Diode Lasers and Photonic Integrated Circuits

WILEY SERIES IN MICROWAVE AND OPTICAL ENGINEERING

KAI CHANG, Editor
Texas A & M University

FIBER-OPTIC COMMUNICATION SYSTEMS • *Govind P. Agrawal*

COHERENT OPTICAL COMMUNICATIONS SYSTEMS • *Silvello Betti,
Giancarlo De Marchis and Eugenio Iannone*

HIGH FREQUENCY ELECTROMAGNETIC TECHNIQUES: RECENT ADVANCES
AND APPLICATIONS • *Asoke K. Bhattacharyya*

COMPUTATIONAL METHODS FOR ELECTROMAGNETICS AND MICROWAVES •
Richard C. Booton, Jr.

MICROWAVE SOLID-STATE CIRCUITS AND APPLICATIONS • *Kai Chang*

DIODE LASERS AND PHOTONIC INTEGRATED CIRCUITS • *Larry A. Coldren
and Scott W. Corzine*

MULTICONDUCTOR TRANSMISSION-LINE STRUCTURES • *J. A. Brandao Faria*

MICROSTRIP CIRCUITS • *Fred Gardiol*

HIGH-SPEED VLSI INTERCONNECTIONS: MODELING, ANALYSIS AND
SIMULATION • *A. K. Goel*

HIGH FREQUENCY ANALOG INTEGRATED CIRCUIT DESIGN •
Ravender Goyal, Editor

OPTICAL COMPUTING: AN INTRODUCTION • *Mohammad A. Karim
and Abdul Abad S. Awwal*

MICROWAVE DEVICES, CIRCUITS AND THEIR INTERACTION • *Charles A. Lee
and G. Conrad Dalman*

ANTENNAS FOR RADAR AND COMMUNICATIONS: A POLARIMETRIC
APPROACH • *Harold Mott*

SOLAR CELLS AND THEIR APPLICATIONS • *Larry D. Partain, Editor*

ANALYSIS OF MULTICONDUCTOR TRANSMISSION LINES • *Clayton R. Paul*

INTRODUCTION TO ELECTROMAGNETIC COMPATIBILITY • *Clayton R. Paul*

NEW FRONTIERS IN MEDICAL DEVICE TECHNOLOGY • *Arye Rosen
and Harel Rosen, Editors*

FREQUENCY SELECTIVE SURFACE AND GRID ARRAY • *T. K. Wu, Editor*

OPTICAL SIGNAL PROCESSING, COMPUTING AND NEURAL NETWORKS •
Francis T. S. Yu and Suganda Jutamulia

Diode Lasers and Photonic Integrated Circuits

L. A. COLDREN

S. W. CORZINE

University of California
Santa Barbara, California

A WILEY-INTERSCIENCE PUBLICATION

JOHN WILEY & SONS, INC.

NEW YORK / CHICHESTER / BRISBANE / TORONTO / SINGAPORE

Library of Congress Cataloging in Publication Data:

Coldren, L. A. (Larry A.)
Diode lasers and photonic integrated circuits / L. A. Coldren, S. W.
Corzine.
p. cm.—(Wiley series in microwave & optical engineering)
"A Wiley-Interscience publication."
Includes index.
ISBN 0-471-11875-3 (cloth; acid-free paper)
1. Semiconductor lasers. 2. Integrated circuits. I. Corzine, S.
W. (Scott W.) II. Title. III. Series: Wiley series in microwave
and optical engineering.
TA1700.C646 1995
621.36′6—dc20   94-39383

To Donna and Reena for their never-ending support

Contents

APPENDICES

Preface

Diode lasers have become an important commercial component. They are used in a wide variety of applications ranging from the readout sources in compact disk players to the transmitters in optical fiber communication systems. New applications in local area data communications and telecommunications networks as well as in consumer products continue to emerge as the devices become more reliable and manufacturable. Although the "short" wavelength (~ 0.7–$0.9\ \mu\text{m}$) GaAs based lasers and the "long" wavelength (~ 1.3–$1.6\ \mu\text{m}$) InP based lasers continue to fill most application needs, there is expanding interest in developing viable sources in the still shorter wavelength visible range as well as in the longer IR range. Learning the underlying basics as well as the desired advanced details in such a fast-moving field is difficult for graduate students as well as experienced engineers.

The book has been written to provide a textbook for teaching the subject of diode lasers and related photonic integrated circuits to students with a wide variety of backgrounds. The depth of coverage is relatively advanced in most areas, but most of the elementary background material is also provided. Appendices are used both to provide review of background material as well as some of the details of the more advanced topics. Thus, by appropriate use of the appendices, the text can support teaching the material at different academic levels, but it remains self-contained.

The text is intended for use at the graduate level, and it is assumed that students have been exposed to elementary quantum mechanics, solid-state physics, and electromagnetic theory at the undergraduate level, and it is recommended that they have had an introductory optoelectronics course. However, Appendices 1 and 3 review most of the necessary background in just about all of the required detail. Thus, it is possible to use the book with less background, provided these review appendices are covered with some care. In fact, Chapters 1 through 3 together with Appendices 1 through 7 provide a fairly comprehensive introduction to most kinds of diode lasers, and they can be used for a relatively elementary course with students who have not had all of the recommended background material. That is, it would be possible to use this material even for an advanced undergraduate course.

On the other hand, for use in a more advanced graduate class, it would not be necessary to cover the material in the first seven appendices. (Of course, it would still be there for reference, and the associated homework problems could still be assigned to insure its understanding. Nevertheless, it is still recommended that Appendix 5, which covers the definitions of modal gain and loss, be reviewed, since this is not well understood by the average worker in the field.) The coverage could then move efficiently through the first three chapters and into Chapters 4 and 5, which deal with the details of gain and laser dynamics in a first course. For more focus on the gain physics some of Appendices 8 through 12 could be included in the coverage. In any event, their inclusion provides for a very self-contained treatment of this important subject matter.

Chapters 6 through 8 deal more with the electromagnetic wave aspects of diode lasers. This material is essential for understanding the more advanced types of devices used in modern communication links and networks. However, keeping this material to last allows the student to develop a fairly complete understanding of the operation of lasers without getting bogged down in the mathematical techniques necessary for the lateral waveguide analysis. Thus, a working understanding and appreciation of laser operation can be gained in only one course. Chapter 6 deals with perturbation and coupled-mode theory and Chapter 7 with dielectric waveguide analysis. Putting Chapter 6 first emphasizes the generality of this material. That is, one really does not need to know the details of the lateral mode profile to develop these powerful techniques. Using the coupled-mode results, gratings and DFB lasers are again investigated. Historically, these components were primarily analyzed with this theory. However, in this text grating based DFB and DBR lasers are first analyzed in Chapter 3 using exact matrix multiplication techniques, from which approximate formulas identical to those derived with coupled mode theory result. The proliferation of computers and the advent of lasers using complex grating designs with many separate sections has led the authors to assert that the matrix multiplication technique should be the primary approach taught to students. The advent of the vertical-cavity laser also supports this approach. Nevertheless, it should be realized that coupled-mode theory is very important to reduce the description of the properties of complex waveguide geometries to simple analytic formulae, which are especially useful in design work. Chapter 8 pulls together most of the material in the first seven chapters by providing a series of design examples of relatively complex photonic integrated circuits.

Chapters 7 and 8 also introduce some basic numerical techniques. These are becoming increasingly more useful with the availability of good low-cost workstations and software for solving complex matrix equations. In Chapter 7 the finite-difference technique is introduced for optical waveguide analysis and in Chapter 8 the beam-propagation method for analyzing real PIC structures is reviewed.

Unlike many books in this field, this book is written as an engineering text. The student is first trained to be able to solve problems on real diode lasers, based upon a phenomenological understanding, before going into the complex physical details such as the material gain process or mode-coupling in dielectric waveguides. This provides motivation for learning the underlying details as well as a toolbox of techniques to immediately apply each new advanced detail in solving real problems. Also, attention has been paid to accuracy and consistency. For example, a careful distinction between the internal quantum efficiency in LEDs and lasers is made, and calculations of gain not only illustrate an analysis technique, but they actually agree with experimental data. Finally, by maintaining consistent notation throughout all of the chapters and appendices, a unique self-contained treatment of all of the included material emerges.

L. A. COLDREN
S. W. CORZINE

Acknowledgments

This text grew out of lecture notes developed for a graduate-level course on diode lasers and guided-wave optics (Coldren) as well as a Ph.D. dissertation written on vertical-cavity lasers with strained quantum-well active regions at UC-Santa Barbara (Corzine). Both benefited from the contributions of a large number of people. Although the lecture notes were initially intended to supplement books in the area, over time the lectures became more dependent on the notes than the available books. Thus, the need for this text emerged. Much is owed to the many outstanding students who took the course and graded homework problems over the years. Many of the important contributions of this book have derived from classroom questions that could not be easily answered.

Outside of the classroom, and generally independent of the course in question, interactions with other faculty and students at UC-Santa Barbara have contributed greatly to our understanding of the subject on which the book expounds. Faculty include Professors Kroemer, Dagli, Bowers, Yeh, and Suemune (now at Hokkaido University), and students include Drs. Ran-Hong Yan, Randy Geels, Jeff Scott, Bruce Young, Dubravko Babic, Zuon-Min Chuang, Vijay Jayaraman, and Radha Nagarajan. We are particularly indebted to Professor Dagli for providing much of the material on numerical analysis, and Professor Yeh for finding a few errors and omissions in the text.

Lastly, we are heavily indebted to Sherene Strobach for her crucial assistance in preparing the manuscript. We would never have completed the project without her help.

L.A.C.
S.W.C.

List of Fundamental Constants

$$\pi = 3.141\,592\,653\,59$$

$$c = 2.997\,924\,58 \times 10^{10} \text{ cm/s}$$

$$\mu_0 = 4\pi \times 10^{-9} \text{ H/cm}$$

$$\varepsilon_0 = 8.854\,187\,82 \times 10^{-14} \text{ F/cm or C}^2/(\text{J}\cdot\text{cm}) \; (= 1/\mu_0 c^2)$$

$$q = 1.602\,189\,2 \times 10^{-19} \text{ C or J/eV}$$

$$m_0 = 9.109\,534 \times 10^{-31} \text{ kg}$$

$$h = 6.626\,176 \times 10^{-34} \text{ J}\cdot\text{s}$$

$$\hbar = 1.054\,588\,7 \times 10^{-34} \text{ J}\cdot\text{s}$$

$$k_B = 1.380\,662 \times 10^{-23} \text{ J/K} = 8.617\,347 \times 10^{-5} \text{ eV/K}$$

$$k_B T = 25.852\,04 \text{ meV (at } T = 300 \text{ K)}$$

$$\lambda_{ph} \cdot E_{ph} = 1.239\,85 \; \mu\text{m}\cdot\text{eV}$$

Source: *CRC Handbook of Chemistry and Physics*, 62nd Edition, 1981–1982.

Diode Lasers and Photonic Integrated Circuits

CHAPTER ONE

Ingredients

1.1 INTRODUCTION

Diode lasers, like most other lasers incorporate an optical gain medium in a resonant optical cavity. The design of both the gain medium and the resonant cavity are critical in modern lasers. The gain medium consists of a material which normally absorbs incident radiation over some wavelength range of interest. But, if it is *pumped* by inputting either electrical or optical energy, the electrons within the material can be excited to higher, nonequilibrium energy levels, so the incident radiation can be amplified rather than absorbed by stimulating the de-excitation of these electrons along with the generation of additional radiation. If the resulting gain is sufficient to overcome the losses of some resonant optical mode of the cavity, this mode is said to have reached *threshold*, and relatively coherent light will be emitted. The resonant cavity provides the necessary positive feedback for the radiation being amplified, so that a lasing oscillation can be established and sustained above threshold pumping levels. As in any other oscillator, the output power level saturates at a level equal to the input minus any internal losses.

Diode lasers represent one class of many different types of lasers today. Two other important classes are *gas* lasers and *solid-state* lasers. The helium–neon gas laser and the Nd-doped YAG (yttrium–aluminum–garnet) solid-state laser are two popular examples. Diode lasers are distinguished from these other types primarily by their ability to be pumped directly by an electrical current. Generally, this results in a much more efficient operation. Overall power conversion efficiences of $\sim 50\%$ are not uncommon for a diode laser, whereas efficiencies on the order of 1% are common for gas and solid-state lasers, which generally are pumped by plasma excitation or an incoherent optical flashlamp source, respectively. However, efficiencies can be somewhat higher, such as in the case of the CO_2 gas laser which has a typical efficiency of over 10%.

Because of their longer cavities and more narrow gain bandwidth, gas and solid-state lasers also tend to have more coherent outputs than simple semiconductor lasers. However, more sophisticated single-frequency diode lasers can have comparable linewidths in the low megahertz range. Net size is another striking difference between semiconductor and other lasers. While gas and solid-state lasers are typically tens of centimeters in length, diode laser chips are generally about the size of a grain of salt, although the mounting and packaging hardware increases the useful component size to the order of a cubic centimeter or so.

A final attribute of diode lasers that has led to their widespread use in important applications such as fiber-optic communications systems is their high reliability or useful lifetime. While the useful life of gas or solid-state lasers is typically measured in thousands of hours, that of carefully qualified diode lasers is measured in hundreds of years. Recent use of diode lasers to pump solid-state lasers, e.g., diode-pumped Nd-YAG, may, however, provide the best advantages of both technologies.

In this chapter, we shall attempt to introduce some of the basic ingredients needed to understand semiconductor diode lasers. First, energy levels and bands in semiconductors are described starting from background given in Appendix 1. The interaction of light with these energy levels is next introduced. Then, the enhancement of this interaction by carrier and photon confinement using heterostructures is discussed. Materials useful for diode lasers and how epitaxial layers of such materials can be grown is briefly reviewed. And finally, the lateral patterning of these layers to provide lateral current, carrier, and photon confinement for practical lasers is introduced.

1.2   ENERGY LEVELS AND BANDS IN SOLIDS

In order to begin to understand how gain is accomplished in lasers, we must have some knowledge of the energy levels that electrons can occupy in the gain medium. The allowed energy levels are obtained by solving Schrödinger's equation using the appropriate electronic potentials. Appendix 1 gives a brief review of this important solid-state physics, as well as the derivation of some other functions that we shall need later. Figure 1.1 schematically illustrates the

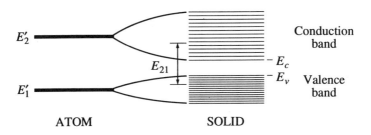

FIGURE 1.1   Illustration of how two discrete energy levels of an atom develop into bands of many levels in a crystal.

energy levels that might be associated with optically induced transitions in both an isolated atom and a semiconductor solid. Electron potential is plotted vertically.

In gas and solid-state lasers, the energy levels of the active atomic species are only perturbed slightly by the surrounding gas or solid host atoms, and they remain effectively as sharp as the original levels in the isolated atom. For example, lasers operating at the 1.06 μm wavelength transition in Nd-doped YAG, use the $^4F_{3/2}$ level of the Nd atom for the upper laser state #2, and the $^4I_{11/2}$ level for the lower laser state #1. Because only these atomic levels are involved, emitted or absorbed photons need to have almost exactly the correct energy, $E_{21} = hc/1.06\ \mu m$.

On the other hand, in a covalently bonded solid like the semiconductor materials we use to make diode lasers, the uppermost energy levels of individual constituent atoms each broaden into bands of levels as the bonds are formed to make the solid. This phenomenon is illustrated in Fig. 1.1. The reason for the splitting can be realized most easily by first considering a single covalent bond. When two atoms are in close proximity, the outer valence electron of one atom can arrange itself into a low-energy *bonding* (symmetric) charge distribution concentrated between the two nuclei, or into a high-energy *antibonding* (antisymmetric) distribution devoid of charge between the two nuclei. In other words, the isolated energy level of the electron is now split into two levels due to the two ways the electron can arrange itself around the two atoms.[1] In a covalent bond, the electrons of the two atoms both occupy the lower energy bonding level (provided they have opposite spin), while the higher energy antibonding level remains empty.

If another atom is brought in line with the first two, a new charge distribution becomes possible that is neither completely bonding nor antibonding. Hence, a third energy level is formed between the two extremes. When N atoms are covalently bonded into a linear chain, N energy levels distributed between the lowest-energy bonding state and the highest-energy antibonding state appear, forming a band of energies. In our linear chain of atoms, spin degeneracy allows all N electrons to fall into the lower half of the energy band, leaving the upper half of the band empty. However in a three-dimensional crystal, the number of energy levels is more generally equated with the number of *unit cells*, not the number of atoms. In typical semiconductor crystals, there are two atoms per primitive unit cell. Thus, the first atom fills the lower half of the energy band (as with the linear chain) while the second atom fills the upper half, such that the energy band is entirely full.

The semiconductor *valence* band is formed by the multiple splitting of the

---

[1] The energy level splitting is often incorrectly attributed to the Pauli exclusion principle which forbids electrons from occupying the same energy state (and thus forces the split, as the argument goes). In actuality, the splitting is a fundamental phenomenon associated with solutions to the wave equation involving two coupled systems, and applies equally to probability, electromagnetic, or any other kind of waves. It has nothing to do with the Pauli exclusion principle.

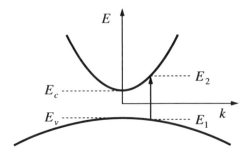

FIGURE 1.2 Electron energy vs. wavenumber in a semiconductor showing a transition of an electron from a bound state in the valence band (E_1) to a free carrier state in the conduction band (E_2). The transition leaves a hole in the valence band. The lowest and highest energies in the conduction and valence bands are E_c and E_v, respectively.

highest occupied atomic energy level of the constituent atoms. In semiconductors, the valence band is by definition entirely filled with no external excitation at $T = 0$ K. Likewise, the next higher-lying atomic level splits apart into the *conduction* band which is entirely empty in semiconductors without any excitation. When thermal or other energy is added to the system, electrons in the valence band may be excited into the conduction band analogous to how electrons in isolated atoms can be excited to the next higher energy level of the atom. In the solid then, this excitation creates holes (missing electrons) in the valence band as well as electrons in the conduction band, and both can contribute to conduction.

Although Fig. 1.1 suggests that many conduction–valence band state pairs may interact with photons of energy E_{21}, Appendix 1 shows that the imposition of momentum conservation in addition to energy conservation limits the interaction to a fairly limited set of state pairs for a given transition energy. This situation is illustrated on the electron energy vs. k-vector (E–k) plot shown schematically in Fig. 1.2. (Note that momentum $\equiv \hbar\mathbf{k}$.) Since the momentum of the interacting photon is negligibly small, transitions between the conduction and valence band must have the same k-vector, and only vertical transitions are allowed on this diagram. This fact will be very important in the calculation of gain.

1.3 SPONTANEOUS AND STIMULATED TRANSITIONS: THE CREATION OF LIGHT

With a qualitative knowledge of the energy levels that exist in semiconductors, we can proceed to consider the electronic transitions that can exist and the interactions with lightwaves that are possible. Figure 1.3 illustrates the different kinds of electronic transitions that are important, emphasizing those that involve the absorption or emission of photons (lightwave quanta).

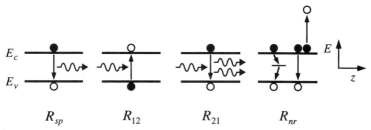

FIGURE 1.3 Electronic transitions between the conduction and valence bands. The first three represent radiative transitions in which the energy to free or bind an electron is supplied by or given to a photon. The fourth illustrates two nonradiative processes.

Although we are explicitly considering semiconductors, only a single level in both the conduction and valence bands is illustrated. As discussed above and in Appendix 1, momentum conservation selects only a limited number of such pairs of levels from these bands for a given transition energy. In fact, if it were not for a finite bandwidth of interaction owing to the finite state lifetime, a single pair of states would be entirely correct. In any event, the procedure to calculate gain and other effects will be to find the contribution from a single state pair and then integrate to include contributions from other pairs; thus, the consideration of only a single conduction–valence band state pair forms an entirely rigorous basis.

As illustrated, four basic electronic recombination/generation (photon emission/absorption) mechanisms must be considered separately: (1) spontaneous recombination (photon emission), (2) stimulated generation (photon absorption), (3) stimulated recombination (coherent photon emission), and (4) nonradiative recombination. The open circles represent unfilled states (holes) and the solid circles represent filled states (electrons). Since electron and hole densities are highest near the bottom or top of the conduction or valence bands, respectively, most transitions of interest involve these carriers. Thus, photon energies tend to be only slightly larger than the bandgap, i.e., $E_{21} = hv \sim E_g$. The effects involving electrons in the conduction band are all enhanced by the addition of some pumping means to increase the electron density to above the equilibrium value there. Of course, the photon absorption can still take place even if some pumping has populated the conduction band somewhat.

The first case (R_{sp}) represents the case of an electron in the conduction band recombining spontaneously with a hole (missing electron) in the valence band to generate a photon. Obviously, if a large number of such events should occur, relatively incoherent emission would result, since the emission time and direction would be random, and the photons would not tend to contribute to a coherent radiation field. This is the primary mechanism within a light-emitting diode (LED), in which photon feedback is not provided. The second illustration (R_{12}) outlines photon absorption which stimulates the generation of an electron in the conduction band while leaving a hole in the valence band. The third process (R_{21}) is exactly the same as the second, only the sign of the interaction is reversed. Here an incident photon perturbs the system, stimulating the recombination of an electron and hole, and simultaneously generating a new photon. Of course, this is the all-important positive gain mechanism that is necessary for lasers to operate. Actually, it should be realized that the net combination of stimulated absorption and emission of photons, effects (R_{12}) and (R_{21}), will represent the net gain experienced by an incident radiation field.

Because spontaneous recombination requires the presence of an electron–hole pair, the recombination rate tends to be proportional to the product of the density of electrons and holes, NP. In undoped active regions, charge neutrality requires that the hole and electron densities be equal. Thus, the spontaneous recombination rate becomes proportional to N^2. In a similarly undoped active region, net stimulated recombination (photon emission) depends upon the existence of photons in addition to a certain value of electron density to overcome the photon absorption. Thus, as we shall later show more explicitly, the net rate of stimulated recombination is proportional to the photon density, N_p, multiplied by $(N - N_{tr})$, where N_{tr} is a *transparency* value of electron density (i.e., where $R_{21} = R_{12}$).

Finally, the fourth schematic in Fig. 1.3 represents the several nonradiative ways in which a conduction band electron can recombine with a valence band hole without generating any useful photons. Instead, the energy is dissipated as heat in the semiconductor crystal lattice. Thus, this schematic represents the ways in which conduction band electrons can escape from usefully contributing to the gain, and as such these effects are to be avoided if possible. In practice, there are two general nonradiative mechanisms for carriers that are important. The first involves nonradiative recombination centers, such as point defects, surfaces, and interfaces, in the active region of the laser. In order to be effective, these do not require the simultaneous existence of electrons and holes or other particles. Thus, the recombination rate via this path tends to be directly proportional to the carrier density, N. The second mechanism is Auger recombination, in which the electron–hole recombination energy, E_{21}, is given to another electron or hole in the form of kinetic energy. Thus, again for undoped active regions in which the electron and hole densities are equal, Auger recombination tends to be proportional to N^3, since we must simultaneously have the recombining electron–hole pair and the third particle that receives the ionization energy. Appendix 2 gives techniques for calculating the carrier density from the density of electronic states and the probability that they are occupied, generally characterized by a Fermi function.

1.4   TRANSVERSE CONFINEMENT OF CARRIERS AND PHOTONS IN DIODE LASERS: THE DOUBLE HETEROSTRUCTURE

In order for the gain material in a semiconductor laser to function, it must be pumped or excited with some external energy source. A major attribute of diode

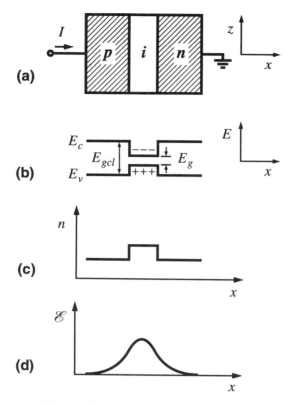

FIGURE 1.4 Aspects of the double-heterostructure diode laser: (a) a schematic of the material structure; (b) an energy diagram of the conduction and valence bands vs. transverse distance; (c) the refractive index profile; (d) the electric field profile for a mode traveling in the z-direction.

lasers is their ability to be pumped directly with an electrical current. Of course, the active material can also be excited by the carriers generated from absorbed light, and this process is important in characterizing semiconductor material before electrical contacts are made. However, we shall focus mainly on the more technologically important direct current injection technique in most of our analysis.

Figure 1.4 gives a schematic of a broad-area *pin* double-heterostructure (DH) laser diode, along with transverse sketches of the energy gap, index of refraction and resulting optical mode profile across the DH region. As illustrated, a thin slab of undoped active material is sandwiched between *p*- and *n*-type cladding layers which have a higher conduction–valence band energy gap. Typical thicknesses of the active layer for this simple three-layer structure are ~0.1–0.2 µm. Because the bandgap of the cladding layers is larger, light generated in the active region will not have sufficient photon energy to be absorbed in them, i.e., $E_{21} = hv < E_{gcl}$.

For this DH structure, a transverse (x-direction) potential well is formed for electrons and holes that are being injected from the n- and p-type regions, respectively, under forward bias. As illustrated in part (b), they are captured and confined together, thereby increasing their probability of recombining with each other. In fact, unlike in most semiconductor diodes or transistors that are to be used in purely electronc circuits, it is desirable to have all of the injected carriers recombine in the active region to form photons in a laser or LED. Thus, simple pn-junction theory, which assumes that all carriers entering the depletion region are swept through with negligible recombination, is totally inappropriate for diode lasers and LEDs. In fact, a better assumption for lasers and LEDs is that all carriers recombine in the i-region [1]. Appendix 2 also discusses a possible "leakage current" which results from some of the carriers being thermionically emitted over the heterobarriers before they can recombine.

To form the necessary resonant cavity for optical feedback, simple cleaved facets can be used, since the large index of refraction discontinuity at the semiconductor–air interface provides a reflection coefficient of $\sim 30\%$. The lower bandgap active region also usually has a higher index of refraction, n, than the cladding, as outlined in Fig. 1.4(c), so that a transverse dielectric optical waveguide is formed with its axis along the z-direction. The resulting transverse optical energy density profile (proportional to the photon density or the electric field magnitude squared, $|\mathscr{E}|^2$) is illustrated in Fig. 1.4(d). The derivation of this optical mode shape is given in Appendix 3. Thus, with the in-plane waveguide and perpendicular mirrors at the ends, a complete resonant cavity is formed. Output is provided at the facets, which only partially reflect. Later on we shall consider more complex reflectors which can provide stronger feedback and wavelength selective feedback. One should also realize that if the end facet reflections are suppressed by antireflection coatings, the device would then function as an LED. When we analyze lasers in the next chapter, the case of no feedback will also be considered.

The carrier-confining effect of the double-heterostructure is one of the most important features of modern diode lasers. After many early efforts that used homojunctions or single heterostructures, the advent of the DH structure made the diode laser truly practical for the first time. It turns out that many modern diode lasers involve a little more complexity in their transverse carrier and photon confinement structure as compared to Fig. 1.4, but the fundamental concepts remain valid. For example, with in-plane lasers, where the light propagates parallel to the substrate surface, a common departure from Fig. 1.4 is to use a thinner *quantum-well* carrier-confining active region ($d \sim 10$ nm), and a surrounding intermediate bandgap *separate confinement* region to confine the photons. Figure 1.5 illustrates transverse bandgap profiles for such separate-confinement heterostructure, single quantum-well (SCH-SQW) lasers. The transverse optical energy density is also overlaid to show that the photons are confined primarily by the outer heterointerfaces and the carriers by the inner quantum well. The advantages of the quantum-well active region are introduced in Appendix 1 and discussed in detail in Chapter 4.

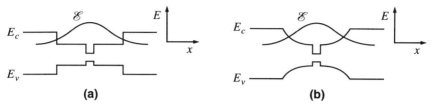

FIGURE 1.5   Transverse band structures for two different separate-confinement hetero-structures (SCHs): (a) standard SCH; (b) graded-index SCH (GRINSCH). The electric field (photons) are confined by the outer step or graded heterostructure; the central quantum well confines the electrons.

1.5   SEMICONDUCTOR MATERIALS FOR DIODE LASERS

The successful fabrication of a diode laser relies very heavily upon the properties of the materials involved. There is a very limited set of semiconductors that possess all of the necessary properties to make a good laser. For the desired double heterostructures at least two compatible materials must be found, one for the cladding layers and another for the active region. In more complex geometries, such as the SCH mentioned above, three or four different bandgaps may be required within the same structure. The most fundamental requirement for these different materials is that they have the same crystal structure and nearly the same lattice constant, so that single-crystal, defect-free films of one can be *epitaxially* grown on the other. Defects generally become nonradiative recombination centers which can steal many of the injected carriers that otherwise would provide gain and luminescence. In a later section we shall discuss some techniques for performing this epitaxial growth, but first we need to understand how to select materials that meet these fundamental boundary conditions.

Figure 1.6 plots the bandgap vs. lattice constant for several families of III–V semiconductors. These III–V compounds (which consist of elements from columns III and V of the periodic table) have emerged as the materials of choice for lasers that emit in the 0.7–1.6 μm wavelength range. This range includes the important fiber-optic communication bands at 0.85, 1.31, and 1.55 μm, the pumping bands for fiber amplifiers at 1.48 and 0.98 μm, the window for pumping Nd-doped YAG at 0.81 μm, and the wavelength currently used for optical disk players at 0.78 μm. Most of these materials have a *direct gap* in E–k space, which means that the minimum and maximum of the conduction and valence bands, respectively, fall at the same k-value, as illustrated in Fig. 1.2. This facilitates radiative transitions because momentum conservation is naturally satisfied by the annihilation of the equal and opposite momenta of the electron and hole. (The momentum of the photon is negligibly small.)

The lines on this diagram represent ternary compounds which are alloys of the binaries labeled at their end-points. The dashed lines represent regions of indirect gap. The triangular areas enclosed by lines between three binaries

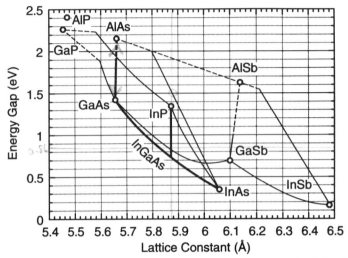

FIGURE 1.6   Energy gap vs. lattice constant of ternary compounds defined by curves that connect the illustrated binaries.

represent quaternaries, which obviously have enough degrees of freedom that the energy gap can be adjusted somewhat without changing the lattice constant. Thus, in general, a quaternary compound is required in a DH laser to allow the adjustment of the energy gap while maintaining lattice matching. Fortunately, there are some unique situations which allow the use of more simple ternaries. As can be seen, the AlGaAs ternary line is almost vertical. That is, the substitution of Al for Ga in GaAs does not change the lattice constant very much. Thus, if GaAs is used as the substrate, any alloy of $Al_xGa_{1-x}As$ can be grown and it will naturally lattice match, so that no misfit dislocations or other defects should form. As suggested by the formula, the x-value determines the percentage of Al in the group III half of the III–V compound. The AlGaAs/GaAs system provides lasers in the 0.7–0.9 μm wavelength range. For DH structures in this system, about two-thirds of the band offset occurs in the conduction band.

The most popular system for long-distance fiber optics is the InGaAsP/InP system. Here the quaternary is specified by an x and y value, i.e., $In_{1-x}Ga_xAs_yP_{1-y}$. This is grown on InP to form layers of various energy gap corresponding to wavelengths in the 1.0–1.6 μm range, where silica fiber has minima in loss (1.55 μm) and dispersion (1.3 μm). Using InP as the substrate, a range of lattice-matched quaternaries extending from InP to the InGaAs ternary line can be accommodated, as indicated by the vertical line in Fig. 1.6. Fixing the quaternary lattice constant defines a relation between x and y. It has been found that choosing x equal to ∼$0.46y$ results in approximate lattice matching to InP. The ternary endpoint is $In_{0.53}Ga_{0.47}As$. For DH structures in this system, only about 40% of the band offset occurs in the conduction band.

For a little more precision, the lattice constants of quaternaries can be calculated from Vegard's law, which gives a value equal to the weighted average of all of the four possible constituent binaries. For example, in $In_{1-x}Ga_xAs_yP_{1-y}$, we obtain

$$a(x, y) = xya_{GaAs} + x(1 - y)a_{GaP} + (1 - x)ya_{InAs} + (1 - x)(1 - y)a_{InP}. \quad (1.1)$$

Similarly, the lattice constants for other alloys can be calculated. Table 1.1 lists the lattice constants, bandgaps, effective masses, and indices of refraction for some common materials. (Subscripts on effective masses, C, HH, LH, and SH denote values in the conduction, heavy-hole, light-hole, and split-off bands, respectively.) Other parameters, for example bandgap, can also be interpolated in a similar fashion to Eq. (1.1), however a second-order *bowing* parameter must oftentimes be added to improve the fit. In addition, one must be careful if different bands come into play in the process. For example, in AlGaAs, the values for GaAs and $Al_{0.2}Ga_{0.8}As$ can be linearly extrapolated for direct gap AlGaAs up to $x \sim 0.45$, but fail to describe the indirect gap at higher x values.

In addition to the usual III–V compounds discussed above, Table 1.1 also lists some of the nitride compounds. These have gained attention primarily because of some recent successes in demonstrating LEDs emitting at high energies in the visible spectrum. Whereas the InAlGaAsP based compounds are limited to emission in the red and near infrared regions, the nitrides have demonstrated blue and UV emission. Although they possess the primary feature of providing a higher-energy-gap system, there are problems in obtaining high quality epitaxial layers since substrates with similar lattice constants and thermal properties are not available. Thus, the nitrides remain an active area of research.

Lattice matching is generally necessary to avoid defects which can destroy the proper operation of diode lasers. However, it is well known that a small lattice mismatch ($\Delta a/a \sim 1\%$) can be tolerated up to a certain thickness (~ 20 nm) without any defects. Thus, for a thin active region, one can move slightly left or right of the lattice matching condition illustrated in Fig. 1.6 or by Eq. (1.1). In this case, the lattice of the deposited film distorts so as to fit the substrate lattice in the plane, but it also must distort in the perpendicular direction to retain approximately the same unit cell volume it would have without distortion. Figure 1.7 shows a cross section of how unit cells might distort to accommodate a small lattice mismatch. After a critical thickness is exceeded, misfit defects are generated to relieve the integrated strain. However, up to this point, it turns out that such *strained layers* may have more desirable optoelectronic properties than their unstrained counterparts. Also, as we have already mentioned, quantum-well active regions, which are thinner than typical critical thicknesses, are desirable in diode lasers for reduced threshold and improved thermal properties. Thus, these quantum wells can also be *strained-layer quantum wells* without introducing

TABLE 1.1 Material Parameters for III–V Compounds.

III–V Compounds	a(Å)	E_g (eV) 0 K	E_g (eV) 300 K	m_C	m_{HH}	m_{LH}	m_{SH}	ε @ dc	n @ E_g	n @ (λ μm)
GaAs	5.6533	1.519	1.424	0.067	0.38	0.09	0.15	13.2	3.62	3.52(0.98)
AlGaAs (0.2)	5.6548	1.769	1.673	0.084	0.39	0.1	0.16	12.5	3.64	3.46(0.87); 3.39(0.98)
AlAs*	5.660	2.228	2.153	0.19	0.48	0.2	0.29	10.06	3.2	2.98(0.87); 2.95(0.98)
InGaAs (0.2) comp. strained on GaAs	5.6533	1.296	1.215	0.059	0.37 / 0.078†	0.062 / 0.16†	0.11	13.6	3.6	
InP	5.8688	1.424	1.351	0.077	0.61	0.12	0.20	12.4	3.41	3.21(1.3); 3.17(1.55); 3.40(1.55)
InGaAsP (1.3 μm)	5.8688	1.029	0.954	0.056	0.42	0.055	0.1	13.3	3.52	
InGaAsP (1.55 μm)	5.8688	0.874	0.800	0.045	0.37	0.044	0.08	13.75	3.55	
InGaAs (1.65 μm)	5.8688	0.818	0.748	0.046	0.36	0.041	0.07	13.9	3.56	
InAs	6.0583	0.418	0.359	0.027	0.34	0.027	0.05	15.15	3.52	
GaP*	5.4505	2.35	2.272	0.254	0.67	0.17	0.46	11.1	3.5	
AlP*	5.4635	2.505	2.41	0.21	0.51	0.21	0.3	9.8	2.97	
AlSb*	6.1355	1.696	1.63	0.33	0.47	0.16	0.24	12.0	3.5	
GaSb	6.0959	0.811	0.70	0.041	0.27	0.05	0.08	15.69	3.92	
InSb	6.4794	0.237	0.175	0.014	0.34	0.016	0.03	16.8	3.5	
GaN (hexagonal)	a = 3.189 c = 5.185	3.50	3.39	0.20				8.9	2.67	2.33(1 eV)
AlN (hexagonal)	a = 3.112 c = 4.982	6.28	6.20					8.5	2.15	2.15(3 eV)

* Indirect gap
† In-plane masses

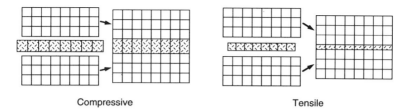

Compressive   Tensile

FIGURE 1.7   Schematic of sandwiching quantum wells with either a larger or smaller lattice constant to provide either compressive or tensile strain, respectively.

any undesired defects. These structures will be analyzed in some detail in Chapter 4.

1.6   EPITAXIAL GROWTH TECHNOLOGY

In order to make the multilayer structures required for diode lasers, it is necessary to grow single-crystal lattice-matched layers with precisely controlled thicknesses over some suitable substrate. We have already discussed the issue of lattice matching and some of the materials involved. Here we briefly introduce several techniques to perform epitaxial growth of the desired thin layers.

We shall focus on the three most important techniques in use today: liquid-phase epitaxy (LPE), molecular beam epitaxy (MBE), and organometallic vapor-phase epitaxy (OMVPE). OMVPE is often also referred to as metal-organic chemical vapor deposition (MOCVD), although purists do not like the omission of the word 'epitaxy'. As the names imply, the three techniques refer to growth either in liquid, vacuum, or a flowing gas, respectively. The growth under liquid or moderate pressure gas tends to be done near equilibrium conditions, so that the reaction can proceed in either the forward or reverse direction to add or remove material, whereas the MBE growth tends to be more of a physical deposition process. Thus, the near-equibrium processes, LPE and MOCVD, tend to better provide for the removal of surface damage at the onset of growth, and they are known for providing higher quality interfaces generally important in devices. MBE on the other hand provides the ultimate in film uniformity and thickness control.

Figure 1.8 gives a cross section of a modern LPE system. In this system the substrate is placed in a recess in a graphite slider bar which forms the bottom of a sequence of bins in a second graphite housing. The bins are filled with solutions from which a desired layer will grow as the substrate is slid beneath that bin. This entire assembly is positioned in a furnace, which is accurately controlled in temperature. There are several different techniques of controlling the temperature and the dwell time under each melt, but generally the solutions are successively brought to saturation by reducing the temperature very slowly as the substrate wafer is slid beneath alternate wells. In modern systems, the process of slider positioning and adjusting furnace temperature is done by

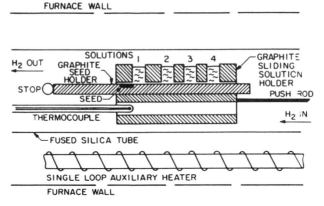

FIGURE 1.8 Schematic of liquid-phase epitaxy (LPE) system [2]. (Reprinted, by permission, from *Applied Physics Letters*.)

computer control for reproducibility and efficiency. However, LPE is rapidly being replaced by MOCVD for the manufacture of most diode lasers.

The melts typically consist mostly of one of the group III metals with the other constituents dissolved in it. For InGaAsP growth, In metal constitutes most of the melt. For an $In_{0.53}Ga_{0.47}As$ film only about 2.5% of Ga and 6% As is added to the melt for growth at 650°C. For InP growth only about 0.8% of P is added. Needless to say, the dopants are added in much lesser amounts. Thus, LPE growth requires some very accurate scales for weighing out the constituents, and an operator with a lot of patience.

Figure 1.9 shows a schematic of an MOCVD system. As can be seen, a large part of the system is devoted to gas valving and manifolding to obtain the proper mixtures for insertion into the growth reactor chamber. The substrate is positioned on a susceptor which is heated typically by rf induction, or in some cases, by resistive heaters. Both low-pressure and atmospheric-pressure systems are being used. While the atmospheric-pressure system uses the reactant gases more effectively, the layer uniformity and the time required to flush the reactor before beginning a new layer is long. Low pressure is more popular where very abrupt interfaces between layers are desired, and this is very important for quantum-well structures.

The sources typically used for MOCVD consist of a combination of hydrides such as arsine (AsH_3) and phosphine (PH_3), and organometallic liquids which are used to saturate an H_2 carrier gas. Example organometallics are triethyl-indium and triethyl-gallium. Dopants can be derived from either other hydrides or liquid sources. For example, H_2S or triethyl-zinc can be used for *n*- or *p*-type dopants, respectively.

One of the key concerns with MOCVD is safety. The problems are primarily with the hydrides, which are very toxic. Thus, much of the cost of an MOCVD facility is associated with elaborate gas handling, monitoring, and emergency

SCHEMATIC VIEW OF AN MOCVD REACTOR

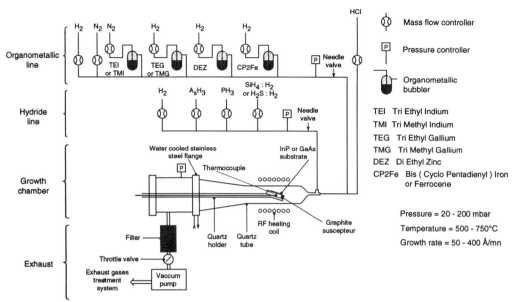

FIGURE 1.9 Schematic of metal-organic chemical vapor deposition (MOCVD) system [3]. (From *GaInAsP Alloy Semiconductors*, T. P. Pearsall, Ed., Copyright © John Wiley & Sons, Inc. Reprinted by permission of John Wiley & Sons, Inc.)

disposal techniques. Recently, however, there has been considerable work with less toxic liquid sources for As and P; e.g., *t*-butyl-arsine and *t*-butyl-phosphine. Although still toxic, liquid sources give off only modest amounts of poisonous gases due to their vapor pressure. Such quantities could be accommodated by conventional fume hoods. The hydrides, on the other hand, are contained in high-pressure gas cylinders which conceivably can fail and release large quantities of concentrated toxic gas in a short time.

Figure 1.10 shows a cross section of a solid-source MBE growth chamber. As illustrated, MBE is carried out under ultrahigh vacuum (UHV) conditions. Constituent beams of atoms are evaporated from effusion cells, and these condense on a heated substrate. Liquid nitrogen cryoshields line the inside of the system to condense any stray gases. For stoichiometry control MBE makes use of the fact that the group V elements are much more volatile than the group III elements. Thus, if the substrate is sufficiently hot, the group V atoms will reevaporate unless there is a group III atom with which to form the compound. At the same time, the substrate must be sufficiently cool so that the group III atoms will stick. Therefore, the growth rate is determined by the group III flux, and the group V flux is typically set to several times that level. Typical growth temperatures for the AlGaAs system are in the 600–650°C range. However, because Al tends to oxidize easily, and such oxides create nonradiative recombination centers, AlGaAs lasers may be grown over 700°C.

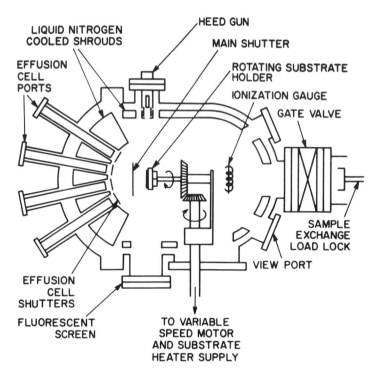

LIQUID NITROGEN
COOLED SHROUDS

HEED GUN

MAIN SHUTTER

EFFUSION
CELL
PORTS

ROTATING SUBSTRATE
HOLDER

IONIZATION GAUGE

GATE VALVE

SAMPLE
EXCHANGE
LOAD LOCK

VIEW PORT

EFFUSION
CELL
SHUTTERS

FLUORESCENT
SCREEN

TO VARIABLE
SPEED MOTOR
AND SUBSTRATE
HEATER SUPPLY

FIGURE 1.10 Schematic of molecular beam epitaxy (MBE) system [4]. (Reprinted, by permission, from *Journal of Applied Physics.*)

One of the key features of MBE is that UHV surface analysis techniques can be applied to the substrate either before or during growth in the same chamber. One of the most useful tools is reflection high-energy electron diffraction (RHEED) which is an integral part of any viable MBE system. It is particularly useful in monitoring the growth rate *in situ*, since the intensity of the RHEED pattern varies in intensity as successive monolayers are deposited.

Some hybrid forms of the last two techniques have also been developed, i.e., gas source MBE, metal-organic MBE (MOMBE), and chemical beam epitaxy (CBE) [5, 6]. These techniques are particularly interesting for the phosphorous-containing compounds, such as the important InGaAsP. Basically, gas-source MBE involves using the hydrides for the group V sources. Generally, these gases must be cracked by passing them through a hot cell prior to arriving at the substrate. MOMBE uses the metal-organics for the group III sources; again, some cracking is necessary. CBE is basically just ultralow-pressure MOCVD since both the group III and V sources are the same as in an MOCVD system. However, in the CBE case one still retains access to the UHV surface analysis techniques that have made MBE viable.

1.7   LATERAL CONFINEMENT OF CURRENT, CARRIERS, AND PHOTONS FOR PRACTICAL LASERS

Practical diode lasers come in two basic varieties: those with *in-plane* cavities and those with *vertical* cavities. The in-plane (or edge-emitting) types have been in existence since the late 1960s whereas the vertical cavity types have been viable only since about 1990. As mentioned earlier, feedback for the in-plane type can be accomplished with a simple cleaved-facet mirror; however, for vertical-cavity lasers a multilayer reflective stack must be grown below and above the active region for the necessary cavity mirrors. Figure 1.11 illustrates both types.

As suggested by this figure, practical lasers must emit light in a narrow beam, which implies that a *lateral* patterning of the active region is necessary. In the case of the in-plane types, a stripe laser is formed which typically has lateral dimensions of a few microns. Similarly, the vertical-cavity types typically consist of a circular dot geometry with lateral dimensions of a few microns. This emitting aperture of a few microns facilitates coupling to optical fibers or other simple optics, since it is sufficiently narrow to support only a single lateral mode of the resulting optical waveguide, but sufficiently wide to provide an emerging optical beam with a relatively small diffraction angle.

Figure 1.12 shows cross-sectional scanning electron micrographs (SEMs) of both the in-plane and vertical-cavity lasers. The reference coordinate systems, also introduced in Fig. 1.11, are somewhat different for these two generic types of diode lasers. The difference arises from our insistence on designating the optical propagation axis as the z-axis. We shall also refer to this direction as the *axial* direction. For both types the *lateral* y-direction is in the plane of the substrate. For in-plane lasers the vertical to the substrate is the *transverse* x-direction, as illustrated in Fig. 1.4, whereas for vertical-cavity lasers the x-direction lies in the plane and is deemed a second lateral direction.

Once we have decided that a lateral patterning of the active region is desirable for lateral carrier and photon confinement, we also must consider lateral *current* confinement. That is, once the active region is limited in lateral extent, we must insure that all of the current is injected into it rather than

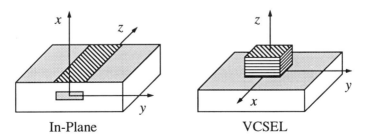

In-Plane   VCSEL

FIGURE 1.11   Schematic of in-plane and vertical-cavity surface-emitting lasers showing selected coordinate systems.

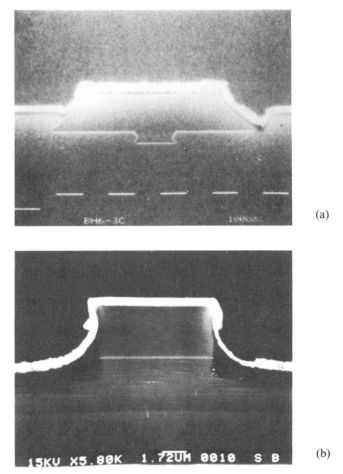

(a)

(b)

FIGURE 1.12 Cross-sectional SEMs of (a) in-plane and (b) vertical-cavity semi-conductor lasers.

finding some unproductive shunt path. In fact, current confinement is the first and simplest step in moving from a broad-area laser to an in-plane stripe or vertical-cavity dot laser. For example, current can be channeled to some degree simply by limiting the contact area. However, in the best lasers current confinement is combined with techniques to laterally confine the carriers and photons in a single structure.

Lateral confinement of current, carriers, and photons has been accomplished in literally dozens of ways, and there are even more acronyms to describe all of these. For brevity's sake, we will focus on only a few generic types illustrated in Fig. 1.13. The first two types only provide current confinement; the third adds a weak photon confinement; and the last three provide all three types

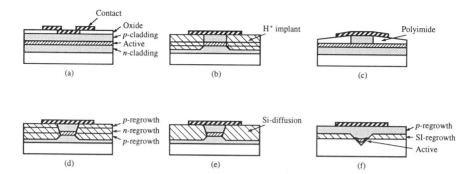

FIGURE 1.13   Lateral confinement structures for heterostructure lasers: (a) oxide-stripe provides current confinement; (b) proton-implant provides current confinement; (c) ridge structure provides current plus photon confinement; (d) etched mesa buried-heterostructure (BH) provides current, photon, and carrier confinement; (e) impurity-induced disordered BH provides current, photon, and carrier confinement; (f) channeled substrate BH provides current, photon, and carrier confinement.

of confinement. The examples are explicit for in-plane lasers, but many are also applicable to vertical-cavity lasers.

Figure 1.13(a) illustrates a simple oxide stripe laser. This stripe laser is the simplest to make, since the area of current injection is limited simply by limiting the contact area. This laser has some current confinement, but no carrier or photon confinement. The proton-implanted configuration of Fig. 1.13(b) has essentially the same attributes, although the current confinement is extended to the active region. It uses the fact that implanted hydrogen ions (protons) create damage and trap out the mobile charge, rendering the implanted material nearly insulating.

The configurations of Fig. 1.13(a) and (b) are described as *gain-guided* stripe lasers, since the current is apertured, but there is no lateral heterobarrier to provide a potential well for carriers or photons. Thus, carriers injected into the active region can diffuse laterally, decreasing the laser's efficiency. Also, there is no lateral index change to guide photons along the axis of the cavity, so optical losses tend to be high. Although these two configurations were of some commercial importance in the early days of diode lasers before the advent of viable etching or regrowth techniques, currently their use is limited. The oxide stripe laser is still used in the lab to characterize material, since the processing is so simple, and proton isolation is still used to limit current leakage in practical lasers, but the implant areas are usually kept a few microns away from the active region for reduced loss and improved reliability.

Figure 1.13(c) illustrates a *ridge* laser which combines current confinement with a weak photon confinement. The efficiency of current injection can be high, but since the processing involves etching down to just above the active

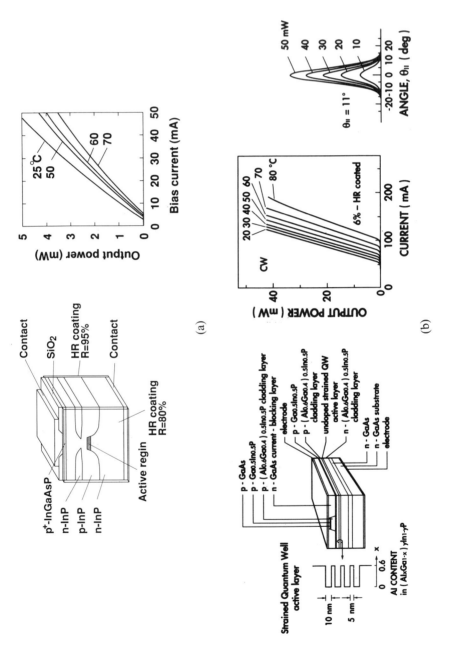

(a)

(b)

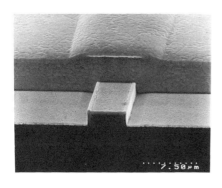
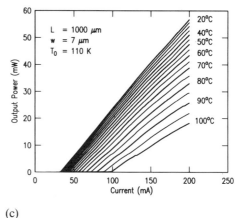

(c)

FIGURE 1.14 Practical examples of index-guided in-plane lasers—schematics and measured light-out vs. bias current. (a) 300 μm long, 1.2 μm wide, 1.5 μm wavelength InGaAsP/InP BH structure with 80 and 95% facet coatings and a compressively strained 5QW active region contained in a 1.2 μm wavelength SCH region [7]. Data taken for the indicated temperatures. (© 1993 IEEE) (b) 690 nm wavelength strained-QW AlGaInP/GaAs buried ridge structure with 6%—HR facet coatings for high output power [8]. Far-field emission angle also indicated for various powers. (© 1993 IEEE) (c) Dry-etched ridge-waveguide InGaP/GaAs strained-QW laser with gold heat spreader shown in SEM photo [9]. Device length, L, ridge width, w, and characteristic temperature, T_0 indicated with data. (© 1993 IEEE)

region, carriers are still not confined laterally. They are free to diffuse laterally and recombine without contributing to the gain. The etching depth is adjusted to provide just enough effective lateral index change to provide a single lateral mode optical waveguide. If a deeper etch is used, e.g., down to the active layer, the device will support more than one lateral mode and coupling to fibers or other optics is spoiled. In fact, even if we ignore the optical waveguide problem, etching through the active region to laterally confine the carriers is not productive, because the nonradiative recombination associated with surface states can be as bad or even worse than the carrier loss due to lateral diffusion. Ridge lasers continue to be important because they can be fabricated following a single epitaxial crystal growth. Since optical guiding is provided by the dielectric loading of the ridge, optical losses can be low. Thus, this structure is still considered a good compromise by many looking for a simple device fabrication process.

Figures 1.14 and 1.15 give cross-sectional schematics and illustrative light power out versus drive current characteristics of a number of experimental in-plane and vertical-cavity lasers [7–12]. The figure captions contain some of the relevant descriptive details. In the chapters to follow, the operating principles of such lasers will be detailed.

Effective lateral carrier or photon confinement requires the creation of a

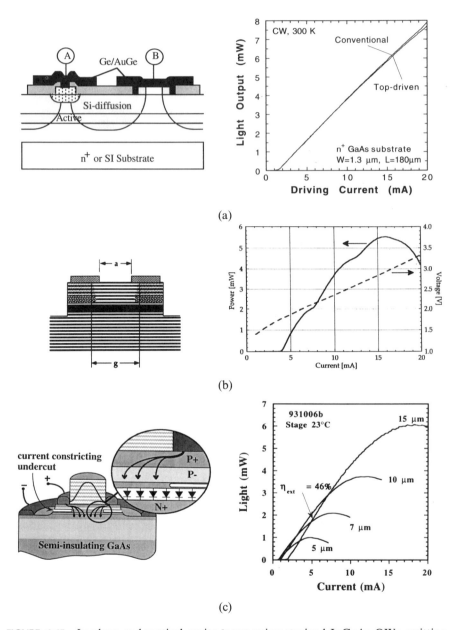

FIGURE 1.15 In-plane and vertical-cavity lasers using strained InGaAs QWs emitting at 980 nm on GaAs substrates—schematics and measured light-out vs. bias current. (a) Top-contacted impurity-induced disordered BH laser using a single strained QW active layer and uncoated facets [10]. Elemental Si is diffused from the top surface at 850°C to disorder the active region. (© 1993 IEEE) (b) Gain-guided VCSEL structure using a 3QW active region and proton implantation above the active region to aperture the current [11]. Data is from a device with $a = g = 15$ μm diameter contact and implant apertures. (© 1994 IEEE) (c) Index-guided 3QW VCSEL with an etched top mesa and intracavity contacts on a semi-insulating substrate [12]. Data is given for several different device diameters. (© 1994 IEEE)

potential well for these particles. Just as in the discussion of transverse carrier and photon confinement in the one-dimensional diode (Fig. 1.4), we recognize that formation of a lateral heterostructure would be most desirable. Figure 1.13(d) through (f) give examples of three different *buried heterostructure* (BH) configurations. These are clearly the most popular lateral configurations since they can combine all the desired features of current, carrier, and photon confinement. There are many variations on this theme beyond Fig. 1.13(d) through (f), but all require a second growth or solid-state intermixing step, unless a complex initial growth on a pre-patterned substrate is performed.

Explicitly shown in Fig. 1.13(d) is an etched mesa BH (EMBH), which is formed by first growing a planar DH configuration, next etching a narrow mesa stripe down through the active region, and finally regrowing additional lattice-matched semiconductor material with a higher bandgap around this mesa. This regrown material must also block current flow, either by in-corporating reverse biased junctions as shown, or by utilizing material which is doped to be semi-insulating. This latter configuration has been dubbed the semi-insulating planar buried heterostructure (SIPBH). Another interesting variation on this theme, which works well in the InGaAsP/InP system, involves stopping the first growth after just completing the active layer, etching a channel on each side of what is to be the active stripe, and then regrowing with several layers of LPE, which tends to fill in the channels before any growth proceeds on the active stripe. If the doping type is switched a couple of times as the channels are being filled, blocking junctions can be formed there but not on the active stripe. This double-channel planar BH (DCPBH) has the additional advantage of having a larger contact area for reduced series resistance.

Figure 1.13(e) gives another variation on the BH theme. In this case only one epitaxial growth is performed, but the active layer to the left and right of the desired active stripe is *modified to increase its bandgap after growth*. This laterally selective modification has been induced by the diffusion of impurities or vacancies implanted or otherwise created in the desired regions. The diffusing species cause an intermixing of the original cladding and active lattice atoms by requiring them to hop from site to site as the diffusing species move through. As indicated in Fig. 1.13(e), the result is a BH structure which provides lateral carrier and photon confinement. Current confinement is also facilitated by the increased turn-on voltage of the larger-bandgap heterojunction in the inter-mixed regions. This technique works particularly well in the AlGaAs/GaAs system, since the Al and Ga ions have the same size, and they can interchange without changing the lattice constant. Difficulties with the technique result from the relatively high temperatures required ($>800°C$ for several hours with AlGaAs) to accomplish the intermixing.

Figure 1.13(f) illustrates the final type of BH laser that we shall introduce. Here the device is formed by *growing on a prepatterned substrate*. That is, the laterally patterned active region is completely defined by a single growth. However, it may be desirable to grow a first epilayer before patterning the substrate to provide for better current confinement. This is the specific embodiment illustrated. Again, there are many variations on this general theme, but the V-groove laser shown in Fig. 1.13(f) will make the point. In this case the substrate is first prepared by growing an epilayer of semi-insulating material, which for InP is accomplished by Fe doping the layer. V-grooves are then etched so that their bottoms extend through the semi-insulating layer to the conducting substrate. Finally, the DH laser layers are grown, but again, due to the tendency of the growth to planarize (especially by LPE or MOCVD), a thicker and separate active stripe is formed in the V-groove where the current is constrained to flow. As in the other case, many variations on this general theme are possible. For example, the initial SI growth could be skipped in favor of using a final proton implant around the active stripe to improve current confinement. This type of laser is desired because the active region does not have to be exposed to air or damaged in the definition of the lateral heterostructure. Thus, nonradiative interface recombination should be avoided. However, forming a reproducible active region is a challenge, since such patterned growth tends to be very sensitive to the detailed shape of the substrate pattern as well as the growth system parameters.

REFERENCES

[1]   S.M. Sze, *Physics of Semiconductor Devices*, Ch. 12, Wiley-Interscience, New York (1981).

[2]   I. Hayashi, M.B. Panish, P.W. Foy, and S. Sumski, *Appl. Phys. Lett.*, **17**, 109 (1970).

[3]   J.P. Hirtz, M. Razeghi, M. Bonnet, and J.P. Duchemin, in *GaInAsP Alloy Semiconductors*, ed. T.P. Pearsall, Ch. 3, Wiley, New York (1982).

[4]   K.Y. Chang and A.Y. Cho, *J. Appl. Phys.*, **53**, 441, (1982).

[5]   M.B. Panish, H. Temkin, and S. Sumski, *J. Vac. Sci. Technol.*, **B3**, 657, (1985).

[6]   W.T. Tsang, in *Beam Processing Technologies*, ed. N.G. Einspruch, S.S. Cohen, and R.N. Singh, "Chemical Beam Epitaxy," Academic Press, New York, (1989).

[7]   T. Odagawa, K. Nakajima, K. Tanaka, T. Inoue, N. Okaaki, and K. Wakao, *IEEE J. Quantum Electron.*, **29**, 1682, (1993).

[8]   Y. Ueno, H. Fujii, H. Sawano, K. Kobayashi, K. Hara, A. Gomyo, and K. Endo, *IEEE J. Quantum Electron.*, **29**, 1851, (1993).

[9]   P. Unger, G.-L. Bona, R. Germann, P. Roentgen, and D.J. Webb, *IEEE J. Quantum Electron.*, **29**, 1880, (1993).

[10]   W.-X. Zou, K.-K. Law, L.-C. Wang, J.L. Merz, H.E. Hager, and C.-S. Hong, *IEEE J. Quantum Electron.*, **29**, 2097, (1993).

[11]   K.L. Lear, S.P. Kilcoyne, and S.A. Chalmers, *IEEE Photon. Technol. Lett.*, **6**, 778, (1994).

[12]   J.W. Scott, B.J. Thibeault, D.B. Young, L.A. Coldren, and F.H. Peters, *IEEE Photon. Technol. Lett.*, **6**, 678, (1994).

READING LIST

P. Bhattacharya, *Semiconductor Optoelectronic Devices*, Chs. 1 and 2, Prentice Hall, Englewood Cliffs, NJ (1994).

B.E.A. Saleh and M.C. Teich, *Fundamentals of Photonics*, Chs. 15 and 16, Wiley, New York (1991).

A. Yariv, *Optical Electronics*, 4th ed., Ch. 15, Saunders College Publishing, Philadelphia (1991).

G.P. Agrawal and N.K. Dutta, *Semiconductor Lasers*, 2d ed., Chs. 4 and 5, Van Nostrand Reinhold, New York (1993).

PROBLEMS

These problems draw on material from Appendices 1 through 3.

1.1 List three advantages and three disadvantages diode lasers have relative to gas or solid-state lasers.

1.2 Why are III–V materials better than Si for LEDs and lasers?

1.3 An electron is trapped in a one-dimensional potential well 5 nm wide and 100 meV deep.

   (a) How many bound energy states exist?

   (b) What are the energy levels of the first three measured relative to the well bottom?

   (c) If the well energy depth were doubled, how many states would be confined?

   (Assume the free electron mass.)

1.4 Repeat Problem 1.3 for a 10 nm wide GaAs well and AlGaAs barriers.

1.5 Ten potential wells that each have two bound states are brought together so that their wavefunctions overlap slightly. How many bound energy states exist in this system?

1.6 A very long one-dimensional chain consists of atoms covalently bonded together with a resulting center-to-center spacing of 0.3 nm. The band structure of this system can be determined from the overlap of the individual atomic wavefunctions. The coupling energy given by Eq. (A1.21) for a particular atomic energy level, E_a, is 0.2 eV.

   (a) Calculate the band structure over the first two Brillouin zones.

   (b) Calculate the electron effective mass at the band extrema.

1.7 A light source emits a uniform intensity in the wavelength range 0.4–2.0 μm. A polished wafer of GaAs with antireflection coatings on both surfaces is placed between the source and an optical spectrum analyzer.

(a) Sketch the wavelength spectrum received.

(b) Which processes in Fig. 1.3 are significant in forming this spectrum?

1.8   The light source in Problem 1.7 is replaced by a GaAs laser emitting at 850 nm, and it is found that 99.5% of the incident light is absorbed in the GaAs wafer. Now an Ar-ion laser emitting at 488 nm is trained to the same place on the wafer.

(a) As the power of the Ar-ion laser is increased to 3 W, the absorption of the GaAs laser beam is reduced to 50%. Assuming the heat is conducted away, explain what might be happening.

(b) The power in the Ar-ion laser is further increased, and it is found that at about 10 W the GaAs laser beam passes through the wafer unattenuated. Again, neglecting heating effects, explain why it requires 10 W rather than ~ 6 W to reach transparency.

1.9   For good carrier confinement it has been found that the quasi-Fermi levels should remain at least $5\,kT$ below the top of a quantum well at operating temperature. In a particular GaAs quantum-well SCH laser, the operating active region temperature is found to be 125°C. If the quantum well is 80 Å wide, how much Al should be in the separate confinement region to provide the desired $5\,kT$ margin in the conduction band at a carrier density of 4×10^{18} cm^{-3}?

1.10   (a) Plot the carrier density vs. the quasi-Fermi level for the conduction band in bulk GaAs and InGaAsP (1.3 µm) at 300 K. Cover the carrier density range from 1×10^{17} cm^{-3} to 1×10^{19} cm^{-3}, and use a logarithmic scale for the carrier density axis.

(b) With this result answer Problem 1.9 for a AlGaAs/GaAs bulk DH structure.

1.11   Calculate the density of states vs. energy for a "quantum wire" potential well in which two dimensions are relatively small. That is, assume a large dimension ($\gg 10$ nm) in the z-direction and quasi-continuous state energies only for k_z.

1.12   Calculate the density of states vs. momentum for a quantum well.

1.13   Derive Eq. (A3.3).

1.14   Photons are transversely confined in a simple three-layer waveguide in a DH laser consisting of an InGaAsP active region 0.2 µm thick sandwiched between InP cladding layers. The bandgap wavelength of the active region is 1.3 µm.

(a) How many transverse TE modes can exist in this slab waveguide?

(b) Plot the transverse electric field for the lowest-order TE mode.

(c) What is the energy density 0.5 µm above the active-cladding interface relative to the peak value in the active region?

(d) What is the effective index of the guided mode?

(e) What is the transverse confinement factor?

1.15 Suppose the DH laser of Problem 1.14 is now used to form a BH laser with an active region 2 µm wide and InP lateral cladding regions, as in Fig. 1.13(d)

(a) What is the effective index for the fundamental two-dimensionally guided mode?

(b) How many lateral modes are possible?

(c) What is the lateral confinement factor for the fundamental mode?

1.16 VCSELs have been formed by etching 5 µm square pillars through the entire laser structure, creating rather large index discontinuities at the lateral surfaces. Assuming the axial propagation constant, β, is fixed at the same value for all resonant modes, and that the lowest-order mode has a wavelength of 1.0 µm, plot the mode spectrum including the first six lateral modes.

1.17 Derive Eq. (A3.14) and verify Eq. (A3.15).

1.18 It has been proposed that if the lateral dimensions of VCSELs or in-plane lasers become sufficiently small, the density of states for electrons and holes can be modified by the lateral size effect. In VCSEL material with an 80 Å thick GaAs quantum-well active region and high barriers, devices of various lateral widths are formed. How narrow must the device be before the lateral size effect shifts the lowest state energy up by 10 meV (about $\frac{1}{2}kT$ at room temperature)? Neglect any indirect surface-state pinning effect.

CHAPTER TWO

A Phenomenological Approach to Diode Lasers

2.1 INTRODUCTION

In this chapter we attempt to develop an engineering toolbox of diode laser properties based largely upon phenomenological arguments. In the course of this development, we make heavy reference to several appendices for a review of some of the underlying physics.

The chapter begins by developing a rate equation model for the flow of charge into double-heterostructure active regions and its subsequent recombination. Some of this electron–hole recombination generates photons by spontaneous emission. This incoherent light is important in LEDs, and a section is devoted to deriving the relevant equations governing LED operation.

Sections 2.4 through 2.6 provide a systematic derivation of the dc light-current characteristics of diode lasers. First, the rate equation for photon generation and loss in a laser cavity is developed. This shows that only a small portion of the spontaneously generated light contributes to the lasing mode. Most of it comes from the stimulated recombination of carriers. All of the carriers that are stimulated to recombine by light in a certain mode contribute more photons to that same mode. Thus, the stimulated carrier recombination/ photon generation process is a *gain* process. The threshold gain for lasing is studied next, and it is found to be the gain necessary to compensate for cavity losses. The current required to reach this gain is called the threshold current, and it is shown to be the current necessary to supply carriers for the unproductive nonradiative and spontaneous recombination processes, which clamp at their threshold value as more current is applied. Above threshold, all additional injected carriers recombining in the active region are shown to contribute to photons in the lasing mode. A fraction escape through the mirrors; others are absorbed by optical losses in the cavity.

The next section deals with the modulation of lasers. Here for the first time we solve the rate equations for a modulated current. Under small-signal modulation, the rate equations for carriers and photons are found to be analogous to the differential equations that describe the current and voltage in an *RLC* circuit. Thus, the optical modulation response is found to have a resonance and to fall off rapidly above this frequency.

Finally, this chapter reviews techniques for characterizing real lasers. These techniques can be used to extract the important device parameters used in the theoretical derivations. They also provide practical terminal parameters that are useful in the design of optoelectronic circuits.

2.2 CARRIER GENERATION AND RECOMBINATION IN ACTIVE REGIONS

The text accompanying Figs. 1.4, 1.5, and 1.13 considers the current injected into the terminals of a diode laser or LED, and suggests it is desirable to have all of it contribute to electrons and holes which recombine in the active region. However, in practice only a fraction, η_i, of the injected current, I, does contribute to such carriers. In Fig. 2.1 we again illustrate the process of carrier injection into a double-heterostructure active region using a somewhat more accurate sketch of the energy gap vs. depth into the substrate.

Since the definitions of the active region and the internal quantum efficiency, η_i, are so critical to further analysis, we highlight them here for easy reference.

Active region: the region where recombining carriers contribute to useful gain and photon emission.

The active region is usually the lowest bandgap region within the depletion region of a *pin* diode for efficient injection. However, it occasionally is convenient to include some of the surrounding intermediate bandgap regions. Also in this definition, *useful* is the operative word. There may be photon emission and even gain at some undesired wavelength elsewhere in the device.

Internal quantum efficiency, η_i: the fraction of terminal current that generates carriers in the active region.

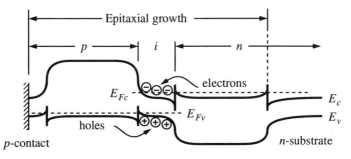

FIGURE 2.1 Band diagram of forward biased double-heterostructure diode.

It is important to realize that this definition includes *all* of the carriers that are injected into the active region, not just carriers that recombine radiatively at the desired transition energy. This definition is oftentimes misstated in the literature.

We also will specifically analyze active regions that are undoped or lightly doped, so that under high injection levels relevant to LEDs and lasers, charge neutrality dictates that the electron density equals the hole density, i.e., $N = P$ in the active region. Thus, we can greatly simplify our analysis by specifically tracking only the electron density, N.

The carrier density in the active region is governed by a dynamic process. In fact, we can compare the process of establishing a certain steady-state carrier density in the active region to that of establishing a certain water level in a reservoir which is being simultaneously filled and drained. This is shown schematically in Fig. 2.2. As we proceed, the various filling (generation) and drain (recombination) terms illustrated will be defined. The current leakage illustrated in Fig. 2.2 contributes to reducing η_i and is created by possible shunt paths around the active region. The carrier leakage, R_l, is due to carriers "splashing" out of the active region (by thermionic emission or by lateral diffusion if no lateral confinement exists) before recombining. Thus, this leakage contributes to a loss of carriers in the active region that could otherwise be used to generate light.

For the DH active region, the injected current provides a generation term, and various radiative and nonradiative recombination processes as well as carrier leakage provide recombination terms. Thus, we can write the rate equation,

$$\frac{dN}{dt} = G_{gen} - R_{rec}, \tag{2.1}$$

where G_{gen} is the rate of injected electrons and R_{rec} is the rate of recombining electrons per unit volume in the active region. Since there are $\eta_i I/q$ electrons per second being injected into the active region,

$$G_{gen} = \frac{\eta_i I}{qV}, \tag{2.2}$$

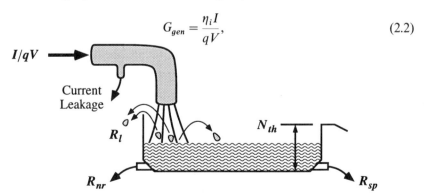

FIGURE 2.2   Reservoir with continuous supply and leakage as an analog to a DH active region with current injection for carrier generation and radiative and nonradiative recombination (LED or laser below threshold).

where V is the volume of the active region. For example, if a current of $I = 20$ mA is flowing into the laser's terminals, a fraction $\eta_i = 80\%$ of the carriers are injected into the active region, and if the active volume is 100 μm^3, then $G_{gen} = 10^{27}$ electrons/s-cm^3. Or, 10^{18} cm^{-3} electrons are injected in 1 ns.

The recombination process is a bit more complicated, since several mechanisms must be considered. As introduced in Fig. 1.3, there is a spontaneous recombination rate, R_{sp}, and a nonradiative recombination rate, R_{nr}. And as depicted in Fig. 2.2, a carrier leakage rate, R_l, must sometimes be included if the transverse and/or lateral potential barriers are not sufficiently high (see Appendix 2 for a discussion of R_l). Finally, under the right conditions, a net stimulated recombination, R_{st}, including both stimulated absorption and emission, is important. Thus, we can write

$$R_{rec} = R_{sp} + R_{nr} + R_l + R_{st}. \tag{2.3}$$

The first three terms on the right refer to the natural or unstimulated carrier decay processes. The fourth one, R_{st}, requires the presence of photons. It is common to describe the natural decay processes by a *carrier lifetime*, τ. In the absence of photons or a generation term, the rate equation for carrier decay is just, $dN/dt = N/\tau$, where $N/\tau \equiv R_{sp} + R_{nr} + R_l$, by comparison to Eq. (2.3). This rate equation defines τ. Also, as mentioned in Chapter 1, this natural decay can be expressed in a power series of the carrier density, N, since each of the terms depends upon the existence of carriers. Thus, we can rewrite Eq. (2.3) in several ways.

$$R_{rec} = R_{sp} + R_{nr} + R_l + R_{st}, \tag{2.3a}$$

$$R_{rec} = \frac{N}{\tau} + R_{st}, \tag{2.3b}$$

$$R_{rec} = BN^2 + (AN + CN^3) + R_{st}, \tag{2.3c}$$

where as the grouping suggests in (2.3c), it has been found that $R_{sp} \sim BN^2$ and $R_{nr} + R_l \sim (AN + CN^3)$. The coefficient B is called the *bimolecular recombination coefficient*, and it has a magnitude, $B \sim 10^{-10}$ cm^3/s for most AlGaAs and InGaAsP alloys of interest. We also note that the carrier lifetime, τ, is not independent of N in most circumstances.

Thus, so far we can write our carrier rate equation in several equivalent ways. We shall deal with R_{st} a little later, but using Eq. (2.3b), our carrier rate equation may be expressed as

$$\frac{dN}{dt} = \frac{\eta_i I}{qV} - \frac{N}{\tau} - R_{st}. \tag{2.4}$$

In the absence of a large photon density, such as in a laser well below threshold or in most LEDs, it can be shown that R_{st} can be neglected. Figure 2.2 illustrates each of these terms in our reservoir analogy, explicitly showing "leaks" R_{sp}, R_{nr}, and R_l for N/τ.

2.3 SPONTANEOUS PHOTON GENERATION AND LEDs

Before proceeding to the consideration of lasers, where R_{st} will become a dominant term above threshold, let us first try to gain some understanding of the situation where the photon density is relatively low, such as in an LED where no feedback is present to provide for the build-up of a large photon density. This case is actually similar to a laser below threshold, in which the gain is insufficient to compensate for cavity losses, and generated photons do not receive net amplification.

The spontaneous photon generation rate per unit volume is exactly equal to the spontaneous electron recombination rate, R_{sp}, since by definition every time an electron–hole pair recombines radiatively, a photon is generated. (Again, N equals the density of electron–hole pairs as well as electrons for relatively light doping). Under *steady-state* conditions ($dN/dt = 0$), the generation rate equals the recombination rate, i.e., from Eqs. (2.2) and (2.3), with $R_{st} \approx 0$,

$$\frac{\eta_i I}{qV} = R_{sp} + R_{nr} + R_l. \tag{2.5}$$

The spontaneously generated optical power, P_{sp}, is obtained by multiplying the number of photons generated per unit time per unit volume, R_{sp}, by the energy per photon, hv, and the volume of the active region, V. We could solve Eq. (2.5) for R_{sp}. but since the exact dependence of $R_{nr} + R_l$ on I is unknown, this leads only to a parametric equation. The conventional approach is to bury this problem by defining a *radiative efficiency*, η_r, where

$$\eta_r = \frac{R_{sp}}{R_{sp} + R_{nr} + R_l}. \tag{2.6}$$

We must not forget that η_r usually depends upon carrier density somewhat. Then, from Eqs. (2.5) and (2.6),

$$P_{sp} = hvVR_{sp} = \eta_i \eta_r \frac{hv}{q} I. \tag{2.7}$$

The product of $\eta_i \eta_r$ is sometimes referred to as the LED internal efficiency. However, we shall *not* use this definition here, since it can lead to serious confusion when we move on to lasers. As we shall see, only η_i appears in the laser output power, and we have called it alone the internal efficiency.

If we are interested in how much power the LED emits into some receiving aperture, P_{LED}, we must further multiply P_{sp} by the *net collection efficiency*, η_c, experienced in transmitting photons out of the semiconductor and into this aperture. This is typically relatively low ($< 10\%$) for most LEDs, because the light is emitted in all directions, and much of it is totally reflected at the semiconductor–air interface. This situation is illustrated in Fig. 2.3.

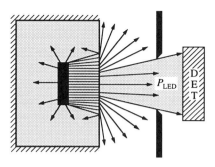

FIGURE 2.3   Schematic of LED showing how only a small portion of the generated light reaches a desired detector.

As indicated by Fig. 2.3 much of the light is reflected back toward the active region rather than being coupled out of the semiconductor chip. A possible consequence is the regeneration of new carriers by the reabsorption of this light. In properly designed LEDs this "photon recycling" can greatly increase their efficiency, yielding an effective $\eta_c > 10\%$.

In any event, the product of the three efficiencies (fraction of carriers injected into the active region, fraction of these recombining radiatively, and the fraction of those usefully coupled out) gives the *external LED quantum efficiency*, η_{ex}. That is,

$$P_{LED} = \eta_c \eta_i \eta_r \frac{h\nu}{q} I = \eta_{ex} \frac{h\nu}{q} I. \qquad (2.8)$$

Thus, ignoring the slight dependence of η_{ex} on I, we see that the power coupled from an LED is directly proportional to the drive current. The external LED quantum efficiency, η_{ex}, is the number of photons coupled to the receiving aperture per electron flowing into the LED.

The frequency response of the LED can also be derived from the carrier rate equation (2.4), with $R_{st} \approx 0$. We shall use the theorem that the Fourier transform of the impulse response in the time domain gives the frequency response. An impulse of current is simply a quantity of charge, which will establish an initial condition of $N(t = 0^+) = N_i$. For $t > 0$, the rate equation can be written as

$$\frac{dN}{dt} = -\frac{N}{\tau} = -AN - BN^2 - CN^3. \qquad (2.9)$$

With the polynomial expansion of the recombination rate, we are reminded that the carrier lifetime, τ, is generally a function of the carrier density. If it were independent of N, the solution would be a simple exponential decay, and the frequency response would be analogous to that of a simple RC circuit in which the 3 dB cutoff frequency, $\omega_c = 1/\tau$. In order for τ to be constant: (1) the cubic term must be negligible and (2) either the linear term, AN, must dominate (not good, since this represents nonradiative recombination) or the active region must be heavily doped, such that the BN^2 term which really equals BNP, can be written as $(BP_d)N$. That is, the p-type doping level, P_d, must be greater than the injection level, N, so that $P_d + P \approx P_d$. Under these conditions, then, the time response is just a simple exponential decay,

$$N(t) = N_i e^{-t/\tau}, \tag{2.10}$$

and the frequency response is a Lorentzian function,

$$N(\omega) = \frac{N(0)}{1 + j\omega\tau}, \tag{2.11}$$

which drops to 0.707 $N(0)$ at $\omega\tau = 1$. For $R_{sp} \approx (BP_d)N$, the power out, P_{LED}, which is proportional to R_{sp}, will also have the same frequency response. The other cases are left as exercises for the reader, but it should be clear that the cutoff frequency will be reduced if the carrier lifetime is increased.

2.4   PHOTON GENERATION AND LOSS IN LASER CAVITIES

For the diode laser, we must now further investigate the nature of the net stimulated recombination rate, R_{st}, in generating photons as well as the effect of the resonant cavity in storing photons. In analogy with Section 2.2, we wish to construct a rate equation for the *photon density*, N_p, which includes the photon generation and loss terms. We shall use the subscript p to indicate that variables are referring to photons.

A main difference between the laser and LED, discussed in Section 2.3 above, is that we only consider light emission into a single mode of the resonant cavity in the laser. Since there are typically thousands of possible optical modes in a diode laser cavity, only a small fraction of R_{sp} contributes to the photon generation rate for a particular mode. Appendix 4 discusses the possible optical modes of a resonant cavity using some of the results of Appendix 3. Note that the number of effective modes in a small vertical-cavity laser can be much fewer, typically dozens rather than thousands.

The main photon generation term above threshold (the regime of interest in lasers) is R_{st}. Every time an electron–hole pair is stimulated to recombine, another photon is generated. However, as indicated in Fig. 2.4, since the cavity volume occupied by photons, V_p, is usually larger than the active region volume occupied by electrons, V, the photon *density* generation rate will be $[V/V_p]R_{st}$ not just R_{st}. This electron–photon overlap factor, V/V_p, is generally referred to as the *confinement factor*, Γ. Sometimes it is convenient to introduce an effective thickness, width, and length that contains the photons, d_{eff}, w_{eff}, and L,

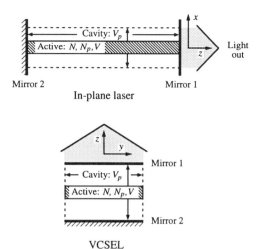

FIGURE 2.4 Schematics of in-plane and vertical-cavity lasers illustrating the active (cross-hatched) and cavity (within dashed lines) volumes as well as the coordinate systems.

respectively. That is, $V_p = d_{eff} w_{eff} L$. Then, if the active region has dimensions, d, w, and L_a, the confinement factor can be expressed as, $\Gamma = \Gamma_x \Gamma_y \Gamma_z$, where $\Gamma_x = d/d_{eff}$, $\Gamma_y = w/w_{eff}$, and $\Gamma_z = L_a/L$. Appendix 5 puts the derivation of Γ on a more rigorous foundation, pointing out that Γ_z is subject to an enhancement factor for $L_a \lesssim \lambda$.

Photon loss occurs within the cavity due to optical absorption and scattering out of the mode, and it also occurs at the output coupling mirror where a portion of the resonant mode is usefully coupled to some output medium. These losses will be quantified in the next section, but for now we can characterize the net loss by a *photon (or cavity) lifetime*, τ_p, analogous to how we handled electron losses above. A first version of the photon rate equation takes the form:

$$\frac{dN_p}{dt} = \Gamma R_{st} + \Gamma \beta_{sp} R_{sp} - \frac{N_p}{\tau_p}, \tag{2.12}$$

where β_{sp} is the *spontaneous emission factor*. As indicated in Appendix 4 for uniform coupling to all modes, β_{sp} is just the reciprocal of the number of optical modes in the bandwidth of the spontaneous emission. As also indicated by Eq. (2.12), in the absence of generation terms, the photons decay exponentially with a decay constant of τ_p. Again, this is really the definition of τ_p.

Equations (2.4) and (2.12) are two coupled equations that can be solved for the steady-state and dynamic responses of a diode laser. However, in their present form there are still several terms that need to be written explicitly in terms of N and N_p before such solutions are possible. First, we shall consider R_{st}.

R_{st} represents the photon-stimulated net electron–hole recombination which generates more photons. This is a *gain* process for photons. As illustrated in Fig. 1.3 and discussed more fully in Appendix 6, the net effect of the upward and downward electronic transitions, corresponding to stimulated absorption and emission of photons, respectively, are included. In Fig. 2.5 we show the growth of a photon density from an incoming value of N_p to an exiting value of $N_p + \Delta N_p$ as it passes through a small length, Δz, of active region. Without loss of generality, but for simplicity, we assume full overlap between the active region and the photon field, i.e., $\Gamma = 1$. As shown, we can also describe this growth in terms of a *gain per unit length, g*, by

$$N_p + \Delta N_p = N_p e^{g\Delta z}. \tag{2.13}$$

If Δz is sufficiently small, $\exp(g\Delta z) \approx (1 + g\Delta z)$. Also, using the fact that $\Delta z = v_g \Delta t$, where v_g is the group velocity, we find that, $\Delta N_p = N_p g v_g \Delta t$. That is, the generation term for dN_p/dt is given by

$$\left(\frac{dN_p}{dt}\right)_{gen} = R_{st} = \frac{\Delta N_p}{\Delta t} = v_g g N_p. \tag{2.14}$$

Thus, we can now rewrite the carrier and photon density rate equations,

$$\frac{dN}{dt} = \frac{\eta_i I}{qV} - \frac{N}{\tau} - v_g g N_p, \tag{2.15}$$

$$\frac{dN_p}{dt} = \Gamma v_g g N_p + \Gamma \beta_{sp} R_{sp} - \frac{N_p}{\tau_p}. \tag{2.16}$$

Of course, we still have not made all the substitutions necessary to directly solve the two equations simultaneously. In Appendix 6 it is suggested that the

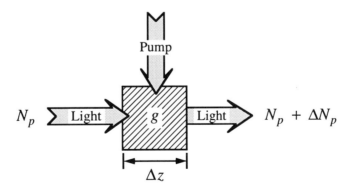

FIGURE 2.5   Definition of gain in terms of the increase in photon number across a small segment of gain material.

gain as a function of carrier density can be approximated by a straight line, at least under small-signal conditions. That is,

$$g \approx a(N - N_{tr}), \tag{2.17}$$

where a is the *differential gain*, $\partial g / \partial N$, and N_{tr} is a *transparency carrier density*. (Actually, a logarithmic function fits the gain better over a wider range of N, as we shall detail in Chapter 4.) Of course, we also know that N/τ can be replaced by the polynomial $AN + BN^2 + CN^3$, where the terms estimate defect, spontaneous (R_{sp}), and Auger recombination, respectively. Nevertheless, we shall leave the rate equations in the general form of Eqs. (2.15) and (2.16) for future reference.

2.5   THRESHOLD OR STEADY-STATE GAIN IN LASERS

In Section 2.4, we characterized the cavity loss by a photon decay constant or lifetime, τ_p. Here, we wish to explicitly express τ_p in terms of the losses associated with optical propagation along the cavity and the cavity mirrors. Also, we wish to show that the net loss of some mode gives the value of net gain required to reach the lasing threshold.

As shown in Appendix 3 and discussed in Chapter 1, the optical energy of a modern diode laser propagates in a dielectric waveguide mode which is confined both transversely and laterally as defined by a normalized transverse electric field profile, $U(x, y)$. In the axial direction this mode propagates as $\exp(-j\tilde{\beta}z)$, where $\tilde{\beta}$ is the complex propagation constant which includes any loss or gain. Thus, the time- and space-varying electric field can be written as

$$\mathscr{E} = \hat{\mathbf{e}}_y E_0 U(x, y) e^{j(\omega t - \tilde{\beta}z)}, \tag{2.18}$$

where $\hat{\mathbf{e}}_y$ is the unit vector indicating TE polarization and E_0 is the magnitude of the field. The complex propagation constant, $\tilde{\beta}$, includes the incremental *transverse modal gain*, $\langle g \rangle_{xy}$ and *internal modal loss*, $\langle \alpha_i \rangle_{xy}$. That is,

$$\tilde{\beta} = \beta + j\beta_i = \beta + \frac{j}{2}(\langle g \rangle_{xy} - \langle \alpha_i \rangle_{xy}), \tag{2.19}$$

where the real part of $\tilde{\beta}$; $\beta = 2\pi\bar{n}/\lambda$, and $\bar{n}$ is an effective index of refraction for the mode, also defined in Appendix 3. As shown in Appendix 5, the transverse modal gain, $\langle g \rangle_{xy}$, and loss, $\langle \alpha_i \rangle_{xy}$, are found from weighted averages of the gain and loss, respectively, across the mode shape, $U(x, y)$. Both are related to power; thus, the factor of $\frac{1}{2}$ in this equation for the amplitude propagation coefficient. From Appendix 5, we can let $\langle g \rangle_{xy} = \Gamma_{xy} g$, where Γ_{xy} is the transverse confinement factor, if $g(x, y)$ is constant across the active region and

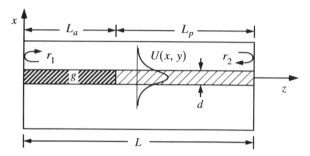

FIGURE 2.6 Generic laser cavity cross section showing active and passive sections (no impedance discontinuity assumed) and the guided-mode profile.

zero elsewhere. This is generally valid for in-plane lasers, but not for VCSELs. Also, for notational convenience, we shall let $\langle \alpha_i \rangle_{xy} = \alpha_i$.

As illustrated in Fig. 2.6, most laser cavities can be divided into two general sections: an active section of length L_a and a passive section of length L_p. Also, g and α_i will clearly be different in these two sections. In the passive section, by definition $g = 0$, and α_i can be given a second subscript to designate its location. The propagating mode is reflected by end mirrors, which have amplitude reflection coefficients of r_1 and r_2, respectively, to provide a resonant cavity. The amount transmitted is potentially useful output.

In order for a mode of the laser to reach threshold, the gain in the active section must be increased to the point where all the propagation and mirror losses are compensated, so that the electric field exactly replicates itself after one round-trip in the cavity. Equivalently, we can unravel the round-trip to lie along the z-axis and require that $\mathscr{E}(z = 2L) = \mathscr{E}(z = 0)$, provided we insert the mode reflection coefficients at $z = 0$ and $z = L$. As a consequence of inserting these boundaries into Eq. (2.18), we obtain

$$r_1 r_2 e^{-2j\tilde{\beta}_{ath}L_a} e^{-2j\tilde{\beta}_{pth}L_p} = 1. \tag{2.20}$$

The subscript th denotes that this characteristic equation only defines the threshold value of $\tilde{\beta}$. (In Chapter 3 we shall take a more basic approach to obtain this same characteristic equation.) Using Eq. (2.19), we can break the complex Eq. (2.20) into two equations for its magnitude and phase. For the magnitude,

$$r_1 r_2 e^{(\Gamma_{xy}g_{th} - \alpha_{ia})L_a} e^{-\alpha_{ip}L_p} = 1, \tag{2.21}$$

where we have chosen reference planes to make the mirror reflectivities real. Solving for $\Gamma_{xy}g_{th}L_a$ we obtain

$$\Gamma_{xy}g_{th}L_a = \alpha_{ia}L_a + \alpha_{ip}L_p + \ln\left(\frac{1}{R}\right), \tag{2.22}$$

where the mean mirror intensity reflection coefficient, $R = r_1 r_2$. For cleaved-facet lasers based upon GaAs or InP, $R \sim 0.32$. Dividing Eq. (2.22) by the total cavity length, L, realizing that $\Gamma_{xy} L_a / L \approx \Gamma_{xy} \Gamma_z = \Gamma$ (exact for $L_a \gg \lambda$), and defining the average internal loss $(\alpha_{ia} L_a + \alpha_{ip} L_p)/L$ as $\langle \alpha_i \rangle$ we have

$$\langle g \rangle_{th} = \Gamma g_{th} = \langle \alpha_i \rangle + \frac{1}{L} \ln\left(\frac{1}{R}\right). \tag{2.23}$$

For convenience the mirror loss term is sometimes abbreviated as, $\alpha_m \equiv (1/L) \ln(1/R)$. Noting that the photon decay rate, $1/\tau_p = 1/\tau_i + 1/\tau_m = v_g(\langle \alpha_i \rangle + \alpha_m)$, we can also write

$$\Gamma g_{th} = \langle \alpha_i \rangle + \alpha_m = \frac{1}{v_g \tau_p}. \tag{2.24}$$

As noted in Appendix 5, if the averaging is initially done over the whole volume, the three-dimensional modal gain and loss used in Eqs. (2.23) and (2.24) are obtained directly. However, this obscures the physics of the re-circulating mode in the cavity, so we have chosen to show the longitudinal weighting separately here. As suggested above, the general $\langle g \rangle_{th}$ form is always valid, but the Γg_{th} form only should be used for in-plane lasers. The limitation that $L_a \gg \lambda$ for $L_a/L = \Gamma_z$ listed above is also discussed in Appendix 5. As explained there, the axial averaging of gain and loss must also use a weighted average over the axial standing wave pattern in the general case. In fact, for very short active regions ($L_a \ll \lambda$), such as in many vertical-cavity lasers, it is possible for $\Gamma_z \approx 2L_a/L$, if the active segment is placed at the peak of the electric-field standing wave (see Appendix 5).

It is important to realize that Eqs. (2.23) and (2.24) give only the cavity loss parameters necessary to calculate the threshold gain. They have nothing to do with the stimulated emission physics which determines what the gain is for a given injection current. This physics is briefly summarized in Appendix 6, and it will be the primary subject of Chapter 4.

For the phase part of Eq. (2.20), $\exp(2j\beta_{tha} L_a) \exp(2j\beta_{thp} L_p) = 1$, requires that $\beta_{tha} L_a + \beta_{thp} L_p = m\pi$, which gives a condition on the modal wavelength,

$$\lambda_{th} = \frac{2}{m} [\bar{n}_a L_a + \bar{n}_p L_p], \tag{2.25}$$

where m is the longitudinal mode number. It should also be realized that $\bar{n}$ varies with wavelength ($\partial \bar{n}/\partial \lambda$, dispersion), and it generally is also dependent upon the carrier density ($\partial \bar{n}/\partial N$, plasma loading). Thus, when making computations these dependences must be included. That is, to determine $\bar{n}$ at a wavelength $\lambda = \lambda_0 + \Delta\lambda$ and a carrier density, $N = N_0 + \Delta N$, we use

$$\bar{n}(\lambda, N) = \bar{n}(\lambda_0, N_0) + \frac{\partial\bar{n}}{\partial\lambda}\Delta\lambda + \frac{\partial\bar{n}}{\partial N}\Delta N. \qquad (2.26)$$

Typically, $\partial\bar{n}/\partial\lambda \sim -1\ \mu m^{-1}$, and $\partial\bar{n}/\partial N \approx \Gamma_{xy}\partial n_A/\partial N \sim -\Gamma_{xy}10^{-20}\ cm^3$, where n_A is the index in the active region. Using Eqs. (2.25) and (2.26) we can find the wavelength separation between two modes, m and $m + 1$, to be

$$\delta\lambda = \frac{\lambda^2}{2(\bar{n}_{ga}L_a + \bar{n}_{gp}L_p)}, \qquad (2.27)$$

where the group effective index for the jth section, $\bar{n}_{gj} = \bar{n}_j - \lambda(\partial\bar{n}/\partial\lambda) = \bar{n}_j + \omega(\partial\bar{n}/\partial\omega)$. The group index in semiconductors is typically 20–30% larger than the index of refraction, depending on the specific wavelength relative to the band edge. From experiments, the values of $\bar{n}_g$ for the active sections of GaAs and InGaAsP DH in-plane lasers are near 4.5 and 4, respectively.

Finally, it is important to note that the steady-state gain in a laser operating above threshold must also equal its threshold value as given by Eq. (2.23). That is, *in a laser cavity,*

$$g(I > I_{th}) = g_{th}. \qquad \text{(steady state)} \qquad (2.28)$$

If the gain were higher than g_{th}, then the field amplitude would continue to increase without bound, and this clearly cannot exist in the steady state. Furthermore, since the gain is monotonically related to the carrier density, this implies that the carrier density must also *clamp* at its threshold value. That is,

$$N(I > I_{th}) = N_{th}. \qquad \text{(steady state)} \qquad (2.29)$$

In fact, what happens when the current is increased to a value above threshold is that the carrier density and gain initially (for on the order of a nanosecond) increase to values above their threshold levels, and the photon density grows. But then, the stimulated recombination term R_{st} also increases, reducing the carrier density and gain until a new steady-state dynamic balance is struck where Eqs. (2.28) and (2.29) are again satisfied. Put another way, the stimulated recombination term in (2.15) uses up all additional carrier injection above threshold. In terms of our reservoir analogy depicted in Fig. 2.2, the water level has reached the spillway and any further increase in input goes over the spillway without increasing the water depth. Of course, the spillway represents simulated recombination. Figure 2.7 shows the analogy in this case.

Figure 2.8 summarizes this carrier clamping effect in a laser cavity. The physics of the g vs. N curve never changes. The feedback effect causes the carrier density to clamp, in order to keep the gain at its threshold value.

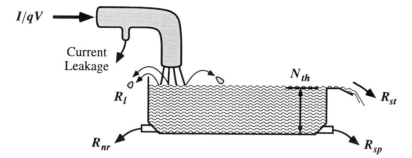

FIGURE 2.7 Reservoir analogy above threshold where water level has risen to the spillway so that an increased input results in an increased output (R_{st}) but no increase in carrier density (water level). The flows R_{nr} and R_{sp} do not change above threshold.

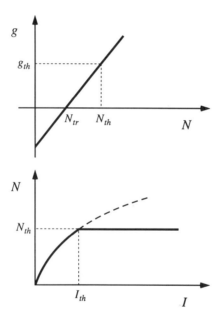

FIGURE 2.8 Gain vs. carrier density and carrier density vs. input current. The carrier density clamps at threshold causing the gain to clamp also.

2.6 THRESHOLD CURRENT AND POWER OUT VS. CURRENT

2.6.1 Basic P–I Characteristics

Although the rate equations (2.15) and (2.16) are valid both above and below threshold, we shall piece together a below-threshold LED characteristic with an above-threshold laser characteristic to construct the power out vs. current in for a diode laser. The LED part is already largely complete with Eq. (2.8). Thus, we shall here concentrate on the above threshold laser part. The first step is to use the below threshold steady-state carrier rate equation, Eq. (2.5) almost at threshold. That is,

$$\frac{\eta_i I_{th}}{qV} = (R_{sp} + R_{nr} + R_l)_{th} = \frac{N_{th}}{\tau}. \qquad (2.30)$$

Then, recognizing that $(R_{sp} + R_{nr} + R_l) = AN + BN^2 + CN^3$ depends monotonically on N, we observe from Eq. (2.29) that above threshold $(R_{sp} + R_{nr} + R_l)$ will also clamp at its threshold value, given by Eq. (2.30). Thus, we can substitute Eq. (2.30) into the carrier rate equation, Eq. (2.15), to obtain a new above-threshold carrier rate equation,

$$\frac{dN}{dt} = \eta_i \frac{(I - I_{th})}{qV} - v_g g N_p, \qquad (I > I_{th}) \qquad (2.31)$$

where we have assumed η_i is not a function of current above threshold. From Eq. (2.31) we can now calculate a steady-state photon density above threshold where $g = g_{th}$. That is,

$$N_p = \frac{\eta_i(I - I_{th})}{qv_g g_{th} V}. \qquad \text{(steady state)} \qquad (2.32)$$

Now with some relatively straightforward substitutions, we can calculate the power out, since it must be proportional to N_p. To obtain the power out, we first construct the *stored optical energy in the cavity*, E_{os}, by multiplying the photon density, N_p, by the energy per photon, hv, and the cavity volume, V_p. That is, $E_{os} = N_p hv V_p$. Then, we multiply this by the *energy loss rate through the mirrors*, $v_g \alpha_m = 1/\tau_m$, to get the optical power output from the mirrors,

$$P_0 = v_g \alpha_m N_p hv V_p. \qquad (2.33)$$

Substituting from Eqs. (2.32) and (2.24), and using $\Gamma = V/V_p$, in Eq. (2.33),

$$P_0 = \eta_i \left(\frac{\alpha_m}{\langle \alpha_i \rangle + \alpha_m}\right) \frac{hv}{q} (I - I_{th}). \qquad (I > I_{th}) \qquad (2.34)$$

Now, by defining

$$\eta_d = \frac{\eta_i \alpha_m}{\langle \alpha_i \rangle + \alpha_m}, \qquad (2.35)$$

we can simplify Eq. (2.34) to be

$$P_0 = \eta_d \frac{hv}{q} (I - I_{th}). \qquad (I > I_{th}) \qquad (2.36)$$

Equation (2.36) represents the total power out of both mirrors. If the mirrors have equal reflectivity, then exactly half will be emitted out of each. If one is totally reflecting, then all will be emitted out the other. On the other hand, if the mirrors have partial but unequal reflectivity, the fraction emitted from each is a nontrivial function which we shall derive in Chapter 3. Equation (2.36) also shows that the power out above threshold is a linear function of the current above threshold. This is true regardless of our assumptions about the form of the gain–current relationship or the nature of the nonradiative recombination mechanisms. The assumptions necessary for this P–I linearity are that the gain–current relationship, the internal efficiency, the confinement factor, and the cavity losses remain constant. As shown in Appendix 5, by confinement factor, we really mean that the *modal gain* must remain constant.

To determine what we should call η_d, we can compare the calculated result of Eq. (2.36) to a measurement. Postulating that it might be related to a quantum efficiency, we calculate a differential quantum efficiency, defined as the number of photons out per electron in from a measured P–I characteristic. As shown in Fig. 2.9, the differential quantum efficiency would be found by measuring the slope $[\Delta P_0/\Delta I]$ in watts/amp above threshold (including output from both ends) and then multiplying this number by $[q/hv]$ in Coulombs/joule to get an empirical number of photons per electron equal to $[\Delta P_0/\Delta I][q/hv]$. Now, if we take the derivative with respect to current of Eq. (2.36), and solve for η_d, we get the same result. This shows that η_d is indeed *the differential quantum efficiency*. To repeat then,

$$\eta_d = \left[\frac{q}{hv}\right]\frac{dP_0}{dI}. \qquad (I > I_{th}) \qquad (2.37)$$

The region in Fig. 2.9 below threshold ($I < I_{th}$) can be approximated by neglecting the stimulated emission term in Eq. (2.16) and solving for N_p, again

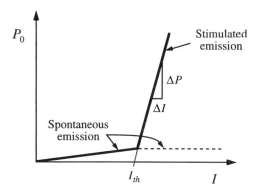

FIGURE 2.9 Illustration of output power vs. current for a diode laser. Below threshold only spontaneous emission is important; above threshold the stimulated emission power increases while the spontaneous emission is clamped at its threshold value.

under steady-state conditions. In this case we find that

$$N_p = \Gamma \beta_{sp} R_{sp} \tau_p. \qquad (I < I_{th}) \qquad (2.38)$$

Using Eqs. (2.38), (2.23), (2.6), and (2.5) in Eq. (2.33), we get the spontaneous emission into the laser mode as

$$P_0(I < I_{th}) = \eta_r \eta_i \left(\frac{\alpha_m}{\langle \alpha_i \rangle + \alpha_m} \right) \frac{h\nu}{q} \beta_{sp} I. \qquad (2.39)$$

Comparing this to the LED expression of Eq. (2.8) shows that $\eta_c = \alpha_m \beta_{sp}/(\langle \alpha_i \rangle + \alpha_m)$ as might have been expected.

At threshold the spontaneous emission clamps as the carrier density clamps since R_{sp} depends upon N. Thus, as the current is increased above threshold, the spontaneous emission *noise* remains constant at the value of Eq. (2.39) with $I = I_{th}$, while the coherent stimulated emission power grows according to Eq. (2.34). As we shall find in Chapter 5, this results in a gradual reduction in the linewidth of the output wavelength as the power is increased.

2.6.2   Relation of Laser Drive Current to Mirror Reflectivity and Cavity Length

Equation (2.36) gives the output power in terms of the additional current applied above threshold. The proportionality factors are constants involving the cavity losses, the lasing wavelength, and the internal efficiency. To design lasers for minimum current at a given output power, we also need an analytic expression for the threshold current. We have the threshold modal gain in terms of the cavity losses, Eq. (2.24), and we have suggested that the gain can be related to the carrier density by either an approximate linear, Eq. (2.17), or more accurate logarithmic (Chapter 4) relationship. The threshold current is also related to the threshold carrier density via the recombination rates, which can be expressed as a polynomial in N, e.g., Eq. (2.30).

In Chapter 4 it will be shown that the gain vs. carrier density can be well approximated by a simple three-parameter logarithmic formula.

$$g = g_0' \ln \frac{N + N_s}{N_{tr} + N_s}. \qquad (2.40)$$

In this approximation, g_0' is an empirical gain coefficient, N_{tr} is the transparency carrier density, and N_s is a shift to force the natural logarithm to be finite at $N = 0$ such that the gain equals the unpumped absorption. However, if we restrict our attention to positive gains, $g \geq 0$, Eq. (2.40) can be further approximated as

$$g = g_0 \ln \frac{N}{N_{tr}}, \qquad (g \geq 0) \qquad (2.41)$$

provided that we use a new gain coefficient g_0. In this case the differential gain, $\partial g/\partial N = g_0/N$. Generally, N_{tr} and $\partial g/\partial N$ will be quite different for bulk, quantum-well, and strained-layer quantum-well active regions. This is the basis

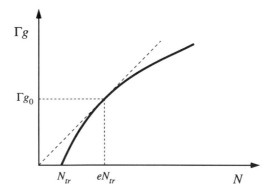

FIGURE 2.10   Schematic illustration of modal gain versus injected carrier density with values labeled from the two-parameter logarithmic fit of Eq. (2.41).

for many of the arguments for and against certain of these structures. Figure 2.10 illustrates schematically the modal gain vs. carrier density with some of the relevant parameters labeled. The point where a line from the origin is tangent, which represents the maximum gain per unit carrier density injected, is simply given by the coordinates, Γg_0, eN_{tr}, with the assumed analytic approximation, Eq. (2.41).

Fitting Eq. (2.41) to numerical gain plots to be found in Chapter 4, a strained 80Å InGaAs/GaAs quantum well yields $g_0 \sim 2100 \text{ cm}^{-1}$ and $N_{tr} \sim 1.8 \times 10^{18} \text{ cm}^{-3}$; and an 80Å GaAs quantum well gives $g_0 \sim 2400 \text{ cm}^{-1}$ and $N_{tr} \sim 2.6 \times 10^{18} \text{ cm}^{-3}$. For InP substrate cases, a strained 30Å InGaAs/InP gives $g_0 \sim 4000 \text{ cm}^{-1}$ and $N_{tr} \sim 3.3 \times 10^{18} \text{ cm}^{-3}$; and an unstrained 60Å InGaAs quantum well gives $g_0 \sim 1800 \text{ cm}^{-1}$ and $N_{tr} \sim 2.2 \times 10^{18} \text{ cm}^{-3}$.

Now we can combine Eqs. (2.24) and (2.41) to get the threshold carrier density,

$$N_{th} = N_{tr} e^{g_{th}/g_0} = N_{tr} e^{(\langle \alpha_i \rangle + \alpha_m)/\Gamma g_0}. \tag{2.42}$$

Using the polynomial fit for the recombination rates in Eq. (2.30), and recognizing that for the best laser material the recombination at threshold is dominated by spontaneous recombination, we have, $I_{th} \cong BN_{th}^2 qV/\eta_i$. Thus,

$$I_{th} \cong \frac{qVBN_{tr}^2}{\eta_i} e^{2(\langle \alpha_i \rangle + \alpha_m)/\Gamma g_0}, \tag{2.43}$$

where for most III–Vs of interest the bimolecular recombination coefficient, $B \sim 10^{-10} \text{ cm}^3/\text{s}$.

Equations (2.36) and (2.43) can now be used for a closed-form expression of output power vs. applied current. However, since we are usually trying to minimize the current needed for a given required power from one mirror, P_{01}, we solve for I.

$$I \cong \frac{qP_{01}(\langle\alpha_i\rangle + \alpha_m)}{F_1\eta_i h\nu\alpha_m} + \frac{qVBN_{tr}^2}{\eta_i}e^{2(\langle\alpha_i\rangle + \alpha_m)/\Gamma g_0}, \tag{2.44}$$

where the first term is the additional current required above threshold to obtain power P_{01} from Eq. (2.36), and the second term is the threshold current, or Eq. (2.43). The factor F_1 is the fraction of the total output power coming out of mirror 1. An exact analytic formula for it will be derived in Chapter 3, but clearly, it is one-half for equal mirrors and unity if mirror 2 has a reflectivity of one.

Equations (2.43) and (2.44) give reasonable accuracy in simple analytic expressions which correctly show that it is always desirable to reduce the transparency value and increase the differential gain of the active material. Both points argue in favor of using quantum-well, especially strained-layer quantum-well, active regions. Relative to the cavity design, the equations also indicate that it is desirable to reduce the cavity loss ($\langle\alpha_i\rangle + \alpha_m$) and volume, V, subject to retaining a reasonably large confinement factor, Γ. Thus, the merits of using vertical-cavity surface emitters or short-cavity in-plane lasers with coated facets are also suggested. These points are quantified in the example given in Fig. 2.11, which plots Eq. (2.44) for a typical set of assumed parameters. Note that for any given values of internal loss and power out there is a trough in required drive current which slopes monotonically downward as R approaches unity and L tends to zero. In practice, this $R = 1$, $L = 0$ minimum can not be approached too closely, since high current densities lead to device heating. For higher powers out or internal losses, the bottom of the current trough moves to smaller Ls and Rs. Figure 2.11 (b) isolates two specific lengths from part (a) and adds other power levels to illustrate this point. Further discussion of optimum laser design is left until Chapter 8 and Appendix 17.

Due to the exponential dependence on g_{th}/g_0 in Eqs. (2.43) and (2.44), it may be beneficial to use more than one quantum well to increase Γ in a quantum-well laser. This dependence is a result of the saturation of the gain as the carrier density is increased to nearly fill the lowest set of states, as discussed in Appendix 6 and Chapter 4. Thus, by distributing the carriers over N_w wells, the gain per well is reduced by less than N_w times, but the modal gain is still multiplied by nearly N_w times this value. For such multiple quantum-well (MQW) lasers Eqs. (2.43) and (2.44) are still valid but one must be sure to multiply the single-well confinement factor, Γ_1, and volume, V_1, by the number of wells, N_w. That is, for an MQW laser, from Eq. (2.43) or the second term in Eq. (2.44), one can explicitly write

$$I_{th_{MQW}} \cong \frac{qN_wV_1BN_{tr}^2}{\eta_i}e^{2(\langle\alpha_i\rangle + \alpha_m)/N_w\Gamma_1 g_0}. \tag{2.45}$$

Here, we have assumed a separate confinement waveguide, so that the optical mode does not change significantly as more wells are added. Also, the number

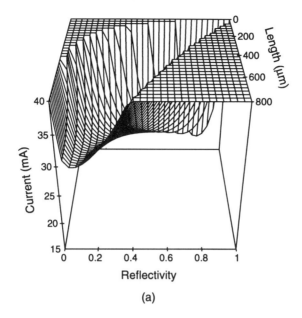

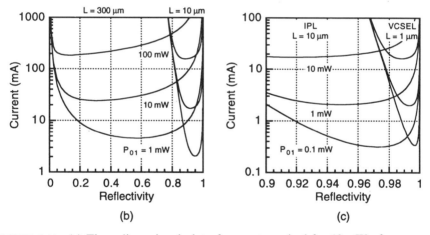

FIGURE 2.11 (a) Three-dimensional plot of current required for 10 mW of power out for lasers of variable length, L, and mean mirror reflectivity, R. Other parameters are: $w = 2.0 \, \mu m$; $d = 10 \, nm$; $\langle \alpha_i \rangle = 20 \, cm^{-1}$; $\Gamma g_0 = 50 \, cm^{-1}$; $F_1 = 0.5$; $N_{tr} = 2 \times 10^{18} \, cm^{-3}$; $\eta_i = 1$; $\Gamma_{xy} = 0.033$; $\Gamma_z = 1$. (b) I vs. R for two lengths from (a) with two additional power levels shown. (c) Comparison of small in-plane laser and VCSEL. $\Gamma_{xy} = 1$; $\Gamma_z = 0.06$ for VCSEL. Active area ($A = 20 \, \mu m^2$) is the same in both. The VCSEL uses three quantum wells rather than one, tripling the active volume. This reduces the value of the optimum R for the VCSEL and broadens the current minimum. However, the VCSEL minimum current is nearly the same as the IPL laser here, since three wells are used.

of wells is limited to the number that can be placed near the maximum of the optical mode. The optimum number of wells is the number that minimizes Eq. (2.45), neglecting nonradiative recombination. With the increased confinement factor we also see that higher powers can be obtained efficiently without moving too far up the gain curve.

If nonradiative recombination is important at threshold, an additional nonradiative threshold current component must be added as outlined earlier. For the long-wavelength InGaAsP/InP materials, nonradiative recombination is known to be very important. In fact, were it not for such recombination, the threshold current densities of lasers using such materials would be lower than those using GaAs quantum wells, as indicated by the gain parameters listed after Eq. (2.41). If such higher-order nonradiative carrier recombination is important at threshold, one must add another component to the threshold current due to the CN_{th}^3 term in the recombination rate. Then, Eqs. (2.43) and (2.44) should be increased by

$$I_{nr_{th}} = \frac{qVCN_{tr}^3}{\eta_i} e^{3(\langle \alpha_i \rangle + \alpha_m)/\Gamma g_0},$$ (2.46)

where for 1.3 μm InGaAsP material, the Auger coefficient, $C \sim 3 \times 10^{-29}$ cm⁶/s, and for 1.55 μm material it is about two or three times larger. The cubic dependence on N_{th} places more importance on reducing the threshold carrier density in this material system. In fact, this additive Auger term dominates Eq. (2.43) for carrier densities above $N_{th} \sim 3 \times 10^{18}$ or 1.5×10^{18} cm⁻³ at 1.3 and 1.55 μm, respectively. This fact focuses more attention on reducing cavity losses, $(\langle \alpha_i \rangle + \alpha_m)$, and maintaining a large confinement factor, Γ. With the use of strained-layer InGaAs/InGaAsP or InGaAs/InGaAlAs quantum wells on InP, a considerable improvement is possible, since all the parameters affecting N_{th} move in the right direction. In fact, the Auger coefficient, C, may also be reduced due to the splitting of the valence bands.

2.7 RELAXATION RESONANCE AND FREQUENCY RESPONSE

Chapter 5 will discuss dynamic effects in some detail. Here, we wish to use Eqs. (2.15) and (2.16) to briefly outline the calculation of relaxation resonance frequency and its relationship to laser modulation bandwidth. As shown in Chapter 5, because of gain compression with increasing photon density and possible transport effects, the calculations are a bit oversimplified, particularly with respect to quantum-well structures. However, these simple equations do seem to work well for standard DH structures, and the method of attack for calculating resonance frequency is also instructive for the more complex calculations to follow in Chapter 5.

Consider the application of an above-threshold dc current, I_0, superimposed with a small ac current, I_1, to a diode laser. Then, under steady-state conditions the laser's carrier density and photon density would respond similarly, with some possible harmonics of the drive frequency, ω, that we shall ignore. Using complex frequency domain notation,

$$I = I_0 + I_1 e^{j\omega t}, \tag{2.47a}$$

$$N = N_0 + N_1 e^{j\omega t}, \tag{2.47b}$$

$$N_p = N_{p0} + N_{p1} e^{j\omega t}. \tag{2.47c}$$

Before applying these to Eqs. (2.15) and (2.16), we first rewrite the rate equations using Eq. (2.17) for the gain. This is valid since small-signal conditions are assumed and the gain can be well approximated as a straight line over some distance, *provided the local slope is used*. We also assume the dc current is sufficiently far above threshold that the spontaneous emission can be neglected. That is,

$$\frac{dN}{dt} = \frac{\eta_i I}{qV} - \frac{N}{\tau} - v_g a (N - N_{tr}) N_p, \tag{2.48}$$

$$\frac{dN_p}{dt} = \Gamma v_g a (N - N_{tr}) N_p - \frac{N_p}{\tau_p}. \tag{2.49}$$

Now, after plugging in Eqs. (2.47) for I, N, and N_p, we recognize that the dc components satisfy the steady-state versions of Eqs. (2.48) and (2.49), i.e., with $d/dt \to 0$; and they can be grouped together and set to zero. Next, we recognize that the steady-state gain factors, $a(N_0 - N_{tr})$ are just equal to g_{th}, and can be replaced by $[\Gamma v_g \tau_p]^{-1}$ according to Eq. (2.24). Finally, we delete the second-harmonic terms that involve $e^{j2\omega t}$, and divide out an $e^{j\omega t}$ common factor. Then,

$$j\omega N_1 = \frac{\eta_i I_1}{qV} - \frac{N_1}{\tau} - \frac{N_{p1}}{\Gamma \tau_p} - v_g a N_1 N_{p0}, \tag{2.50}$$

$$j\omega N_{p1} = \Gamma v_g a N_1 N_{p0}. \tag{2.51}$$

With the above manipulations we have generated frequency domain equations which can easily be solved for the transfer function $N_{p1}(\omega)/I_1(\omega)$.

Before doing this however, let's briefly examine the coupling between the small-signal photon density, N_{p1}, and the small-signal carrier density, N_1. The carrier density depends on N_{p1} through the third term in Eq. (2.50), while the photon density depends on N_1 through Eq. (2.51). If we view the left-hand sides of these two equations as time derivatives, then we observe from Eq. (2.51) that as N_1 increases and becomes positive, N_{p1} increases in time due to increased gain in the laser. However, from the third term in Eq. (2.50), once N_{p1} becomes positive, it serves to decrease N_1 through increased stimulated emission. As N_1 decreases and becomes negative, N_{p1} begins to fall, and once it becomes negative, it again produces an increase in N_1. At this point, the cycle repeats itself. This phenomenon produces a natural resonance in the laser cavity which shows up as a ringing in the output power of the laser in response to sudden changes in the input current. The natural frequency of oscillation associated with this mutual dependence between N_1 and N_{p1} can be found by multiplying Eqs. (2.50) and (2.51) together, ignoring all but the third term on the right-hand side of the first equation:

$$\omega_R^2 = \frac{v_g a N_{p0}}{\tau_p}. \tag{2.52}$$

This natural resonance frequency is commonly referred to as the *relaxation resonance frequency*, ω_R (where *relaxation* refers to an attempt by the photons and carriers to relax to their steady-state values). It is directly proportional to the square root of the differential gain and average photon density in the cavity (output power), and inversely proportional to the square root of the photon lifetime in the cavity.

The relaxation resonance of the laser cavity is much like the natural oscillation of an *LC* circuit. However, the additional terms present in Eq. (2.50) lead to more of an *RLC* circuit behavior, dampening the resonant response. The overall modulation frequency response including these terms is governed by the small-signal transfer function, $N_{p1}(\omega)/I_1(\omega)$. Solving for N_1 in Eq. (2.51), we have $N_1 = j\omega N_{p1}/\Gamma v_g a N_{p0}$. Then eliminating N_1 from Eq. (2.50), and using $P_{ac} = v_g \alpha_m N_{p1} h\nu V_p$, we obtain

$$\frac{P_{ac}(\omega)}{I_1(\omega)} = \frac{\eta_i h\nu}{q} \frac{v_g \alpha_m (v_g a N_{p0})}{v_g a N_{p0}/\tau_p - \omega^2 + j\omega[v_g a N_{p0} + 1/\tau]}. \tag{2.53}$$

Setting $v_g a N_{p0} \equiv \omega_R^2 \tau_p$ and using Eqs. (2.24) and (2.35), the transfer function can be written in a more normalized form:

$$\frac{P_{ac}(\omega)}{I_1(\omega)} = \frac{\eta_d h\nu/q}{1 - (\omega/\omega_R)^2 + j(\omega/\omega_R)[\omega_R \tau_p + 1/\omega_R \tau]}. \tag{2.54}$$

For sufficiently low modulation frequencies, the denominator reduces to one and Eq. (2.54) reduces to the ac equivalent of Eq. (2.36). For higher modulation frequencies, the $1 - (\omega/\omega_R)^2$ term in the denominator creates a strong resonance in the response. Figure 2.12 illustrates the frequency dependence for a wide range of output powers. Note that the resonance is damped at low and high output powers. This occurs because the imaginary damping term in Eq. (2.54) depends on both ω_R and $1/\omega_R$. In Chapter 5, we will find that inclusion of gain compression and transport effects creates significantly more damping than predicted here. In fact, on real laser devices the resonance is typically limited to 5–10 dB (as opposed to the peak ~ 25 dB suggested in Fig. 2.12).

Beyond the strong resonance, the transfer characteristics degrade significantly. Thus, effective modulation of the output power can only be achieved over a

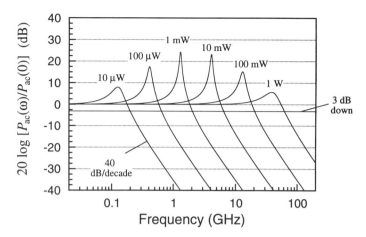

FIGURE 2.12 Frequency response of an idealized diode laser for several different output powers. The active region is characterized by: $hv = 1.5\,\text{eV}$, $a = 5 \times 10^{-16}\,\text{cm}^2$, $\tau = 3 \times 10^{-9}\,\text{s}$, $\eta_i = 86.7\%$, and $v_g = 3 \times 10^{10}/4$ cm/s. The laser cavity is characterized by: $\tau_p = 2 \times 10^{-12}\,\text{s}$ (with $\alpha_m = 60\,\text{cm}^{-1}$ and $\alpha_i = 5\,\text{cm}^{-1}$), $\eta_d = 80\%$, and $V_p = 5\,\mu\text{m} \times 0.25\,\mu\text{m} \times 200\,\mu\text{m}$. The $20\log[P_{ac}(\omega)/P_{ac}(0)]$ is used because photodetection generates an electrical *current* in direct proportion to the optical *power*. Thus, for a power ratio in the electrical circuit, this current must be squared.

modulation bandwidth of $\sim \omega_R$. When the damping is small, the electrical 3 dB down frequency (i.e. the frequency which reduces the received *electrical* power to one-half its dc value) is given by $\omega_{3\,dB} = \sqrt{1 + \sqrt{2}}\,\omega_R$. Expanding Eq. (2.52) using Eqs. (2.24), (2.33), and (2.35), we can express this result in terms of the output power:

$$f_{3\,dB} \approx \frac{1.55}{2\pi}\left[\frac{\Gamma v_g a}{hvV}\frac{\eta_i}{\eta_d}\right]^{1/2}\sqrt{P_0}. \qquad \text{(small damping)} \qquad (2.55)$$

The modulation bandwidth of the laser can be steadily enhanced by increasing the output power. However, increased damping of the resonance at high powers, thermal limitations, and high-power mirror facet damage set practical limits on the maximum average operating power we can use.

Since thermal limits are usually associated with the drive current, it is also convenient to express ω_R in terms of current. Using Eq. (2.32) for N_{p0}, with g_{th} given by Eq. (2.24), Eq. (2.52) becomes

$$\omega_R = \left[\frac{\Gamma v_g a}{qV}\eta_i(I - I_{th})\right]^{1/2}. \qquad (2.56)$$

In this form we observe that it is desirable to enhance the differential gain, minimize the volume of the mode ($\Gamma/V = 1/V_p$), and maximize the current relative to threshold for maximum bandwidth. If we want to keep the overall drive current low, then we should also try to minimize the threshold current, perhaps by increasing the facet reflectivity. If however, we are more concerned about keeping the photon density low (for example, to reduce the risk of facet damage), then from (2.52) we should try to decrease the cavity lifetime instead, perhaps by *decreasing* the facet reflectivity. Thus, the optimum cavity design for a high-speed laser depends on what constraints we place on the device operation. In Chapter 5 we will find that at very high powers, the maximum bandwidth actually becomes independent of ω_R, and is more fundamentally related to the damping factor (the K-factor) which is affected by gain compression and transport effects.

2.8   CHARACTERIZING REAL DIODE LASERS

In this section we wish to review some of the common measurements that are made on diode lasers. We shall emphasize those which can be used to extract internal parameters that we have used in the rest of this chapter. More complex characterization techniques will be delayed until after the discussion of dynamic effects and the introduction of more complex cavity geometries.

2.8.1   Internal Parameters for In-Plane Lasers: $\langle \alpha_i \rangle$, η_i, and g vs. J

Perhaps the most fundamental characteristic of a diode laser is the P–I characteristic as has been illustrated in Fig. 2.9. From a measured P–I characteristic one can immediately determine the experimental threshold current, I_{th}, from the intercept of the above-threshold curve with the abscissa. The differential quantum efficiency, η_d, can be calculated from Eq. (2.36), provided the wavelength is known. Usually the mean mirror reflectivity, $R = r_1 r_2$, can be calculated with good accuracy, and the length, L, can be measured. Thus, the mirror loss, $\alpha_m \equiv (1/L) \ln(1/R)$ can be calculated. However, the net internal optical loss $\langle \alpha_i \rangle$ and quantum efficiency, η_i, cannot be determined from a single device.

To determine these important internal parameters, one commonly uses two or more lasers of different length fabricated from the same material with identical mirrors. This is relatively straightforward for in-plane lasers, since the length can be varied at the final cleaving step. From Eq. (2.35) it can be seen that by measuring the differential efficiency of two such lasers, one is left with two equations containing two unknowns, $\langle \alpha_i \rangle$ and η_i. That is,

$$\eta_d = \frac{\eta_i \ln\left(\dfrac{1}{R}\right)}{L\langle \alpha_i \rangle + \ln\left(\dfrac{1}{R}\right)},$$

and

$$\eta_d' = \frac{\eta_i \ln\left(\dfrac{1}{R}\right)}{L'\langle\alpha_i\rangle + \ln\left(\dfrac{1}{R}\right)}, \tag{2.57}$$

where L and L' are the lengths of the two different lasers. Solving, we find

$$\langle\alpha_i\rangle = \frac{\eta_d' - \eta_d}{L\eta_d - L'\eta_d'} \ln\left(\frac{1}{R}\right),$$

and   (2.58)

$$\eta_i = \eta_d\eta_d' \frac{L - L'}{L\eta_d - L'\eta_d'}.$$

If indeed one can make two identical lasers except for their lengths, then Eqs. (2.58) will give the desired internal parameters. However, experimental data usually have some uncertainty which limits the utility of these expressions. For more reliability, it is generally better to plot a number of data points on a graph and determine the unknowns by fitting a curve to the data. In the present case it is most convenient to plot the reciprocal of the measured differential efficiencies vs. L. Then a straight line through the data has a slope and intercept from which $\langle\alpha_i\rangle$ and η_i can be determined. More specifically,

$$\frac{1}{\eta_d} = \frac{\langle\alpha_i\rangle}{\eta_i \ln(1/R)} L + \frac{1}{\eta_i}. \tag{2.59}$$

Thus, the intercept gives η_i, and this can be used in the slope to get $\langle\alpha_i\rangle$.

Figure 2.13 shows such a plot for some data taken from broad-area in-plane InGaAs/GaAs quantum-well lasers. Single (SQW) and double (DQW) quantum-well cases are included [1].

For shorter cavity lengths, the data in Fig. 2.13 will fall above the line indicated. This data was ignored when determining the line fit because it represents a region where higher-order effects result in an incomplete clamping of the carrier density above threshold. The result is an apparent decrease in η_i (see Appendix 2 for details). If one assumes that the net internal loss does not change in this process, it is possible to estimate the decrease in η_i by repeated use of Eq. (2.59) or (2.35) for these high-gain points.

In the process of taking the above data, one can also generate a table of threshold current densities in the active region, $J_{th} = (\eta_i I_{th}/wL)$ vs. L. These are usually taken from broad-area devices so that lateral current and carrier leakage can be neglected. From Eq. (2.23) we also see that corresponding threshold modal gains Γg_{th} can be calculated for each length once the internal loss is found. Thus, it is possible to construct the modal gain vs. current density

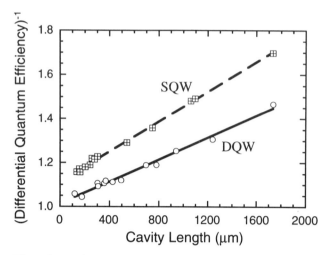

FIGURE 2.13   Plot of experimental reciprocal external differential efficiencies vs. laser cavity length for 50 μm wide $In_{0.2}Ga_{0.8}As$ GRINSCH quantum well lasers. For both single quantum-well (SQW) and double quantum-well (DQW) cases, the $In_{0.2}Ga_{0.8}As$ well(s) was 80Å wide. For the DQW a 12 nm GaAs separation barrier was used, and in both cases 40 nm of GaAs was used on each side of the well(s). On each side of this active region, the barrier stepped to $Al_{0.2}Ga_{0.8}As$ for 8 nm and then tapered to $Al_{0.8}Ga_{0.2}As$ over 80 nm to form the graded-index GRINSCH structure [1]. From these data the SQW had $\alpha_i = 3.2$ cm^{-1} and $\eta_i = 89.6\%$; and the DQW had $\alpha_i = 2.6$ cm^{-1} and $\eta_i = 98.6\%$.

characteristic for the laser from these threshold values. Since the confinement factor, Γ, can usually be calculated as discussed in Appendix 5, one can ultimately determine the basic material gain vs. current density characteristic for the active material. Figure 2.14 gives the result for the example in Fig. 2.13.

2.8.2   Internal Parameters for VCSELs: η_i and g vs. J, $\langle \alpha_i \rangle$, and α_m

In vertical-cavity lasers the above procedure is a little difficult to carry out since the cavity length is set by the crystal growth. Multiple growths may result in other changes in the material besides the cavity length. Therefore, it has been proposed that the desired information can be determined by making in-plane cleaved lasers fabricated from the vertical-cavity laser material. However, a somewhat different approach is followed. Clearly, the internal loss determined for the in-plane laser will not be the same as for the VCSEL since the optical mode travels through a different cross section of materials. Nevertheless, if the electrical pumping current follows the same path and the threshold current densities covered in the in-plane diagnostic lasers includes the VCSEL values, the measured internal quantum efficiency should be the same.

The most valuable piece of information provided by the diagnostic lasers is the gain vs. current density characteristic. Combining this characteristic and

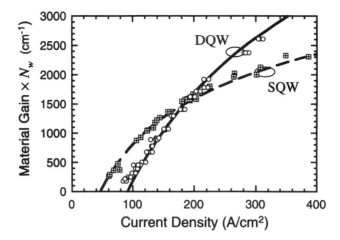

FIGURE 2.14 Experimentally determined gain vs. current density for an InGaAs/GaAs quantum well laser described in Fig. 2.13 [1]. For the ordinate the modal gain Γg is divided by the confinement factor for one well Γ_w to give the material gain g times the number of wells, N_w. The solid curves are from a calculation based upon the theory to be developed in Chapter 4.

the internal quantum efficiency from the in-plane lasers together with the measured threshold current density and differential quantum efficiency from the VCSEL, we now have enough information to unambiguously determine the VCSEL internal loss and mirror loss (and thus, reflectivity). That is, Eqs. (2.35) and (2.23) can be solved for $\langle \alpha_i \rangle$ and α_m, since Γg_{th}, η_d, and η_i are known. The results are

$$\alpha_m = \Gamma g_{th} \frac{\eta_d}{\eta_i},$$

and

(2.60)

$$\langle \alpha_i \rangle = \Gamma g_{th} 1 - \frac{\eta_d}{\eta_i} \rightarrow \langle \alpha_i \rangle = \Gamma g_{th} \left[1 - \frac{\eta_d}{\eta_i} \right]$$

As before, the confinement factors for both the in-plane diagnostic lasers and the VCSELs must be calculated.

2.8.3 Efficiency and Heat Flow

Just as the differential efficiency is important in determining the electrical to optical modulation efficiency, the overall net power conversion efficiency is also important in determining the achievable optical power out as well as the circuit heating and system power requirements. This so-called *wall-plug* efficiency is simply the optical power out relative to the electrical power in, $\eta = P_0/P_{in}$. The optical power out is given by Eq. (2.34), and the electrical power in is the product of the drive current and the total voltage across the diode's terminals. We can express this as

$$P_{in} = I^2 R_s + I V_d + I V_s, \qquad (2.61)$$

where R_s is the series resistance, V_s is a current-independent series voltage, and V_d is the ideal diode voltage, which is equal to the quasi-Fermi level separation. This voltage is clamped at its threshold value above threshold.

The power dissipated in the laser is

$$P_D = P_{in} - P_0 = P_{in}[1 - \eta], \qquad (2.62)$$

and the temperature rise is

$$\Delta T = P_D Z_T, \qquad (2.63)$$

where Z_T is the thermal impedance. Analytic expressions for Z_T, which are approximately valid for several practical cases of interest, exist. Three are illustrated in Fig. 2.15. For a heat sink plane positioned much closer than the lateral dimensions of the regions generating the heat (Fig. 2.15(a)), a one-dimensional heat flow can be assumed. In this case,

$$Z_T = \frac{h}{\xi A}, \qquad \text{(1-D flow)} \qquad (2.64)$$

where ξ is the thermal conductivity of the material separating the source of area A a distance h from the ideal sink. For GaAs and AlAs, $\xi \sim 0.45$ and 0.9 W/cm-°C. For $Al_x Ga_{1-x} As$, alloy scattering reduces ξ to a minimum of 0.11 W/cm-°C at $x \approx 0.5$. For a linear stripe heat source of length l and width w on a thick substrate (thickness h), which is somewhat wider (width $2w_s$) than this thickness (Fig. 2.15(b)), a quasi two-dimensional heat flow results. Then,

$$Z_T \approx \frac{\ln(4h/w)}{\pi \xi l}. \qquad \text{(line: } w \ll h < w_s) \qquad (2.65)$$

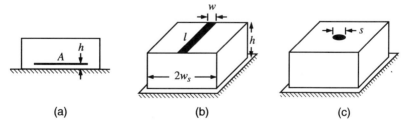

(a)   (b)   (c)

FIGURE 2.15 Schematics of heat flow geometries relevant to lasers: (a) planar or one-dimensional flow for a heat sink relatively near the heat source; (b) a line source on a thick substrate; and (c) a disk source on a half-space.

A narrow stripe in-plane laser mounted active region up on a relatively thick substrate approximates this case. For a disk heat source of diameter s on a half-space (Fig. 2.15(c)), a three-dimensional flow into the half-space can be assumed. Then,

$$Z_T = \frac{1}{2\xi s}. \quad \text{(disk)} \quad (2.66)$$

This is approximately valid for a small-diameter VCSEL mounted on the top side of a relatively thick substrate.

2.8.4 Temperature Dependence of Drive Current

The required drive current for a given power out of a laser is given by Eq. (2.44), in which the first and second terms give the needed current above threshold and the threshold current, respectively. In this equation it is assumed that the recombination below threshold is dominated by spontaneous emission events. If significant nonradiative recombination exists, an additional threshold term, such as Eq. (2.46), must be added. For both in-plane and vertical cavity lasers these expressions are functions of temperature. Generally, more current is required both for threshold and the increment above threshold as the temperature is increased, and we can estimate the nature of this dependence by exploring the temperature dependence of each of the factors in the terms of Eqs. (2.44) and (2.46).

However, for VCSELs as well as single axial-mode in-plane lasers, the situation is complicated by the integrated mode selection filter (e.g., Bragg mirrors), which can force the lasing mode to be well off the wavelength where the gain is a maximum. Thus, such lasers can be designed to have anomolous temperature behavior, since the wavelength of the cavity mode and the gain peak shift at different rates versus temperature. In fact, by deliberately misaligning the mode from the gain peak at room temperature, it is even possible to make the threshold go down with increasing temperature as the gain moves into alignment with the mode [2]. In this section, we will not consider these relative mode–gain alignment issues. Rather, we shall assume that a spectrum of modes exist, as in a simple in-plane laser, so that lasing always can occur at the gain peak. Thus, we again can focus only on the temperature dependence of the various factors in Eqs. (2.44) and (2.46).

For the threshold current in Eq. (2.43), there are three factors which generally have a significant temperature dependence: N_{tr}, g_0, and $\langle \alpha_i \rangle$. From the gain calculations of Chapter 4, it may be shown that over some range of temperatures, $N_{tr} \propto T$, $g_0 \propto 1/T$, and $\langle \alpha_i \rangle \propto T$. The transparency carrier density is increased and the gain parameter is reduced because injected carriers spread over a wider range in energy with higher temperatures. The increased internal loss results from the required higher carrier densities for threshold. From Eq. (2.43), we conclude that both the gain and the internal loss variations result in an exponential temperature dependence of the threshold current, while the linear dependence of the transparency carrier density is not significant over small temperature ranges. Additional threshold components such as Eq. (2.46) will introduce further temperature dependencies. For example, in Chapter 4 and Appendix 2 it is shown that $C \propto \exp(\gamma_C T)$ and $R_l \propto \exp(\gamma_l T)$. Thus, Auger recombination and carrier leakage both contribute additional exponential increases in the threshold current. These observations suggest that the threshold current can be approximately modeled by

$$I_{th} = I_0 \, e^{T/T_0}, \tag{2.67}$$

where T_0 is some overall characteristic temperature, and both temperatures are given in degrees Kelvin, K. Note that small values of T_0 indicate a larger dependence on temperature (since $dI_{th}/dT = I_{th}/T_0$). It should also be noted that any minor temperature dependence of other parameters can easily fit into this model over some limited temperature range. For example, the internal efficiency can decrease at higher temperatures due to increased leakage currents and/or higher-order effects discussed in Appendix 2. This decrease in η_i will show up as a reduction of T_0 over a limited temperature range, regardless of the exact dependence of η_i on temperature.

For good near-infrared (~ 850 nm) GaAs/AlGaAs DH lasers, observed values of T_0 tend to be greater than 120 K near room temperature. For quantum-well GaAs/AlGaAs the values are somewhat higher (~ 150–180 K), and for strained-layer InGaAs/AlGaAs quantum wells, $T_0 \geq 200$ K have been observed. For 1.3–1.55 μm InGaAsP/InP DH and quantum-well lasers the characteristic temperature is generally quite a bit lower as expected. Measured values tend to fall in the 50–70 K range, due to Auger recombination as well as possible carrier leakage and intervalence band absorption effects. Thus, the threshold tends to change significantly between room temperature and 100°C, usually resulting in relatively poor performance at the higher temperatures, and generally requiring the use of thermoelectric coolers. Shorter wavelength (600–800 nm) AlGaAs/GaAs and AlInGaP/GaAs lasers also tend to have a smaller T_0 than the near-infrared variety, presumably due to increased carrier leakage.

The above-threshold current required to obtain a desired output power is also temperature dependent, although the dependence is usually smaller than for the threshold current. This dependence results from a reduction in the differential quantum efficiency. As suggested by the constituent factors in the first term in Eq. (2.44), an increase in $\langle \alpha_i \rangle$ as well as a drop in η_i are usually the cause of the increase in $I - I_{th}$. In analogy with Eq. (2.67), we can write

$$I - I_{th} = I_{p0} \, e^{T/T_\eta}, \tag{2.68}$$

where T_η is the characteristic temperature for the above-threshold current increment. As can be found from Fig. 1.14, T_η is generally two or three times larger than T_0, as might be expected from the above discussion. That is, T_0

includes several effects in addition to those in T_η. In summary, for the total drive current for a given power out, we need four parameters to express the temperature dependence. That is,

$$I = I_0 e^{T/T_0} + I_{p0} e^{T/T_\eta}. \tag{2.69}$$

2.8.5 Derivative Analysis

Real diode lasers do not always have perfectly linear P–I characteristics above threshold, and they have parasitic series resistance as well as a possible series voltage as outlined in Eq. (2.61) above. Derivatives of the P–I and V–I characteristics can be useful in sorting out these nonidealities. The dP/dI characteristic in an ideal laser would only provide a good measure of the threshold current and a slope to determine η_d above threshold. However, actual P–I characteristics can have kinks, and they tend to be nonlinear. The kinks

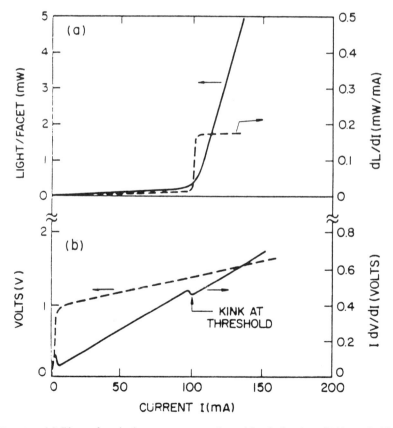

FIGURE 2.16   (a) Plots of optical output power, P, and its derivative, $dP/dI \equiv dL/dI$, vs. drive current, I, for a stripe-geometry gain-guided InGaAsP laser. (b) Plots of the terminal voltage, V, and IdV/dI vs. current for the same device [3]. (Reproduced, by permission, from *Semiconductor Lasers*.)

can indicate a switching between lateral or axial modes or an additional parasitic mirror in the device. These are obviously emphasized in a derivative curve. Premature saturation of the output power may indicate the existence of current leakage paths that "turn-on" at higher current levels or excessive heating of the gain material. The derivative curve gives a good quantitative measure of these symptoms. Figure 2.16(a) gives example plots of $P-I$ and dP/dI for an in-plane laser.

In addition to the $V-I$ characteristic, it is common to plot IdV/dI vs. I. This latter characteristic gives a sensitive measure of the series resistance, and it is particularly useful in identifying shunt current paths. Since the voltage across the junction clamps at threshold with the carrier density, a kink in the curve occurs at that point. Figure 2.16(b) shows example plots of both V and IdV/dI vs. I. The information contained in the plot can be derived by

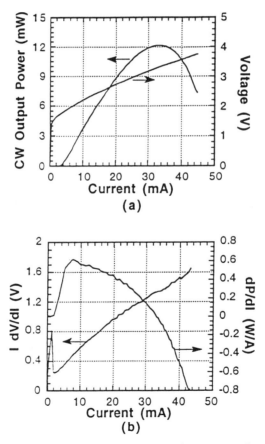

FIGURE 2.17   (a) Output power and terminal voltage for a 20 μm diameter VCSEL with three InGaAs strained quantum wells and AlAs/GaAs DBR mirrors. (b) Derivative curves for the same device [4].

considering an equivalent circuit with a parasitic resistance in series with an ideal heterojunction diode. The diode $V–I$ is described by

$$I = I_0[e^{qV_d/nkT} - 1].\tag{2.70}$$

Taking the derivative of the terminal voltage, $V = V_d + IR$, and solving for dV_d/dI from Eq. (2.70), we obtain for $I \gg I_0$ but below threshold,

$$I\frac{dV}{dI} = \frac{nkT}{q} + IR.\tag{2.71}$$

Above threshold V_d is constant, so

$$I\frac{dV}{dI} = IR.\tag{2.72}$$

Thus, we see that the slope above and below threshold should be R, but there is a positive offset of nkT/q below threshold, which provides a kink of this magnitude at threshold. Now if a shunt resistance is added to the equivalent circuit, it turns out that an additional term must be added to Eq. (2.71). This provides a peak in the $I(dV/dI)$ characteristic below threshold. For common DH structures the diode ideality factor $n \sim 2$.

Figure 2.17 gives plots analogous to Fig. 2.16 for an InGaAs/GaAs VCSEL. Here, significant local heating causes the $P–I$ curve to roll over at relatively low powers. This results in a negative dP/dI beyond this point. In addition, significant series resistance makes it difficult to discern the nkT/q kink in the IdV/dI characteristic at threshold. Thus, the derivative analysis is not always very effective for VCSELs.

REFERENCES

[1] S.Y. Hu, D.B. Young, S.W. Corzine, A.C. Gossard, and L.A. Coldren, *J. Appl. Phys.*, **76**, 3932 (1994).

[2] D.B. Young, J.W. Scott, F.H. Peters, M.G. Peters, M.L. Majewski, B.J. Thibeault, S.W. Corzine, and L.A. Coldren, *IEEE J. Quantum Electron.*, **29**, 2013 (1993).

[3] G.P. Agrawal and N.K. Dutta, *Semiconductor Lasers*, 2d ed. Ch. 4, Van Nostrand Reinhold, New York (1993).

[4] M.G. Peters, PhD dissertation, Electrical and Computer Engineering Dept., University of California, Santa Barbara (1995).

READING LIST

G.P. Agrawal and N.K. Dutta, *Semiconductor Lasers*, 2d ed., Ch. 2, Van Nostrand Reinhold, New York (1993).

K.J. Ebeling, *Integrated Opto-electronics*, Ch. 10, Springer-Verlag, Berlin (1993).

J.T. Verdeyen, *Laser Electronics*, 2d ed., Chs. 7 and 11, Prentice-Hall, Englewood Cliffs, NJ (1989).

A. Yariv, *Optical Electronics*, 4th ed., Ch. 15, Saunders College Publishing, Philadelphia (1991).

PROBLEMS

These problems draw on material from Appendices 4 through 6.

2.1   In a diode laser, the terminal current is I, the current bypassing the active region is I_b, the current due to carriers leaking out of the active region before they recombine is I_l, the current contributing to nonradiative recombination in the active region is I_{nr}, the current contributing to spontaneous emission in the active region is I_{sp}, the currents contributing to spontaneous emission and nonradiative recombination outside the active region are I'_{sp} and I'_{nr}, respectively, and the current contributing to stimulated emission in the active region is I_{st}.

   (a) What is the internal efficiency?

   (b) If the measured external differential efficiency above threshold is η_d, what is the ratio of the mirror loss to the total cavity loss?

   (c) For below-threshold operation, what is the radiative efficiency?

2.2   A reservoir of area A is filled at a rate of R_f (in ft^3/min.) and simultaneously drained from two pipes which have flow rates that depend upon the height of water, h. The drain rates are, $R_{d1} = C_1 h$, and $R_{d2} = C_2 h^2$, respectively.

   (a) Write a rate equation for the water height.

   (b) What is the steady-state water height?

   (c) If $A = 100$ ft^2, $R_f = 10$ ft^3/min, and $C_1 \approx 0$, what is C_2 for a steady-state depth of 5 ft?

2.3   What is the approximate intrinsic cutoff frequency of an LED with a p-type active region doping of 6.3×10^{18} cm^{-3}?

2.4   The relative increase in photons in passing through a piece of GaAs is found to be, $[1/N_p][dN_p/dt] = 10^{13}$ s^{-1}. What is the material gain in cm^{-1}?

2.5   A 1.3 μm wavelength InGaAsP/InP diode laser cavity is found to have an optical loss rate of 4×10^{12}/s.

   (a) What is the photon lifetime?

   (b) What is the threshold modal gain?

2.6   In a cleaved-facet 1.55 μm InGaAsP/InP multiple quantum-well laser

400 µm in length, it is known that the internal efficiency and losses are 80% and 10 cm^{-1}, respectively.

(a) What is the threshold modal gain?

(b) What is the differential efficiency?

(c) What is the axial mode spacing?

2.7   A cleaved-facet, DH GaAs laser has an active layer thickness of 0.1 µm, a length of 300 µm, and a threshold current density of 1 kA/cm^2. Assume unity internal efficiency, an internal loss of 10 cm^{-1}, a confinement factor of 0.1, and only radiative recombination.

(a) What is the threshold carrier density in the active region?

(b) What is the power out of one cleaved facet per micrometer of width at a current density of 2 kA/cm^2?

(c) What are the photon and carrier densities at 2 kA/cm^2?

2.8   In the device of Problem 2.7, gain transparency ($g = 0$) is found to occur at 0.5 kA/cm^2 and the transverse confinement factor is 0.15. What is the relaxation resonance frequency at 2 kA/cm^2?

2.9   Two broad-area DH 1.3 µm InGaAsP/InP lasers are cleaved from the same material. One is 200 µm long and the other is 400 µm long. The threshold current densities are found to be 3 kA/cm^2 and 2 kA/cm^2, respectively, and the differential efficiencies including both ends are measured to be 60% and 50%, respectively.

(a) What are the internal quantum efficiency and internal loss for this material?

(b) For a $\pm 1\%$ error in each of the measured differential efficiencies, what are the errors in the calculated internal loss and quantum efficiency?

2.10   For the material of Problem 2.9, the relaxation resonance frequency for the 200 µm laser biased at twice threshold is found to be 3 GHz. What is the resonance frequency for the 400 µm device also biased at twice threshold?

2.11   A VCSEL is formed with multilayer AlGaAs mirrors and a 3-quantum-well GaAs active region. Current is injected through the mirrors. At a terminal current density of 1 kA/cm^2 the active region provides 1% of one-pass gain for the propagating axial mode. The internal efficiency is assumed to be 80% and the average internal loss is 25 cm^{-1}. The effective cavity length is 1.5 µm.

(a) What mean mirror reflectivity is necessary for the device to reach threshold at 1 kA/cm^2?

(b) For this case, plot the output power density vs. terminal current density.

(c) If we assume the gain is linear with carrier density, and that only spontaneous recombination is important below threshold, plot the threshold current density vs. mean mirror reflectivity for $0.98 < R < 1.0$. On the opposite axis label the differential efficiency at each 0.005 reflectivity increment.

2.12 With the VCSEL material of Problem 2.11, etched square mesas are now formed measuring s on each side. Assuming a spontaneous bandwidth of 30 nm, an axial confinement factor of $2L_a/L$, lateral confinement factors of unity, and that the approximations of Appendix 4 are valid, plot the spontaneous emission factor vs. s.

2.13 Again with the VCSEL material of Problem 2.11, square mesas are formed measuring s on each side by etching down to the active region, as illustrated in Fig. 1.11. The GaAs substrate may be assumed to be thick and wide. If we assume a threshold current density of 1 kA/cm^2 independent of area, a series voltage of 1 V, and a series resistance that is inversely proportional to device area, $R_s = 20$ kΩ-μm^2/s^2.

(a) Plot the temperature at the base of the mesa (active region location) vs. s for a current of twice threshold. Cover $1 < s < 20$ μm.

(b) Assuming a differential efficiency of 50%, plot the power out and required current vs. s for the conditions of (a).

2.14 In a 1.55 μm InGaAsP/InP BH laser, the active region is 0.2 μm thick, 3 μm wide and 300 μm long. The internal efficiency is 70%. In addition, there is a 400 μm long passive waveguide channel with the same lateral and transverse dimensions butted to the end of the active region. The transverse and lateral confinement factors are 0.2 and 0.8, respectively. Cleaved mirrors form a 700 μm long cavity, and other internal reflections can be neglected. The material losses are 80 cm^{-1}, 20 cm^{-1}, and 5 cm^{-1} in the active, passive, and cladding regions, respectively. The gain vs. carrier density characteristic for the active material is linear with a transparency carrier density of 2×10^{18} cm^{-3} and a differential gain of 5×10^{-16} cm^2. Assume a spontaneous emission bandwidth of 100 nm. At transparency the Auger recombination rate equals the spontaneous recombination rate, and other nonradiative terms can be neglected.

(a) Plot the P–I characteristic, labeling the threshold current, the spontaneous emission power into the mode at threshold, and the differential efficiency above threshold.

(b) Plot the small-signal frequency response for a bias current of twice threshold.

2.15 Using Eq. (A6.25) calculate the gain 50 meV above the band edge in GaAs as a function of $(f_2 - f_1)$. (Assume $\tau_{sp}^{21} = 0.3$ ns, and consider only the heavy-hole band.)

CHAPTER THREE

Mirrors and Resonators for Diode Lasers

3.1 INTRODUCTION

Modern diode lasers use a variety of cavity structures. In Chapter 1 and Appendix 3 we introduced the transverse and lateral guiding structures that are generally used. However, the axial structure has not been discussed in much detail. In gas and solid-state lasers the entire cavity is defined by the axial mirrors, since no lateral guiding structure tends to be employed. So far, our analysis has assumed a simple two-mirror Fabry–Perot cavity for axial photon confinement and some waveguide for lateral and transverse confinement. In this chapter we shall focus on the axial dimension.

First, we develop a scattering matrix formalism so the various structures can be analyzed rigorously and easily. The use of the associated transmission matrices reduces the analysis of axial structures with numerous impedance discontinuities to a mathematical exercise in matrix multiplication.

Next, we successively explore several different axial geometries for diode lasers. Three- and four-mirror cavities are treated first. The mathematical condensation of the additional cavities and mirrors into a single effective complex mirror allows many of the basic formulae for Fabry–Perot lasers to be used. Because the mirror loss varies with wavelength, the loss of a single axial mode can be less than others. Thus, lasing in a single axial mode is possible. Also, if the phase in various laser sections can be changed independently, a tuning of the laser's wavelength can result.

Third, we introduce the concept of grating mirrors. These mirrors are interesting for both in-plane and vertical-cavity lasers. They consist of a series of relatively small impedance discontinuities along the axial propagation direction, phased so that the reflections add constructively at some frequency. Thus, the grating mirrors can provide a high level of net reflection when only small impedance differences exist, and because these gratings may be many wavelengths in length, the desired phasing can only occur over a narrow band of wavelengths to provide single-frequency operation. If substituted for a discrete mirror of a diode laser, a distributed Bragg reflector (DBR) laser results. In such cases the net complex grating reflection can replace the discrete mirror reflectivity in the Fabry–Perot formulae of Chapter 2. Multiple section DBR lasers have emerged as the most practical for wide wavelength tunability.

Fourth, we consider gratings with gain. Lasers made with such gratings are called distributed feedback (DFB) lasers. They can be simpler to fabricate than DBRs, since no transitions from active to passive regions are necessary, but their analysis is a little more complex. Fortunately, the transmission matrix formalism still works for complex propagation constants, so the threshold gains and wavelengths can still be obtained numerically. Because of their relative fabrication simplicity, DFB lasers have emerged as the best choice for single-frequency operation. Wavelength tuning is possible, but the range is limited as compared to the DBR.

The final section of this chapter deals with the spectral purity of axial mode selection that is possible with the above compound cavity lasers. The mode suppression ratio (MSR) is the ratio of the power out of the primary mode to the next largest mode. Experiments have shown that an MSR of at least 30 dB is necessary for single-frequency system applications.

3.2   SCATTERING THEORY

In working with complex laser cavities it is convenient to work with normalized amplitudes, a_j, which have a magnitude equal to the square root of the power flow and a phase equal to a selected observable such as the electric field. If we choose to reference the phase to the electric field, which we have written as

$$\mathscr{E}(x, y, z, t) = \hat{\mathbf{e}} E_0 U(x, y) e^{j(\omega t - \tilde{\beta} z)}, \tag{3.1}$$

we define

$$a_j = \frac{E_0}{\sqrt{2\eta_j}} e^{-j\tilde{\beta} z}, \tag{3.2}$$

where $\eta_j = 377\Omega/\bar{n}_j$ is the mode impedance (the ratio of the transverse electric to transverse magnetic field magnitudes of the mode). Thus, provided $\int |U|^2 \, dx \, dy = 1$, we see that $a_j a_j^* = P_j^+$, the power flowing in the positive z-direction in the mode.

At some waveguide reference plane, in general, there are incident and reflected powers. We characterize the incident waves (or inputs) by a normalized amplitude, a_j, and the reflected waves (or outputs) by a normalized amplitude, b_j, where the j's refer to the reference plane or port in question. Thus, at port j, the net power flowing into the port is

$$P_j = a_j a_j^* - b_j b_j^*. \tag{3.3}$$

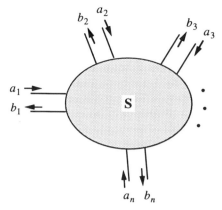

FIGURE 3.1   Generic scattering junction illustrating the inputs, a_j, and outputs, b_j, for the various ports.

Note that the impedance can be different at each port, but the definitions are unchanged. This is one of the most important features of the normalized amplitudes. Figure 3.1 shows a multiport scattering junction with inputs and outputs at each port for reference.

If the outputs can be linearly related to the inputs, a matrix formalism can be developed to express the outputs as a weighted combination of the inputs:

$$b_i = \sum_j S_{ij} a_j, \tag{3.4}$$

where the S_{ij} are called the *scattering coefficients*. Note from Eq. (3.4) that to determine a particular S_{ij}, all inputs except a_j must be set to zero. That is,

$$S_{ij} = \frac{b_i}{a_j}\bigg|_{a_k = 0, k \neq j}. \tag{3.5}$$

This is equivalent to terminating all ports in their characteristic impedance to prevent reflections back into the network.

More generally, $\mathbf{b} = \mathbf{Sa}$, where $\mathbf{a}$ and $\mathbf{b}$ are column vectors and $\mathbf{S}$ is a matrix. For example, for a two-port scattering junction, such as a partially transmissive mirror, we have

$$\begin{bmatrix} b_1 \\ b_2 \end{bmatrix} = \begin{bmatrix} S_{11} & S_{12} \\ S_{21} & S_{22} \end{bmatrix} \begin{bmatrix} a_1 \\ a_2 \end{bmatrix}. \tag{3.6}$$

The scattering coefficients are particularly useful because they have direct physical significance. All represent the ratio of a normalized output amplitude to a normalized input amplitude. The diagonal elements of the matrix are the respective complex amplitude reflection coefficients. For example in the two-port, we have previously referred to S_{11} and S_{22} as simply r_1 and r_2. The power reflection coefficients in this case are $|S_{11}|^2$ and $|S_{22}|^2$, respectively. The off-diagonal terms represent the complex (amplitude and phase) output at one port due to the input at another. Thus, they are really *transfer functions*. In all cases the magnitude squared of a scattering coefficient, $|S_{ij}|^2$, represents the fraction of power appearing at the port i due to the power entering port j.

Scattering matrices may have a number of interesting properties if the networks they describe satisfy certain criteria. For example, the scattering matrix of a linear reciprocal system is symmetric. That is for a two-port, $S_{12} = S_{21}$. For a lossless two-port, power conservation yields, $|S_{11}|^2 + |S_{21}|^2 = 1$, and $|S_{22}|^2 + |S_{12}|^2 = 1$. Also, the scattering matrix of a lossless system is unitary.

Another important matrix that relates the normalized amplitudes is the *transmission matrix*. The transmission matrix expresses the inputs and outputs at a given port in terms of those at the others. In the case of a two-port, it is used most often to cascade networks together, since simple matrix multiplication can be used (as we will see in a moment). Referring to Fig. 3.2(a), the transmission matrix of a two-port is defined as

$$\begin{bmatrix} A_1 \\ B_1 \end{bmatrix} = \begin{bmatrix} T_{11} & T_{12} \\ T_{21} & T_{22} \end{bmatrix} \begin{bmatrix} A_2 \\ B_2 \end{bmatrix}, \tag{3.7}$$

where instead of using the input and output amplitudes, a_i and b_i, defined above, we have chosen to denote right or forward-going waves as A_i, and left or backward-going waves as B_i. From Fig. 3.2(a), the correspondence between the **T**-matrix and **S**-matrix amplitudes is as follows: $A_1 = a_1$, $B_1 = b_1$, $A_2 = b_2$, and $B_2 = a_2$. This change of notation is convenient when cascading two-port networks in a serial chain. For example, in Fig. 3.2(b), port #1 of the second is connected to port #2 of the first. By equating $A_2 = A_1'$ and $B_2 = B_1'$, we can relate the fields on the left side of the overall structure, A_1 and B_1, to the fields on the right side, A_2' and B_2', as follows:

$$\begin{bmatrix} A_1 \\ B_1 \end{bmatrix} = \begin{bmatrix} T_{11} & T_{12} \\ T_{21} & T_{22} \end{bmatrix} \begin{bmatrix} A_2 \\ B_2 \end{bmatrix} = \begin{bmatrix} T_{11} & T_{12} \\ T_{21} & T_{22} \end{bmatrix} \begin{bmatrix} T_{11}' & T_{12}' \\ T_{21}' & T_{22}' \end{bmatrix} \begin{bmatrix} A_2' \\ B_2' \end{bmatrix}. \tag{3.8}$$

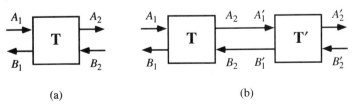

(a)   (b)

FIGURE 3.2   (a) Single two-port network. (b) Two networks cascaded together.

This process can be continued to obtain the net transmission matrix of arbitrarily complex multisection waveguide devices. The **T**-matrix can be obtained directly from the **S**-matrix using the following:

$$T_{11} = \frac{1}{S_{21}}, \qquad T_{12} = -\frac{S_{22}}{S_{21}},$$

$$T_{21} = \frac{S_{11}}{S_{21}}, \qquad T_{22} = -\frac{S_{11}S_{22} - S_{12}S_{21}}{S_{21}}.$$

(3.9)

Definitions and relations between the **S**- and **T**-matrices are summarized in Table 3.1.

TABLE 3.1   Relations Between Scattering and Transmission Matrices.

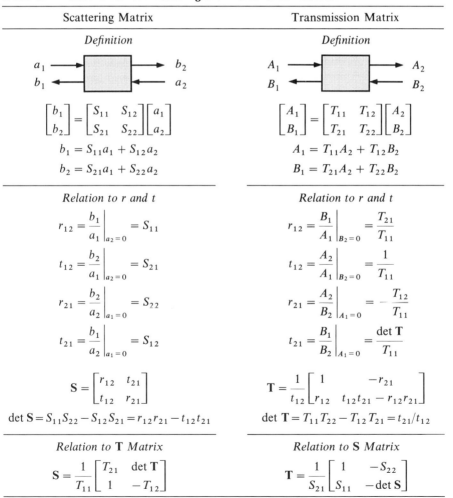

Scattering Matrix	Transmission Matrix
Definition	*Definition*
$\begin{bmatrix} b_1 \\ b_2 \end{bmatrix} = \begin{bmatrix} S_{11} & S_{12} \\ S_{21} & S_{22} \end{bmatrix}\begin{bmatrix} a_1 \\ a_2 \end{bmatrix}$	$\begin{bmatrix} A_1 \\ B_1 \end{bmatrix} = \begin{bmatrix} T_{11} & T_{12} \\ T_{21} & T_{22} \end{bmatrix}\begin{bmatrix} A_2 \\ B_2 \end{bmatrix}$
$b_1 = S_{11}a_1 + S_{12}a_2$	$A_1 = T_{11}A_2 + T_{12}B_2$
$b_2 = S_{21}a_1 + S_{22}a_2$	$B_1 = T_{21}A_2 + T_{22}B_2$
Relation to r and t	*Relation to r and t*
$r_{12} = \left.\dfrac{b_1}{a_1}\right|_{a_2=0} = S_{11}$	$r_{12} = \left.\dfrac{B_1}{A_1}\right|_{B_2=0} = \dfrac{T_{21}}{T_{11}}$
$t_{12} = \left.\dfrac{b_2}{a_1}\right|_{a_2=0} = S_{21}$	$t_{12} = \left.\dfrac{A_2}{A_1}\right|_{B_2=0} = \dfrac{1}{T_{11}}$
$r_{21} = \left.\dfrac{b_2}{a_2}\right|_{a_1=0} = S_{22}$	$r_{21} = \left.\dfrac{A_2}{B_2}\right|_{A_1=0} = -\dfrac{T_{12}}{T_{11}}$
$t_{21} = \left.\dfrac{b_1}{a_2}\right|_{a_1=0} = S_{12}$	$t_{21} = \left.\dfrac{B_1}{B_2}\right|_{A_1=0} = \dfrac{\det \mathbf{T}}{T_{11}}$
$\mathbf{S} = \begin{bmatrix} r_{12} & t_{21} \\ t_{12} & r_{21} \end{bmatrix}$	$\mathbf{T} = \dfrac{1}{t_{12}}\begin{bmatrix} 1 & -r_{21} \\ r_{12} & t_{12}t_{21} - r_{12}r_{21} \end{bmatrix}$
$\det \mathbf{S} = S_{11}S_{22} - S_{12}S_{21} = r_{12}r_{21} - t_{12}t_{21}$	$\det \mathbf{T} = T_{11}T_{22} - T_{12}T_{21} = t_{21}/t_{12}$
*Relation to **T** Matrix*	*Relation to **S** Matrix*
$\mathbf{S} = \dfrac{1}{T_{11}}\begin{bmatrix} T_{21} & \det \mathbf{T} \\ 1 & -T_{12} \end{bmatrix}$	$\mathbf{T} = \dfrac{1}{S_{21}}\begin{bmatrix} 1 & -S_{22} \\ S_{11} & -\det \mathbf{S} \end{bmatrix}$

As mentioned earlier, various network properties allow us to specify relationships *between* the matrix coefficients, allowing us to reduce the total number of independent parameters. For example, the *normalized* fields of a system which satisfy Maxwell's equations with scalar ε and μ are known to obey reciprocity, which simply put, means that the scattering matrix is equal to its transpose, or that $S_{12} = S_{21}$.

In addition to being reciprocal (a property of any linear network), the network might also be lossless. If this is the case, then other simplifying relations can be derived. For example, it can be shown that $T_{12} = T_{21}^*$ and $T_{22} = T_{11}^*$, which simplifies the determination of the **T**-matrix significantly. Table 3.2 summarizes these relations for various network properties.

TABLE 3.2   Network Properties and Their Consequences on the Matrix Coefficients.

Reciprocal Network (valid for normalized fields with and without loss)

$$\mathbf{S}_t = \mathbf{S} \to \begin{array}{c} S_{12} = S_{21} \\ \det \mathbf{T} = 1 \end{array}$$

$$\mathbf{S} = \begin{bmatrix} S_{11} & S_{21} \\ S_{21} & S_{22} \end{bmatrix} = \frac{1}{T_{11}} \begin{bmatrix} T_{21} & 1 \\ 1 & -T_{12} \end{bmatrix}$$

$$\mathbf{T} = \begin{bmatrix} T_{11} & T_{12} \\ T_{21} & (T_{12}T_{21} + 1)/T_{11} \end{bmatrix} = \frac{1}{S_{21}} \begin{bmatrix} 1 & -S_{22} \\ S_{11} & S_{21}^2 - S_{11}S_{22} \end{bmatrix}$$

Lossless Reciprocal Network

$$\mathbf{S}_t^* \mathbf{S} = 1 \to \begin{array}{cc} |S_{11}|^2 + |S_{21}|^2 = 1 & |T_{21}|^2 + 1 = |T_{11}|^2 \\ |S_{12}|^2 + |S_{22}|^2 = 1 \to 1 + |T_{12}|^2 = |T_{11}|^2 \\ S_{11}^* S_{12} + S_{21}^* S_{22} = 0 & T_{21}^* - T_{12} = 0 \end{array}$$

$$\mathbf{S} = \begin{bmatrix} S_{11} & S_{21} \\ S_{21} & -S_{11}^*(S_{21}/S_{21}^*) \end{bmatrix} = \frac{1}{T_{11}} \begin{bmatrix} T_{21} & 1 \\ 1 & -T_{21}^* \end{bmatrix}$$

$$\mathbf{T} = \begin{bmatrix} T_{11} & T_{21}^* \\ T_{21} & T_{11}^* \end{bmatrix} = \begin{bmatrix} 1/S_{21} & S_{11}^*/S_{21}^* \\ S_{11}/S_{21} & 1/S_{21}^* \end{bmatrix}$$

Lossless Reciprocal Network with r and t Phase Shifts of 0 or π

$$\begin{array}{c} S_{22} = -S_{11} \\ S_{11} = S_{11}^* \\ S_{21} = S_{21}^* \end{array} \to \det \mathbf{S} = -1 \\ T_{22} = T_{11}, \quad T_{12} = T_{21}$$

$$\mathbf{S} = \begin{bmatrix} S_{11} & S_{21} \\ S_{21} & -S_{11} \end{bmatrix} = \frac{1}{T_{11}} \begin{bmatrix} T_{21} & 1 \\ 1 & -T_{21} \end{bmatrix}$$

$$\mathbf{T} = \begin{bmatrix} T_{11} & T_{21} \\ T_{21} & T_{11} \end{bmatrix} = \frac{1}{S_{21}} \begin{bmatrix} 1 & S_{11} \\ S_{11} & 1 \end{bmatrix}$$

3.3   S AND T MATRICES FOR SOME COMMON ELEMENTS

The utility of the **S** and **T** matrices should become clearer as we consider a few common "scattering junctions." As we shall see, these form the basis of many more complex waveguide networks encountered in diode lasers. The further development of these matrices for more complex photonic integrated circuits, such as ones including directional couplers, will be left to a later chapter.

3.3.1   The Dielectric Interface

Figure 3.3 illustrates the normalized amplitudes at a dielectric interface. The media are characterized by indices of refraction n_1 and n_2.

In any such problem, one has the freedom to select reference planes for each port. The phase of the scattering coefficients will clearly depend upon the location of such planes. Some feel that it is important to make such an asymmetric problem have a symmetric scattering matrix by carefully selecting the reference planes, so that **S** will satisfy certain mathematical niceties. However, we believe that tends to create confusion, and it certainly obscures the physics of the problem. Thus, we shall always attempt to select "natural" reference planes at physical boundaries.

In the present case we select both reference planes at the physical interface between the two dielectrics, so that the scattering junction has zero length. Using Eq. (3.5), this leads to

$$S_{11} = \left.\frac{b_1}{a_1}\right|_{a_2=0} = -r_1 = \frac{n_1 - n_2}{n_1 + n_2}, \tag{3.10}$$

where for the last equality, we have assumed normally incident plane waves. This is approximately true for weakly guided dielectric waveguide modes, but n_1 and n_2 should be replaced by the effective indices of the modes, $\bar{n}_1$ and $\bar{n}_2$. Note that we have chosen to label the reflection from the n_1 side $-r_1$ rather than $+r_1$ in Fig. 3.3. The reason is for compatibility with other calculations in this book. Also, if $n_2 > n_1$, r_1 would be a positive real number.

Similarly, for the second port

$$S_{22} = \left.\frac{b_2}{a_2}\right|_{a_1=0} = r_2 = -(-r_1), \tag{3.11}$$

and

$$S_{12} = S_{21} = t = \sqrt{1 - r_1^2}, \tag{3.12}$$

where we have used power conservation for Eq. (3.12). This is clearly valid for plane waves, since there can be no loss in zero length, but for waveguide modes, power conservation also implies that the transverse mode profiles are equal. (If

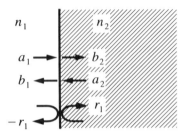

FIGURE 3.3 Interface between two dielectrics illustrating reference planes at interface for both ports.

the waveguide modes do not match, there can be scattering loss. This will be taken up in a later chapter.) For the normally incident plane wave case, $t = 2(n_1 n_2)^{1/2}/(n_1 + n_2)$.

Thus, the complete scattering matrix for the dielectric interface can be written as

$$\mathbf{S} = \begin{bmatrix} -r_1 & t \\ t & r_1 \end{bmatrix}, \tag{3.13}$$

where again, the sign of r_1 follows the convention in Fig. 3.3. The corresponding **T**-matrix is

$$\mathbf{T} = \frac{1}{t} \begin{bmatrix} 1 & -r_1 \\ -r_1 & 1 \end{bmatrix}. \tag{3.14}$$

3.3.2 Transmission Line with no Discontinuities

Figure 3.4 shows a network which is a length of waveguide, L, in which there are no discontinuities. In fact, this network consists only of two reference planes on a waveguide. Thus, the problem is to find how to relate variables

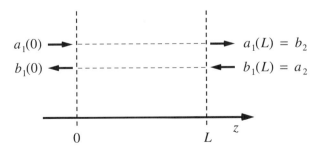

FIGURE 3.4 Transmission line section of length L.

from one reference plane to another where there are no scattering junctions in between.

In such cases it is common to express the normalized amplitudes as a function of distance as defined by Eq. (3.2). Thus, if the origin is put at port #1, then $b_2 = a_1(L)$ and $a_2 = b_1(L)$, as indicated in Fig. 3.4. Since there is no coupling between the waves propagating in the forward and backward directions, $S_{11} = S_{22} = 0$. From Eq. (3.2), we have $a_1(z) = a_1(0)e^{-j\tilde{\beta}z}$ and $b_1(z) = b_1(0)e^{j\tilde{\beta}z}$, for the forward and backward modes. Thus,

$$b_2 = a_1(L) = a_1(0)e^{-j\tilde{\beta}L} = a_1 e^{-j\tilde{\beta}L},$$

and

$$a_2 = b_1(L) = b_1(0)e^{j\tilde{\beta}L} = b_1 e^{j\tilde{\beta}L}. \tag{3.15}$$

Forming S_{12} and S_{21}, we find that

$$S_{12} = S_{21} = e^{-j\tilde{\beta}L}. \tag{3.16}$$

This is a very important result because it verifies that a propagation delay is the same for both forward and backward waves. In other words, the waveguide modes do not know how you have chosen the coordinate system. Mode propagation a distance L in any direction results in a phase shift of $-\beta L$ and a growth rate $\beta_i L$ for that mode, assuming that $\tilde{\beta} = \beta + j\beta_i$ as in Eq. (2.19). The scattering matrix is summarized as

$$\mathbf{S} = \begin{bmatrix} 0 & e^{-j\tilde{\beta}L} \\ e^{-j\tilde{\beta}L} & 0 \end{bmatrix}. \tag{3.17}$$

The corresponding **T**-matrix found using Eqs. (3.9) is given by

$$\mathbf{T} = \begin{bmatrix} e^{j\tilde{\beta}L} & 0 \\ 0 & e^{-j\tilde{\beta}L} \end{bmatrix}. \tag{3.18}$$

The basic component matrices are summarized in Table 3.3. For the **T**-matrix, lossless mirrors have been assumed in the first and third cases. To include the possibility of a lossy mirror, T_{22} for these two cases should be multiplied by $r_{12}^2 + t_{12}^2$. An advantage of the **T**-matrices is that more complicated structures can be constructed simply by matrix multiplying together the basic components shown in the table. For example, the third matrix in the list is easily constructed by matrix multiplying the first by the second.

3.3.3   Dielectric Segment and the Fabry–Perot Etalon

Figure 3.5 shows a dielectric block of length L and index n_2. To the left is a region of index, n_1, and to the right is a region of index n_3. We shall use r_1

TABLE 3.3   Summary of S- and T-matrices for Simple "Building-Block" Components.

Scattering Matrix	Structure	Transmission Matrix
$\begin{bmatrix} r_{12} & t_{12} \\ t_{12} & -r_{12} \end{bmatrix}$	1 2 t_{12} r_{12} $r_{21} = -r_{12} \quad t_{21} = t_{12}$	$\dfrac{1}{t_{12}}\begin{bmatrix} 1 & r_{12} \\ r_{12} & 1 \end{bmatrix}$ $r_{12}^2 + t_{12}^2 = 1$
$\begin{bmatrix} 0 & e^{-j\phi} \\ e^{-j\phi} & 0 \end{bmatrix}$	2 2 2 L $\phi = \tilde{\beta}_2 L$	$\begin{bmatrix} e^{j\phi} & 0 \\ 0 & e^{-j\phi} \end{bmatrix}$
$\begin{bmatrix} r_{12} & t_{12}e^{-j\phi} \\ t_{12}e^{-j\phi} & -r_{12}e^{-j2\phi} \end{bmatrix}$	1 2 2 L t_{12} r_{12}	$\dfrac{1}{t_{12}}\begin{bmatrix} e^{j\phi} & r_{12}e^{-j\phi} \\ r_{12}e^{j\phi} & e^{-j\phi} \end{bmatrix}$ $r_{12}^2 + t_{12}^2 = 1$

and t_1 for the left interface and r_2 and t_2 for the right interface viewed from the central medium. (Thus, choosing the reference planes at the physical interfaces and assuming that $n_2 > n_1$, r_1 would be a positive real number.) Such a structure is known as a Fabry–Perot etalon.

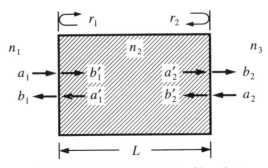

FIGURE 3.5   Dielectric block of length L.

As indicated in the diagram, we can look at this problem as three scattering networks cascaded—two dielectric interfaces and one transmission line. Thus, the problem could be solved by multiplying the **T**-matrices. This is to be shown in one of the problems at the end of the chapter. Here, however, we would like to demonstrate how to solve the system of normalized amplitudes to obtain the **S**-matrix directly. As we shall see as this chapter proceeds, once we have the **S** and **T**-matrices of such a dielectric segment, most multisection diode laser problems can be solved by their repetitive use.

Referring to Fig. 3.5 where primes are used for internal variables, and initially assuming that $a_2 = 0$, we can write the following relationships for the outputs, b_j:

$$b_1 = -a_1 r_1 + a_1' t_1,$$
$$b_1' = a_1 t_1 + a_1' r_1,$$
$$b_2 = a_2' t_2,$$
$$b_2' = a_2' r_2.$$

Also, we can express

$$a_1' = b_2' e^{-j\tilde{\beta}L},$$
$$a_2' = b_1' e^{-j\tilde{\beta}L}. \tag{3.19}$$

Solving this system of equations for $S_{11} = b_1/a_1$ and $S_{21} = b_2/a_1$ (since $a_2 = 0$), we obtain

$$S_{11} = -r_1 + \frac{t_1^2 r_2 e^{-2j\tilde{\beta}L}}{1 - r_1 r_2 e^{-2j\tilde{\beta}L}}, \tag{3.20}$$

$$S_{21} = \frac{t_1 t_2 e^{-j\tilde{\beta}L}}{1 - r_1 r_2 e^{-2j\tilde{\beta}L}}. \tag{3.21}$$

Similarly, with $a_1 = 0$,

$$S_{22} = -r_2 + \frac{t_2^2 r_1 e^{-2j\tilde{\beta}L}}{1 - r_1 r_2 e^{-2j\tilde{\beta}L}}, \tag{3.22}$$

$$S_{12} = S_{21}. \tag{3.23}$$

The common factor of $1 - r_1 r_2 e^{-2j\tilde{\beta}L}$ gives rise to the characteristic resonances and antiresonances associated with Fabry–Perot etalons.

The corresponding **T**-matrix for the Fabry–Perot etalon can be obtained in a number of ways: (1) we can use the **S**-matrix coefficients and apply Eqs. (3.9); (2) we can multiply the third and first matrices in Table 3.3, using the appropriate refractive indices; or (3) we can solve for the relevant ratios,

$T_{11} = a_1/b_2$ and $T_{21} = b_1/b_2$, (with $a_2 = 0$); $T_{12} = -T_{11}b_2/a_2$ and $T_{22} = b_1/a_2 - T_{21}b_2/a_2$, (with $a_1 = 0$). With $a_2 = 0$,

$$T_{11} = \frac{1}{t_1 t_2} [e^{j\tilde{\beta}L} - r_1 r_2 e^{-j\tilde{\beta}L}], \qquad (3.24)$$

$$T_{21} = -\frac{1}{t_1 t_2} [r_1 e^{j\tilde{\beta}L} - r_2 e^{-j\tilde{\beta}L}]. \qquad (3.25)$$

Similarly, with $a_1 = 0$,

$$T_{12} = -\frac{1}{t_1 t_2} [r_1 e^{-j\tilde{\beta}L} - r_2 e^{j\tilde{\beta}L}]. \qquad (3.26)$$

$$T_{22} = \frac{1}{t_1 t_2} [e^{-j\tilde{\beta}L} - r_1 r_2 e^{j\tilde{\beta}L}], \qquad (3.27)$$

For the latter three T-parameters we have assumed lossless interfaces, $t_1^2 = 1 - r_1^2$ and $t_2^2 = 1 - r_2^2$, to simplify the expressions. For the S-parameters, Eqs. (3.20) to (3.23), the reflection and transmission at each mirror has been left in general form. Thus, these apply to any Fabry–Perot etalon of length L. For example, the dielectric interfaces could include loss, or they could be coated to enhance the reflectivities, r_1 and r_2, to values greater than that indicated by Eq. (3.10).

The absolute squares of S_{11} and S_{21} give the amount of power reflected by and transmitted through the Fabry–Perot etalon as a function of the wavelength of the incident field. Using Eqs. (3.20) and (3.21), and assuming $t_i^2 = 1 - r_i^2$, we obtain

$$|S_{11}|^2 = \frac{(r_1 - r_2 e^{2\beta_i L})^2 + 4R \sin^2 \beta L}{(1 - R)^2 + 4R \sin^2 \beta L} \rightarrow \frac{4R \sin^2 \beta L}{(1 - R)^2 + 4R \sin^2 \beta L},$$

$$|S_{21}|^2 = \frac{(1 - r_1^2)(1 - r_2^2)e^{2\beta_i L}}{(1 - R)^2 + 4R \sin^2 \beta L} \rightarrow \frac{(1 - R)^2}{(1 - R)^2 + 4R \sin^2 \beta L},$$

where $R = r_1 r_2 e^{2\beta_i L}$ and $\tilde{\beta} = \beta + j\beta_i$. The arrows indicate the special case of a symmetric ($r_1 = r_2$), lossless ($\beta_i = 0$) Fabry–Perot cavity. Note that without loss, $|S_{11}|^2 + |S_{21}|^2 = 1$, as required by power conservation. Figure 3.6 plots both the magnitude and phase of S_{11} and S_{21} vs. $2\beta L$ using three different reflectivities. The periodic maxima in the transmission spectrum of S_{21} go to unity for this symmetric, lossless case, and become extremely sharp at high values of r. These maxima occur at the axial resonances or *modes* of the Fabry–Perot cavity where $e^{-2j\beta L} = |r_1 r_2|/r_1 r_2 = 1$. By power conservation, the minima of the net reflection, S_{11}, go to zero at the resonances in this case. If the cavity has loss, the maxima of S_{21} cannot reach unity. Of course, if the cavity

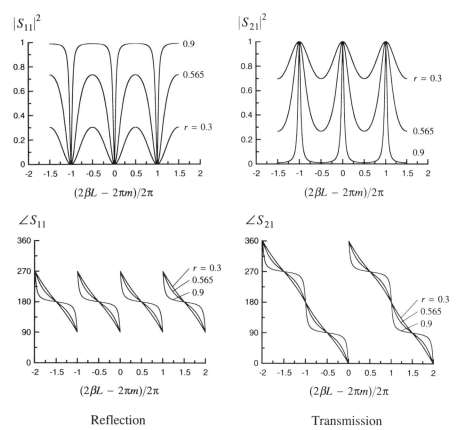

Reflection   Transmission

FIGURE 3.6 Magnitude and phase of the reflection (S_{11}) and transmission (S_{21}) coefficients of a Fabry–Perot etalon as a function of the cavity round-trip phase relative to an integer multiple of 2π (in the lower-right plot, the phase is relative to an *even* multiple of 2π). Three mirror reflectivities are plotted assuming $r = r_1 = r_2$ and zero loss (a cleaved-facet cavity corresponds to $r = 0.565$).

has gain as in a laser, the maxima can be larger than unity. In the lossy case, it is also worth noting that S_{11} can still be adjusted to zero by reducing the reflectivity of the input mirror to be lower than the output mirror such that $r_1 = r_2 e^{-\alpha L}$. This is equivalent to setting the first term equal to the second in Eq. (3.20) on resonance. This asymmetric Fabry–Perot is useful in efficient optical modulators and detectors.

3.3.4   Fabry–Perot Laser

In Chapter 2 we analyzed the use of a Fabry–Perot cavity to form a diode laser. Based upon an intuitive argument, we labeled threshold as the point where the net round-trip gain equaled the net round-trip loss for an axial mode.

We now can argue this point more rigorously using the S-parameters. For lasing we must have coherent light being emitted for no optical inputs. Thus, the threshold for lasing must correspond to a pole of the S-parameters, which give the ratio of outputs over inputs. Referring to Eqs. (3.20) to (3.23), we observe that all of the S-parameters have a factor $[1 - r_1 r_2 e^{-2j\tilde{\beta}L}]$ in the denominator of one term. Setting this factor to zero gives the needed pole. This definition of threshold is therefore equivalent to Eq. (2.20).

For lossless mirrors the differential efficiency is given by Eq. (2.35), and the power out is given by Eq. (2.36). However, we have not solved for the relative power out of each end of the laser. We now can determine the ratio of the powers out of each end, P_{01}/P_{02}, since it equals $|b_1/b_2|^2$. Looking at Fig. 3.5 with $a_1 = a_2 = 0$, we can see that $b_1 = a_1' t_1$, $b_2 = a_2' t_2$, and $a_1' = a_2' r_2 e^{-j\tilde{\beta}L}$. For a mode above threshold, $\text{Im}\{\tilde{\beta}\} = (\Gamma_{xy}g_{th} - \alpha_i)/2$. Thus,

$$\frac{P_{01}}{P_{02}} = \left|\frac{b_1}{b_2}\right|^2 = \frac{t_1^2}{t_2^2} r_2^2 e^{(\Gamma_{xy}g_{th} - \alpha_i)L}. \tag{3.28}$$

But, $\exp[\Gamma_{xy}g_{th} - \alpha_i]L = 1/(r_1 r_2)$ from Eq. (2.21). Therefore,

$$\frac{P_{01}}{P_{02}} = \frac{t_1^2 r_2}{t_2^2 r_1}. \tag{3.29}$$

The more general result for more complex cavities can be obtained directly from either the T-matrix or S-matrix coefficients using $P_{01}/P_{02} = |T_{21}|^2 = |S_{11}/S_{21}|^2$ evaluated at threshold.

Now if the mirrors are *lossy*, such that $r_j^2 + t_j^2 \neq 1$, we are more interested in the fraction, F_1, of *power delivered* from end 1, P_{01}, relative to the *total coupled out of the cavity* by the mirrors, P_m. This is because it is F_1 that must multiply the differential quantum efficiency and power expressions of Chapter 2, which assumed that all mirror loss was delivered to the outside. Referring again to Fig. 3.5 with $a_1 = a_2 = 0$, we construct the desired ratio, $F_1 = P_{01}/P_m = |b_1|^2/[|a_1'|^2(1 - r_1^2) + |a_2'|^2(1 - r_2^2)]$, which gives

$$F_1 = \frac{t_1^2}{(1 - r_1^2) + \dfrac{r_1}{r_2}(1 - r_2^2)}. \tag{3.30}$$

Therefore, using Eq. (2.35), the differential quantum efficiencies for light delivered out of end 1 and end 2 of the general Fabry–Perot laser are

$$\eta_{d1} = F_1 \eta_i \frac{\alpha_m}{\langle\alpha_i\rangle + \alpha_m},$$

and

$$\eta_{d2} = F_2 \eta_i \frac{\alpha_m}{\langle \alpha_i \rangle + \alpha_m},$$
(3.31)

where F_1 is given by Eq. (3.30) and F_2 is found by switching subscripts in F_1. From the above and Eq. (2.36) the power out of the jth port is

$$P_{0j} = \eta_{dj} \frac{hv}{q} (I - I_{th}).$$
(3.32)

For a symmetrical laser cavity with $r_1 = r_2 = r$ and $t_1 = t_2 = t$, the fractions reduce to

$$F_1 = F_2 = \frac{1}{2} \frac{t^2}{1 - r^2}.$$
(3.33)

Therefore, any loss occurring at the mirrors (other than what is coupled as useful output) is taken into account through F_1 and F_2.

3.4   THREE- AND FOUR-MIRROR LASER CAVITIES

Many modern diode laser configurations have at least one additional discontinuity within their cavities. In Fig. 2.6 and the associated derivations, we neglected any reflection from the interface between the active and passive sections. Here we shall discuss first one and then two discontinuities within the laser. The additional discontinuities may be within the semiconductor material or one may be at the semiconductor—air interface in an external cavity configuration.

3.4.1   Three-Mirror Lasers

Figure 3.7 gives a schematic of a three-mirror laser. The transmission coefficient across the interface, t_2, includes any scattering or coupling loss, so $r_2^2 + t_2^2 \neq 1$, in general. Generally, our analysis will hold for active–active as well as active–passive devices, but the results presented are only meaningful if the first cavity (labeled active) is the *dominant* cavity, which provides most of the gain.

Also shown in Fig. 3.7 is an equivalent two-mirror cavity which replaces the passive section by an *effective mirror* with reflectivity, r_{eff}. This substitution is valid for steady-state analyses, but it will not necessarily properly model the compound cavity for dynamic operation. The value of r_{eff} was derived above, since it is really S_{11} for the passive section as viewed from the active section.

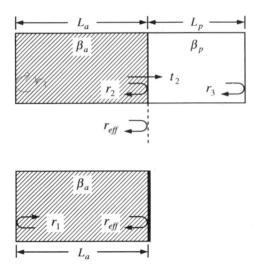

FIGURE 3.7   External cavity laser and equivalent cavity with effective mirror to model the external section.

Using the reference planes and mirror reflectivities defined in Fig. 3.7 and Eq. (3.20) we find that

$$r_{eff} = r_2 + \frac{t_2^2 r_3 e^{-2j\tilde{\beta}_p L_p}}{1 + r_2 r_3 e^{-2j\tilde{\beta}_p L_p}}. \tag{3.34}$$

The complete characteristic equation for a three-mirror laser may be constructed by replacing r_2 by r_{eff} in Eq. (2.20), (or by solving for the poles of an active-cavity S-parameter with this substitution). However, this leads to a fairly complex equation. We can obtain most of the information we want by using the equivalent cavity in Fig. 3.7, and carrying r_{eff} along in the threshold calculation. This results in only a slight modification to the threshold gain expression of Eq. (2.23). That is, the threshold gain of our three-mirror laser can be written as

$$\Gamma g_{th} = \langle \alpha_i \rangle_a + \frac{1}{L_a} \ln \left[\frac{1}{r_1 |r_{eff}|} \right], \tag{3.35}$$

where Γ and $\langle \alpha_i \rangle_a$ average over the active section of the cavity only (any losses encountered in the passive section are contained in r_{eff}). To complete the model, we need to specify the threshold condition for the round-trip phase. With $r_{eff} = |r_{eff}| e^{j\phi_{eff}}$ and r_1 positive and real, the round-trip phase must satisfy $e^{-2j\beta_a L_a} e^{j\phi_{eff}} = 1$, which translates into $2\beta_a L_a - \phi_{eff} = 2\pi m$. Taking derivatives of all variables dependent on frequency, we obtain

$$d\beta_a L_a - \tfrac{1}{2} d\phi_{eff} = \pi dm. \tag{3.36}$$

The spacing between adjacent modes is found by setting $dm = 1$ and solving for $d\beta_a$:

$$d\beta_a = \frac{\pi}{L_a - \frac{1}{2}d\phi_{eff}/d\beta_a}. \tag{3.37}$$

In this expression, the cavity length which defines the mode spacing includes the active section length plus an additional factor dependent on how the effective mirror phase changes with frequency, motivating us to consider the second quantity as an effective length. However, because ϕ_{eff} explicitly depends on β_p, not β_a, we choose to define the effective length as

$$L_{eff} = -\frac{1}{2}\frac{d\phi_{eff}}{d\beta_p}. \tag{3.38}$$

The mode spacing can be defined in terms of wavelength, $d\beta_a = -d\lambda(2\pi/\lambda^2)\bar{n}_{ga}$; or in terms of frequency, $d\beta_a = dv(2\pi/c)\bar{n}_{ga}$, where $\bar{n}_g = \bar{n} - \lambda\,\partial\bar{n}/\partial\lambda$. Furthermore, it follows that $d\beta_p/d\beta_a = \bar{n}_{gp}/\bar{n}_{ga}$. Using these expressions in Eq. (3.37), the mode spacing in either wavelength or frequency becomes

$$d\lambda = \frac{\lambda^2}{2(\bar{n}_{ga}L_a + \bar{n}_{gp}L_{eff})} \quad \text{or} \quad dv = \frac{c}{2(\bar{n}_{ga}L_a + \bar{n}_{gp}L_{eff})}. \tag{3.39}$$

If no reflection exists at the active–passive interface ($r_2 = 0$) and r_3 is positive and real, then $\phi_{eff} = -2\beta_p L_p$, and $L_{eff} = L_p$, reducing Eq. (3.39) to Eq. (2.25) given earlier. For the more general Fabry–Perot etalon, the slope of the phase will be dependent on whether we are near a resonance or an antiresonance of the etalon (as shown earlier in Fig. 3.6). Thus, L_{eff} can be larger or smaller than L_p. However, if the phase varies rapidly and nonlinearly within the range of one mode spacing (for example, near the Fabry–Perot resonances ($\beta L = m\pi$) in Fig. 3.6), then Eq. (3.39) will most likely not be very accurate, since this derivation assumes that ϕ_{eff} varies linearly over at least one mode spacing.

The differential quantum efficiency and power out of end 1 are given by Eqs. (3.31) and (3.32), respectively, using Eq. (3.30). The second mirror reflectivity, r_2, in these single-section laser expressions should be replaced by r_{eff} wherever it shows up, and the mirror loss, α_m, is given by the second term in Eq. (3.35) above.

Figure 3.6 gives plots of the magnitude of r_{eff} in the special case where $r_2 = -r_3$ and loss can be neglected. As can be seen, the magnitude of the reflectivity of this mirror can vary significantly, and this will provide a filtering effect on the cavity modes. As indicated by Eq. (3.35), the modes with the lowest loss or highest mirror reflectivity will tend to lase first. Thus, such a second section or etalon can be used to filter out unwanted modes. However, a point often confused is that the maxima of r_{eff} always occur at the antiresonances of the etalon. Thus, in the three-mirror configuration, having a high-Q external cavity actually leads to worse mode selectivity, because the maxima become very flat in this case. In fact, there is an optimum value of net external cavity loss that provides the largest curvature at the maxima of r_{eff}. Figure 3.6 is actually not a very practical case, since there is usually some loss both in traversing the passive cavity and in coupling back into the active section, and generally $r_2 \neq r_3$. Thus, the minima do not tend to be as deep, and the maxima have more shape. This is one case in nature where loss seems to help.

External cavities are not very useful for axial mode selection in the VCSEL case, since their short cavities together with the finite gain bandwidth usually provides single-axial mode operation. Here lateral modes are the larger problem. However, for in-plane lasers ≥ 100 μm in length several axial modes will exist near the gain maximum, and the loss modulation caused by the external cavity is useful for axial mode selection.

The length of the external cavity (etalon) determines how modes will be selected. Ideally, the periodic loss modulation will combine with the gain roll-off to provide a single net gain maximum where the nearest axial mode will be selected. Figure 3.8 illustrates the three cases where the relative length, L_p/L_a is (a) $\ll 1$, (b) ~ 1 (but $\neq 1$), and (c) $\gg 1$.

Here we show the variations of α_m, from Eq. (3.35), and a generic net gain curve, ($\Gamma_{xy}g - \alpha_i$) vs. wavelength along with indications of the mode locations. (Note that the maxima in r_{eff} correspond to minima in α_m.) The active cavity length and the width of the gain peak are held constant for all three cases. When the gain reaches the loss at some point, the mode at that wavelength reaches threshold. Also, we know that once one mode reaches threshold, the gain becomes clamped, and if the loss margin (the loss minus the gain) is large enough, the other modes are suppressed.

If the external cavity is somewhat shorter than the active section, as in case (a), the modes of the active cavity will be more closely spaced than the minima in α_m. In this case a single loss minimum can effectively select a single axial mode of the active cavity, if α_m varies enough. That is, the period of the loss modulation cannot be so large that the minimum of α_m is as wide as the gain peak, or no additional filtering will be provided. On the other hand, it must be large enough so that the next minima of α_m fall sufficiently far off the gain peak. (If too close, these secondary minima may select unwanted repeat modes.)

If the lengths are comparable as in Fig. 3.8(b), the resonances of both cavities are spaced by about the same amount, and the active cavity modes will slowly slide across the minima of α_m providing an action similar to a vernier scale. Again, relatively good mode suppression is possible if the beat period is not too large or too small. In the third case, Fig. 3.8(c), good mode suppression is generally not possible unless the external cavity mirror itself is a filter. In fact, a grating mirror is sometimes used to provide for single-frequency operation of the laser with a long external cavity.

One of the uses of a two-section (three-mirror) laser is to provide a tunable single-frequency source. The repeat modes can make this somewhat problematic,

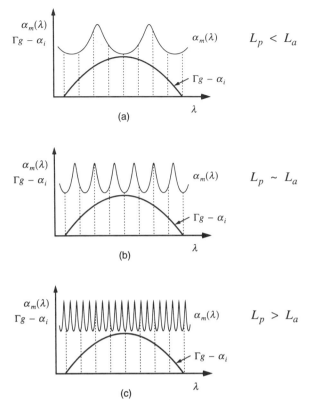

FIGURE 3.8 Schematic illustration of net propagation gain, $\Gamma_{xy}g - \alpha_i$, and the net mirror loss, α_m, as a function of wavelength for external cavities with length (a) shorter, (b) about the same, and (c) longer than the active section.

but the tuning mechanism is still worth reviewing. If the passive section shown in Fig. 3.7 is formed of electro-optic material, it would be possible to change its index and the round-trip phase, $-\beta_p L_p$, by applying an electric field across the material. Then, according to Eq. (3.34), the r_{eff} and α_m characteristics would tune in wavelength. This would cause a successive selection of different axial modes, as can be envisioned from Fig. 3.8. Moreover, because the phase of r_{eff} also varies across each period, some continuous tuning of the cavity mode is possible prior to the shift to a new mode. Perhaps one of the key reasons to have the tunability is to provide an active mechanism to optimally align the loss minimum with an axial mode for best spurious mode suppression.

3.4.2   Four-Mirror Lasers

If the r_{eff} characteristic could be flipped over so that its maxima (α_m minima) also occurred at the resonances of the external cavity, the filtering action would be much better. As stated above this is not possible in a three-mirror cavity, but with a *four-mirror cavity* it is. However, the relative positions of the reflectors must provide the desired phasing. The best known example of a four-mirror laser is the coupled-cavity laser, which incorporates a narrow space to separate *two active sections.* This same effect can also be achieved by using coatings on the facet of the active section in an external-cavity laser.

To model the four-mirror laser, we replace the interface between the active and passive sections in Fig. 3.7 by another section, which can be fully described by another complex scattering matrix. (Again, this analysis also holds for the active–active situation, in which the highest gain cavity is referred to as the *active* cavity.) Physically, this interface section might be another dielectric region with an index different from either the active or passive sections. Figure 3.9 shows the model and a possible implementation.

For this case we only need to modify the expression for r_{eff} slightly. Replacing r_2, $-r_2$, t_2, and t_2 with the more general terms, S_{s11}, S_{s22}, S_{s21}, and S_{s12}, respectively, in Eq. (3.34), we obtain

$$r_{eff} = S_{s11} + \frac{S_{s21}S_{s12}r_3 e^{-2j\tilde{\beta}_p L_p}}{1 - S_{s22}r_3 e^{-2j\tilde{\beta}_p L_p}}, \tag{3.40}$$

or rearranging,

$$r_{eff} = S_{s11}\left[1 + \frac{\sigma R'}{1 - R'}\right], \tag{3.41}$$

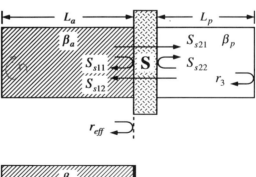

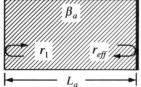

FIGURE 3.9 Generic four-mirror or three-section laser and equivalent mirror representation.

where $R' = S_{s22}r_3 e^{-2j\tilde{\beta}_p L_p}$, and $\sigma = (S_{s21}S_{s12}/S_{s11}S_{s22})$. Equation (3.41) shows that for resonance R' is real and positive. Thus, for the second term to add to the first for a maximum in r_{eff}, the ratio σ must also be real and positive. As can be verified by reference to Eqs. (3.20) to (3.23), this occurs when the space is a multiple of a half-wavelength wide, if its index is either lower or higher than both end cavities (e.g., a simple air gap between two semiconductors). Other situations will be left as exercises for the reader, but it should be clear that the phase of this *gap factor*, σ, will determine whether the maxima of r_{eff} will occur at passive cavity resonances ($\angle \sigma = 0$), or antiresonances ($\angle \sigma = \pi$), as in Fig. 3.6. In the case of the three-mirror cavity, the phase of the gap factor is always π.

For the optimum case of $\angle \sigma = 0$ the minima in α_m are sharpest and the mode selection best. Also, if the index can be adjusted in the second cavity, the resulting tunability is somewhat improved. In particular, since the phase of r_{eff} varies most rapidly near the resonance of the second cavity, this optimum case leads to better continuous tunability.

3.5 GRATINGS

3.5.1 Introduction

Many important diode lasers use gratings or distributed Bragg reflectors (DBRs) for one or both cavity mirrors. With in-plane lasers the reason is to use their frequency selectivity for single-axial mode operation, and with vertical-cavity lasers the reason is to obtain a very high value of reflectivity. Gratings consist of a periodic array of index (sometimes gain) variations. At the Bragg frequency, the period of the grating is half of the average optical wavelength in the medium. Significant reflections can also occur at harmonics of this frequency. In the vertical-cavity case, quarter wavelength thick layers of two different index materials are alternated during growth. In the in-plane case, corrugations are typically etched on the surface of the waveguide, and these are refilled with a different index material during a second growth. Figure 3.10 illustrates the two cases. As indicated, the in-plane case usually has many more grating periods than the VCSEL.

In either case, the concept of the grating is that many small reflections can add up to a large net reflection. At the Bragg frequency the reflections from each discontinuity add up exactly in phase. For the rectangular gratings shown, there are two discontinuities per period, each of reflectivity r. Thus, the net reflection from m grating periods is $r_g \sim 2mr$ when the net reflection is weak, so that each discontinuity sees nearly the same incident field. For a significant reflection the field will fall off into the grating and the problem becomes more difficult. As the frequency is deviated from the Bragg condition, the reflections from discontinuities further into the grating return with progressively larger

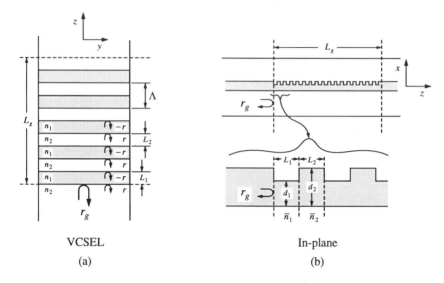

FIGURE 3.10   (a) Schematic of a DBR mirror for a vertical-cavity laser. (b) Schematic of a DBR mirror for an in-plane laser.

phase mismatch. This causes a roll-off in the net reflection which occurs more rapidly in longer gratings.

Since the dielectric interfaces extend uniformly across the mode in the vertical-cavity case, the results from Section 3.3.3 above apply directly. That is, we can obtain the net grating reflectivity, r_g, exactly by cascading the T-matrices, Eqs. (3.24) to (3.27). For the in-plane case, we must assume some effective reflectivity at each discontinuity in order to use these results. This is indicated by the inset in Fig. 3.10(b). For the rectangular gratings shown, this effective modal reflectivity at each discontinuity can be estimated by using the effective indices in each waveguide segment in Eq. (3.10). That is, if $\bar{n}_1$ and $\bar{n}_2$ are the effective indices in waveguide segments 1 and 2 of width d_1 and d_2, respectively, the reflectivity in going from segment 2 to segment 1 is approximately

$$\bar{r} \approx \frac{\bar{n}_2 - \bar{n}_1}{\bar{n}_2 + \bar{n}_1}. \tag{3.42}$$

The effective indices can be calculated following the procedure outlined in Appendix 3. The approximation is necessary because the transverse grating mode will be different from (actually something in between) the modes in uniform waveguides of width d_1 or d_2. The approximation is best for small impedance discontinuities. The reflectivity in going from segment 1 to segment 2 is $-\bar{r}$ by symmetry. As we shall discuss more in Chapter 6, for grating profiles other than rectangular, the effective reflectivity at each segment interface can be found by multiplying Eq. (3.42) by the relative Fourier coefficients for the harmonic of interest.

3.5.2   Transmission Matrix Theory

Figure 3.11 shows how the periodic gratings of Fig. 3.10 can be represented using T-matrices. If the output fields are known, the input fields (and hence the reflectivity, $r_g = B_m/A_m$) can be determined by matrix multiplying the individual components of the grating, starting from the output and proceeding to the input. Now, for the simple uniform gratings depicted in Fig. 3.10, in which only two indices are involved, the matrix multiplication to obtain r_g can be simplified by realizing that each period is the same. (However, we may need to use a different T-matrix for the output segment.)

To determine the T-matrix for a single period, we must matrix multiply *four* simple T-matrices together. Starting at the reference plane and moving in the positive z-direction in Fig. 3.10, we encounter (1) a 2–1 dielectric interface, (2) a propagation delay of length L_1, (3) a 1–2 dielectric interface, and (4) a propagation delay of length L_2. At this point another 2–1 interface is encountered, marking the beginning the next period. The results of multiplying the first three of these matrices together has already been derived in Eqs. (3.24) to (3.27). To apply those equations to Fig. 3.10 we need to change notation slightly by setting $r_1 = r_2 \equiv -r$, $t_1 = t_2 \equiv t$ and $L = L_1$. Now, the associative property is maintained under matrix multiplication implying that $T_1 T_2 T_3 T_4 = (T_1 T_2 T_3)T_4$. Thus, by multiplying the composite of three matrices by the fourth propagation delay matrix defined in Eq. (3.18), we obtain the single period T-matrix.

Alternatively, we can group the T-matrices as follows: $(T_1 T_2)(T_3 T_4)$, and identify each group as being equivalent to the third structure listed in Table 3.3. Multiplying two such matrices (using the appropriate refractive indices) also gives us the single period T-matrix. Either way, the general

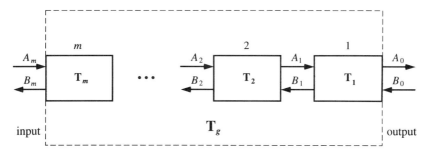

FIGURE 3.11   Cascaded scattering junctions characterized by transmission matrices. The net transmission matrix of the cascade is T_g.

forms, and corresponding forms for a lossless DBR at the Bragg frequency (right of arrow), become

$$T_{11} = \frac{1}{t^2} [e^{j\phi_+} - r^2 e^{-j\phi_-}] \rightarrow -\frac{1+r^2}{t^2},$$

$$T_{21} = \frac{r}{t^2} [e^{j\phi_+} - e^{-j\phi_-}] \rightarrow -\frac{2r}{t^2},$$

$$T_{12} = \frac{r}{t^2} [e^{-j\phi_+} - e^{j\phi_-}] \rightarrow -\frac{2r}{t^2},$$

$$T_{22} = \frac{1}{t^2} [e^{-j\phi_+} - r^2 e^{j\phi_-}] \rightarrow -\frac{1+r^2}{t^2},$$

(3.43)

where $\phi_\pm \equiv \tilde{\beta}_1 L_1 \pm \tilde{\beta}_2 L_2$, which becomes either π or 0 at the Bragg condition. For convenience, we define an *average* complex propagation constant of one grating period: $\tilde{\beta}_\pm \Lambda \equiv \tilde{\beta}_1 L_1 \pm \tilde{\beta}_2 L_2$, where $\tilde{\beta}_\pm = \beta_\pm - j\alpha_\pm/2$. Setting $\Lambda = L_1 + L_2$, $L_1 = \lambda_0/4n_1$, and $L_2 = \lambda_0/4n_2$ (where λ_0 is the Bragg wavelength), we find that

$$\beta_\pm = \frac{\beta_1/n_1 \pm \beta_2/n_2}{1/n_1 + 1/n_2} \rightarrow \frac{1}{\beta_+} \equiv \frac{1}{\beta} = \frac{1}{2}\left[\frac{1}{\beta_1} + \frac{1}{\beta_2}\right], \qquad \beta_- = 0,$$

$$\alpha_\pm = \frac{\alpha_1/n_1 \pm \alpha_2/n_2}{1/n_1 + 1/n_2} \rightarrow \alpha_+ = \alpha, \qquad \alpha_- = 0 \quad (\text{if } \alpha_1 = \alpha_2 = \alpha).$$

(3.44)

In the first line, β_+ is defined as the average propagation constant of the grating, β, while β_- is always zero regardless of the frequency. In the limit of small index differences, $\beta \approx \beta_1 \approx \beta_2$. In the second line, the average loss reduces to that shown on the right only if the loss in each layer is the same.

With these definitions we can write the phase terms in Eqs. (3.43) as $j\phi_+ = j\beta\Lambda + \alpha_+\Lambda/2$ and $j\phi_- = \alpha_-\Lambda/2$. Furthermore, at the Bragg frequency, the phase delay of each layer is $\beta_1 L_1 = \beta_2 L_2 = \pi/2$, and $\beta_0\Lambda = \beta_1 L_1 + \beta_2 L_2 = \pi$. Defining a *detuning parameter*,

$$\delta \equiv \beta - \beta_0,$$

(3.45)

the phase terms with no loss become simply

$$\phi_+ = \pi + \delta\Lambda, \qquad \phi_- = 0.$$

(3.46)

The detuning parameter can alternatively be expressed as $\delta\Lambda = \pi(\nu - \nu_0)/\nu_0$. The arrows in Eqs. (3.43) indicate this no loss case at the Bragg frequency ($\delta \rightarrow 0$).

For a cascade of m such matrices for the m grating segments of length Λ, we have

$$\mathbf{T}_g = \begin{bmatrix} T_{11} & T_{12} \\ T_{21} & T_{22} \end{bmatrix}^m, \tag{3.47}$$

which can be simplified using a mathematical identity derived in Appendix 7. The only restriction for the form used here is that the system must be *reciprocal* such that $T_{11}T_{22} - T_{12}T_{21} = 1$. In other words, transmission through the dielectric stack must be equivalent for light incident from either side of the stack. For all **T**-matrices we have discussed so far, this condition is satisfied even with loss or gain in the layers. Using the additional subscript g for the T-parameters of the entire grating, and assuming reciprocity, we can write

$$T_{g11} = (1 + jm_{eff}\Delta) \cosh m\xi,$$

$$T_{g21} = \frac{T_{21}}{T_{11}} m_{eff}(1 + j\Delta) \cosh m\xi,$$

$$\tag{3.48}$$

$$T_{g12} = \frac{T_{12}}{T_{22}} m_{eff}(1 - j\Delta) \cosh m\xi,$$

$$T_{g22} = (1 - jm_{eff}\Delta) \cosh m\xi,$$

where

$$\pm\xi = \ln\{\tfrac{1}{2}[T_{11} + T_{22}] \pm \sqrt{\tfrac{1}{4}[T_{11} + T_{22}]^2 - 1}\},$$

$$\Delta = j\frac{T_{22} - T_{11}}{T_{22} + T_{11}}, \tag{3.49}$$

$$m_{eff} = \frac{\tanh m\xi}{\tanh \xi}.$$

The equation $\pm\xi = \ln\{\pm\}$ implies that the negative root $\{-\} = 1/\{+\}$ or $\{-\}\{+\} = 1$, which is easily verified. As will be discussed below, the three parameters, ξ, Δ, and m_{eff}, all have a physical significance and are important in the analysis of dielectric stacks.

The first parameter is the *discrete* propagation constant, ξ. Its value is very dependent on the wavelength of the incident light, and is in general complex. It represents a discrete propagation constant because the fields (A_j, B_j) shown in Fig. 3.11 are multiplied by $e^{\pm\xi}$ upon passing to the next period. (As shown more exactly in Appendix 7, the fields when decomposed into the two eigenvectors of the matrix are multiplied by the eigenvalues $e^{+\xi}$ and $e^{-\xi}$ upon passing to the next period.) For example, if ξ is purely imaginary at some wavelength, the incident field will only encounter a phase shift of $m\xi$ upon passing through m periods, suffering no attentuation and hence, providing perfect transmission through the stack. Wavelength regimes for which this occurs are referred to as *passbands* of the dielectric stack. If ξ is purely real at some wavelength, the field will be attenuated by $e^{\pm m\xi}$ as it propagates through m periods, which can lead to very low transmission and hence, high reflection. These wavelength regimes are referred to as *stopbands* of the dielectric stack. Using Eqs. (3.43) at the Bragg frequency, it is shown below that $\xi \approx j\pi + 2r$, revealing that the field experiences a π phase shift through each period, and is attenuated by e^{-2r}. If r is high enough, and enough periods are used, the attentuation can be very high, leading to extremely low transmission and extremely high reflection of the field.

The second parameter is defined as the generalized detuning parameter, Δ, and is a measure of how far away we are from the Bragg condition. For small reflectivities, using Eqs. (3.43) together with Eq. (3.46), we find $\Delta \approx \tan \delta\Lambda \approx \delta\Lambda$. At the Bragg condition, $T_{11} = T_{22}$ in a lossless dielectric stack, and $\Delta = 0$ exactly, from its definition in Eq. (3.49).

The third parameter defines the effective number of periods, m_{eff}, seen by the incident field. For very weak attenuation ($\mathrm{Re}\{\xi\} \ll 1$), and a small number of periods, the tanh functions reduce to their arguments and $m_{eff} = m$. For large attenuation, m_{eff} as a function of m saturates at a value of $m_{eff} = 1/\tanh \xi$ which when multiplied by Λ, determines the *penetration depth* of the field into the dielectric stack to be discussed later.

As an example in evaluating the above three parameters, we examine the special case of when the wavelength satisfies the Bragg condition, and there is no loss or gain. From Eqs. (3.43), we observe that at the Bragg frequency, $T_{11} = T_{22}$ and $T_{21} = T_{12}$. The reciprocity condition for this case reduces to $T_{11}^2 - T_{21}^2 = 1$. Using these relations in the first equation of (3.49), we obtain $e^{\pm\xi} = T_{11} \pm T_{21}$, which allows us to set $\tanh \xi = T_{21}/T_{11}$. The three parameters defined in Eq. (3.49) then reduce to

$$\xi = j\pi + \ln(-T_{11} - T_{21}),$$

$$\Delta = 0,$$

$$m_{eff} = \frac{T_{11}}{T_{21}} \tanh[m \ln(-T_{11} - T_{21})].$$

(3.50)

In the first equality, the identity: $\ln(-1) = j\pi$, was used to introduce a minus sign into the argument of the ln function. Inserting Eqs. (3.43), we have

$$\mathrm{Re}\{\xi\} = \ln\left(\frac{1+r}{1-r}\right) \approx 2r,$$

$$m_{eff} = \frac{1+r^2}{2r} \tanh\left[m \ln\left(\frac{1+r}{1-r}\right)\right] \approx \frac{1}{2r} \tanh(2mr),$$

(3.51)

where the latter relations are obtained by neglecting second- and higher-order terms in r (in the ln function expansion, the second-order terms actually cancel making it an excellent approximation). Thus, at the Bragg wavelength, the attenuating portion of the discrete propagation constant is roughly equal to the sum of the reflectivities encountered within one period, as one might intuitively expect. The effective number of periods is also inversely related to $2r$. In other words, as r increases, fewer and fewer periods are effectively seen by the field, which also agrees with intuition. Note that as m increases toward infinity, m_{eff} saturates at $1/2r$.

The desired reflectivity of the overall stack in the general case is given by $r_g = S_{g11} = T_{g21}/T_{g11}$. With Eqs. (3.48), the reflectivity becomes

$$r_g = \frac{T_{21}}{T_{11}} m_{eff} \frac{1 + j\Lambda}{1 + jm_{eff}\Delta}. \tag{3.52}$$

At the Bragg frequency with no loss or gain, the second term disappears using Eqs. (3.50) and the reflectivity reduces to

$$r_g = \tanh\left[m \ln\left(\frac{1+r}{1-r}\right)\right] \approx \tanh(2mr), \qquad (\Delta = 0) \tag{3.53}$$

where the approximation is good to second order in r. Expressing r in terms of the refractive indices, the reflectivity at the Bragg frequency can also be written as

$$r_g = \frac{1 - (n_1/n_2)^{2m}}{1 + (n_1/n_2)^{2m}}. \qquad (\Delta = 0) \tag{3.54}$$

It is shown in Appendix 7 that Eq. (3.54) can more generally be applied to dielectric stacks with different values of index (for example at the input or output of the stack) by replacing $(n_1/n_2)^{2m}$ with the product of the low-to-high index ratios of every interface in the stack.

Figure 3.12 shows example plots of the magnitude and phase of r_g from Eq. (3.52) for several values of $2mr$. For low reflection magnitudes it approaches a $\sin(\delta L_g)/(\delta L_g)$ function, while for high reflection magnitudes, the top flattens out (saturating at a value of unity) and the stopband broadens. In the phase spectrum, increasing $2mr$ has the effect of suppressing the phase slope over the range of the stopband. We also see that the phase jumps by π every time the reflectivity passes through a null. A phasor diagram of the reflectivity would reveal that as δL_g approaches a zero crossing, the reflectivity phasor aligns itself with the negative real axis, shrinks to zero, and then increases again, pointing along the positive real axis.

In the low reflection limit the peak net reflection should intuitively approach $2mr$, since there are two discontinuities per grating period, and multiple reflections should be negligible. Also from Fourier transform theory, the net

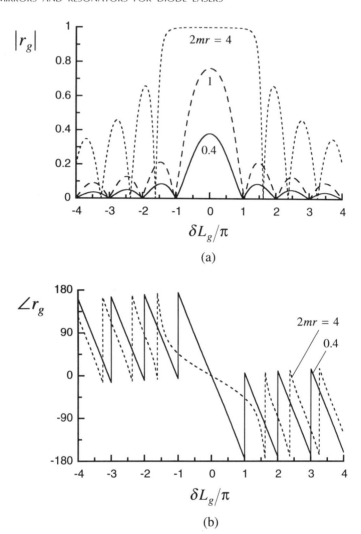

FIGURE 3.12 (a) Magnitude and (b) phase of the grating reflection coefficient vs. the normalized frequency deviation from the Bragg condition for different values of the reflection parameter, $\kappa L_g \equiv 2mr$. For small r and large m, the reflection spectrum plotted above is only dependent on the product $2mr$. However, for larger values of r and/or smaller values of m, there is some dependence on the individual values of r and m. For example, $r = 0.01$ and $m = 100$ would be roughly equivalent to $r = 0.1$ and $m = 10$. However, some changes in the spectrum would occur with $r = 0.2$ and $m = 5$, even though the $2mr$ product is equivalent in all three cases. The above plots use $r = 0.1$, 0.025, 0.01 with $m = 20$ for all cases. In the phase plot, the $2mr = 1$ case is very similar to the 0.4 case and hence is not plotted. Here $\delta \equiv \beta - \beta_0$ where β is the average propagation constant of the grating.

reflection from m elements, equally spaced by half a wavelength, and each causing a reflection of $2r$, should produce the following spectral response:

$$|r_g| \approx 2mr\, \frac{\sin(\delta L_g)}{\delta L_g}, \qquad (mr < 0.2) \qquad (3.55)$$

where δ is the deviation of the average propagation constant from the Bragg frequency. The qualifier reminds us that Eq. (3.55) is only valid in the weak reflection limit. Practical lasers usually require mirror power reflectivities of greater than 15% where multiple reflections cannot be ignored. Thus, Eq. (3.55) is of limited utility for diode laser work unless a relatively low reflectivity mirror is desired.

Historically, researchers working with long grating reflectors in in-plane lasers have chosen to use a different dimensionless parameter to quantify the net grating reflection rather than $2mr$, which is the reflection per grating segment, $2r$, times the number of segments, m. The parameter of choice is the reflection per unit length, κ, times the grating length, L_g. Thus, for the square wave grating, we see that the *coupling constant, $\kappa\!\!\underset{\varepsilon}{L}$* is given by

$$\kappa L_g \equiv 2mr = \frac{m\Delta\bar{n}}{\bar{n}} = \frac{L_g}{\Lambda}\left(\frac{\Delta\bar{n}}{\bar{n}}\right), \qquad (3.56)$$

where $\Delta\bar{n} \equiv |\bar{n}_2 - \bar{n}_1|$ and $\bar{n} \equiv (\bar{n}_2 + \bar{n}_1)/2$. For small index differences, $\Lambda = (\lambda_0/4)(1/\bar{n}_1 + 1/\bar{n}_2) \approx \lambda_0/2\bar{n}$ which gives $\kappa = 2\Delta\bar{n}/\lambda_0$. Later in Chapter 6 when we introduce coupled mode theory, we shall see that this form also naturally results (however, for the sinusoidal gratings analyzed there, κ is reduced by $\pi/4$). Therefore, the approximation for the grating reflection in Eq. (3.53) can also be written as, $r_g \approx \tanh(\kappa L_g)$. However, for very short gratings, such as in VCSELs, our original $2mr$ form seems more natural.

3.5.3 Effective Mirror Model for Gratings

From Fig. 3.12 we note that the phase varies relatively linearly near the reflection maximum. Such a reflection can be well approximated by a discrete mirror reflection equal to the magnitude of the grating's reflection, $|r_g|$, but placed a distance L_{eff} away as shown in Fig. 3.13.

From Eqs. (3.15) and the associated discussion in Section 3.3.2 above, we know that the incident and reflected wave amplitudes each experience a phase shift of $-\beta L_{eff}$ in traversing the distance to the effective mirror and back. Thus, knowing that the reflection phase is zero at the Bragg frequency, we can express r_g as

$$r_g \approx |r_g|e^{-2j(\beta - \beta_0)L_{eff}}. \qquad (|\delta L_g| \ll \pi) \qquad (3.57)$$

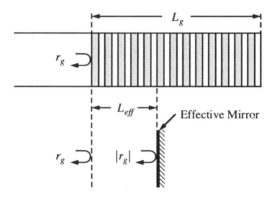

FIGURE 3.13   Definition of an effective mirror for a grating reflector.

Expanding the true DBR reflection phase in a Taylor series about the Bragg frequency: $j\phi \approx j\phi_0 + j(\beta - \beta_0)(\partial\phi/\partial\beta) + \cdots$, and equating the linear $(\beta - \beta_0)$ coefficient with the exponent in Eq. (3.57), we find that the effective length is given by

$$L_{eff} = -\frac{1}{2}\frac{\partial\phi}{\partial\beta}, \qquad (3.58)$$

which is the same result we found earlier in Eq. (3.38) by different means. From Fig. 3.12(b), it is clear that as $2mr$ increases, L_{eff} decreases over the range of the stopband. Using Eqs. (3.52) and (3.43), it can be shown (after a bit of math) that if third- and higher-order terms in r are neglected, the effective length becomes

$$L_{eff} = \frac{1}{2\kappa}\tanh(\kappa L_g), \qquad (|\delta L_g| \ll \pi) \qquad (3.59)$$

where $\kappa L_g \equiv 2mr$ as given by Eq. (3.56); the exact expression for L_{eff} is given in Appendix 7. For weakly reflecting gratings ($\tanh \kappa L_g \to \kappa L_g$), the effective mirror plane is at the center of the grating, and for strongly reflecting gratings ($\tanh \kappa L_g \to 1$), $L_{eff} \to \Lambda/(4r) \approx \lambda_0/4\Delta\bar{n}$.

Although L_{eff} was defined to give the proper mirror phase, and thus, can be used to locate cavity modes, it also gives the approximate optical energy penetration depth into the grating mirror. As mentioned earlier, m_{eff} gives the effective number of periods seen by the incident field. The optical power is the square of the field and hence it penetrates half as far into the mirror. Therefore, the energy penetration depth is given by $L_{pen} = \Lambda m_{eff}/2$ (see Appendix 7 for more details). Substituting Eq. (3.51) for m_{eff} and setting $2r = \kappa\Lambda$, this definition reduces to the one given in Eq. (3.59) and hence $L_{pen} \approx L_{eff}$. Therefore, the total energy stored in the mirror is approximately equal to the energy density at its input multiplied by L_{eff}.

A small propagation loss can be approximately added back into a lossless calculation of r_g by multiplying it by a factor $e^{-\alpha_i L_{eff}}$. In this case we are still using the effective mirror approximation, but β has been replaced by its complex form $\tilde{\beta}$, in Eq. (3.57). This perturbation technique is not valid for significant levels of loss or gain, because if such levels exist, the rate of decay of energy into the grating will be significantly affected. In this case, r_g should be recalculated using the transmission matrix method incorporating complex propagation constants, $\tilde{\beta}$, throughout. But, for many practical cases of interest we find that it is possible to model the grating near its reflection maximum by using the effective mirror concept, for which we need only know (1) the lossless reflection magnitude at the Bragg frequency, $|r_g|_{max}$; (2) the effective mirror location, given by L_{eff}; and (3) the propagation loss over the entire grating length, given by α_i in the grating.[1]

In the general case, $r_g = T_{g21}/T_{g11} = B_m/A_m|_{(B_0 = 0)}$ (or in terms of the S-parameters, $r_g = S_{g11} = b_m/a_m|_{(a_0 = 0)}$), which can be calculated numerically by performing the operations indicated in Fig. 3.11. The numerical procedure proceeds from the output of the grating backwards, after first assuming some value for A_0, such as unity. The intermediate A_j's and B_j's are evaluated by matrix multiplication moving to the left toward the beginning of the grating. Using this technique each segment in principle could be different, and the loss or gain can be naturally included in the T_{ij}'s, by using the appropriate complex propagation constants in Eqs. (3.24) to (3.27).

3.6 DBR LASERS

3.6.1 Introduction

A distributed Bragg reflector (DBR) laser can be formed by replacing one or both of the discrete laser mirrors with a passive grating reflector. Figure 3.14 shows schematics of both in-plane and VCSEL configurations with one grating mirror.

By definition, the grating reflectors are formed along a passive waveguide section, so one of the issues is how to make the transition between the active and passive waveguides without introducing an unwanted discontinuity. This is of little concern in the VCSEL case, since the axial direction is the growth direction, and switching materials is always done several times during growth.

---

[1] If the propagation loss is not distributed evenly throughout the grating (as is often the case with VCSEL mirrors), we must use an effective propagation loss:

$$\alpha_{i,eff} = \frac{1}{L_{eff}} \int_0^{L_g} \alpha_i(z) e^{-z/L_{eff}} (1 \pm \cos 2\beta z)\, dz.$$

The upper (lower) sign in the standing wave term is for a DBR with 0 (π) reflection phase at the Bragg wavelength. For a constant loss, $\alpha_i(z) = \alpha_{i0}$, and the equation reduces to $\alpha_{i,eff} = \alpha_{i0}$ for $L_g \gg L_{eff} \gg \lambda/4\pi n$.

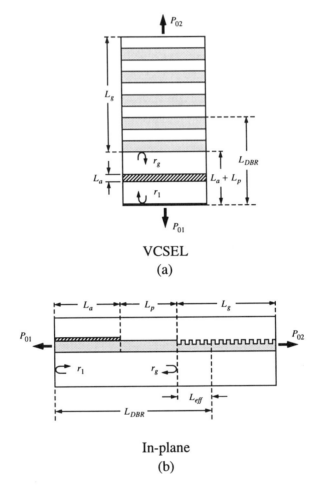

FIGURE 3.14 (a) Vertical-cavity surface-emitting laser schematic illustrating various lengths and reference planes. (b) In-plane laser schematic illustrating various lengths and reference planes.

Thus, forming the mirrors only requires growing more uniform layers. However, for coherence along the axial direction these layers must be very accurately controlled in thickness.

In the in-plane laser case making a DBR laser is relatively complex, since a lot of structure must be created along the surface of the wafer. This generally includes a joint between the active and passive regions as well as grating patterning and regrowth. For this reason in-plane DBR lasers are only formed when their unique properties are required. Besides the single-frequency property provided by the frequency-selective grating mirrors, these attributes can include wide tunability, if the effective index is varied electro-optically in the several sections by separate electrodes.

3.6.2   Threshold Gain and Power Out

The threshold gain of a DBR laser is the same as we have already calculated elsewhere, but we must interpret the parameters properly and consistently. This interpretation, however, can vary depending upon how we choose to model the DBR. If we treat the grating reflector as a separate element characterized by reflection and transmission scattering parameters, Case (a), we logically choose the cavity length to be $L_a + L_p$. However, if we use the effective mirror model outlined in Fig. 3.14, Case (b), we would choose the cavity length to be L_{DBR}, and include the grating passive losses in the penetration depth, L_{eff}, rather than the reflection coefficient. Fortunately, these approaches lead to the same result near the Bragg frequency, provided the losses are not too large. In practice we find that the approximate Case (b) is most useful for design where a minimum of computation is desired. Of course, Case (a) is more exact for detailed analysis.

With reference to the single grating mirror configuration in Fig. 3.14, the threshold gain given by Eq. (2.23) becomes

Case (a)

$$\Gamma g_{th} = \langle \alpha_i \rangle + \frac{1}{L_a + L_p} \ln\left[\frac{1}{r_1 |r_g|}\right], \tag{3.60}$$

where Γ and $\langle \alpha_i \rangle$ average over the active and passive sections of the cavity only (any losses encountered in the DBR section are contained in $|r_g|$, the magnitude of the DBR reflectivity given by Eq. (3.52)). Using the effective mirror model,

Case (b)

$$\Gamma g_{th} = \langle \alpha_i \rangle + \frac{1}{L_{DBR}} \ln\left[\frac{1}{r_1 |r_g'|}\right], \tag{3.61}$$

where Γ and $\langle \alpha_i \rangle$ average over the entire effective cavity length, $L_{DBR} = L_a + L_p + L_{eff}$, and losses encountered in the DBR section are treated as propagation losses in L_{eff}. The prime on r_g denotes the use of its lossless value. In either case, the mode spacing is determined using Eq. (3.39) with the optical length set equal to $\bar{n}_{ga} L_a + \bar{n}_{gp} L_p + \bar{n}_{gDBR} L_{eff}$.

For VCSELs, the confinement factor Γ must generally be calculated by Eq. (A5.10). Even for uniform gain within the active region Eq. (A5.14) must be used for Γ_z. If two grating mirrors are used in a DBR laser, r_1 in Eq. (3.60) must be replaced by $|r_{g1}|$ for the other grating reflection, and the mode spacing needs to include the effective lengths of both DBRs.

Since the distributed mirrors are lossy in general, we must use Eqs. (3.31) and (3.32) for the differential quantum efficiency and the power out, respectively. However, we need to use the proper value for α_m, and we must replace r_2 and

t_2 by the relevant grating S-parameters. Again, we have different expressions to be consistent with the two different models considered. That is, with a single Bragg mirror at end 2 and a discrete mirror at end 1 of the laser, Eq. (3.60) gives

Case (a)

$$\alpha_m = \frac{1}{L_a + L_p} \ln\left[\frac{1}{r_1|r_g|}\right], \tag{3.62}$$

and for the effective mirror model,

Case (b)

$$\alpha_m = \frac{1}{L_{DBR}} \ln\left[\frac{1}{r_1|r_g'|}\right].$$

For the fractional power out of each end in Case (a), we plug the grating S-parameters into Eq. (3.30). Setting $r_2 = S_{g11} \equiv r_g$ and $t_2 = S_{g21} \equiv t_g$, we obtain

Case (a)

$$F_1 = \frac{t_1^2}{(1 - r_1^2) + \dfrac{r_1}{|r_g|}(1 - |r_g|^2)},$$

and

$$F_2 = \frac{|t_g|^2}{(1 - |r_g|^2) + \dfrac{|r_g|}{r_1}(1 - r_1^2)}. \tag{3.63}$$

For the effective mirror model,

Case (b)

$$F_1 = \frac{t_1^2}{(1 - r_1^2) + \dfrac{r_1}{|r_g'|}(1 - |r_g'|^2)},$$

and

$$F_2 = \frac{|t_g'|^2}{(1 - |r_g'|^2) + \dfrac{|r_g'|}{r_1}(1 - r_1^2)}.$$

In this case, r_g is replaced by its lossless value r_g', and $|t_g|^2$ is replaced by the power transmission exclusively through the effective mirror, $|t_g'|^2$, which is found by setting $e^{-\alpha_i L_{eff}}|t_g'|^2 = |S_{g21}|^2$.

For both Case (a) and (b), it is useful to determine $|S_{g21}|^2$ for a lossy DBR. It can be shown using Eqs. (3.43)–(3.49) for a small uniform loss, α_i, at the Bragg condition that $\Delta \approx -j\alpha_i \Lambda/2$ and

$$|S_{g21}|^2 = 1/|T_{g11}|^2 \approx (1 - |r_g'|^2)e^{-2\alpha_i L_{eff}}. \tag{3.64}$$

Hence, we reach the unintuitive conclusion that the transmitted power is $\propto e^{-2\alpha_i L_{eff}}$ just like the reflected power. As a result, the transmission through the effective mirror becomes $|t'_g|^2 = (1 - |r'_g|^2)e^{-\alpha_i L_{eff}}$, which unfortunately does not correspond to the lossless transmission through the DBR (as one might have hoped).

With Eq. (3.64), we can show that the ratio of powers out of the two ends, F_1/F_2, is preserved in Case (a) and (b). Furthermore, we can show that F_1 and F_2 are both larger in Case (b). This increase in F_1 and F_2 compensates for the smaller mirror loss of Case (b), such that η_d is also approximately preserved in Case (a) and (b) (the effective mirror model does tend to overestimate η_d, but not significantly for $\alpha_i L_{eff} < 0.1$, or for larger losses if the grating reflectivity is high).

For a second general lossy mirror at end 1, its S-parameters should also be used for r_1 and t_1 in Eqs. (3.62) and (3.63). In Case (b), the same additional substitutions are made. Finally, as mentioned above, the power out of each end can be obtained from Eq. (3.32).

3.6.3 Mode Selection and Tunability

Figure 3.15 illustrates two plots similar to those shown in Fig. 3.8 for the DBR case. One is for cavity lengths common to in-plane lasers, the other is for VCSELs. As compared to the coupled-cavity cases, the key difference with grating mirrors is that there is only a single loss minimum.

For in-plane lasers, the relatively rapid roll-up in α_m leads to good loss margins at the adjacent axial mode wavelength. As indicated in Fig. 3.15, the net loss margin at the adjacent axial mode is the sum of the roll-off in the net modal gain, $\Delta\Gamma g$, and the roll-up of the mirror loss, $\Delta\alpha_m$. In a later section, we shall show that this net loss can be used to derive an expression for the steady-state mode suppression ratio (MSR) which predicts how much unwanted modes are suppressed.

In the VCSEL case, both the mode spacing and the width of the loss minimum are much larger, since the lengths of the cavity and grating, respectively, are much shorter. Thus, if we scale the wavelength axis to make them look comparable to the in-plane case, the primary effect is to make the gain look much more narrow relative to the mode spacing and mirror loss width. In fact, the roll-off in gain tends to be more important than the roll-up in loss for the VCSEL. That is, the primary cause of single-axial mode operation is just the short cavity.

The potential tunability of DBR lasers is one of the main reasons they are of great importance. As indicated in Fig. 3.14, there are usually three sections to a DBR, one active, one passive, and one passive grating. For the in-plane DBR it is convenient to place three separate control electrodes over these regions as shown in Fig. 3.16. One section provides gain, one allows independent mode phase control, and one can shift the mode-selective grating filter, respectively. By applying a control current or voltage to the grating section, its

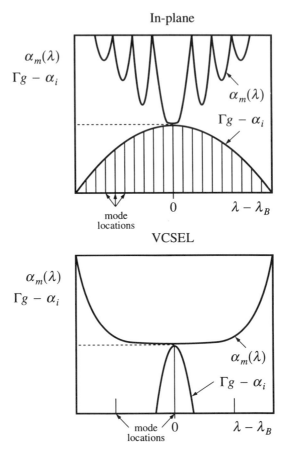

FIGURE 3.15 Schematic illustration of how a single axial mode is selected in an in-plane or vertical-cavity DBR laser. The VCSEL wavelength axis covers a five times larger range, i.e., the gain curve has the same width in both plots.

index, $\bar{n}_{DBR}$, changes, and the center wavelength of the grating, λ_g, moves according to $\Delta\lambda_g/\lambda_g = \Delta\bar{n}_{DBR}/\bar{n}_{DBR}$. Alternate axial modes can be selected as the mirror loss curve, $\alpha_m(\lambda)$, moves relative to the gain and modes. This is referred to as *mode hop* tuning. Also, the modes will move slightly in wavelength, since part of the net cavity length (L_{eff}) is in the grating.

By applying a current or voltage to the phase control electrode, the index of the passive cavity section, $\bar{n}_p$, changes, shifting the axial modes of the cavity. Thus, by applying a combination of control signals to the grating and phase control sections, a broad range of wavelengths are accessible. Since the carrier density is clamped in the active region, changes in current there only have a second-order effect on its index, $\bar{n}_a$, and only small changes in mode wavelength result. We can see more explicitly how the continuous mode shift occurs by

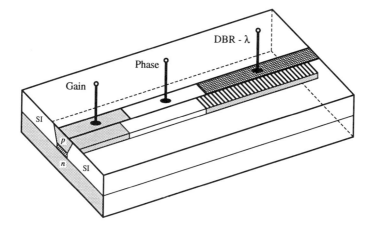

FIGURE 3.16   Illustration of a tunable single-frequency three-section DBR laser.

solving for the relative shift, $\Delta\lambda_m/\lambda_m$, from Eq. (2.25):

$$\frac{\Delta\lambda_m}{\lambda_m} = \frac{\Delta\bar{n}_a L_a + \Delta\bar{n}_p L_p + \Delta\bar{n}_{DBR} L_{eff}}{\bar{n}_a L_a + \bar{n}_p L_p + \bar{n}_{DBR} L_{eff}}. \qquad (3.65)$$

From Eq. (2.26) we can see how the indices are shifted by carrier injection. For example, with a transverse confinement factor, $\Gamma_{xy} = 10\%$, the effective index shifts by $\Delta\bar{n}/\bar{n} \approx -0.1\%$ for $\Delta N = 10^{18}$ cm^{-3}. This can occur in the phase control and grating regions above threshold. Since these lengths typically account for about half of L_{DBR} in an in-plane laser, the wavelength would be continuously tuned by $\sim 0.05\%$, or ~ 8 nm at 1.55 µm. The injected carrier density in the passive regions can be calculated from Eq. (2.15) with $g = 0$ and $dN/dt = 0$. Thus, the effective index change in the jth section can be written as

$$\Delta\bar{n}_j = \frac{\partial\bar{n}}{\partial N}\frac{\eta_i \tau I_j}{q V_j}, \qquad (3.66)$$

where $[\partial\bar{n}/\partial N] \approx -\Gamma_{xy} 10^{-20}$ cm^3.

A reverse bias voltage can also change the index of refraction via linear and quadratic electro-optic effects. Effective index shifts of $\sim 0.1\%$ are possible, but this is a little lower than possible with high injection currents. However, this reverse biased effect can have a much faster response time than current injection, since the carrier lifetime limits changes in the carrier density to a few hundred megahertz, similar to LEDs. Also, current injection leads to local heating which can change the index with time constants in the tens of microseconds range. Reverse bias also is more practical in the multisection VCSEL case.

More complex structures than the basic three-section DBR are possible, and some of these will be discussed in Chapter 8 on photonic integrated circuits.

3.7   DFB LASERS

A distributed feedback laser (DFB) also uses grating mirrors, but gain is included in the gratings. Thus, it is possible to make a laser from a single grating, although it is desirable to have at least a fraction of a wavelength shift near the center to facilitate lasing at the Bragg frequency. Historically, the DFB laser preceded the DBR, primarily because of its simplicity and relative ease of fabrication. Figure 3.17 gives a schematic of in-plane versions. Vertical cavity versions are also possible, but there is no advantage over the DBR and the fabrication is not any easier.

The basic characteristic equation is still the same as for other lasers, i.e., Eq. (2.20), but the gain is now in the complex mirror reflectivities, r_1 and r_2. Also, to avoid the troublesome active–passive transitions there is no passive cavity ($L_p = 0$) and the additional active cavity length (L_a) is typically only a fraction of a wavelength. The complex mirror reflectivities are given by Eq. (3.52) if antireflection (AR) coatings are used at the ends. If no AR coatings are used, one more **T**-matrix must be multiplied by that of Eq. (3.47) before calculating the grating's S_{11}.

For no shift in the gratings, the cavity can be taken to be anywhere within the DFB, since all periods look the same. The active length is then a quarter-wavelength long, since we have chosen mirror reference planes to fall at a downstep in index looking both to the left and right. (As discussed earlier, this yields a zero grating reflection phase at the Bragg frequency.) Thus, at the Bragg frequency, we can see that this DFB is antiresonant. Since the cavity is

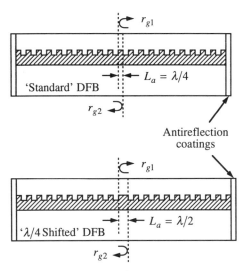

FIGURE 3.17 Standard and quarter-wave shifted DFB lasers. The entire length is filled with active material embossed with a grating.

so short, we can neglect any phase change in it over the reflection band of the mirrors. Thus, inserting $L_a = \lambda/4$ in Eq. (2.20) and assuming uniform pumping, the threshold condition becomes,

$$r_{g1}(\tilde{\beta}_{th})r_{g2}(\tilde{\beta}_{th}) = -1. \qquad \text{(unshifted)} \qquad (3.67)$$

Since there is no solution at the Bragg frequency, the wavelength must be scanned for each gain until Eq. (3.67) is satisfied.

The exception to this case is the *gain-coupled* DFB, in which the deviations in refractive index are purely imaginary, as for the case of added gain or loss. For example, the grating could consist of alternate sections of index n_1 and index $n_2 = n_1 + jn_i$; (by definition, $n_i = g\lambda/(4\pi)$ for added gain). Then, the reflection at each discontinuity, $r = jn_i/(2n_1)$, and the net grating reflection, r_g, at the Bragg frequency would be purely imaginary (reflection phase of $\pi/2$) for the selected reference planes indicated in Fig. 3.10. Thus, the fundamental solutions to (3.67) do occur at the Bragg frequency in this case.

If we again consider only real index perturbations and the cavity is half a wavelength long, we can see that the device is resonant at the Bragg frequency where the reflection phase is zero. Actually, as can be seen in Fig. 3.17, a half-wavelength mirror spacing corresponds to a quarter-wave shift between the two gratings. Thus, this configuration is usually referred to as a *quarter-wave shifted* DFB. In this case we use $L_a = \lambda/2$ in Eq. (2.20), and the threshold condition for uniform pumping becomes

$$r_{g1}(\tilde{\beta}_{th})r_{g2}(\tilde{\beta}_{th}) = 1. \qquad \text{(quarter-wave shifted)} \qquad (3.68)$$

The threshold gain and wavelength solutions to the quarter-wave shifted case are a little easier, since we know they occur near the Bragg condition. In both cases, if the pumping is different on the two sides of the cavity, the complex propagation constants, $\tilde{\beta}_{th}$, will be different in each grating section. In such cases, there will be pairs of threshold gains (one in each section) that satisfy the threshold condition.

Figure 3.18 shows calculated threshold gain results for DFB lasers with and without the quarter-wave shift in the center of the cavity. Antireflection (AR) coatings are assumed.

For the unshifted DFB, it should be realized that two modes equally spaced on each side of the Bragg wavelength reach threshold simultaneously, if there exist no additional perturbing reflections, such as from uncoated cleaves at the end. This simplest of DFBs must rely upon such additional reflections to destroy the unwanted degeneracy. In practice, at least one cleaved mirror will do the job if the gratings are not too strong ($\kappa L_g \leq 1$) so the net reflection phase from one end is shifted from that of the grating alone. However, there is still a yield problem, since the reflection from the cleave will have a random relative phase, but optimally it should be in quadrature to shift the net phase from that laser end by the maximum amount. One more T-matrix must be multiplied by that

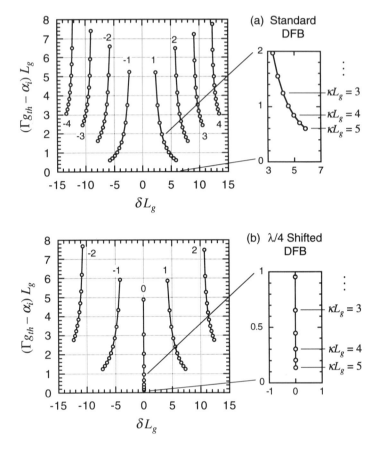

FIGURE 3.18   Normalized plots of threshold modal gain and threshold wavelength for different modes of standard and quarter-wave shifted DFB lasers with κL_g ($\equiv 2mr$) ranging from 5 to 0.5 in 0.5 increments. Here $\delta = \beta - \beta_0$, where β is the average propagation constant of the grating.

of Eq. (3.47) to obtain the net grating S_{11} in this case. Figure 3.19 replots the threshold gain and wavelength for a DFB laser with an AR coating on one end and a cleave at the other, assuming the cleave is in quadrature phase (i.e., cleaved exactly between two interfaces in the grating).

The threshold gain and wavelength can also be calculated by observing the net transfer function, $S_{21}(\omega) = 1/T_{11}(\omega)$, through the DFB laser (or any other laser) as the gain is increased, rather than solving Eq. (3.67) or (3.68). The poles of one of the S-parameters (or zeros of the T_{11} parameter) *for the entire system* give another form of the characteristic equation. From Eq. (3.48), the characteristic equation of a DFB with AR coated facets can be written as

$$m_{eff}\Delta = j,$$ (3.69)

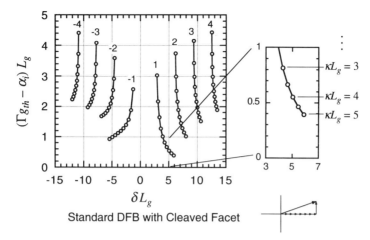

Standard DFB with Cleaved Facet

FIGURE 3.19 Normalized plot of threshold modal gain and threshold wavelength for different modes of a standard DFB laser with κL_g ($\equiv 2mr$) ranging from 5 to 0.5 in 0.5 increments. One end of the laser is AR coated and the other end is cleaved such that the facet reflection (with a field magnitude of 0.565) is 90° out of phase with the small grating reflections (as illustrated in the lower right corner). Here $\delta = \beta - \beta_0$, where β is the average propagation constant of the grating.

where the effective number of mirror periods, m_{eff}, and the detuning parameter, Δ, are defined in Eq. (3.49).

In a numerical calculation, as the gain is increased, the transmission spectrum of the device under study will develop a strong maximum. The gain required for this maximum to reach some large value and its wavelength are the desired threshold values. This technique is particularly useful to determine the threshold gain margin for spurious modes, since even after one mode reaches threshold, the gain can still be increased to look for the next mode to blow up at another wavelength. In Figs. 3.18 and 3.19 the threshold gain and wavelength for the first few modes are shown.

The corrugations of the grating can also cause significant periodic loss and gain variations. That is, r_1 (and κ) may be complex (have a nonzero phase angle) even with the reference planes chosen above, which makes r_1 real for real index variations. In the extreme of pure gain modulation and no index variation, Eqs. (3.42) through (3.52) show that r_g would have an angle of $\phi = \pi/4$ at the Bragg wavelength. Thus, as already discussed, the unshifted DFB laser characterized by Eq. (3.67) provides a mode at the Bragg wavelength of the gratings. It performs as the quarter-wave shifted DFB with real index variations. For these reasons and other potential benefits of this *gain-coupled* DFB, there has been continued active research in this direction.

3.8   MODE SUPPRESSION RATIO IN SINGLE-FREQUENCY LASERS

As mentioned above, a primary reason that people are interested in coupled-cavity, DBR, and DFB lasers is their potential for single-frequency operation. But, we must realize that "single-frequency" is a relative term. In this section we wish to discuss a measure of single-frequency purity, the mode suppression ratio (MSR). It is simply the ratio of the output power in the primary laser mode to that in the next strongest mode from one end of the laser:

$$MSR = \frac{P(\lambda_0)}{P(\lambda_1)}, \tag{3.70}$$

where we have dropped the subscripts from the output power as given by Eq. (3.32) and labeled the primary mode as the one at λ_0. More fundamentally, the output power from one end of the laser at the nth mode is given by Eq. (2.33) multiplied by the fraction out one end, $F_1(\lambda_n)$ as given by Eq. (3.30). That is,

$$P(\lambda_n) = F_1(\lambda_n)v_g\alpha_m(\lambda_n)N_p(\lambda_n)h\nu V_p. \tag{3.71}$$

From Eq. (2.16) we can express the steady-state ($dP/dt = 0$) photon density as

$$N_p(\lambda_n) = \frac{\Gamma\beta_{sp}R_{sp}(\lambda_n)}{1/\tau_p(\lambda_n) - \Gamma v_g g(\lambda_n)}. \tag{3.72}$$

Note that in Chapter 2, we used Eq. (2.15) to solve for N_p in terms of the terminal current, but here we are interested in expressing it for the various modes in terms of the net gain margin, the denominator of Eq. (3.72). In this form we can see how the noise injected into a particular mode, $\Gamma\beta_{sp}R_{sp}(\lambda_n)$, is amplified to a large steady-state value as the denominator approaches zero. (But it never actually goes to zero for any finite power out.)

Now, we can plug Eq. (3.72) into Eq. (3.71) and form the ratio given in Eq. (3.70) for the desired pair of modes. That is,

$$MSR = \frac{F_1(\lambda_0)\alpha_m(\lambda_0)[\alpha_i + \alpha_m(\lambda_1) - \Gamma g(\lambda_1)]}{F_1(\lambda_1)\alpha_m(\lambda_1)[\alpha_i + \alpha_m(\lambda_0) - \Gamma g(\lambda_0)]}, \tag{3.73}$$

where we have assumed that the spontaneous emission is coupled equally into both modes, and that the modes are similar in frequency, volume, and velocity. We have also used Eq. (2.24) for the cavity lifetime, i.e., $1/\tau_p(\lambda_n) = v_g(\alpha_i + \alpha_m)$. Before trying to simplify this any further, we wish to review a generic schematic of loss and gain vs. wavelength shown in Fig. 3.20, similar to Figs. 3.8, 3.15, and indirectly 3.18 for different types of lasers.

Finally, by reference to the figure we can simplify Eq. (3.73) by calling the denominator bracket, which is the separation between the mirror loss and the

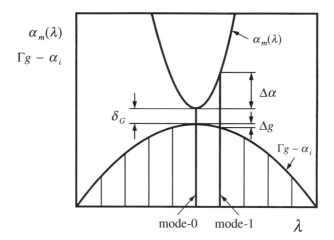

FIGURE 3.20   Definition of gain and loss margins for use in MSR calculations.

net modal gain for the main mode, $\delta_G = \alpha_m(\lambda_0) - [\Gamma g(\lambda_0) - \alpha_i]$, the loss margin, $\Delta\alpha = \alpha_m(\lambda_1) - \alpha_m(\lambda_0)$, and the modal gain margin, $\Delta g = \Gamma g(\lambda_0) - \Gamma g(\lambda_1)$. Furthermore, if the back mirror provides the frequency-dependent loss, then the fraction of light coupled out of the front mirror will be reduced as the mirror loss is increased. As a result, the coupling fraction ratio out of the front mirror times the mirror loss ratio is ~ 1. We are then left with

$$\text{MSR} \approx \frac{\Delta\alpha + \Delta g}{\delta_G} + 1. \tag{3.74}$$

More commonly, the MSR is expressed in terms of decibels (dB) of optical power:

$$\text{MSR(dB)} \approx 10 \log_{10}\left[\frac{\Delta\alpha + \Delta g}{\delta_G} + 1\right]. \tag{3.75}$$

If the spectrum is observed by direct detection, the photodiode current is directly proportional to the optical power, while the electrical power is proportional to its square. Thus, if *electrical* power is displayed in decibels on a spectrum analyzer, the observed MSR will appear twice as large as (3.75) would predict.

The value of δ_G can be calculated in terms of cavity parameters and the drive current by combining a number of existing equations. First, we solve Eqs. (3.71) and (3.72) for δ_G at λ_0,

$$\delta_G = \frac{F_1 \alpha_m h\nu V_p \Gamma \beta_{sp} R_{sp}}{P_{01}}. \tag{3.76}$$

Then, using Eqs. (2.5) and (2.6) for R_{sp}, and Eqs. (3.30) to (3.32) for P_{01}, we get an expression valid for $I > I_{th}$,

$$\delta_G = (\alpha_i + \alpha_m)\beta_{sp}\eta_r \frac{I_{th}}{(I - I_{th})}, \tag{3.77}$$

and the spontaneous emission factor, β_{sp}, is given by Eq. (A4.10). For typical values of the parameters $\delta_G \sim 10^{-3} I_{th}/(I - I_{th}) \, \text{cm}^{-1}$.

READING LIST

G.P. Agrawal and N.K. Dutta, 2d ed. *Semiconductor Lasers*, Chs. 7 and 8, Van Nostrand Reinhold, New York (1993).

H.A. Haus, *Waves and Fields in Optoelectronics*, Ch. 3, Prentice Hall, Englewood Cliffs, NJ (1984).

S. Ramo, J.R. Whinnery, and T. Vanduzer, *Fields and Waves in Communication Eectronics*, Ch. 11, Wiley, New York (1984).

G. Björk and O. Nilsson, *IEEE J. Lightwave Technol.*, **5**, 140 (1987).

PROBLEMS

These problems include material from Appendix 7.

3.1 (a) Write the **S** and **T** matrices between two ports bounding a section of transmission line and a dielectric interface as shown in Fig. 3.21.

(b) For $n_t = 1$, $n_d = 3.5$, $L_t = 10 \, \mu\text{m}$, and $L_d = 5 \, \mu\text{m}$, plot S_{11} and S_{21} vs. wavelength for $900 < \lambda < 1000$ nm.

3.2 Verify Eqs. (3.24) to (3.27) by solving for the appropriate ratios of scattering amplitudes with the appropriate boundary conditions.

3.3 Verify Eq. (3.30) by showing that $F_1 + F_2 = 1$ and that $F_1/F_2 = P_{01}/P_{02}$.

3.4 Write the characteristic equation for a three-mirror laser using only its mirror reflectivities and dimensions (i.e., no r_{eff} in the answer).

3.5 Plot the effective mirror reflectivity, r_{eff}, vs. wavelength near 1.3 μm over two full cycles of oscillation for an external cavity laser of the form illustrated in Fig. 3.7. Assume the external cavity medium has an index of 1.6, a loss of $5 \, \text{cm}^{-1}$, a length $L_p = 200 \, \mu\text{m}$, and that the mirrors, $r_3 = -0.9$, $r_2 = 0.5$, $t_2 = 0.3$ (mode mismatch loss).

3.6 In Problem 3.5, if we tune the index of the external cavity by 1%, by how many nanometers do the maxima of r_{eff} shift?

3.7 In a four-mirror coupled-cavity laser, it is desired to have the maxima of r_{eff} be narrower in wavelength than the minima for best mode selectivity.

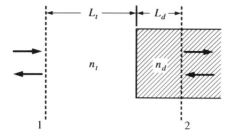

FIGURE 3.21 Two-port scattering junction consisting of a section of transmission line and a section of dielectric.

A device is fabricated by etching a deep and narrow groove across 1.55 μm InGaAsP/InP DH material to form two active sections. Assume the minimum accurately controlled groove width is 1.2 μm.

(a) Find the minimum groove width to accomplish the desired goal.

(b) Assuming a diffraction power loss of 50% per pass in coupling from one section to the other, what are the values of S_{s11}, R', σ, and r_{eff} in Eq. (3.41) for this case?

(c) Plot r_{eff} vs. wavelength over two periods for this case.

3.8   Verify Eqs. (3.43).

3.9   Show that Eq. (3.52) reduces to Eq. (3.55) in the low reflection limit.

3.10   A VCSEL mirror consists of three grating periods backed by a metallic reflector. The position of the metallic reflector is adjusted so that its reflection adds in-phase with the grating's. Assume the amplitude reflection at each discontinuity of the grating is 0.1, and the metallic layer has an amplitude reflection of 0.95.

(a) What is the net amplitude reflectivity at the Bragg frequency?

(b) What is the effective penetration depth measured in wavelengths? Be sure to define a reference plane.

3.11   An $In_{0.2}Ga_{0.8}As/AlGaAs$ VCSEL as in Fig. 3.14 has the following parameters:

$L_a = 0.02$ μm GaAs DH active region placed at standing wave peak.
$L_p + L_a = 1$ wavelength in the medium.
Two DBR mirrors: AlAs/GaAs quarter-wave stacks 18 periods each; top (rear) mirror metalized to give a net mirror reflectivity of 99.9%.
Average internal loss, $\langle \alpha_i \rangle = 20$ cm^{-1}.

(a) What is the bottom (front) mirror reflectivity?

(b) What is the effective penetration depth into each mirror?

(c) What is the threshold modal gain?

(d) What is the differential efficiency, assuming $\eta_i = 100\%$?

3.12 Consider the bottom (front) DBR mirror of Problem 3.11.

(a) Plot its power reflectivity vs. wavelength relative to the Bragg value. Show two minima on each side of the central maximum.

(b) Plot $\ln(1/R)$ for this grating vs. wavelength.

3.13 A tunable three-section DBR as in Fig. 3.16 is constructed to operate near 1.55 µm from InGaAsP/InP materials. Above threshold, the wavelength is tuned by changing the effective indices in the phase and DBR passive sections by injecting current. For no current injection, the operating wavelength is 1.57 µm, the effective index in all sections is 3.4, $\partial \bar{n}/\partial N = 10^{-21}$ cm^{-3}, $\eta_i = 70\%$, and the carrier lifetime is independent of carrier density and equals 3 ns in all sections. The waveguide cross section in all regions is 0.2×3 µm; the gain, phase shift, and grating regions are each 200 µm long, and the grating has a reflectivity per unit length of 100 cm^{-1}. The other mirror is a cleaved facet.

Plot the wavelength vs. current to the grating.

(a) Assume no current is applied to the phase shift region and show at least three axial mode jumps.

(b) Repeat for a phase shift current sufficient to maintain operation at the grating's Bragg wavelength.

In (b) also plot the required phase shift current on the opposite axis. Stop plots when any current reaches 50 mA.

3.14 A 1.55 µm InGaAsP DBR laser consists of an active section 500 µm long butted to a passive grating section 500 µm long. The coupling constant–length product, $\kappa L_g = 1$ for the grating. The active section is terminated with a cleaved facet on the opposite end. The active section internal efficiency is 70%, and the average internal modal losses are 20 cm^{-1} throughout both sections. What are η_{d1} and η_{d2}?

3.15 For the problem of 3.14, what is the MSR out of each end of the laser? If the grating is now made stronger such that $\kappa L_g = 2$, what are the differential efficiencies and MSRs out of both ends?

3.16 A quarter-wave shifted DFB laser has an internal quantum efficiency of 60%, a modal loss of 10 cm^{-1}, AR coated facets, $L_g = 500$µm, and a $\kappa L_g = 1$.

(a) What is the threshold modal gain?

(b) For operation at twice threshold with $\beta_{sp} = 10^{-4}$, what is the MSR?

(c) What is the differential efficiency from each end?

(d) What κL_g gives the best MSR, and what is it?

3.17 Standard DFBs with internal modal losses of 10 cm^{-1} and various κL_g's are fabricated with one end AR coated and one end cleaved. Assume that the cleave provides a reflection in quadrature with that of the grating and that the output is from the AR coated end.

(a) What κL_g gives the best MSR, and what is η_{d1} in this case?

(b) What κL_g gives the worst MSR, and what is η_{d1} in this case?

CHAPTER FOUR

Gain and Current Relations

4.1 INTRODUCTION

In Chapter 1, various transitions responsible for the generation and recombination of carriers within the semiconductor were introduced. In Chapter 2, the rates at which these transitions occur were shown to provide the fundamental description of LED and laser operation through the development of the rate equations. The optical gain, for example, was defined in terms of the difference between the stimulated emission and absorption rates. Radiative efficiency was defined in terms of the spontaneous and nonradiative recombination rates. Simple relationships between these rates and the carrier density were assumed in Chapter 2 to provide a feel for how semiconductor lasers generally behave. In the present chapter, we would like to delve a little deeper into the fundamentals of these transitions.

We will first develop a quantitative description of radiative transitions, from which we will be able to determine both the optical gain and the corresponding radiative current density as a function of injection level. Then, we will consider nonradiative transitions and see how they compare to the radiative transition rates in different material systems. Finally we will provide a set of example gain calculations for common materials in order to quantify the various relationships between the gain, carrier density, and current density.

In Appendix 6, an alternative description of radiative transitions traditionally applied to discrete energy level lasers is adapted for use in semiconductors. The reader is encouraged to examine this appendix, for not only does the analysis bridge the gap between Einstein's approach and the treatment provided here, but it is also hoped that by covering the same material from a different perspective, the reader will gain a deeper understanding of radiative processes in semiconductors.

More in-depth discussions of many of the relations used in this chapter can be found in Appendices 8 through 11. These discussions are presented at a higher level and are not required for the basic understanding of material presented in this chapter. In brief, the envelope function approximation and the calculation of the valence subband structure in quantum wells with and without strain can be found in Appendix 8. Fermi's Golden Rule, a key relation for estimating gain in semiconductors, is derived from first principles in Appendix 9. The resulting transition matrix element and polarization-dependent effects related to it are considered in Appendix 10. Finally, the effects of strain on the bandgap of semiconductors are discussed in Appendix 11. The latter part of Appendix 8 and Appendix 11 are recommended reading for anyone particularly interested in strained materials.

4.2   RADIATIVE TRANSITIONS

4.2.1   Basic Definitions and Fundamental Relationships

In Chapter 1, many types of transitions were discussed in reference to Fig. 1.3. Here we would like to concentrate on the radiative transitions. Specifically, there are three types of radiative transitions between the conduction and valence bands which are important in modern semiconductor lasers. These *band-to-band* radiative transitions are sketched in Fig. 4.1.

In the first diagram, the energy of the photon is transferred to an electron, elevating it from some state 1 in the valence band to some state 2 in the conduction band. Such *stimulated absorption* events generate new carriers and are also responsible for the disappearance of photons. In the second diagram, the incoming photon stimulates the electron to liberate energy in the form of a new photon, lowering it from state 2 in the conduction band to state 1 in the valence band. Such *stimulated emission* events provide a recombination path for carriers and are more importantly the source of new photons. The third diagram is really no different from the second diagram except that the field which stimulates the electron to emit a photon and make a downward transition is not a real field, but a *vacuum-field* (as it is commonly referred to in the quantum world). Because vacuum-field-induced transitions can occur with no classical field stimulation, we refer to them as *spontaneous emission* events. In the absence of classical fields, spontaneous emission serves as one of the

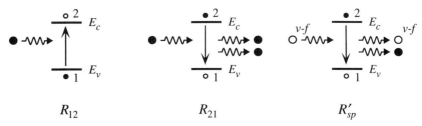

FIGURE 4.1 Band-to-band radiative transitions: stimulated absorption, stimulated emission, and spontaneous emission. (All rates are defined per unit volume.)

dominant recombination paths of carriers in direct bandgap semiconductors, and is by far the most common source of photons provided by nature.

The rates at which the three radiative processes in Fig. 4.1 occur depend on a number of factors. Two primary factors are the density of photons and the density of *available* state pairs. As we will find later, the dependence on the photon density enters through the local electric field strength, $|\mathscr{E}|^2$. Thus, R_{12}, $R_{21} \propto |\mathscr{E}|^2$ and $R'_{sp} \propto |\mathscr{E}^{v\text{-}f}|^2$, where $|\mathscr{E}^{v\text{-}f}|^2$ is the vacuum-field strength. We will have more to say about vacuum-fields and how to evaluate $|\mathscr{E}^{v\text{-}f}|^2$ in Section 4.4.

The dependence of the transition rates on the density of available state pairs can be broken down into two components: one which is strictly material dependent, and the other which depends on the injection levels. The first component is the density of *total* state pairs, which is found by taking the appropriate average between the density of states in the conduction and valence bands. We will learn how to evaluate this *reduced* density of states function, ρ_r, a little later. The second component is the *fraction* of state pairs available to participate in the transition. For upward transitions, this fraction is maximized when all carriers are placed in the valence band. For downward transitions, this fraction is maximized when all carriers are placed in the conduction band. The former population of carriers occurs naturally, while the latter *inverted* population can only be achieved by providing energy which pumps the carriers into the conduction band (for example, by current injection into the center of a *pn* junction). The fraction of available state pairs will be quantified below.

Now let's consider the electromagnetic field a little more carefully. First of all, it is important to appreciate that downward transitions not only create a new photon but they create a new photon *into the same optical mode as the stimulating photon* (whether it is a real or vacuum-field photon). For this reason, it is important to distinguish photons in one optical mode from photons in another (see Appendices 3 and 4 for a discussion of optical modes). By associating $|\mathscr{E}|^2$ with the field strength of one optical mode, we can interpret the transition rates in Fig. 4.1 as *single-mode* transition rates (in fact, the prime on R'_{sp} is used to distinguish this single-mode spontaneous emission rate from the total band-to-band spontaneous emission rate, R_{sp}). The total transition rates are then found by summing over all optical modes.

Another interesting feature of downward transitions is that in addition to appearing in the same optical mode, the newly created photon also contributes to the existing field constructively. This feature allows the optical mode to build up a very coherent field. Unfortunately, the vacuum-field phase is not correlated with the phase of the real fields in the optical mode. As a result, new photons introduced into the mode through spontaneous emission have random phases relative to the coherent fields created through stimulated emission. And although the number of photons introduced into the mode through spontaneous emission can be made small relative to the photons introduced through stimulated emission, they can never be removed completely, implying that perfect coherence

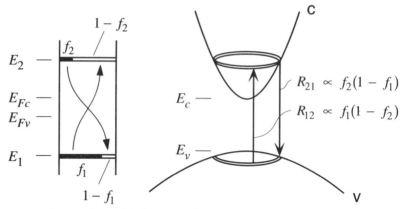

FIGURE 4.2   State pairs which interact with photons at E_{21}. Energy and momentum conservation reduce the set of state pairs to the annulus shown in the plot of energy vs. momentum in two dimensions. The occupation probabilities, f_1 and f_2, reduce this set even further.

in a laser can never be achieved. Chapter 5 considers the implications of spontaneous emission as a phase noise source in more detail.

So far we have not specified the electron states 1 and 2 in any detail. As will be shown below, photons with energy $h\nu$ induce upward and downward transitions only between those electron state pairs which conserve both energy and momentum in the course of the transition. In other words, we must have $E_2 - E_1 = E_{21} = h\nu$ and $\mathbf{k}_2 = \mathbf{k}_1$. These conservation laws reduce the interaction to a very particular region of the E–k diagram of the semiconductor, as illustrated in Fig. 4.2 for two dimensions of k-space. Furthermore, within this region only vertical transitions are allowed.

Now electrons typically only spend about 0.1 ps in any given state due to collisions with phonons and other electrons. As a result, their energy is uncertain, making the annulus shown in Fig. 4.2 appear fuzzy. To properly account for this, the total transition rates should include an integration over the energy uncertainty. In Appendix 6, this integration is included right from the start. However, because the integral tends to clutter the math, we will defer this procedure to the very end, where a more thorough discussion will be included.

Another restriction we must consider is that transitions occur only between filled initial states and empty final states. Figure 4.2 illustrates the fraction of state pairs which satisfy this criterion for both upward and downward transitions. Writing out the Fermi factors explicitly, the three radiative transition rates become

$$\begin{aligned}
R_{12} &= R_r \cdot f_1(1 - f_2), \\
R_{21} &= R_r \cdot f_2(1 - f_1), \\
R'_{sp} &= R_r^{v-f} \cdot f_2(1 - f_1).
\end{aligned} \tag{4.1}$$

In these equations, R_r represents the radiative transition rate that would exist if all state pairs were available to participate in the transition. For the spontaneous emission rate, we must use $R_r^{v-f} = R_r$ with $|\mathscr{E}|^2 \rightarrow |\mathscr{E}^{v-f}|^2$. We will derive an explicit expression for R_r later, but from the above discussions we already know that R_r is proportional to the field strength and the reduced density of states function.

Because R_{21} and R_{12} are competing effects in that one generates new photons and the other takes them away, we would also like to know the *net* generation rate of photons in the semiconductor, or

$$R_{st} \equiv R_{21} - R_{12} = R_r \cdot (f_2 - f_1). \tag{4.2}$$

We will show in Section 4.3 as we have by a more phenomenological route in Chapter 2, that the *net stimulated emission rate*, R_{st}, is directly proportional to the optical gain in the material.

The occupation probabilities in Eqs. (4.1) and (4.2) can usually be described using Fermi statistics even under nonequilibrium conditions by using a separate Fermi level for the conduction and valence bands:

$$f_1 = \frac{1}{e^{(E_1 - E_{Fv})/kT} + 1} \quad \text{and} \quad f_2 = \frac{1}{e^{(E_2 - E_{Fc})/kT} + 1}, \tag{4.3}$$

where E_{Fc} and E_{Fv} are the conduction and valence band *quasi-Fermi levels*. Under nonequilibrium forward bias conditions, E_{Fc} and E_{Fv} are separated by slightly less than the applied voltage to the junction.

Simple relations between the transition rates are easily derived by substituting Eq. (4.3) for the occupation probabilities:

$$\frac{R_{21}}{R_{12}} = \frac{f_2(1 - f_1)}{f_1(1 - f_2)} = e^{(\Delta E_F - E_{21})/kT}, \tag{4.4}$$

$$\frac{R'_{sp}}{R_{st}} = \frac{|\mathscr{E}^{v-f}|^2}{|\mathscr{E}|^2} \frac{f_2(1 - f_1)}{f_2 - f_1} = \frac{|\mathscr{E}^{v-f}|^2}{|\mathscr{E}|^2} \frac{1}{1 - e^{(E_{21} - \Delta E_F)/kT}}. \tag{4.5}$$

The first ratio (4.4) reveals that the stimulated emission rate will be larger than the absorption rate only when

$$E_{Fc} - E_{Fv} \equiv \Delta E_F > E_{21}. \tag{4.6}$$

Stated another way, the *net* stimulated emission rate (and hence the optical gain) will become positive only when the quasi-Fermi level separation is greater than the photon energy of interest. And because the photon energy must at the very least be equal to the bandgap energy, we conclude that to achieve gain in the semiconductor, we must have

$$\Delta E_F > E_g. \tag{4.7}$$

This condition demands that the voltage across a *pn* junction must be greater than the bandgap to achieve gain in the active region.

The second ratio (4.5) reveals a fundamental relationship between the single-mode spontaneous emission rate and the net stimulated emission rate. This relation will be developed further in Section 4.4. We will find there that the ratio of field strengths is just equal to the reciprocal of the number of photons in the optical mode.

4.2.2 Fundamental Description of the Radiative Transition Rate

To fully quantify all three radiative transition rates, we need only evaluate the one transition rate, R_r, appearing in Eq. (4.1). The treatment in Appendix 6 also concludes that Einstein's stimulated rate constant, B_{21}, is all that is necessary to determine the three radiative transition rates. However, Einstein's approach does not provide a means of determining B_{21} in semiconductors. Fortunately, the transition rate, R_r can be estimated using a relation known as Fermi's Golden Rule, derived in Appendix 9. To evaluate Fermi's Golden Rule, we need to provide an accurate description of the interaction which occurs between the electron in the crystal and the electro-magnetic field.

To describe the electron fully we must provide a model for the electron's wavefunction in both states 1 and 2. To be rigorous, ψ_1 and ψ_2 must be found by solving the Schrödinger equation with the appropriate crystal potential. Such an exact solution, however, would be difficult to find and inconvenient to work with. Fortunately, a useful approximation can be made which decomposes the crystal potential into (1) a complex atomic-scale potential which is periodic with the crystal lattice, and (2) a macroscopic potential which follows the spatial dependence of the conduction or valence band edge (created by either doping or material composition variations) as illustrated in Fig. 4.3. Appendix 8 shows

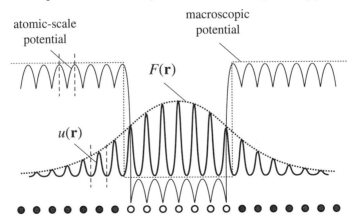

FIGURE 4.3 Illustration of a quantum-well potential and the corresponding lowest energy electron wavefunction.

that the corresponding electron wavefunctions can then be written as the product of two functions:

$$\psi_1 = F_1(\mathbf{r}) \cdot u_v(\mathbf{r}) \quad \text{and} \quad \psi_2 = F_2(\mathbf{r}) \cdot u_c(\mathbf{r}). \tag{4.8}$$

The *envelope* function, $F(\mathbf{r})$, is a slowly varying function satisfying Schrödinger's equation using the macroscopic potential and an appropriate *effective* mass. The *Bloch* function, $u(\mathbf{r})$, is a complex periodic function which satisfies Schrödinger's equation using the atomic-scale potential. Each energy band in the crystal has its own Bloch function. Fortunately, one never really needs to determine $u(\mathbf{r})$ precisely. Only the symmetry properties of these functions are necessary for most calculations, as discussed in Appendix 8. Thus we can concentrate our attention on the simpler envelope function.

In uniformly doped bulk material, the flat energy bands imply a constant macroscopic potential which leads to simple plane wave solutions for the envelope functions. In a quantum-well potential, plane wave solutions exist along directions within the plane of the well. However, along the confinement direction, $F(\mathbf{r})$ takes on either a cosine or sine wave distribution inside the well and decays exponentially outside the well. The bulk, quantum-well, and (by extension) quantum-wire envelope functions therefore take the following form:

$$F(\mathbf{r}) = e^{-j\mathbf{k}\cdot\mathbf{r}}/\sqrt{V}, \qquad \text{(bulk)} \tag{4.9}$$

$$F(\mathbf{r}) = F(z) \cdot e^{-j\mathbf{k}\cdot\mathbf{r}_\parallel}/\sqrt{A}, \qquad \text{(quantum well)} \tag{4.10}$$

$$F(\mathbf{r}) = F(x, y) \cdot e^{-jk_z z}/\sqrt{L}, \qquad \text{(quantum wire)} \tag{4.11}$$

where V (A, L) is the volume (area, length) of the crystal, and appears for normalization purposes. For the quantum well, the position vector $\mathbf{r}_\parallel$ is parallel to the quantum-well plane, and $F(z)$ is the simple one-dimensional solution to the quantum-well potential considered in Appendix 1. For the quantum wire, the length of the wire runs along z, and $F(x, y)$ is the two-dimensional solution within the quantization plane. The *overall* quantum-well electron wavefunction illustrating both envelope and Bloch function components is superimposed over the crystal potential in Fig. 4.3. Because we often need only the envelope function of the electron for many calculations, it is common to associate $F(\mathbf{r})$ with the complete wavefunction of the electron, ψ. This association is oftentimes harmless, however, in the present case it is necessary to emphasize that it represents only the slowly varying envelope of the complete electron wavefunction.

With the electron wavefunctions defined, we can move on to describing the interaction between the electron and the electromagnetic wave. The wave's interaction with the electron enters into Schrödinger's equation through the vector potential:

$$A(\mathbf{r}, t) = \hat{e}\, \text{Re}\{\mathscr{A}(\mathbf{r})e^{j\omega t}\} = \hat{e}\tfrac{1}{2}[\mathscr{A}(\mathbf{r})e^{j\omega t} + \mathscr{A}^*(\mathbf{r})e^{-j\omega t}], \qquad (4.12)$$

where $\hat{e}$ is the unit polarization vector in the direction of A, and ω ($\hbar\omega$) is the angular frequency (energy) of the photon. The vector potential is related to the electric field via $\mathscr{E} = -\partial A/\partial t$. The kinetic energy term of Schrödinger's equation describing the electron in the crystal (given by (A8.1)) is now modified by the substitution

$$\mathbf{p}^2 \rightarrow (\mathbf{p} + q\mathbf{A})^2 \approx \mathbf{p}^2 + 2q\mathbf{A}\cdot\mathbf{p}, \qquad (4.13)$$

where q is the *magnitude* of the electron charge. This modification accounts for the electromagnetic field's ability to accelerate and/or decelerate charged particles (and hence modify the electron's kinetic energy). In expanding the square, we can neglect the squared vector potential term since it does not affect our final results (orthogonality of the wavefunctions ensures us that the operator $\mathbf{A}^2$ does not perturb the system, assuming we can neglect the spatial variation of $\mathbf{A}$ within one unit cell of the crystal). Substituting (4.13) into (A8.1), we can write the new Hamiltonian as

$$H = H_0 + [H'(\mathbf{r})e^{j\omega t} + h.c.], \qquad H'(\mathbf{r}) = \frac{q}{2m_0}\mathscr{A}(\mathbf{r})\hat{e}\cdot\mathbf{p}. \qquad (4.14)$$

The *h.c.* stands for Hermitian conjugate, and simply means that we take the complex conjugate of all terms except the Hermitian momentum operator $\mathbf{p}$. The term in brackets can be viewed as a time-dependent perturbation to the original Hamiltonian, H_0. This perturbation term is the driving force for transitions between the conduction and valence bands.

By studying the time evolution of some electron wavefunction initially in a valence band state, for example, as it makes an upward transition to the conduction band in the presence of the time-harmonic perturbation, it is possible to determine the rate at which such transitions will occur. This procedure is carried out in Appendix 9. The resulting transition rate per unit volume of active material is given (in units of $s^{-1}\,cm^{-3}$) by

$$R_r = \frac{2\pi}{\hbar}|H'_{21}|^2\rho_f(E_{21})|_{E_{21}=\hbar\omega}, \qquad (4.15)$$

$$H'_{21} \equiv \langle\psi_2|H'(\mathbf{r})|\psi_1\rangle = \int_V \psi_2^* H'(\mathbf{r})\psi_1\, d^3\mathbf{r}. \qquad (4.16)$$

Equation (4.15) is known as Fermi's Golden Rule. It reveals that the number of transitions per unit active volume, V, occurring per second is dependent on (1) the density of final states, $\rho_f(E_{21})$, (in units of energy^{-1} cm^{-3}) available to the electron, and (2) the spatial overlap of the initial and final electron wavefunctions with the harmonic perturbation defined in (4.14)—the integral being defined as H'_{21} (in units of energy). It is important to appreciate that the field will only invoke a response from electrons which exist in states which have possible final states separated by $E_{21} \approx \hbar\omega$. That is, the two electron states must be in resonance with the oscillating field. As a result, $\rho_f(E_{21})$ and H'_{21} in (4.15) must be evaluated at $E_{21} = \hbar\omega$. This resonance condition derived in Appendix 9 is a statement of energy conservation.

Using (4.15), the job of determining R_r is reduced to providing explicit relations for both the density of final states and the overlap integral (or *matrix element* as it is commonly called). The next two subsections tackle this chore.

4.2.3   Transition Matrix Element

The matrix element $|H'_{21}|^2$ determines the *strength of interaction* between two states. This interaction can be strong, negligible, or identically zero, all depending on the wavefunctions describing the two electron states. For example, in a quantum well only transitions between subbands with the same quantum number are *allowed*, all others are *forbidden*. The wavefunction overlap also leads to the k-selection rule, which dictates that transitions between plane wave states are forbidden unless the k-vectors of the two states are equal (the two electron states must propagate along the same direction). In addition to these considerations, the interaction strength can also depend on the polarization of the incident light, if the material has some preferential axis of symmetry. For example, the interaction strength between conduction and heavy-hole states in a quantum well is much stronger for electric fields in the plane of the well than perpendicular to the well.

To derive an expression for $|H'_{21}|^2$, we insert the definition of $H'(\mathbf{r})$ in Eq. (4.14) into the definition of $|H'_{21}|^2$ given by Eq. (4.16). We can reduce Eq. (4.16) by expressing the electron and hole wavefunctions in terms of the envelope/Bloch function formalism using Eq. (4.8). Because the momentum operator when operating on a product can be written as $\mathbf{p}AB = B\mathbf{p}A + A\mathbf{p}B$, the overlap integral can be expressed as the sum of two terms

$$H'_{21} = \frac{q}{2m_0} \int_V F_2^* u_c^* (\mathscr{A}(\mathbf{r})\hat{\mathbf{e}} \cdot \mathbf{p}) F_1 u_v \, d^3\mathbf{r}$$

$$= \frac{q}{2m_0} \left[\int_V u_c^* u_v F_2^* (\mathscr{A}(\mathbf{r})\hat{\mathbf{e}} \cdot \mathbf{p}) F_1 \, d^3\mathbf{r} + \int_V [F_2^* \mathscr{A}(\mathbf{r}) F_1] u_c^* \hat{\mathbf{e}} \cdot \mathbf{p} u_v \, d^3\mathbf{r} \right]. \quad (4.17)$$

In transitions from the conduction band to the valence band, the first integral within the brackets vanishes[1] due to the orthogonality condition expressed in Eq. (A8.5) and due to the fact that the other terms in the integrand are, to a good approximation, constant in any one unit cell. To evaluate the second integral, we breakup the integration over the crystal volume into a sum of integrations over each unit cell. The terms collected in brackets in the second integral can again be taken as constant over the dimensions of a unit cell, and we can write

$$H'_{21} = \frac{q}{2m_0} \sum_j [F_2^* \mathscr{A}(\mathbf{r}) F_1]_{\mathbf{r}=\mathbf{r}_j} \int_{\text{unit cell}} u_c^* \hat{\mathbf{e}} \cdot \mathbf{p} u_v \, d^3\mathbf{r}, \qquad (4.18)$$

where j sums over all units cells in the crystal, and $\mathbf{r}_j$ is a position vector to the jth cell. Because the Bloch functions, u, repeat themselves in each unit cell, the integral can be pulled out of the summation to obtain

$$H'_{21} = \frac{q}{2m_0} \left[\frac{1}{V_{uc}} \int_{\text{unit cell}} u_c^* \hat{\mathbf{e}} \cdot \mathbf{p} u_v \, d^3\mathbf{r} \right] \sum_j [F_2^* \mathscr{A}(\mathbf{r}) F_1]_{\mathbf{r}=\mathbf{r}_j} V_{uc}$$

$$= \frac{q}{2m_0} \langle u_c | \hat{\mathbf{e}} \cdot \mathbf{p} | u_v \rangle \int_V F_2^* \mathscr{A}(\mathbf{r}) F_1 \, d^3\mathbf{r}, \qquad (4.19)$$

where, by assuming the volume of a unit cell to be very small, we have converted the summation back into an integral. We have also used Dirac notation to express the Bloch function overlap integral.

The envelope function overlap integral in (4.19) can be further simplified by recognizing that the spatial variation of $\mathscr{A}(\mathbf{r})$ is typically much slower than that of the envelope functions allowing us to pull it out of the integration. Assuming $\mathscr{A}(\mathbf{r})$ to be a plane wave of the form $\mathscr{A}_0 e^{-j\boldsymbol{\kappa} \cdot \mathbf{r}}$, and ignoring the spatial dependence (i.e., the exponential term), we obtain

$$\int_V F_2^* \mathscr{A}(\mathbf{r}) F_1 \, d^3\mathbf{r} \approx \mathscr{A}_0 \int_V F_2^* F_1 \, d^3\mathbf{r} \equiv \mathscr{A}_0 \langle F_2 | F_1 \rangle. \qquad (4.20)$$

We will consider this integral in more detail a little later in Eqs. (4.22) through (4.24). Substituting (4.20) into (4.19), we finally obtain

$$|H'_{21}|^2 = \left(\frac{q\mathscr{A}_0}{2m_0} \right)^2 |M_T|^2, \qquad \text{where} \qquad |M_T|^2 \equiv |\langle u_c | \hat{\mathbf{e}} \cdot \mathbf{p} | u_v \rangle|^2 |\langle F_2 | F_1 \rangle|^2.$$
$$(4.21)$$

---

[1] For transitions within the *same* energy band, the Bloch function overlap is equal to unity and the first integral may or may not be zero, depending on the envelope function overlap.

The prefactor in the first equality comes directly from the perturbation Hamiltonian (4.14). The second term, $|M_T|^2$, is referred to as the *transition matrix element* and is given special attention in Appendix 10. The first component, $|\langle u_c|\hat{e}\cdot\mathbf{p}|u_v\rangle|^2$, contains the polarization dependence of the interaction which will depend on the particular symmetries of the conduction and valence band Bloch functions. Aside from the polarization dependence (which can be a function of photon energy), we can consider this *momentum matrix element* to be a constant, $|M|^2$, for a given material.

As shown in Appendix 8, the constant $|M|^2$ can be determined experimentally. Table 4.1 tabulates the most accurately reported values for several important materials systems. (Note that $2|M|^2/m_0$ has units of energy.) Appendix 10 shows how $|M_T|^2$ can be expressed in terms of $|M|^2$. This involves expanding the dot product as well as considering the overlap of the envelope functions in $|M_T|^2$ given by Eq. (4.21). Table 4.2 summarizes the results for bulk and quantum-well materials for either transverse electric (TE: electric field in the quantum-well plane) or transverse magnetic (TM: electric field perpendicular to quantum-well plane) polarizations. Due to band mixing effects, the values for quantum wells are only valid for small transverse k-vectors, k_t. Figure A10.5 gives values of $|M_T|^2/|M|^2$ as k_t is varied.

In addition to creating a polarization sensitivity, the transition matrix element also restricts the types of states which can interact. For transitions *between two plane wave states* in a "bulk" active medium (i.e. $V \to \infty$), we can use Eq. (4.9) to set

$$\langle F_2|F_1\rangle = \frac{1}{V}\int_V e^{j\mathbf{k}_2\cdot\mathbf{r}}e^{-j\mathbf{k}_1\cdot\mathbf{r}}\, d^3\mathbf{r} = \delta_{\mathbf{k}_1,\mathbf{k}_2}, \tag{4.22}$$

where the Kronecker delta, $\delta_{\mathbf{k}_1,\mathbf{k}_2}$, is zero unless $\mathbf{k}_2 = \mathbf{k}_1$, in which case it equals one. This spatial *phase-matching* condition, known as the **k**-selection rule, is a statement of momentum conservation. Thus, only states propagating along the same direction in the crystal can interact. If the spatial dependence of the field were not ignored in deriving Eq. (4.21), then an additional plane wave term, $e^{-j\kappa\cdot\mathbf{r}}$, would appear in Eq. (4.22), and the **k**-selection rule for upward transitions

TABLE 4.1   Magnitude of $|M|^2$ for Various Material Systems.

Material system	$\dfrac{2|M|^2}{m_0}$ (in eV)	Reference
GaAs	28.8 ± 0.15	1, 2
$Al_xGa_{1-x}As$ $(x < 0.3)$	$29.83 + 2.85x$	3
$In_xGa_{1-x}As$	$28.8 - 6.6x$	1, 2
InP	19.7 ± 0.6	1, 2
$In_{1-x}Ga_xAs_yP_{1-y}\,(x = 0.47y)$	$19.7 + 5.6y$	2, 4

TABLE 4.2   Magnitude of $|M_T|^2/|M|^2$ for Different Transitions and Polarizations.

Polarization	Bulk		Quantum-well $(k_t \sim 0)$	
	C–HH	C–LH	C–HH	C–LH
TE	1/3	1/3	1/2	1/6
TM	1/3	1/3	0	2/3

would become: $\mathbf{k}_2 = \mathbf{k}_1 + \boldsymbol{\kappa}$. However, the wavevector of the field, $\boldsymbol{\kappa}$, is typically orders of magnitude smaller than the electron wavevector, and can usually be ignored (justifying our earlier simplifying step in Eq. (4.20)).

In a quantum well, the envelope functions are given by Eq. (4.10). Transitions between two such quantum-well states are governed by the following overlap:

$$\langle F_2|F_1 \rangle = \frac{1}{A} \int_V F_2^*(z) e^{j\mathbf{k}_2 \cdot \mathbf{r}_\parallel} F_1(z) e^{-j\mathbf{k}_1 \cdot \mathbf{r}_\parallel} \, d^3\mathbf{r}$$

$$= \int_z F_2^*(z) F_1(z) \, dz. \qquad \text{(with } \mathbf{k}_2 = \mathbf{k}_1) \qquad (4.23)$$

Thus, we can assume $\mathbf{k}$-selection in the plane of the quantum well, but we still need to evaluate $|\langle F_2|F_1 \rangle|^2$ perpendicular to the plane, where again F_2 and F_1 are simply the particle-in-a-box envelope functions found for the quantized energy levels of the quantum well in both conduction and valence bands.

Due to orthogonality between the quantum-well wavefunction solutions, the overlap integral in Eq. (4.23) reduces to the following rule for subband transitions:

$$|\langle F_2|F_1 \rangle|^2 \approx \delta_{n_c, n_v}. \qquad (4.24)$$

This means that transitions can only occur between quantum-well subbands which have the same quantum number, $n_c = n_v$. These are referred to as *allowed* transitions. Transitions between subbands with dissimilar quantum numbers are *forbidden* transitions. Both are illustrated in Fig. 4.4. The allowed transitions are usually referred to as the $n = 1$ transition, the $n = 2$ transition, etc. The "nearly equal to" sign is used in (4.24) because the different effective mass and barrier height in the conduction and valence bands means that the wavefunctions of the two bands are not completely orthogonal to each other. Nevertheless, allowed transition overlaps are usually (but not always) close to unity (0.9–1) and forbidden transition overlaps are usually very small (0–0.1).

The above considerations can be extended to quantum wires which have potential barriers in two dimensions. For example, in a quantum wire, F_2 and F_1 are functions of both quantized directions, such that $|\langle F_2|F_1 \rangle|^2$ represents

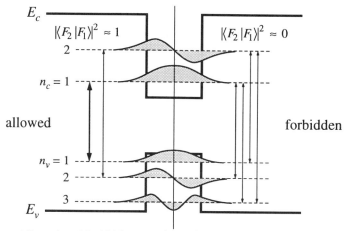

FIGURE 4.4 Allowed and forbidden transitions in a quantum well. The most important "$n = 1$" transition is highlighted in bold.

an integration over the "quantization plane" (see Eq. (4.11)). In this case, **k**-selection is obeyed only along the length of the wire.

In heavily doped materials, electron states which are bound to charged donors or acceptors can exist. If the concentrations are high enough, it is conceivable that many transitions near the band edge will be either *band-to-bound* or *bound-to-bound* transitions. In these cases, the **k**-selection rule cannot be assumed in any direction and $|\langle F_2|F_1\rangle|^2$ must be evaluated explicitly for the envelope functions which correspond to these bound states. In early treatments [5], such band-to-bound transitions were given considerable attention. However, these analyses were provided when most active regions were heavily doped. More recently, heavily doped active regions have faded in popularity in favor of undoped quantum-well active regions (or sometimes *modulation doped* quantum wells, where the doping ions are physically separated from the quantum wells). As a result, most current semiconductor lasers operate on the physics of band-to-band transitions. The rest of this chapter will concentrate on transitions occurring between plane wave states such that the **k**-selection rule can be assumed.

With $|H'_{21}|^2$ given by (4.21) and the **k**-selection rule established, we now need to define the final density of states more carefully.

4.2.4 Reduced Density of States

The derivation of Fermi's Golden Rule in Appendix 9 assumes the electron initially occupies a single state which makes a transition to one of a large number of final states. In a semiconductor, both final *and* initial states of the electron are immersed within a large number of nearby states, as illustrated in Fig. 4.5. For this case, the final density of states appearing in (4.15) should

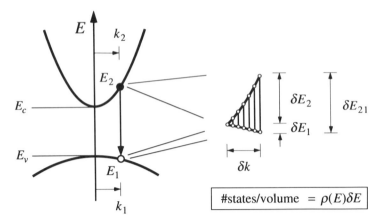

FIGURE 4.5 Relationship between the energy ranges in the conduction and valence bands for a given dk in k-space, assuming **k**-selection applies.

actually be interpreted as the *density of transition pairs per unit transition energy*, δE_{21}. This density of transition pairs is referred to as the *reduced* density of states function, $\rho_r(E_{21})$. One specific form for $\rho_r(E_{21})$ is derived in Appendix 6, however, here we would like to determine the general form.

If we assume the **k**-selection rule applies, then only states with identical k-vectors can form a transition pair, and only vertical transitions in k-space can occur. Because of this restriction, the number of transition pairs within δk is equal to the number of states in either the conduction or valence band, and $\rho_r \delta E_{21} = \rho_c \delta E_2 = \rho_v \delta E_1$. This allows us to set $\delta E_2 = (\rho_r/\rho_c)\delta E_{21}$ and $\delta E_1 = (\rho_r/\rho_v)\delta E_{21}$. Summing these relations and setting $\delta E_{21} = \delta E_2 + \delta E_1$, we immediately obtain

$$\frac{1}{\rho_r} = \frac{1}{\rho_c} + \frac{1}{\rho_v}. \tag{4.25}$$

This is the more general form for the final density of states to be used in Fermi's Golden Rule. Note that as $\rho_v \to \infty$, we have $\rho_r \to \rho_c$. This is the case we solved in Appendix 9, with ρ_c interpreted as the final density of states. For a finite density of states in the valence band, such as $\rho_v = \rho_c$, ρ_r is reduced from ρ_c to $\rho_c/2$ (hence the name *reduced* density of states). In typical semiconductors, $\rho_v \geq 5\rho_c$, or $(5/6)\rho_c \leq \rho_r \leq \rho_c$. Thus, ρ_r is generally very close to ρ_c. However, in strained materials, ρ_v can be reduced significantly, as shown in Appendix 8, bringing ρ_r closer to $\rho_c/2$.

For general use, it turns out that Eq. (4.25) is not that practical. An alternate definition of $\rho_r(E_{21})$ can be found by relating it to the density of states in k-space. From Fig. 4.5, we can set $\rho_r(E_{21})\delta E_{21} = \rho(k)\delta k$. Rearranging, we find

$$\frac{1}{\rho_r(E_{21})} = \frac{1}{\rho(k)}\frac{dE_{21}(k)}{dk} = \frac{1}{\rho(k)}\left[\frac{dE_2(k)}{dk} - \frac{dE_1(k)}{dk}\right]. \tag{4.26}$$

TABLE 4.3 Density of States for Bulk (3D), Quantum-Well (2D), and Quantum-Wire (1D) Structures (Including Spin).

Dimension	$\rho(k)$	$\rho(E)$
3	$\dfrac{k^2}{\pi^2}$	$\dfrac{\sqrt{E}}{2\pi^2}\left[\dfrac{2m}{\hbar^2}\right]^{3/2}$
2	$\dfrac{k}{\pi d_z}$	$\dfrac{m}{\pi\hbar^2 d_z}$
1	$\dfrac{2}{\pi d_x d_y}$	$\dfrac{\rho(k)}{\sqrt{E}}\left[\dfrac{2m}{\hbar^2}\right]^{1/2}$

This definition allows $\rho_r(E_{21})$ to be evaluated at any given point in k-space once the derivatives of the electron and hole energies with respect to k are known at that point. This definition is especially useful when $E_2(k)$ and $E_1(k)$ are not parabolic functions (see Figs. A8.4 and A8.6 for examples of non-parabolic subbands in QWs). The density of states in k-space, $\rho(k)$, for various dimensional structures is summarized in Table 4.3. (Note that the z-direction is taken as the narrow dimension of the quantum well and the axis of the quantum wire to be consistent with Appendices 8 and 10.)

If both bands involved in the transition are parabolic, an even more straightforward definition of $\rho_r(E_{21})$ can be used. We can generally state that the transition energy is equal to the bandgap energy $E_g = E_c - E_v$ plus the kinetic energies of the electron and hole. If the electrons and holes follow parabolic dispersion curves, we have

$$E_{21} = E_g + \frac{\hbar^2 k^2}{2m_c} + \frac{\hbar^2 k^2}{2m_v} = E_g + \frac{\hbar^2 k^2}{2m_r}, \qquad \text{where} \qquad \frac{1}{m_r} \equiv \frac{1}{m_c} + \frac{1}{m_v}. \quad (4.27)$$

In other words, the dispersion of E_{21} with k also follows a parabolic curve with a curvature characterized by a *reduced* mass, m_r. As a result, the density of transition states along E_{21} is entirely analogous to the density of states function in either the conduction or valence band, with the following associations:

$$\rho_r(E_{21}) \leftrightarrow \rho_c(E_2), \rho_v(E_1)$$

$$E_{21} - E_g \leftrightarrow E_2 - E_c, E_v - E_1 \qquad \text{(parabolic bands)} \qquad (4.28)$$

$$m_r \leftrightarrow m_c, m_v$$

The derivation of $\rho_r(E_{21})$ provided in Appendix 6 for bulk material confirms these associations. More generally, the density of states per unit energy given in Appendix 1 and summarized in Table 4.3 for various dimensional structures can be applied directly to $\rho_r(E_{21})$ using (4.28) and (4.27) as long as the energy bands are parabolic.

Finally, in order to evaluate the Fermi occupation probabilities in Eq. (4.3), we need the individual electron and hole energies. With parabolic bands and assuming k-selection, the individual electron and hole energies in terms of the transition energy can be found using (4.27):

$$E_2 = E_c + (E_{21} - E_g)\frac{m_r}{m_c}, \qquad E_1 = E_v - (E_{21} - E_g)\frac{m_r}{m_v}. \qquad (4.29)$$

When the bands are not parabolic, we must in general use $E_2(k)$ and $E_1(k)$ evaluated at the k-vector which yields the desired transition energy, E_{21}.

4.2.5   Correspondence with Einstein's Stimulated Rate Constant

Einstein's original description of radiative transitions outlined in Appendix 6 defines the downward transition rate in terms of a stimulated rate constant, B_{21}, weighted by the radiation spectral density, $W(v)$, and the differential number of state pairs available for downward transitions, dN_2. This rate constant can be related to R_r, allowing us to quantify B_{21} using Fermi's Golden Rule.

Assuming the lineshape broadening function to be a delta function, the stimulated emission rate in Eq. (A6.11) becomes $dR_{21} = B_{21}W(v)\,dN_2$. This differential transition rate must still be integrated over all transition pairs affected by $W(v)$. Using Eq. (A6.8) for dN_2, setting $W(v) \to hvN_p\delta(v - v_0)$, and integrating over the transition energy, the stimulated emission rate becomes

$$R_{21} = B_{21} \cdot hv_0 N_p h\rho_r f_2(1 - f_1). \qquad (4.30)$$

Comparing this to Eq. (4.1), we conclude that

$$B_{21} = \frac{R_r}{hv_0 N_p h\rho_r} = \frac{1}{hv_0 N_p} \frac{|H'_{21}|^2}{\hbar^2}. \qquad (4.31)$$

The second equality uses Fermi's Golden Rule (4.15) to expand R_r. With this and Eq. (4.36) defined below, B_{21} and all relations dependent on B_{21} in Appendix 6 can be quantified.

4.3   OPTICAL GAIN

4.3.1   General Expression for Gain

The explicit relation between the net stimulated emission rate and the optical gain was derived in Chapter 2. For reference, we repeat that derivation here in a slightly different way. As discussed in Chapter 2, we can define the *material gain per unit length* as the proportional growth of the photon density as it propagates along some direction in the crystal. This definition can be related to the transition rates as follows:

$$g = \frac{1}{N_p} \frac{dN_p}{dz} = \frac{1}{v_g N_p} \frac{dN_p}{dt} = \frac{1}{v_g N_p} (R_{21} - R_{12}). \tag{4.32}$$

The second equality uses the group velocity, v_g, to transform the spatial growth rate to the growth rate in time. The growth rate in time is then linked to the net generation rate of photons per unit volume. Finally, using Eq. (4.2) we obtain

$$g = \frac{R_{st}}{v_g N_p} = \frac{R_r}{v_g N_p} (f_2 - f_1). \tag{4.33}$$

Using Fermi's Golden Rule (4.15) for R_r, we have

$$g_{21} = \frac{2\pi}{\hbar} \frac{|H'_{21}|^2}{v_g N_p} \rho_r(E_{21}) \cdot (f_2 - f_1). \tag{4.34}$$

The electromagnetic perturbation is proportional to the field strength. Thus, to evaluate the ratio, $|H'_{21}|^2/N_p$, we need to relate the field strength to the photon density. The energy density in terms of the photon density is $\hbar\omega N_p$. The energy density in terms of the electric field strength is $\frac{1}{2}n^2\varepsilon_0|\mathscr{E}|^2$. If the material is dispersive, this becomes $\frac{1}{2}nn_g\varepsilon_0|\mathscr{E}|^2$. The electric field is related to the vector potential through a time derivative. For time-harmonic fields, we can set $|\mathscr{E}|^2 = \omega^2|\mathscr{A}|^2$. By equating the two versions of the energy density, we obtain the desired relation:

$$\frac{1}{2}nn_g\varepsilon_0\omega^2|\mathscr{A}_0|^2 = \hbar\omega N_p \rightarrow |\mathscr{A}_0|^2 = \frac{2\hbar}{nn_g\varepsilon_0\omega} N_p. \tag{4.35}$$

Using this relation and the definition of the matrix element (4.21), we can set[2]

$$\frac{|H'_{21}|^2}{N_p} = \frac{1}{N_p} \left(\frac{q\mathscr{A}_0}{2m_0}\right)^2 |M_T|^2 = \frac{q^2\hbar}{2nn_g\varepsilon_0 m_0^2\omega} |M_T|^2. \tag{4.36}$$

---

[2] The transition matrix element in the expression for gain is occasionally written instead as the *dipole moment* matrix element, $q^2|x|^2$, where $|x|^2 = |\langle u_c|\hat{e}\cdot\mathbf{x}|u_v\rangle|^2 |\langle F_2|F_1\rangle|^2$, and $\mathbf{x}$ is the position operator. The relationship between the two is given by $q^2|M_T|^2 = m_0^2\omega^2 q^2|x|^2$. Thus, Eq. (4.36) can alternatively be written as

$$\frac{|H'_{21}|^2}{N_p} = \frac{\hbar\omega}{2nn_g\varepsilon_0} q^2|x|^2.$$

This ratio can then be used to define the gain using (4.34), however, we will not make use of this alternate expression for gain in this chapter.

The material gain per unit length (4.34) then becomes

$$g_{21} = g_{max}(E_{21}) \cdot (f_2 - f_1)$$

where

$$g_{max}(E_{21}) = \frac{\pi q^2 h}{n \varepsilon_0 c m_0^2} \frac{1}{h v_{21}} |M_T(E_{21})|^2 \rho_r(E_{21}).$$

(4.37)

The maximum gain, g_{max}, is a property of the material, while the Fermi factor, $f_2 - f_1$, is dependent on the injection level. In GaAs 80–100Å quantum wells, the maximum gain of each subband transition is $g_{max} \sim 10^4$ cm^{-1} (or 1 μm^{-1}).

In reduced-dimensional structures such as a quantum well, ρ_r corresponds to the reduced density of states between two quantized subbands. The total gain at E_{21} is found by summing over all possible subband pairs:

$$g_{21} = \sum_{n_c} \sum_{n_v} g_{21}^{sub}(n_c, n_v). \qquad \text{(quantum well)}$$

(4.38)

The double sum indicates that all subband combinations should be considered. In practice however, the selection rules arising from the envelope function overlap expressed in Eq. (4.24) and illustrated in Fig. 4.4 suggest that the gain from $n_c = n_v$ subband pairs will dominate the gain spectrum. In particular, the $n = 1$ gain is usually the largest, and hence most important transition in quantum-well lasers. We will use the following example to demonstrate this point as well as to highlight many other basic considerations involved with determining the gain for a given injection level in a quantum well.

The left side of Fig. 4.6 illustrates the lowest two energy subbands of a quantum well in both conduction and valence bands (neglecting the light-hole subbands and assuming parabolic subbands for simplicity). Under strong forward-bias conditions, the equilibrium Fermi level is separated into two quasi-Fermi levels, one for all conduction subbands and one for all valence subbands (the quasi-Fermi functions are indicated by the dashed curves). The separation of the quasi-Fermi levels is constrained by the requirement that *charge neutrality* be maintained within the quantum well (if there were a charge imbalance, band-bending would occur in the diode junction in such a way as to neutralize the imbalance). Thus, we must have $N(E_{Fc}) = P(E_{Fv})$ in the quantum well (see Eq. (A6.20) for a more explicit version of this relation). Because the valence band typically has many more states per unit energy, the valence quasi-Fermi function does not have to penetrate nearly as deeply as the conduction quasi-Fermi function to obtain the same overall carrier density. As a result, the quasi-Fermi levels separate asymmetrically as indicated in the figure.

The right side of Fig. 4.6 shows the constant density of states functions of each subband which when added together produce a staircase density of states in each band. When $\rho(E)$ is multiplied by the fraction of filled (empty) states in the band we obtain the electron (hole) distribution as a function of energy

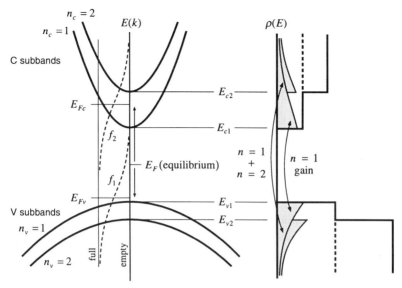

FIGURE 4.6 QW subbands and corresponding density of states illustrating the relationships between the carrier populations, the quasi-Fermi levels, and the gain at the subband edges.

(the contributions from the $n = 1$ and $n = 2$ subbands are indicated separately in the figure). The shaded area under the carrier distribution curves then yields the total carrier density. Since charge neutrality requires that $N = P$, we conclude that the total shaded area must be the same in the conduction and valence bands for any injection level. Thus, the quasi-Fermi levels must always adjust themselves to ensure that this requirement is met (with the larger step heights in the valence band, this again explains why the valence quasi-Fermi function does not have to penetrate as deeply).

With the relationship between the quasi-Fermi levels and the carrier density qualitatively defined, we can now proceed to analyze the gain of the quantum well. Many optical properties of the quantum well depicted in Fig. 4.6 can be qualitatively determined by simple inspection. For example, from Eq. (4.6) we immediately know that at the injection level indicated, the quantum well provides gain at the lowest subband edge, simply because $E_{Fc} - E_{Fv} > E_{c1} - E_{v1}$. Thus, in spite of the fact that the electron density at E_{v1} is slightly larger than the electron density at E_{c1}, a population inversion at the band edge has in fact been achieved. This is because a "population inversion" requires $f_2 > f_1$, *not* $\rho_c f_2 > \rho_v f_1$ (as the name might lead one to conclude). Furthermore, the fact that we have gain at the band edge even though $E_{Fv} > E_{v1}$ reinforces the concept that it is the *relative difference* and not the absolute positions of E_{Fc} and E_{Fv} that determines the gain (i.e., it is not a requirement to have *both* $E_{Fc} > E_{c1}$ and $E_{Fv} < E_{v1}$ to achieve gain).

We can take the analysis a step further by using Eq. (4.37) to estimate the gain at the band edge. Evaluating terms at the appropriate energies, we find the band edge gain to be $g_{max1}(E_{g1}) \cdot (f_2(E_{c1}) - f_1(E_{v1}))$, where g_{max1} uses the envelope function overlap and reduced density of states between the two $n = 1$ subbands. Estimating from the figure that $f_2(E_{c1}) - f_1(E_{v1}) \approx 0.8 - 0.6 = 0.2$, we conclude that the band edge gain is roughly 20% of its maximum possible value. If we assume $g_{max1}(E_{g1}) \sim 10^4$ cm^{-1} (a typical number), then the band edge gain in the figure is roughly 2000 cm^{-1}.

As we move to higher photon energies, the population inversion clearly declines, implying that the gain from $n = 1$ transitions is largest at the band edge. However, at high enough photon energies a second population of carriers starts contributing to the transition process. In principle, this added supply of carriers could increase the gain significantly. However in this case, at photon energies close to the second subband edge, we know the quantum well is absorbing because $E_{Fc} - E_{Fv} < E_{c2} - E_{v2}$. The absorption comes from two contributions: the carrier populations in the $n = 1$ subbands and the carrier populations in the $n = 2$ subbands, as the summation in Eq. (4.38) implies. Estimating from the figure that $f_2(E_{c2}) - f_1(E_{v2}) \approx 0.3 - 0.8 = -0.5$, we find that the absorption from the $n = 2$ subband transitions is 0.5 $g_{max\,2}$ (E_{g2}). For the $n = 1$ subband transitions, it's a little trickier since the Fermi energies must be estimated for electrons and holes separated by E_{g2} in the $n = 1$ subbands. Using Eq. (4.29), we must evaluate $f_2[E_{c1} + (E_{g2} - E_{g1})(m_r/m_c)] - f_1[E_{v1} - (E_{g2} - E_{g1})(m_r/m_v)]$, which from the figure is approximately $\approx 0.2 - 0.7 = -0.5$. The overall absorption at the second subband edge is therefore $0.5(g_{max1}(E_{g2}) + g_{max2}(E_{g2}))$ (cross-population transitions such as g_{max12} and g_{max21} also exist but as mentioned earlier, their contributions are typically small). So we find that the absorption at E_{g2} has not yet been converted into gain but has at least been reduced to half its maximum absorption value with the application of a forward bias. To achieve gain at the $n = 2$ subband edge, the forward bias must be increased to the point where $E_{Fc} - E_{Fv} > E_{g2}$. To surpass the $n = 1$ subband edge gain, the forward bias must be even stronger such that

$$(g_{max1}(E_{g2}) + g_{max2}(E_{g2}))(f_2(E_{c2}) - f_1(E_{v2})) > g_{max1}(E_{g1})f_2(E_{c1}) - f_1(E_{v1})).$$

So for all but very strong forward-bias conditions, the $n = 1$ gain dominates.

To summarize, by invoking charge neutrality to link both quasi-Fermi levels to a given carrier density, evaluating the quasi-Fermi functions at the appropriate energies, and summing over all subband transition pairs, we can determine the gain at any given photon energy relative to the maximum value. In Section 4.3.3, we will return to a more general discussion of the gain spectrum and its properties in bulk and quantum well materials.

4.3.2   Lineshape Broadening

Equation (A6.24) in Appendix 6 expresses the gain in terms of B_{21}. Replacing B_{21} with (4.31) and using (4.36), we find that (A6.24) is indeed equivalent to

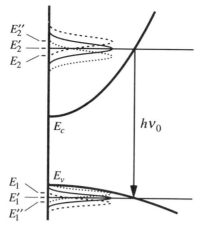

FIGURE 4.7   Three (of many) transition pairs which contribute to gain at hv_0.

(4.37) derived here. However, the gain more generally defined in Eq. (A6.23) has an additional integration which takes into account the energy uncertainty of the electron states. This *energy broadening* of electron states ultimately limits the resolution of features we can observe in the gain spectrum and is therefore particularly important to consider in reduced dimensional structures where the reduced density of states function contains very sharp features. To recover Eq. (A6.23) from Eq. (4.37), we need to consider how the broadening of electron states affects the gain.

Figure 4.7 reveals that when the energy states are broadened, many different transition pairs contribute to gain at a particular photon energy. These transition pairs are primarily clustered within the energy uncertainty width of the *lineshape function* describing the probable energy distribution of each transition pair. To determine the *total* gain at hv_0, we must integrate g_{21} over all transition energies weighted by the appropriate lineshape function, $\mathscr{L}(hv_0 - E_{21})$.[3] The gain including lineshape broadening therefore takes the form:

$$g(hv_0) = \int g_{21}\mathscr{L}(hv_0 - E_{21})\,dE_{21}. \tag{4.39}$$

In Eq. (4.37), hv_{21} should be set equal to hv_0, while all other terms dependent on E_{21} should be considered variables of the integration. This expression for the gain now agrees with Eq. (A6.23) derived in Appendix 6.

The specific form for the lineshape function to be used in Eq. (4.39) can be determined by attempting to study the time evolution of an electron state, taking into account its interaction with phonons and other electrons. In a first-order

---

[3] This lineshape function is actually a combination of the individual electron and hole lineshapes comprising the transition pair:

$$\mathscr{L}(hv_0 - E_{21}) = \int \mathscr{L}_2(E - E_2)\mathscr{L}_1((E - hv_0) - E_1)\,dE.$$

approximation, we might assume that the probability of finding the electron in a given state decays exponentially, as we found for the electron's interaction with photons in Appendix 9. This simplistic time dependence when Fourier transformed to the energy domain immediately leads to a Lorentzian lineshape function:

$$\mathscr{L}(E - E_{21}) = \frac{1}{\pi} \frac{\hbar/\tau_{in}}{(\hbar/\tau_{in})^2 + (E - E_{21})^2}. \tag{4.40}$$

The *intraband relaxation* time, τ_{in}, is the time constant associated with the exponential decay of the electron. The energy full-width is related to τ_{in} via $\Delta E_{21} = 2\hbar/\tau_{in}$ (compare (4.40) to (A6.27)).[4] Early investigations which attempted curve fits of gain and spontaneous emission spectra to measurements in bulk material lead to values of $\tau_{in} \approx 0.1$ ps [6]. However, the gain and emission spectra did not match very well, particularly on the low-energy side of the spectrum near the band edge, where the details of the lineshape function are most apparent. Thus, other more sophisticated theoretical methods of determining the lineshape function have since been employed.

Using a quantum mechanical density matrix approach, Yamanishi and Lee [7] have suggested that the electron state decays initially as a Gaussian but then takes on exponential behavior for larger times. This leads to less energy in the tails of the lineshape function than the Lorentzian function, which is more in line with experimental observations. Asada [8] has also performed a detailed analysis of intraband scattering in quantum wells, arriving at an asymmetrical lineshape function which falls off much faster than a Lorentzian on the low-energy side of the transition, similar to the findings of Yamanishi and Lee. Kucharska and Robbins [9] have kept with a Lorentzian lineshape, but have theoretically derived an energy-dependent lifetime, arguing that the scattering rate out of a state is dependent on where the state is in the band, and on how full the band is.

In an attempt to keep the lineshape function as simple as possible while maintaining some degree of accuracy, Chinn et al. [10] approximated the numerical lineshape function derived by Yamanishi and Lee with a simple curve fit which describes the time dependence of the electron state. This time dependence is given by

$$e^{-l(t)},$$

where

$$\log_{10} l(t[\text{ps}]) = 2 + 1.5 \log_{10} t - 0.5\sqrt{(2 + \log_{10} t)^2 + 0.36}. \tag{4.41}$$

For long times, $l(t) \to t$, reproducing the exponential decay. For short times, $l(t) \to t^2$ implying that the state initially decays as a Gaussian. The most

---

[4] In Appendix 9, it is shown that if a state decays exponentially with a time constant, $\tau (=1/W)$, then the energy uncertainty of the state is $\Delta E_{FWHM} = \hbar/\tau$. The combined energy uncertainty of states 1 and 2 is $\Delta E_{21} = \hbar(1/\tau_2 + 1/\tau_1)$. With τ_{in} defined as the *average* time constant: $1/\tau_{in} = \frac{1}{2}(1/\tau_2 + 1/\tau_1)$, we end up with $\Delta E_{21} = 2\hbar/\tau_{in}$.

efficient way to make use of the Chinn lineshape function is to inverse fast Fourier transform (inverse FFT) the gain spectrum, multiply it by Eq. (4.41), and FFT back to the energy domain. This proves to be the fastest method of evaluating the convolution contained in Eq. (4.39). We will make use of this simplified Chinn lineshape function in later calculations.

Some example lineshape functions are shown in Fig. 4.8. The Chinn lineshape has less energy in the tails than the Lorentzian lineshape. However, the Gaussian lineshape which has been included for comparison has significantly less energy in the tails than either of the other two. The effects of convolving these lineshapes with a typical quantum-well gain spectrum are shown in the lower part of Fig. 4.8. The dramatic smoothing of the sharp features of the gain spectrum can reduce the peak gain substantially. However at higher gains, the reduction is not as significant. Note that due to the energy in the tails, the Lorentzian lineshape function creates absorption of almost 100 cm^{-1} below the band edge. The other two lineshapes do not suffer this problem. Of the three, the Chinn lineshape is perhaps closest to representing the actual

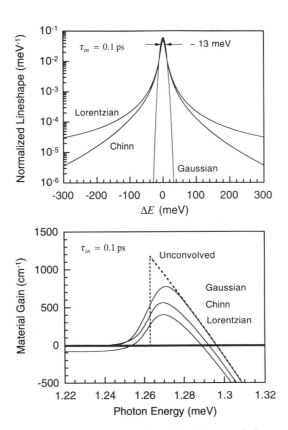

FIGURE 4.8 Comparison of various lineshape functions and the resulting convolved gain spectra.

complex lineshape function of the semiconductor. However, in reality the lineshape function is a complex function of both transition energy and injection level.

4.3.3   General Features of the Gain Spectrum

From the general gain equation (4.37), it is clear that the gain/absorption spectrum is bounded to a maximum value of $|g_{max}|$. With no carrier injection, the material is strongly absorbing with an absorption spectrum equal to $-g_{max}(E_{21})$. With carrier injection, we can invert the carrier population near the band edge and change the Fermi factor, $f_2 - f_1$, from -1 toward $+1$, converting the absorption into gain. As considered earlier in relation to Fig. 4.6, the carrier inversion is highly concentrated near the band edge. Therefore, as the photon energy increases away from the band edge, $f_2 - f_1$ must steadily reduce back to -1, and the gain spectrum must reduce back to the unpumped absorption spectrum. This conversion of absorption into gain is depicted in Fig. 4.9 for both bulk and quantum-well material.

In both bulk and quantum-well cases, the material is transparent below the bandgap. Just above the bandgap, a region of positive gain exists. Beyond this, the material becomes strongly absorbing. The crossing point from positive gain to absorption occurs when $f_2 = f_1$ and was considered earlier in discussions related to Eqs. (4.4) through (4.7). There it was found that the stimulated emission and absorption rates exactly cancel (making the material transparent) when $hv = \Delta E_F$. Thus the region of positive gain extends between the bandgap and the quasi-Fermi level separation:

$$E_g < hv < \Delta E_F. \qquad \text{(positive gain)} \qquad (4.42)$$

In other words, gain is achieved in the material only when the carrier injection is high enough to create a quasi-Fermi level separation exceeding the bandgap. Also, the larger the quasi-Fermi level separation we can create, the wider the gain bandwidth we can achieve in the material.

The shape of the bounding limits in Fig. 4.9 representing g_{max} and $-g_{max}$ is primarily governed by the reduced density of states function[5] from Eq. (4.37). Hence in bulk material, g_{max} follows a square root dependence, while in quantum-well material, g_{max} follows a step-like dependence where each step corresponds to the addition of a new subband transition pair. As a result, the bulk gain spectrum is quite smooth whereas the quantum-well gain spectrum is rather jagged. However in the latter case, lineshape broadening tends to smooth out the discontinuous features into rounded "bumps," one for each subband transition pair.

---

[5] There are other energy-dependent terms comprising $g_{max}(E)$. For example, the $1/\hbar\omega$ dependence modifies the shape of $g_{max}(E)$ slightly, but not noticeably. In reduced-dimensional structures, the matrix element also has an energy dependence, however, we will ignore this dependence for the present discussion.

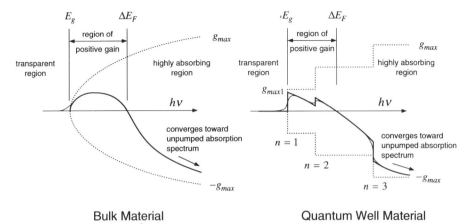

FIGURE 4.9 Gain spectra in bulk and quantum-well materials. The thin smoothed curves indicate the effect of lineshape broadening.

In bulk material, the peak of the gain spectrum increases and shifts to higher photon energies with increasing ΔE_F. In contrast, the peak of the quantum-well gain spectrum remains fixed at the $n = 1$ gain peak near the band edge under most conditions. With increased carrier injection, the $n = 1$ gain saturates at g_{max1} as the states in the first subband reach complete inversion, while the $n = 2$ gain continues to increase to a value twice as high as the $n = 1$ gain (twice as high because it contains contributions from both $n = 1$ and $n = 2$ subband transition pairs). Thus, under very high carrier injection, the overall peak gain can jump to the $n = 2$ gain peak as discussed in relation to Fig. 4.6. Experiments on quantum-well lasers do in fact show a discrete jump in the lasing wavelength from the $n = 1$ to the $n = 2$ peak as the cavity length is reduced (i.e., as the threshold gain is increased).

In drawing the quantum-well gain spectrum in Fig. 4.9, a few features have been idealized. First of all, an additional set of steps spaced differently from the ones shown should be included to account for both conduction-to-heavy hole (C–HH) *and* conduction-to-light hole (C–LH) subband transitions. In other words, there should be an $n = 1$ step for both types of transitions, etc. In practice, for the more common situation where the electric field lies in the plane of the quantum well, the matrix element for C–HH transitions is three times larger than the matrix element for C–LH transitions, implying that the steps related to C–LH transitions are much smaller and not as important. We have also neglected steps due to forbidden transitions. Inclusion of these transitions introduces small peaks in between the major peaks due to the small but finite overlap of forbidden transitions. However, in practice such small peaks are not usually observed.

Another simplification we have made in Fig. 4.9 is that all steps comprising g_{max} have been drawn with equal height. In actuality, the overlap integral defining the matrix element is typically smaller for higher subband transitions, implying that the $n = 1$ step height is usually the largest (sometimes by as much as a factor of 2, depending on the barrier height). Finally, we have completely neglected the energy dependence of the matrix element. This energy dependence is illustrated in Appendix 10. Generally for polarizations of interest, the matrix element is a maximum at the band edge and decays to less than half of its peak value at higher energies. Thus, the flat plateaus in between the steps of g_{max} should actually slope toward zero, modifying the gain spectrum accordingly (the $1/\hbar\omega$ dependence will also contribute to this sloping toward zero). Gain calculations using practical material systems presented at the end of this chapter will reveal these more subtle features in the spectrum.

4.3.4 Many-Body Effects

The above theory of gain involving Fermi's Golden Rule considers each electron in isolation as it interacts with the electromagnetic field. In other words, we have used a single-particle theory to obtain the gain spectrum. In reality, there is a large density of both electrons and holes present in our system. The mutual interactions between these particles are generally referred to as *many-body* effects. We have already considered one consequence of such many-body effects in our discussion of lineshape broadening which is related to collisions between particles and/or phonons in the crystal. In addition to this important effect, there are two other significant consequences of many-body effects: exciton states and bandgap shrinkage. Exciton states exist primarily at low carrier densities and low temperatures, while bandgap shrinkage becomes noticeable at high carrier densities.

Under conditions of low carrier density and low temperature it is possible for an electron and hole to orbit each other for an extended period of time (in analogy with a hydrogen atom), forming what is referred to as an *excition* pair. Such excition paris have a *binding energy* associated with them that is equal to the energy required to separate the electron and hole. As a result, electrons which are elevated from the valence band to one of these exciton states will absorb radiation at energies equal to the bandgap *less the binding energy* (the bandgap will appear to be red-shifted). More significantly however, the overlap integral (and hence the matrix element) of these two-particle states can be quite large. As a result, band-to-exciton transitions tend to dominate the absorption spectrum. However, exciton states are limited to states near $k = 0$, and hence band-to-exciton transitions are clustered at the band edge (or subband edge). The overall effect is the appearance of very strong absorption peaks near the subband edges in quantum-well material, and near the band edge in bulk material.

Exciton absorption peaks are clearly visible in quantum wells at room temperature, as seen in Fig. 4.10 for a typical GaAs QW. The first two steps in the "staircase" absorption spectrum predicted from the density of states (see Fig. 4.9) can be seen, with each step contributing about 10^4 cm^{-1} absorption.

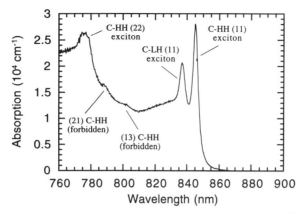

FIGURE 4.10 Absorption spectrum of a GaAs/$Al_{0.3}Ga_{0.7}As$ 100Å QW at room temperature.

However, the exciton peaks riding on top of the steps, particularly the $n = 1$ peaks, dominate the absorption spectrum. Each observed exciton peak corresponds to one of the subband transitions illustrated earlier in Fig. 4.4. The allowed transitions ($n_c = n_v$) are clearly dominant, however, traces of absorption can also be seen from forbidden transitions ($n_c \neq n_v$).

At room temperature, exciton absorption peaks are not nearly as dramatic in bulk material. The reason is that reduced-dimensional structures confine and hold the electron and hole more closely together, producing a higher binding energy. However, the larger QW binding energy is still only on the order of a few meV. Hence exciton states even in QW material are somewhat "fragile," and collisions with phonons and carriers can easily break the exciton apart. For this reason, exciton absorption peaks are strongest and sharpest at low temperatures and low carrier densities. As the temperature and/or carrier density increases, the exciton lifetime diminishes and the exciton absorption peak broadens. Eventually at high enough temperatures and/or carrier densities, the exciton peak disappears altogether. At carrier densities required to achieve gain in the material, exciton states completely vanish and the absorption/emission spectrum becomes dominated by band-to-band transitions. Thus, excitons have little effect on the gain spectrum of the material.

The second many-body effect occurs at high carrier densities, where the charges actually screen out the atomic attractive forces. With a weaker effective atomic potential, the single-atom electron wavefunctions of interest become less localized and the nearest-neighbor electron overlap becomes higher. From discussions in Appendix 1, the larger overlap increases the width of the energy bands (ΔE is larger in Fig. A1.7), reducing the gap between bands. While this description is only qualitative, it does reveal that the bandgap should shrink with increasing carrier density.

It can also be argued theoretically that the *bandgap shrinkage* is inversely related to the average spacing between carriers, or $\Delta E_G \propto -1/r_s$ (the closer the carriers are, the more their own Coulomb potentials screen out the atomic potential). In bulk material, the average volume occupied by one carrier is inversely related to the carrier density, and hence $V \propto 1/N \propto r_s^3$. As a result, we conclude that $\Delta E_G \propto -N^{1/3}$ [5]. In a quantum well, the average area occupied by one carrier in the quantum-well plane is inversely related to the sheet density, or $A \propto 1/N \propto r_s^2$. So in a quantum well, we might expect that $\Delta E_G \propto -N^{1/2}$ [11]. However, taking into account the finite thickness of the quantum well in determining the electron Coulomb potential, the theoretical power dependence on carrier density has been estimated at closer to the one-third power law [12].

Experimental measurements of the power dependence of the bandgap shrinkage on carrier density yield numbers between 0.32 and 0.38 [5, 11, 12] for *both* bulk and quantum well material. As for the absolute shift, at a density of 10^{18} cm^{-3} in bulk or 10^{12} cm^{-2} in a quantum well, the bandgap is reduced by anywhere from 22-32 meV [5, 11, 12], assuming $N = P$ (for measurements on p-doped material, the shift has been doubled assuming electrons would contribute equally to the shift in a laser where $N = P$). In light of the spread in measured data, it is common practice to simply assume the bandgap shrinks with the one-third power of carrier density in both bulk and quantum well material, or

$$\Delta E_g = -cN^{1/3}, \tag{4.43}$$

where N can be either the two- or three-dimensional carrier density. A common value used for the bandgap shrinkage constant in bulk material (assuming $N = P$), which also falls within the measured range for quantum well material, is

$$c \approx 32 \text{ meV}/(10^{18} \text{ cm}^{-3})^{1/3}, \quad \text{(bulk GaAs)} \tag{4.44}$$

$$c \approx 32 \text{ meV}/(10^{12} \text{ cm}^{-2})^{1/3}. \quad \text{(GaAs/AlGaAs QW)} \tag{4.45}$$

Equations (4.43)–(4.45) are not entirely accurate since there is some experimental uncertainty in both the one-third power law and the value of c. However, they do provide a simple and reasonable estimate of the extent of bandgap shrinkage. Less data exists for other material systems, so it is common to assume the same values in the InGaAsP system for example.

The net effect of bandgap shrinkage is that as carrier density increases, the entire gain spectrum redshifts by a noticeable amount. In principle, the shift is accompanied by a slight distortion (i.e., reshaping and enhancement) of the spectrum. However, to first order we can neglect the distortion and simply assume that high carrier densities produce a *rigid* shift of the entire gain spectrum to longer wavelengths. This phenomenon is observable in quantum-well lasers where the high threshold carrier density shifts the lasing wavelength beyond the known band edge wavelength of the quantum well. Bandgap shrinkage is a particularly important factor in situations where there is some critical alignment between a desired cavity mode of the laser and the gain spectrum (as in a short-cavity VCSEL, for example).

4.4   SPONTANEOUS EMISSION

4.4.1   Single-Mode Spontaneous Emission Rate

As we inject a high enough carrier density to achieve gain in the material, we also inevitably increase the spontaneous emission rate. The spontaneous emission rate is important to consider because for every spontaneous photon emitted, a new carrier must be injected into the active region, as discussed in Chapter 2. In short-wavelength materials, this carrier recombination mechanism represents the largest component of the current we must inject into the active region.

To determine the spontaneous emission rate per unit active volume *into one optical mode*, R'_{sp}, we return to Eq. (4.5) which relates the single-mode spontaneous and stimulated emission rates:

$$R'_{sp} = \frac{|\mathscr{E}^{v-f}|^2}{|\mathscr{E}|^2} n_{sp} R_{st}. \tag{4.46}$$

The first term replaces the field strength with the vacuum-field strength of the mode, $|\mathscr{E}^{v-f}|^2$. The second term adjusts the Fermi factor. The last term, R_{st}, can be related to the gain through (4.33). The *population inversion factor*, n_{sp}, discussed in Appendix 6 is defined as

$$n_{sp} \equiv \frac{f_2(1 - f_1)}{f_2 - f_1} = \frac{1}{1 - e^{(E_{21} - \Delta E_F)/kT}}. \tag{4.47}$$

Its value is typically between 1 and 2 at gain thresholds commonly encountered in lasers. Figure 4.11 displays the dependence of n_{sp} on gain for various active materials.

To evaluate $|\mathscr{E}^{v-f}|^2$, we turn to a quantum mechanical description of the optical mode. Without going into the details, it can be shown that an optical mode can be described quantum mechanically using the mathematical formalism developed for harmonic oscillators. One of the basic properties of harmonic oscillators is that the probability of elevating the state n to $n + 1$ (via the creation operator) is proportional to $n + 1$. In describing the optical mode, n refers to the number of photons in the mode. Thus, the probability of adding a new photon to the mode is proportional to the number of photons in the mode *plus one*—as if an imaginary photon were present in the mode. In terms of transition rates, it is the field strength of this imaginary photon which induces "spontaneous" downward transition events. Thus, the vacuum-field strength of

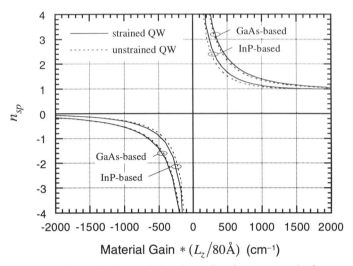

FIGURE 4.11 The theoretical population inversion factor vs. gain for several relevant materials. GaAs based: strained InGaAs/GaAs 80Å QW and unstrained GaAs/AlGaAs 80Å QW. InP based: compressively strained InGaAs/InP 30Å QW and unstrained InGaAs/InP 60Å QW. See Section 4.6 for more details on these structures. In each case, n_{sp} and g are evaluated at a fixed energy slightly larger (11 meV) than the band gap.

the mode is equivalent to the field strength generated by one photon in the mode: $|\mathscr{E}^{v-f}|^2 = |\mathscr{E}_1|^2$. This conclusion is consistent with Einstein's approach which establishes that the equivalent spectral density inducing spontaneous emission is equal to one photon per optical mode (see Appendix 6).

We can express the classical field strength as: $|\mathscr{E}|^2 = N_p V_p |\mathscr{E}_1|^2$, where V_p is the mode volume (i.e. $N_p V_p$ is the number of photons in the mode). Thus, in Eq. (4.46) we can set

$$|\mathscr{E}^{v-f}|^2 = |\mathscr{E}_1|^2,$$
$$|\mathscr{E}|^2 = N_p V_p |\mathscr{E}_1|^2.$$

$$(4.48)$$

Then using Eq. (4.33) to set $R_{st} = v_g g N_p$, Eq. (4.46) becomes

$$R'_{sp} = \frac{v_g g n_{sp}}{V_p} = \frac{\Gamma v_g g n_{sp}}{V}. \qquad (4.49)$$

Setting $1/V_p = \Gamma/V$ in the second equality, we conclude that the spontaneous emission rate into the mode is fundamentally related to the *modal* gain.

4.4.2 Total Spontaneous Emission Rate

To find the total amount of spontaneous emission occurring in the active region, we must sum the single-mode rate (4.49) over all optical modes. Let's denote the *total* spontaneous emission occurring within the energy range dhv, as $R_{sp}^{21} dhv$, where R_{sp}^{21} represents the total spontaneous emission rate *per unit energy* per unit active volume. Equating this with a sum over modes near hv_{21} within dhv, we have

$$R_{sp}^{21} dhv = \sum_{\substack{\text{modes} \\ \text{in } dhv}} R'_{sp} = n_{sp} \sum_{\substack{\text{modes} \\ \text{in } dhv}} \frac{v_g g_{21}}{V_p}. \tag{4.50}$$

Because n_{sp} only depends on hv, it has been pulled out of the sum and evaluated at hv_{21}. However, the remaining three terms depend on the specifics of each mode (either through modal dispersion, polarization dependence, or mode volume). If we define average values for the three terms, the sum reduces to

$$\sum_{\text{modes}} \frac{v_g g_{21}}{V_p} = \frac{\bar{v}_g \bar{g}_{21}}{\bar{V}_p} N_{modes}, \tag{4.51}$$

where N_{modes} is the number of modes within dhv. Defining *individual* average values for each term is justified as long as the variations of each term in the sum are not correlated, and there is no reason to believe otherwise.

Considering that the average is over modes going in all directions, it is probably safe to say that the average group velocity, $\bar{v}_g$, is somewhere close to the material group velocity of the active region, v_g. The dependence of the quantum-well gain is through the polarization state of the mode (in bulk material there is no dependence and $\bar{g}_{21} = g_{21}$). Assuming the polarization states are isotropically distributed over the two in-plane TE and one perpendicular TM polarizations, the average material gain in a quantum well becomes

$$\bar{g}_{21} = \tfrac{1}{3}[2g_{21}^{TE} + g_{21}^{TM}]. \tag{4.52}$$

To handle $1/\bar{V}_p$ (or equivalently, the average confinement factor, $\bar{\Gamma}$), we need to evaluate N_{modes}. In large cavities, N_{modes} is most easily found using the mode density concept. If we assume the cavity is a large rectangular metal box of volume, V_{box}, the density of optical modes per unit frequency per unit volume can be derived. The procedure is outlined in Appendix 4 and the result is given by Eq. (A4.5). We repeat it here for reference:

$$\rho_0(v) \, dv = \frac{8\pi}{c^3} n^2 n_g v^2 \, dv. \tag{4.53}$$

The density of modes per unit *energy* is equal to $\rho_0(v) \, dv/dE = \rho_0(v)/h$. With this definition, the total number of modes within dhv becomes

$$N_{modes} = \rho_0(v)/h \cdot V_{box} \, dhv. \tag{4.54}$$

Inserting this into Eq. (4.51), we are left with the volume ratio $V_{box}/\bar{V}_p$. If the laser cavity were a large metal box then the mode volume would be equivalent to the volume of the box (aside from standing wave effects, which should average out over the spontaneous emission bandwidth), and we could simply set

$$V_{box}/\bar{V}_p = 1. \tag{4.55}$$

Of course, laser cavities are typically much more complex than a simple metal box, and V_p is potentially different for every optical mode. Thus, there is no guarantee that (4.55) holds in real laser cavities, or that we can even define a mode density as given in Eq. (4.53).

Fortunately, if the cavity is much larger than the wavelength of bandgap radiation in the active region, or more specifically if $V_{cav} \gg \lambda^3$, it turns out that a more rigorous treatment usually averages out to the simple metal box treatment [13] (especially when the emission bandwidth is large relative to resonances in the cavity). To get a feel for the numbers, a typical $0.2 \times 4 \times 200\ \mu m^3$ GaAs in-plane laser has $V_{cav} \sim 10{,}000\lambda^3$. Thus, the simple metal box assumption contained in Eqs. (4.53) through (4.55) is expected to work well in this case.

In a VCSEL, the simple box assumption is more questionable. However, the volume of a typical GaAs VCSEL with dimensions of $1 \times 10 \times 10\ \mu m^3$, is smaller than a typical in-plane laser by only a factor of two, and $V_{cav} \sim 5000\lambda^3$. Thus, for VCSELs of this size, the simple box assumption should still hold. However, when the lateral dimensions of the VCSEL are reduced below 1 μm, significant deviations from the simple box assumption are expected to occur. Under these circumstances, the mode density concept must be abandoned and more sophisticated mode-counting techniques must be used to estimate the total spontaneous emission rate. In addition to mode counting, the average value for $1/\bar{V}_p$ (or equivalently $\bar{\Gamma}$) is also required. Numerous researchers interested in microcavity lasers have investigated such numerical exercises for a number of cavity geometries [13, 14].

Combining Eqs. (4.50) through (4.55) assuming the simple box assumption holds, the total spontaneous emission rate per unit energy per unit active volume (in units of $s^{-1}\ cm^{-3}\ eV^{-1}$) becomes

$$R_{sp}^{21} = \frac{1}{h}\rho_0(v_{21}) \cdot v_g n_{sp}\bar{g}_{21}. \tag{4.56}$$

This result is identical to Eq. (A6.32) derived in Appendix 6, with the exception that the gain here is more correctly defined as an average over all polarizations. Using the explicit expression for gain (4.37) and mode density (4.53), the general expression for spontaneous emission becomes

$$R_{sp}^{21} = \frac{4n\pi q^2}{\varepsilon_0 h^2 c^3 m_0^2} hv_{21}|\bar{M}_T(E_{21})|^2 \rho_r(E_{21}) \cdot f_2(1 - f_1), \tag{4.57}$$

where

$$|\bar{M}_T(E_{21})|^2 = \frac{1}{3} \sum_{\substack{\text{all three} \\ \text{polarizations}}} |M_T(E_{21})|^2.$$

The transition matrix element is the only factor dependent on polarization in the expression for gain. This is why the averaging over polarizations only needs to include $|M_T|^2$.

The actual spontaneous emission spectrum will be affected by lineshape broadening in the same way the gain is affected. In analogy with Eq. (4.39), the spontaneous emission spectrum taking lineshape broadening into account is related to R_{sp}^{21} through the following:

$$R_{sp}^{hv}(hv) = \int R_{sp}^{21} \mathcal{L}(hv - E_{21}) \, dE_{21}. \tag{4.58}$$

The spectrum is generally peaked just above the bandgap energy since the electrons and holes are concentrated at the band edges. The spectrum gradually decays to zero at higher energies. To determine the total emission rate, we must integrate over all photon energies. However, in practice integrating over a limited range near the bandgap energy is usually sufficient to account for all of the spontaneous emission. It is shown in Appendix 6 that the integration over $R_{sp}^{hv}(hv)$ is essentially the same as integrating over the simpler R_{sp}^{21}. Thus, the total band-to-band spontaneous emission rate is given by

$$R_{sp} = \int R_{sp}^{21} \, dE_{21} = \eta_i \eta_r \frac{I}{qV}. \tag{4.59}$$

The second equality allows us to determine the radiative component of the current (excluding stimulated emission) required by the active region to obtain a given gain. Example calculations in different material systems will be given later in this chapter.

4.4.3 Spontaneous Emission Factor

With the above description of spontaneous emission we can derive a simple expression for the spontaneous emission factor used in the rate equations to express the fraction of total spontaneous emission which enters the mode of interest. Using Eq. (4.49) for the emission rate into one mode, and Eq. (4.59) for the total emission rate, we have

$$\beta_{sp} \equiv \frac{R'_{sp}}{R_{sp}} = \frac{\Gamma v_g g n_{sp}}{\eta_i \eta_r I/q}, \tag{4.60}$$

where I as used here does not include stimulated emission current.

With Eq. (4.60), parameters readily accessible experimentally can be used to estimate β_{sp}. For example, assume we have a laser with a threshold current of 10 mA, a modal threshold gain of 50 cm^{-1}, and an internal efficiency of 75%. If nonradiative recombination is minimal, then $\eta_r \approx 1$. If we assume the group index is ~ 4 and the population inversion factor is ~ 1.5, we quickly find $\beta_{sp} \approx 1.2 \times 10^{-5}$, typical of experimental values measured with in-plane lasers. To get a better estimate of β_{sp}, we need to determine c/v_g, n_{sp}, and η_r more accurately. However, c/v_g is almost always in the range of 4–5, n_{sp} is usually between 1.25 and 1.75 for gains commonly required in lasers, and η_r is typically between 50 and 80%, implying that simple estimates of c/v_g, n_{sp}, and η_r will get us within a factor of 2 of the correct value of β_{sp}.

In a typical VCSEL, the percent gain per pass is in the range of 0.5–1% and the cavity length is ~ 1 μm. This gives a modal gain of 50–100 cm^{-1}. *Radiative* threshold currents are in the range of 0.5–2.0 mA for 10 μm diameter devices. If we assume a 1 mA radiative threshold current and 100 cm^{-1} threshold modal gain, we find $\beta_{sp} \approx 1.8 \times 10^{-4}$ (assuming again that c/v_g and n_{sp} are equal to 4 and 1.5). This is about an order of magnitude higher than β_{sp} observed in typical in-plane lasers. The main difference lies in the reduction of the threshold current which is in turn predominantly due to the reduction in active-region volume possible with VCSEL structures.

Equation (4.60) also reveals that β_{sp} is not a constant, but is dependent on the injection level. As the injection level increases, the spontaneous emission into one mode saturates at a maximum value just as the gain saturates at g_{max}. Meanwhile the current and total spontaneous emission rate continue to increase as the spectrum broadens to include more modes in the emission process. The net effect is that β_{sp} decreases with increasing injection level. In rate equation analyses, β_{sp} is often approximated as a constant (see Chapters 2 and 5). This is justified for near- and above-threshold analyses, as long as the value assumed for β_{sp} is the actual value that would exist near the threshold injection level.

4.5   NONRADIATIVE TRANSITIONS

With radiative processes defined, we now need to consider nonradiative transitions to determine their relative importance in the overall carrier recombination process. Three common types of nonradiative transitions are depicted in Fig. 4.12. These processes were briefly discussed in Chapter 1 in reference to Fig. 1.3. In the following sections we will provide a more detailed description of each process.

4.5.1   Defect and Impurity Recombination

The first type of nonradiative transition appearing in Fig. 4.12 depicts an energy level in the middle of the gap, which serves to trap an electron from the

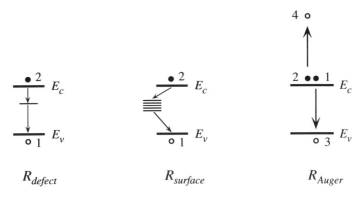

FIGURE 4.12   Various types of nonradiative recombination paths.

conduction band temporarily before releasing it to the valence band. Energy levels of this sort can arise from a variety of causes. Defects in the lattice structure are one source. For example, a void at an atomic site or an extra atom lodged in between the lattice structure can produce insufficient or extra orbitals, which leads to a mismatch in the covalent bonding pattern. Such localized *dangling bonds* give rise to discrete energy levels which can possibly appear in the middle of the bandgap. Another common source of midgap energy levels are impurities. Just as dopants create impurity levels near the band edge, other types of atoms can create impurity levels which are closer to the middle of the gap. Oxygen, for example, is a particularly insidious impurity because it is an abundant element and is known to create large recombination rates in aluminum-containing compounds.

The defect or impurity recombination rate, R_{defect}, or R_d for short, was first analyzed four decades ago in a classic paper by Shockley and Read [15]. Hall simultaneously arrived at a similar result [16]. Thus it is known as the Shockley–Read–Hall recombination theory. The theory begins by writing down the four possible transition rates into and out of the trap (up and down from the conduction band, and up and down to the valence band). Setting the rates equal in thermal equilibrium to determine relationships between the up and down rates (similar to the derivation of radiative transition rates in Appendix 6), an expression for the nonequilibrium recombination rate can be derived. Within the Boltzmann nondegenerate carrier density regime discussed in Appendix 2, the defect recombination rate takes the form

$$R_d = \frac{NP - N_i^2}{(N^* + N)\tau_h + (P^* + P)\tau_e}, \tag{4.61}$$

where N_i is the intrinsic carrier concentration, τ_e is the time required to capture an electron from the conduction band assuming all traps are empty, and τ_h is the time required to capture a hole from the valence band assuming all traps are full. As might be expected, the capture rate is proportional to the density of traps, $1/\tau_{e,h} \propto N_t$, such that the higher the trap density, the shorter the capture times. N^* and P^* appearing in the denominator are the electron and hole densities that would exist if the Fermi level were aligned with the energy level of the trap. The important point here is that if the trap level is close to either band edge, then either N^* or P^* will become large, substantially reducing the recombination rate. Thus, the most effective recombination centers are those with energy levels close to the middle of the gap, so-called *deep-level* traps. Great care must be taken to avoid introducing such deep-level impurities (such as oxygen) into the crystal lattice.

It is apparent that Eq. (4.61) has a nontrivial dependence on both electron and hole densities. However, for laser applications we are primarily interested in the high-level injection regime where $P = N \gg N_i$, N^*, P^*. Under these conditions, Eq. (4.61) simplifies to

$$R_d = \frac{N}{\tau_h + \tau_e}. \quad \text{(high-level injection)} \qquad (4.62)$$

Thus, defect or impurity recombination follows a linear dependence on carrier density in the active region of lasers, and we can define the defect component of the linear recombination constant as: $A_d = 1/(\tau_h + \tau_e)$, using the notation of Chapter 2.

In the low-level injection limit, the form of the recombination rate changes. If we have deep-level traps, then N^* and P^* are negligible ($N^* \approx P^* \approx N_i$), and they can be removed from Eq. (4.61). With $N = N_0 + \delta N$ and $P = P_0 + \delta N$ (assuming equal numbers of excess electrons and holes) and assuming δN is small, the recombination rate becomes

$$R_d = \delta N \frac{N_0 + P_0}{N_0 \tau_h + P_0 \tau_e}. \quad \text{(low-level injection)} \qquad (4.63)$$

The recombination rate is again linear in the excess carrier density, however, the rate depends on whether the material is doped n-type or p-type. For n-type material, $A_d = 1/\tau_h$ (it is limited by the capture of holes), and for p-type material $A_d = 1/\tau_e$ (it is limited by the capture of electrons). Thus, while the defect or impurity recombination rate is linear with the excess carrier density at either low or high injection levels, the lifetime does increase from either τ_h or τ_e to $\tau_h + \tau_e$ as the carrier density is increased.

Recombination via defects and impurities is primarily a problem for excess minority carriers or carriers injected into a region under nonequilibrium conditions. Majority carrier current flow in heavily doped materials is relatively unaffected by defects and impurities (aside from the possibility that defects and impurities can affect the ionization of doping species, which can affect the doping efficiency). In other words, material which contains a very high density of defects or impurities may not be a good choice for a laser's active region, but it still may carry majority carrier current just fine if doped heavily enough.

With modern MBE and MOCVD growth technologies, the crystal quality of semiconductor devices is such that the defect density and the density of impurity atoms are at most 10^{16} cm^{-3}, and are more often below 10^{15} cm^{-3}. With such low trap densities, the defect and impurity recombination rates are negligible in typical laser applications. However, there are instances where such recombination can become large and problematic.

In the early stages of MBE growth technology (in the late 1970s), it was a common belief that growth temperatures should be kept below 580°C to retain good surface morphology. However, a puzzling concern among researchers was that GaAs/AlGaAs lasers grown by MBE invariably had much higher threshold current densities than equivalent structures grown by LPE. Finally, Tsang et al. [17] found that increasing the growth temperature to 650°C led to dramatic reductions in the threshold current density, lower than the best LPE material. Their explanation was that impurities (perhaps oxygen) were being incorporated into the AlGaAs cladding layers at densities high enough to significantly increase the threshold current density. At the higher growth temperatures, the incorporation of impurities was minimized and dramatic improvements in the threshold current were observed. More recently, the purity of aluminum sources used in MBE systems has significantly improved, allowing the growth temperature to be reduced back down to 600–620°C. However, unusually high threshold current densities in just-grown lasers which normally would yield low thresholds can often indicate that the MBE or MOCVD machine is introducing unwanted impurities, either through a leak in the system or corrupted material sources.

An example of where defect recombination is important is in the aging of lasers. As lasers are used they inevitably go through dramatic thermal variations. Such repeated thermal cycles can cause stress to the crystal. This stress can cause small defects within the crystal lattice to spread and grow large, just as a small crack in the windshield of your car can propagate across your entire field of view. A particularly interesting example of this is the emergence of dark-line defects which tend to appear after lasers have been in use for thousands of hours. Such defects show up as dark lines when viewed from the surface due to the absence of carriers and hence spontaneous emission in regions where the "cracks" propagate. If the original quality of the crystal is high, such dark-line defects can be minimized. However, if they do become large with age, the threshold current of the laser will suffer over time and eventually the laser will die.

Another common example of where defect recombination can become important is in the area of strained-layer research. This research is interested in growing epitaxial layers of materials which have a different "native" lattice constant than the substrate material. In a lattice-mismatched growth, the epitaxial layer will attempt to deform to the substrate lattice structure. However, as the layer becomes thicker, atomic forces building up within the epitaxial lattice structure will at some critical thickness break discontinuously with the deformed lattice and begin to force the lattice back to its native form. Strained layers with thicknesses below the critical thickness can be of very high quality (and are indeed superior in many respects to their unstrainted counterparts). However, strained layers with thicknesses much greater than the critical thickness will inevitably be laden with severe lattice defects.

As the strained layer is grown thicker still, the lattice defects unfortunately tend to propagate along the growth direction. As a result, thick strained layers have very high defect recombination rates and are unusable for any type of carrier injection applications such as lasers. However, techniques do exist which can block defect propagation, allowing the upper portion of a thick epitaxial layer to provide a high-quality lattice structure with low defect densities and hence low defect recombination rates. For example, good-quality epitaxially grown GaAs/AlGaAs lasers have been successfully grown on silicon substrates using such techniques even though the lattice constants of the two semiconductors are very different. Furthermore, high doping of the defect-laden transition region can allow high currents to pass across the interface, allowing for the integration and interconnection of silicon circuits with GaAs lasers.

4.5.2 Surface and Interface Recombination

The second type of nonradiative transition in Fig. 4.12 depicts electrons recombining via *surface states* of the crystal. The two-step recombination mechanism is analogous to the defect and impurity recombination mechanism. However, in this case the number of traps is characterized by a *two-dimensional* sheet density at an exposed surface of the crystal or at an interface between two materials. These surface states primarily arise from the termination of the lattice, which inevitably leaves a few unmatched bonds on one side of every exposed unit cell. Such *dangling bonds* occur in very high densities, forming a miniband as opposed to individual energy levels, as depicted in the figure. Surface recombination is most damaging when the exposed surface-to-volume ratio is large—in other words, when the device size is reduced. Interface recombination is damaging when the interface quality is poor. With modern growth technologies, interface recombination is minimal in common materials, but can be very detrimental in more experimental materials research efforts. In addition, devices which make use of regrowth technology (i.e., devices which are put back into the growth chamber after etching or some other processing step is performed) can suffer from poor interfaces and hence high interface recombination.

The description of surface recombination is accomplished via the Shockley–Read–Hall theory in analogous manner to defect and impurity recombination theory. However, instead of defining a caputre *time*, τ, of carriers distributed throughout a volume of material, we define a capture *rate* of carriers located within some capture length of the surface: L_c/τ. A larger capture length allows surface states to capture more carriers per unit time and hence leads to a higher

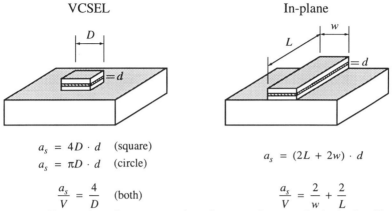

VCSEL

In-plane

$$a_s = 4D \cdot d \quad \text{(square)}$$
$$a_s = \pi D \cdot d \quad \text{(circle)}$$

$$a_s = (2L + 2w) \cdot d$$

$$\frac{a_s}{V} = \frac{4}{D} \quad \text{(both)}$$

$$\frac{a_s}{V} = \frac{2}{w} + \frac{2}{L}$$

FIGURE 4.13   Exposed surface area and surface-to-volume ratio in both pillar-type VCSEL and in-plane ridge laser geometries.

capture rate. Because of the units, this capture rate is referred to as a velocity. The association with velocity does have a physical significance. For example, if the capture velocity of the trap is larger than the average thermal velocity of the carriers, then the capture rate will be limited by the thermal velocity. This maximum arrival rate of carriers at the surface sets an upper limit on the capture velocity of somewhere near 10^7 cm/s. In most semiconductors, the capture velocity is at least an order of magnitude smaller than this.

Redefining the electron and hole capture times as capture *velocities* (i.e., setting $\tau_e \to 1/v_e$ and $\tau_h \to 1/v_h$), the analogous version of Eq. (4.61), neglecting N^* and P^*, for surface recombination becomes[6]

$$R_{sr} = \frac{a_s}{V} \cdot \frac{NP - N_i^2}{N/v_h + P/v_e}. \tag{4.64}$$

The first term effectively distributes the exposed surface area, a_s, over the volume of the active region, V, since we have defined R_{sr} as the rate per unit active volume, not rate per unit surface area. This geometrical factor is shown in Fig. 4.13 for two common laser geometries. As can be seen, a_s/V makes surface recombination important when the pillar diameter or stripe width is small.

Under high-level injection in the active region, $P = N \gg N_i$, and Eq. (4.64) reduces to

$$R_{sr} = \frac{a_s}{V} v_s N \quad \text{(high-level injection)} \tag{4.65}$$

---

[6] A more rigorous version of Eq. (4.64) would integrate over all surface states within the miniband, with possibly energy-dependent capture velocities. However, for our purposes, we will assume a discrete energy level for the surface states.

with

$$\frac{1}{v_s} = \frac{1}{v_h} + \frac{1}{v_e}. \tag{4.66}$$

The linear relationship to the carrier density allows us to define the surface recombination component of the linear recombination constant as $A_{sr} = (a_s/V)v_s$. The *surface recombination velocity*, v_s, which controls the surface recombination rate, is seen to be an average of the individual electron and hole capture velocities. The use of inverse velocities in (4.66) reflects the fact that if, for example, the electron capture velocity is very high, then surface recombination will be limited by the capture rate of holes and $v_s \approx v_h$.

Using (4.65), the surface recombination current and current density in the laser can be written as

$$I_{sr} = q a_s v_s N \quad \text{and} \quad J_{sr} = q v_s N\left(\frac{a_s}{V} d\right). \tag{4.67}$$

To estimate I_{sr}, it is important to realize that we need to know the carrier density *which exists at the surface*. As a first-order, upper-bound estimate of I_{sr} or J_{sr}, we can naively use the carrier density in the center of the active region (which can be roughly estimated from the threshold gain, if we know the gain as a function carrier density). However, in reality the heavy recombination at the surface will deplete the surface carrier density to some level which balances the recombination rate with the gradient-driven lateral diffusion current directed from the center to the surface of the active region (see Problem 4.6 for more details). To determine this carrier density requires solving the carrier diffusion equation.[7]

Under low-level injection conditions, the surface recombination rate changes just as (4.62) and (4.63) are different. For laser applications we are not particularly interested in this case. However, experimental measurements of the surface recombination velocity are often made under such conditions. Using Eq. (4.65) to define v_s, we find that (4.64) evaluated under small carrier density perturbations away from the equilibrium values, N_0 and P_0, gives

$$v_s \equiv \frac{V}{a_s} \cdot \frac{R_{sr}}{\delta N} = \frac{N_0 + P_0}{N_0/v_h + P_0/v_e}. \quad \text{(low-level injection)} \tag{4.68}$$

[7] Such a procedure can take on various degrees of complexity. The simplest approach is to assume an ambipolar diffusion coefficient representing the effective diffusion of both electrons and holes (which within the model sets the carrier densities equal). The recombination then follows (4.65) using the surface recombination velocity given in (4.66), and a single carrier diffusion equation can be used. A more extensive model would allow the electron and hole densities to be different, each with their own diffusion constant. The recombination rate at the surface for both electrons and holes would then take on the form (4.64), where the individual electron and hole capture velocities must be known or fitted. The solution would then have to satisfy the drift-diffusion equations for electrons and holes. To be complete, the drift-diffusion equations must then be coupled with Poisson's equation taking into account the charge distribution of the surface states [18].

If the surface recombination velocity is measured under low-level injection conditions, then the measured value will be dependent on the doping present. For n-type material, $v_s = v_h$ (it is limited by the capture rate of holes), and for p-type material, $v_s = v_e$ (it is limited by the capture rate of electrons). Equation (4.68) also assumes that $\delta N = \delta P$, which may not necessarily be true, yielding still different values for v_s. The most reliable values for v_s relevant for laser applications must be made under strong high-level injection conditions. For example, measuring the threshold current dependence on the laser geometry is the most direct method. However, geometry-dependent optical losses which change the threshold gain can skew these results as well.

Surface recombination tends to be much more of a problem in the short-wavelength GaAs system than in the long-wavelength InGaAsP system. This is because v_s is two orders of magnitude larger in GaAs. Values measured under low-level injection conditions on GaAs and GaAs/AlGaAs quantum wells give [19]

$$v_s \approx 4\text{–}6 \times 10^5 \text{ cm/s.} \quad \text{(GaAs, bulk and QW)} \quad (4.69)$$

In n-type GaAs, values closer to $2\text{–}3 \times 10^6$ cm/s have been reported [20], suggesting that v_h is greater than v_e by close to an order of magnitude. Measurements on strained InGaAs/GaAs quantum-well lasers show a reasonable improvement over GaAs [21]:

$$v_s \approx 1\text{–}2 \times 10^5 \text{ cm/s.} \quad \text{(InGaAs/GaAs, QW)} \quad (4.70)$$

In InGaAsP materials, surface recombination tends to be lower. Measured values on InP give [22]

$$v_s \leq 10^4 \text{ cm/s.} \quad \text{(InP, bulk)} \quad (4.71)$$

For quaternary InGaAsP, such low values are also expected.

In GaAs/AlGaAs and InGaAs/GaAs lasers, surface recombination can be quite severe for small devices. To get a feeling for how small, Fig. 4.14 plots the threshold current density of a typical InGaAs QW VCSEL and an in-plane laser for different values of v_s (using the gain as a function of carrier density and current density given later in the chapter). At $v_s = 1 \times 10^5$ cm/s, the VCSEL threshold current density doubles from ~ 500 A/cm^2 to ~ 1000 A/cm^2 when the pillar diameter is reduced to 10 μm. This characteristic is common for InGaAs QW VCSELs. If we were to use GaAs QWs with v_s perhaps equal to 5×10^5 cm/s, the threshold current density would double at 50 μm! In long-wavelength materials, $v_s \leq 10^4$ cm/s and pillar diameters could be reduced to 1 μm before the threshold current density would double. Similarly for the in-plane laser, problems become severe for stripe widths less than 10 μm using InGaAs QWs (with $v_s = 1 \times 10^5$ cm/s). However, it is important to realize that these estimates assume the carrier profile is flat across the active region. A more

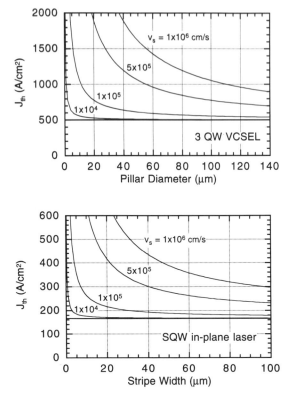

FIGURE 4.14 Threshold current density vs. lateral device dimensions for different surface recombination velocities (assuming $D_{np} = \infty$). Both lasers use $In_{0.2}Ga_{0.8}As/GaAs$ 80Å QWs and have a 1500 cm^{-1} threshold *material* gain (Tables 4.4 and 4.5 are used to obtain $N_{th} = 3.62 \times 10^{18}$ cm^{-3} and $J_{th} = 166.4$ A/cm^2 per QW). The length of the in-plane laser is assumed to be 250 μm.

realistic carrier profile will produce threshold current densities somewhat lower than indicated, particularly at the smaller dimensions (see Problem 4.6).

Because there is much interest in reducing the active volume well below the limits imposed in Fig. 4.14, much attention has focused on ways to reduce or eliminate surface recombination. One obvious method would be to define the stripe or pillar *without etching through* the active region. However, one must now contend with lateral outdiffusion of carriers at the stripe or pillar edges. In Problem 4.6, it is shown that the ambipolar diffusion of carriers out of the active region is equivalent to a surface recombination velocity of magnitude

$$v_{sD} \equiv \frac{D_{np}}{L_{np}} = \sqrt{\frac{D_{np}}{\tau_{np}}} = 1 \times 10^5 \text{ cm/s} \cdot \left[\frac{D_{np}}{20 \text{ cm}^2/\text{s}} \frac{2 \text{ μm}}{L_{np}} \right]. \qquad (4.72)$$

D_{np} and L_{np} are the ambipolar diffusion constant and related diffusion length, and τ_{np} is the average carrier lifetime. The last equality evaluates v_{sD} for common values, revealing that carrier outdiffusion "velocities" and surface recombination velocities are unfortunately of similar magnitude. To minimize carrier out-diffusion, we must bury the active region in a higher-bandgap material which provides a lateral potential barrier to carriers. However, such regrowth technologies, while popular and effective in long-wavelength systems, have remained problematic in AlGaAs-containing compounds.

Other techniques for reducing the effects of surface recombination are also being considered. One interesting method involves surface passivation. As might be expected, the density and character of surface states is very dependent on the surface chemistry. For example, an oxidized surface provides a different recombination velocity than a freshly etched surface. If a GaAs surface is soaked for a while in a water solution containing Na_2S, $(NH_4)_2S$, or any one of various salts that yield aqueous sulfur species, it has been demonstrated that the surface recombination velocity can be reduced dramatically, yielding $v_s \approx 10^3$ cm/s under high-level injection conditions [23]! Unfortunately, the sulfur-containing compounds which tie up and passivate the dangling bonds are somewhat volatile and the surface passivation effect can fade within an hour or two of exposure to air. Thus, efforts are under way to make the passivation more permanent by placing an oxide or nitride cap over the surface immediately after it has been passivated to lock-in and retain the passivating layer.

4.5.3 Auger Recombination

The last type of transition in Fig. 4.12 depicts what is essentially a collision between two electrons which knocks one electron down to the valence band and the other to a higher energy state in the conduction band. The high-energy electron eventually thermalizes back down to the bottom of the conduction band, releasing the excess energy as heat to the crystal lattice. An analogous collision can occur between two holes in the heavy hole (HH) band; in this case, the hole which is knocked deeper into the valence band is transferred to either the split-off (SO) or light hole (LH) band. In all, there are three types of transitions, collectively referred to as *Auger* processes, which are relevant in III–V semiconductors. These are shown in Fig. 4.15. For quantum-well material, additional types of subband-to-subband and bound-to-unbound transitions can be defined, but they still fall into these three general categories.

Because Auger processes depend on carriers colliding with one another, the Auger recombination rate, R_A, should increase rapidly as carrier density increases. The CCCH process involves three electron states and one heavy-hole state, and hence is expected to become important when the electron density is high. The CHHS and CHHL processes involve one electron state, two heavy-hole states, and one split-off or light-hole state. Thus,

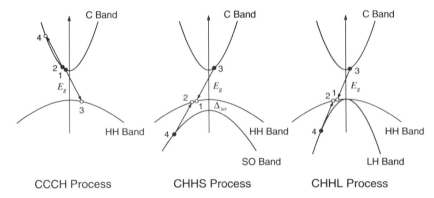

FIGURE 4.15 Auger processes in III–V semiconductors.

they are expected to become important when the hole density is high. In lasers, the electron and hole densities in the active region are equal (if the active region is not heavily doped) implying that all three processes are potentially important.

The Auger recombination rate, like the radiative transition rates, is dependent on the probability of finding the various states occupied or empty, as the case may be. For the Auger processes in Fig. 4.15, the relevant Fermi factors are

$$P_{1 \to 3} = f_{c1} f_{c2} (1 - f_{v3})(1 - f_{c4}), \qquad \text{(CCCH)} \qquad (4.73a)$$

$$P_{3 \to 1} = (1 - f_{v1})(1 - f_{v2}) f_{c3} f_{v4}. \qquad \text{(CHHS and CHHL)} \qquad (4.73b)$$

The subscript on the probability, P, indicates the significant electron recombination path. The additional c and v subscripts on the Fermi functions identify the quasi-Fermi level to be used in each case: E_{Fc} or E_{Fv}.

To obtain simple indicators of how these Fermi factors vary with carrier density, we use the Boltzmann approximation (which is strictly valid only for low carrier densities). From Table A2.1 we have

$$\frac{N}{N_c} \approx e^{-(E_c - E_{Fc})/kT} \qquad \text{and} \qquad \frac{P}{N_v} \approx e^{-(E_{Fv} - E_v)/kT}, \qquad (4.74)$$

which hold as long as $E_{Fc} \ll E_c$ and $E_{Fv} \gg E_v$. Under these conditions, the 1 in the denominator of the Fermi functions defined in (4.3) can be neglected, and

$$f_c \approx e^{-(E - E_{Fc})/kT} = \frac{N}{N_c} e^{-(E - E_c)/kT},$$

$$ \qquad (4.75)$$

$$1 - f_v \approx e^{-(E_{Fv} - E)/kT} = \frac{P}{N_v} e^{-(E_v - E)/kT}.$$

With these relations, the transition probabilities can be approximated by

$$P_{1 \to 3} \approx \frac{N^2 P}{N_c^2 N_v} e^{-(\Delta E_1 + \Delta E_2 + \Delta E_3)/kT}, \quad \text{(CCCH)} \tag{4.76a}$$

$$P_{3 \to 1} \approx \frac{N P^2}{N_c N_v^2} e^{-(\Delta E_1 + \Delta E_2 + \Delta E_3)/kT}, \quad \text{(CHHS and CHHL)} \tag{4.76b}$$

where $\Delta E_i = E_i - E_c$ for conduction band states, and $\Delta E_i = E_v - E_i$ for valence band states. The fourth, high-energy state is assumed to be completely empty (CCCH) or completely full (CHHS and CHHL). These probability factors suggest that CCCH processes are $\propto N^2 P$ while CHHS and CHHL processes are $\propto N P^2$. Furthermore, for a given electron and hole density, the probability of Auger recombination increases exponentially with temperature. The strong dependence on both carrier density and temperature makes Auger recombination a potentially devastating recombination path for carriers in laser applications.

The Auger recombination rate is also strongly dependent on the bandgap of the material. To understand why this is so, we need to examine the consequences of energy and momentum conservation. With Auger transitions, as with radiative transitions, the initial energy and momentum of the system must be conserved. This constrains Auger transitions to specific regions of the bands. In particular, the lowest possible energy configuration does not occur right at the band edge, as it does with radiative transitions, but at a slightly higher energy. This "lowest energy" configuration for the CCCH process is depicted in Fig. 4.16 for two different bandgaps. The shading illustrates the location of electrons assuming a zero temperature distribution. The presence of holes at state 3 is greatly enhanced for the smaller bandgap material; this allows for a much higher Auger transition rate *for the same overall carrier density*. In fact at zero temperature, such lowest-energy Auger transitions would not occur in the higher-bandgap material since state 3 would be fully occupied. The two initial electron states, 1 and 2, are also closer to the band edge in the smaller-bandgap material, further enhancing the Auger transition rate, since at finite temperatures more electrons can be found there.

To gain a more quantitative feel for how the bandgap affects the Auger transition rate, we need to estimate how the transition probabilities in Eq. (4.76) are affected by the bandgap. Figure 4.16 demonstrates how states 1, 2, and 3 move closer to the band edge as the bandgap is reduced. In mathematical terms, the sum of energy offsets from the band edge, $\Delta E_1 + \Delta E_2 + \Delta E_3$, is reduced as the bandgap is reduced. This sum can be described more conveniently in terms of ΔE_4 using energy conservation. Setting the initial energy equal to the final energy for the CCCH process in Fig. 4.15 or 4.16, we must have

$$\Delta E_1 + \Delta E_2 = -(E_g + \Delta E_3) + \Delta E_4, \tag{4.77}$$

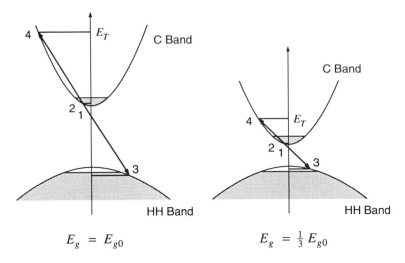

FIGURE 4.16 Lowest-energy CCCH Auger transition for two different bandgaps. The four states are drawn roughly to scale, with $\mu = m_C/m_H \approx 5$.

where E_c is used as the energy reference. Rearranging, we obtain

$$\Delta E_1 + \Delta E_2 + \Delta E_3 = \Delta E_4 - E_g. \tag{4.78}$$

Thus, the probabilities in Eq. (4.76) are maximized when ΔE_4 is minimized. This minimum value for ΔE_4 is referred to as the *threshold* energy, E_T, of the Auger process. The lowest-energy CCCH configuration in Fig. 4.16 corresponds to this threshold energy process and is therefore the most probable configuration for CCCH Auger transitions.

Appendix 12 details the math involved with minimizing (4.78) subject to momentum conservation. This exercise provides expressions for all energies and k-vectors associated with the threshold Auger process. In particular, the threshold energy for all three Auger processes in Fig. 4.15 are found to be

$$E_T = \frac{2m_C + m_H}{m_C + m_H} E_g, \qquad \text{(CCCH)} \tag{4.79}$$

$$E_T = \frac{2m_H + m_C}{2m_H + m_C - m_S} (E_g - \Delta_{so}), \qquad \text{(CHHS)} \tag{4.80}$$

$$E_T = \frac{2m_H + m_C}{2m_H + m_C - m_L} E_g. \qquad \text{(CHHL)} \tag{4.81}$$

For the CHHS process, E_T is measured from the SO band edge. The effective mass prefactors typically fall in the range of 1.1–1.2.[8] The important point to notice is that the threshold energy is proportional to the bandgap. Using Eq. (4.78) in Eq. (4.76), the maximum probability for CCCH Auger transitions becomes

$$P_{1 \to 3} \approx \frac{N^2 P}{N_c^2 N_v} e^{-(E_T - E_g)/kT}, \quad \text{(CCCH)} \quad (4.82)$$

and similarly for the other two processes. With $E_T \propto E_g$, it is clear that the maximum probability for Auger transitions increases exponentially as the bandgap is decreased. For example, if we were to reduce the bandgap of a given material by a factor of two, the probability for CCCH Auger recombination would increase by

$$\frac{P(E_g/2)}{P(E_g)} \approx \frac{e^{-(1/2)(E_T - E_g)/kT}}{e^{-(E_T - E_g)/kT}} = e^{(1/2)aE_g/kT}, \quad (4.83)$$

where

$$a = E_T/E_g - 1 \quad (\approx 0.1\text{–}0.2). \quad (4.84)$$

In GaAs, $a = 0.15$ for the CCCH process, giving an increase of $e^{4.14} \approx 63$ at room temperature, if the bandgap were reduced by a factor of 2. With $a = 0.2$, this factor would be $e^{5.52} \approx 250$, indicating a strong dependence on $E_T/E_g - 1$. This example, while not quantitatively accurate (in that it only considers the transition probability of one set of states, and assumes the Boltzmann approximation), nevertheless serves to illustrate the severity of the problem for long-wavelength lasers, which must deal with Auger recombination rates much larger than in short-wavelength lasers.

A type of Auger transition we have not yet considered involves an additional particle, a phonon of the crystal lattice which can absorb much of the momentum in the transition. Such *phonon-assisted* transitions are familiar from indirect gap materials such as silicon where strong phonon-assisted absorption across the indirect gap can occur, even though the electron's momentum changes dramatically (it is transferred from the phonon). With phonon-assisted *Auger* processes, the final momentum of the two electrons does not have to equal the initial momentum (but the *overall* momentum including the phonon is still conserved). In reference to Fig. 4.16, this implies that states 1, 2, and 3 are free to move closer to the band edge than allowed with momentum conservation. Thus, they become more probable and less sensitive to the bandgap and temperature, since a threshold energy (which depends on bandgap) no longer exists. However, phonon-assisted Auger processes do involve an additional particle and hence are less likely to occur overall.

In general, phonon-assisted Auger transitions become important in

---

[8] For GaAs (0.87 μm):   (CCCH, CHHS, CHHL) = (1.15, 1.22, 1.12).
  For InGaAsP (1.3 μm): (CCCH, CHHS, CHHL) = (1.12, 1.13, 1.07).
  For InGaAsP (1.55 μm): (CCCH, CHHS, CHHL) = (1.11, 1.11, 1.06).

situations where normal Auger recombination is minimal: that is, in large-bandgap materials and/or at low temperatures. For example, Auger recombination in GaAs is dominated by phonon-assisted processes, whereas InGaAsP long-wavelength materials are dominated by normal Auger processes at room temperature. However, at low temperatures below 100–150°C, normal Auger processes are suppressed, and the less temperature sensitive, phonon-assisted processes begin to dominate even in long-wavelength materials.

To quantify the *total* Auger recombination rate, one would in principle use Fermi's Golden Rule to estimate the transition rate for a given set of states. This would involve evaluating the overlap integral of the four states with a coulombic potential perturbation, in addition to evaluating the Fermi factors defined in Eq. (4.73). With Fermi's Golden Rule known for each set of four states, one would then have to sum over all possible sets of states which obey energy and momentum conservation. With four states involved in the transition, this would involve summing independently over four k-vectors. However, energy and momentum conservation constrain the sums to two independent k-vectors.

Theoretical models attempting the above procedure were first considered by Beattie and Landsberg in their pioneering 1959 paper on Auger recombination [24]. Since that time researchers have applied more refined versions of the theory to various material systems. Dutta and Nelson [25] analyzed the InGasP system, while Takeshima [26] also analyzed AlGaAs and other systems. Taylor et al. [27] as well as others have attempted to extend the theory to quantum wells and quantum-wire material. Unfortunately, the difficulty with theories of Auger recombination is that information of the band structure at more than a bandgap away from the band edge must be known accurately. Overlap intergrals of "k-space distant" Bloch functions must also be known. Such experimental information is sparse, and theories are inevitably led to making very simplifying assumptions. In contrast, the spontaneous emission rate considered in Section 4.4 can be obtained from the band edge Bloch function overlap and band edge curvatures, data which is experimentally abundant. Hence, spontaneous emission rate calculations can be quite accurate.

In general, theories can predict the Auger rate to within an order of magnitude. That is not to say they are not important. On the contrary, they remain valuable for predicting trends in the Auger recombination rate, such as the temperature and bandgap dependence. Also, the *relative* effects of material composition variations and of reduced dimensionality can be estimated theoretically. In addition, the relative importance of the three Auger processes can be determined. Most Auger theories predict that the CHHS process dominates in common III–V semiconductors, with the CCCH process almost an order of magnitude smaller (however, some theories estimate comparable magnitudes for these two processes). The CHHL process is orders of magnitude smaller than either the CHHS or CCCH process and its contributions are negligible.

Continued refinements in the theory can hopefully produce more accurate predictions. For example, Takeshima [28] has enhanced the accuracy of his Auger theory by using realistic band structures. His model extends deep into both conduction and valence bands, accounting for the nonparabolicity in all bands. In addition the band model includes the anisotropy of the band structure (the change in the band curvature along different directions of the crystal), which he considers a very important factor. His predictions include phonon-assisted Auger processes implicitly, which he finds also to be important in InGaAsP material, contrary to the results of other researchers. His predictions, in particular, appear to match well with experimentally obtained results on InGaAsP material.

In light of the above theoretical challenges, the most common method of estimating Auger recombination is to use experimentally obtained Auger coefficients in combination with the calculated or experimentally measured carrier density, assuming a recombination rate per unit volume of the form:

$$R_A = C_n N^2 P + C_p N P^2,\tag{4.85}$$

where the first term is due to the CCCH Auger process and the second term is due to the CHHS process. These dependencies are strictly valid only under nondegenerate conditions where Eq. (4.76) can be used. For degenerate carrier densities ($\geq 10^{18}$ cm^{-3}), this functionality overestimates the Auger recombination rate somewhat. However, use of Eq. (4.85) is widespread and convenient.

In laser applications with lightly doped active regions, $N = P$ at high injection levels and the Auger recombination rate simplifies to

$$R_A = CN^3 = \frac{I_{Auger}}{qV},\tag{4.86}$$

where C is a generic experimentally determined Auger coefficient which lumps together CCCH, CHHS, and phonon-assisted Auger processes. In long-wavelength InGaAsP materials, various carrier lifetime measurements that can extract the cubic dependence on carrier density place C anywhere from 10^{-29} cm^6/s to 10^{-28} cm^6/s at room temperature, depending on the particular method of measurement and the material used. Representative values at room temperature are [29, 30]:

$$C \approx 2\text{--}3 \times 10^{-29} \text{ cm}^6/\text{s}, \quad \text{(bulk 1.3 } \mu\text{m InGaAsP)}\tag{4.87}$$

$$C \approx 7\text{--}9 \times 10^{-29} \text{ cm}^6/\text{s}. \quad \text{(bulk 1.55 } \mu\text{m InGaAsP)}\tag{4.88}$$

Equation (4.87) agrees well with the predictions of Takeshima [28]. For GaAs, less data exists. Experimental data from Takeshima [31] gives an Auger coefficient that is about an order of magnitude less than in long-wavelength systems at room temperature:

$$C \approx 4\text{--}5 \times 10^{-30} \, \text{cm}^6/\text{s}. \quad \text{(bulk GaAs)} \quad (4.89)$$

For other material systems, the Auger recombination rate is not well characterized, but is expected to be similar to the values given in Eqs. (4.87) through (4.89), depending on the bandgap of the material.

In quantum wells, the band structure is converted into subbands and it is expected that the Auger rate is modified. Discussions by Smith et al. [32] and Taylor et al. [33] predict that the Auger coefficient should be reduced by $\sim \sqrt{kT/aE_g}$ in quantum-well material, where a is defined in Eq. (4.84), for $aE_g > kT$. For long-wavelength materials, this factor is on the order of 1.5–2.0. Experiments by Hausser et al. [34] indicate a reduction of about 3 in the Auger coefficient when comparing measurements of quantum-well and bulk material. Thus, reduced dimensionality appears to help with Auger recombination, but does not completely solve the problem.

Another possible method of minimizing Auger recombination involves using strain, which it has been suggested should reduce the Auger coefficient (see Problem 4.7). Strained quantum-well lasers in long-wavelength material systems in fact have approached threshold current densities comparable to larger-bandgap quantum-well lasers, suggesting that the Auger recombination is much lower in these devices.

4.6 ACTIVE MATERIALS AND THEIR CHARACTERISTICS

In this last section, we are going to run through a number of example calculations in bulk and quantum-well material in the two most popular material systems, GaAs/AlGaAs and InGaAs/InP. The examples will serve to quantitatively illustrate the various dependencies between the gain, current, and carrier density in different materials. The order of presentation will proceed as follows. We will first look at various gain spectra for different carrier densities. The peak gain and differential gain will then be determined as a function of carrier density. The spontaneous emission spectrum and its relation to carrier density will then be explored. From this, the current as a function of carrier density and ultimately the gain as a function of current density will be obtained. Experimental comparisons will then be made to estimate the accuracy of the model. Finally, we will explore the parameter space by varying the width of the quantum well, the doping, and the temperature.

4.6.1   Strained Materials and Doped Materials

Before getting started with the examples, we need to briefly consider the concept of strained QWs, because a number of the examples considered include them. Strained QWs use a material which has a different native latice constant than the surrounding barrier material. If the QWs native lattice constant is larger than the surrounding lattice constant, the QW lattice will compress in the plane, and the lattice is said to be under *compressive* strain. If the opposite is true, the QW is under *tensile* strain. In the GaAs system, adding a little indium to GaAs to form InGaAs can increase the native lattice constant, allowing for the construction of InGaAs/GaAs *compressively* strained QWs. In the InP system, the InGaAs ternary (or InGaAsP quaternary) can be either smaller or larger than the lattice constant of InP. Thus, both tensile and compressive strain can be achieved with InGaAs/InP QWs. However, in any lattice-mismatched system, it is important to realize that there is a critical thickness beyond which the strained lattice will begin to revert back to its native state, causing high densities of lattice defects. For typical applications, this critical thickness is on the order of a few hundred angstroms, thus limiting the thickness of strained active layers to a few QWs.

The effects of strain on the bandgaps are considered in detail in Appendix 11. The corresponding effects on the band curvature are discussed in the latter part of Appendix 8. Essentially strain of either type increases the curvature of the valence band structure, greatly reducing the effective mass. However, compressive strain is better at doing this. In relation to the gain, the implications are many. For example, Fig. 4.17 shows a typical band structure in the center. Due to charge neutrality under high-injection conditions, and the asymmetry in the effective masses, the quasi-Fermi levels separate more toward the conduction band, as discussed in relation to Fig. 4.6. Strain can reduce the valence band effective mass allowing the quasi-Fermi levels to separate more symmetrically, as shown to the left.

The advantage of the left plot can be seen in the following. Concentrating

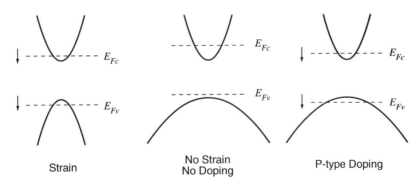

FIGURE 4.17   Illustration of how the quasi-Fermi levels are affected by strain and *p*-type modulation doping.

on the conduction band, it is clear that *for a given quasi-Fermi level separation*, the electron density is always lower in the left plot than in the center plot. Thus, at transparency when the quasi-Fermi levels are separated by the bandgap in both cases, the left plot inevitably has a lower carrier density. In other words, the transparency carrier density, N_{tr}, can be reduced substantially in strained materials. In addition, since the current goes roughly as N^2, the transparency current, J_{tr}, can also be reduced. In practical structures, the reduction can be as much as a factor of 2!

An additional advantage of the arrangement on the left in Fig. 4.17 is that the differential gain is also enhanced. So not only does the material reach transparency faster, but the gain also increases faster with carrier density. To understand this, it is important to realize that the differential gain, dg/dN, depends on how quickly the *band edge carrier density* changes in response to movements in the quasi-Fermi levels. Because the slope of the Fermi occupation probability function with energy is maximized at the location of the quasi-Fermi level, the band edge carrier density will be affected most when the quasi-Fermi level is aligned with the band edge. Thus, for increased differential gain, it is critical to bring *both* quasi-Fermi levels as close to the band edges as possible. The left plot clearly accomplishes this task (particularly near transparency). The improvement in differential gain in strained materials can be as much as a factor of 2.

Another way to improve the differential gain is by *p*-type doping of the active region. As shown in the plot on the right in Fig. 4.17, the addition of doping can pull the quasi-Fermi levels down to a more symmetrical position just as strain does. Similarly, because of the alignment of the quasi-Fermi levels with the band edges, the differential gain can again be improved. However, in this case, while the *electron* density can be reduced at transparency for similar reasons as with the strained example, the *hole* density is increased. Now because the electron density is degenerate while the hole density is nondegenerate, the downward shift in quasi-Fermi levels increases the hole density faster than it decreases the electron density (P increases approximately exponentially while N decreases approximately linearly). Thus, the NP product actually increases in this case, resulting in a higher transparency current density. So unlike strained materials, where both transparency and differential gain are improved, *p*-type doping increases the differential gain at the expense of increased transparency current densities. Later we will examine this relationship more quantitatively.

4.6.2 Gain Spectra of Common Active Materials

The following gain spectra were calculated using Eqs. (4.37) and (4.38) with the reduced density of states functions derived in Appendix 8. Lineshape broadening which smooths the spectral features has been included using Eq. (4.39) with Eq. (4.41) as the lineshape function. The two quasi-Fermi levels defining the gain have been connected by assuming charge neutrality in the

QW region including filling of barrier states directly above the QW states. The carrier density therefore refers to all carriers within the QW region. Also for the QWs, the polarization is assumed to be in the plane of the well (TE polarization), since the matrix element is much larger for this polarization. And unless otherwise specified, all active materials are undoped.

Figure 4.18 shows gain spectra for three common active materials in the GaAs system: (1) bulk GaAs, (2) an unstrained $GaAs/Al_{0.2}Ga_{0.8}As$ 80Å QW, and (3) a compressively strained $In_{0.2}Ga_{0.8}As/GaAs$ 80Å QW. Figure 4.19 shows gain spectra for three common active materials in the InP system: (1) lattice-matched bulk $In_{0.53}Ga_{0.47}As$, (2) an unstrained $In_{0.53}Ga_{0.47}As/(Q1.08)$ 60Å QW, and (3) a compressively strained $In_{0.68}Ga_{0.32}As/(Q1.08)$ 30Å QW. The strained QW has a $+1\%$ lattice mismatch to InP. Both QWs have bandgap wavelengths of 1.5 μm. The (Q1.08) notation implies quaternary InGaAsP with 1.08 μm bandgap wavelength.

The bulk gain spectra in both Figs. 4.18 and 4.19 are much smoother than the staircase QW gain spectra. However, the QW gain spectra tend to provide higher gain near the band edge. In comparing the QW gain spectra, the unstrained QWs have a higher maximum gain, g_{max}, due to the larger density of states in the valence band available to participate in transitions. However, it takes more carriers to reach near complete inversion in the unstrained QWs. Other features in the QW gain spectra include various subband transitions. These features are to be compared with the idealized version in Fig. 4.9. At shorter wavelengths in the GaAs based QWs, a bulk contribution from the 100Å AlGaAs or GaAs barrier layers on either side of the QW appears. In fact, the barriers also become inverted for very high carrier densities, particularly in the InGaAs/GaAs QW. In the InP based QWs, the bulk contributions are not within the wavelength range shown. The sheet carrier density (NL_z) required to achieve gain in all cases appears to be slightly less than 2×10^{12} cm^{-2}.

4.6.3 Gain vs. Carrier Density

To get a better feel for how the gain varies with carrier density, the peak of the gain spectrum is plotted as a function of carrier density in Fig. 4.20. The upper plot includes the GaAs based active materials. Generally, N_{tr} is clustered near 2×10^{12} cm^{-2}. However, the strained QW has the lowest N_{tr} and increases the fastest with N, as expected from earlier discussions in Section 4.6.1. Curiously, the bulk GaAs and the GaAs QW have similar characteristics. In other words, it would appear that an 80Å bulk GaAs layer would behave similarly to an 80Å GaAs QW, and that the major improvement to be gained from QWs is simply a volume effect rather than a "quantum size" effect.

The lower plot in Fig. 4.20 includes the InP based active materials with the addition of two strained QWs: (1) a -0.37% tensile strained $In_{0.48}Ga_{0.52}As/$ (Q1.08) 120Å QW, and (2) a -1% tensile strained $In_{0.38}Ga_{0.62}As/(Q1.08)$ 150Å QW, both with bandgap wavelengths close to 1.5 μm. For these QWs, the C–LH transition (which provides higher gain for polarizations perpendicular

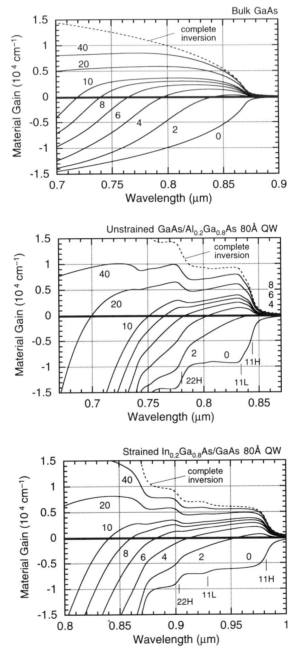

FIGURE 4.18 TE gain spectrum vs. carrier density in GaAs based materials. Indicated values are the sheet carrier densities: $\times 10^{12}$ cm^{-2} (the bulk "sheet" density assumes an 80Å width).

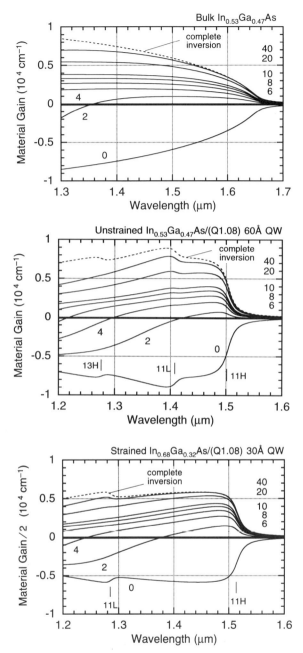

FIGURE 4.19 TE gain spectrum vs. carrier density in InP based materials. Indicated values are the sheet carrier densities: $\times 10^{12}$ cm^{-2} (the bulk "sheet" density assumes a 60Å width). The material gain in the 30Å QW is divided by 2 to account for the smaller optical confinement in comparison to the 60Å QW.

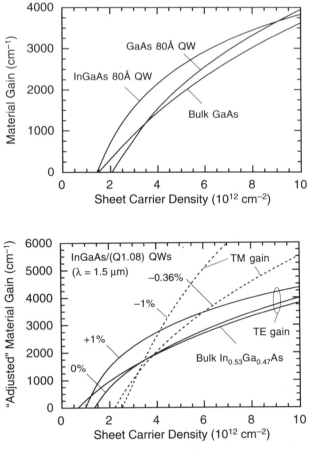

FIGURE 4.20   Peak TE gain vs. sheet carrier density in GaAs (upper) and InP (lower) based materials (bulk "sheet" densities assume an 80Å width for GaAs and a 60Å width for In$_{0.53}$Ga$_{0.47}$As). In the lower plot, the strained QWs have 1.5 μm bandgaps where: 1% = (30Å, x_{ind} = 0.68), 0% = (60Å, x_{ind} = 0.53), −0.37% = (120Å, x_{ind} = 0.48), and −1% = (150Å, x_{ind} = 0.38). Also in the lower plot, the dashed curves are TM gain and the "adjusted" material gain is the material gain multiplied by $L_z/60$Å to account for the difference in confinement for QWs of different width.

to the well plane—TM polarization) is dominant, and hence for these two cases, the peak TM gain is plotted. In this system, N_{tr} for the unstrained and compressive materials is clustered closer to 1×10^{12} cm^{-2}. Again, the +1% strained QW increases the quickest with carrier density. However, it saturates faster than the tensile strained QWs. Bulk In$_{0.53}$Ga$_{0.47}$As has the lowest N_{tr} and remains comparable to the unstrained QW performance. The two tensile strained QWs have a higher N_{tr}, but do not saturate nearly as quickly, allowing for lower threshold carrier densities at high threshold gains.

TABLE 4.4 Three- and Two-Parameter Gain vs. Carrier Density Curve Fits.

Active Material	$g = g_0 \ln\left[\dfrac{N + N_s}{N_{tr} + N_s}\right]$			$g = g_0 \ln[N/N_{tr}]$	
	N_{tr}	N_s	g_0	N_{tr}	g_0
Bulk GaAs	1.85	6	4200	1.85	1500
GaAs/Al$_{0.2}$Ga$_{0.8}$As 80Å QW	2.6	1.1	3000	2.6	2400
In$_{0.2}$Ga$_{0.8}$As/GaAs 80Å QW	1.8	−0.4	1800	1.8	2100
Bulk In$_{0.53}$Ga$_{0.47}$As	1.1	5	3000	1.1	1000
InGaAs 30Å QW (+1%)	3.3	−0.8	3400	3.3	4000
InGaAs 60Å QW (0%)	2.2	1.3	2400	2.2	1800
InGaAs 120Å QW (−0.37%)	1.85	0.6	2100	1.85	1800
InGaAs 150Å QW (−1%)	1.7	0.6	2900	1.7	2300

Inverse Relation	*Differential Gain*
$N = (N_{tr} + N_s)e^{g/g_0} - N_s$	$\dfrac{dg}{dN} = \dfrac{g_0}{N + N_s}$

$[N] = 10^{18}$ cm^{-3}, $[g] = $ cm^{-1}.

Curve fits have been applied to all gain curves in Fig. 4.20. The results appear in Table 4.4. The two-parameter (N_{tr}, g_0) logarithmic functionality works reasonably well. However, for more linear gain curves (such as the bulk gain curves), the fit is not so good. To correct for this, a third linearity parameter, N_s, has been added. For this three-parameter model, the logarithmic functionality converts to linear functionality as $N_s \to \infty$ and $g_0/N_s \to$ constant. Thus, the larger N_s is in the table, the more linear the curve is, and the worse the two-parameter fit. In general, the strained materials follow much closer to a pure logarithmic functionality. Both two- and the more accurate three-parameter fits appear in the table. All carrier densities in the table have been converted to volume carrier densities, and all material gains are for each particular well width ("adjusted" material gains as used in Fig. 4.20 are not used here).

The differential gain, dg/dN, is an important parameter in high-speed laser applications. This is primarily because the relaxation resonance frequency of the laser depends on the square root of the differential gain as we learned in Chapter 2 and will discuss more fully in Chapter 5. Thus, higher differential gains can ideally improve the modulation response of the laser (in practice, however, other damping factors are just as important in determining the overall modulation bandwidth). Figure 4.21 plots dg/dN for the gain curves in Fig. 4.20, dramatically revealing the importance of working close to transparency when designing high-speed lasers. In general, numbers in the mid-10^{-16} cm^2 are expected and this is what is generally observed in QW lasers (however, there can be a large variation in the measured differential gain depending on the threshold carrier density of the laser). In both material systems, the strained QWs yield a $\sim 50\%$ improvement over unstrained QWs. The bulk values in both cases are the lowest, coming in at below 5×10^{-16} cm^2. In the InP system, dg/dN is highest at both extremes of strain and dips by about one-third in between these extremes. The secondary peak on the GaAs QW dg/dN curve corresponds to the peak gain switching to the C–LH (11) transition which rides on top of the C–HH (11) transition at slightly shorter wavelengths (see Fig. 4.18).

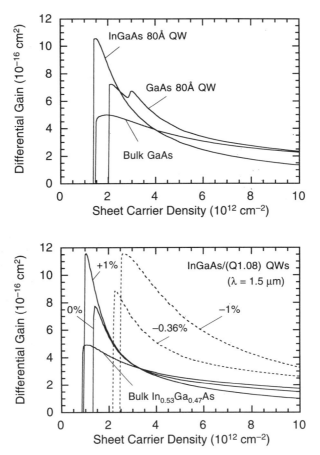

FIGURE 4.21   Peak differential gain, dg/dN, vs. sheet carrier density in GaAs (upper) and InP (lower) based materials (bulk "sheet" densities assume an 80Å width for GaAs and a 60Å width for In$_{0.53}$Ga$_{0.47}$As). The compositions of the strained QWs in the lower plot can be found in the caption of Fig. 4.20.

4.6.4   Spontaneous Emission Spectra and Current vs. Carrier Density

Having determined the gain as a function of carrier density, we now need to understand how the spontaneous emission spectrum which generates radiative recombination varies with carrier density. The spontaneous emission spectrum is found using Eqs. (4.57) and (4.58). An example spectrum is shown in Fig. 4.22 for a strained InGaAs/GaAs QW for different sheet carrier densities. The staircase spectrum can be observed and the saturation of the $n = 1$ transition at very high carrier densities is noticeable. However, the second step is much smaller than the first step. The reason is that the second electron state in the QW is barely confined, leading to a poor overlap between the C2 state and the strongly confined HH2 state (this is also visible in the gain spectrum in Fig. 4.18). The result is a smaller matrix element and hence smaller step height.

The third step indicated by the dashed curves in the spectrum of Fig. 4.22 is actually spontaneous emission occurring in the GaAs barrier layers. Such barrier contributions to the radiative current density are particularly large in this active material. In all other QWs considered, barrier recombination is minimal. The reason is that in this material, the QW valence band states allow E_{Fv} to penetrate deeper than normal due to the strained light band structure. As a result, both N and P in the barrier regions can become large, as opposed to other QWs where P in the barriers usually remains very small. The NP product in the barriers can therefore become large leading to a high recombination there. The barrier heights are also relatively small, allowing the barrier regions to populate quickly as carrier density increases. The barrier recombination could be reduced by increasing the GaAs barriers to AlGaAs,

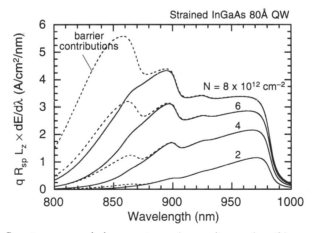

FIGURE 4.22 Spontaneous emission spectrum (per unit wavelength) vs. sheet carrier density in a strained $In_{0.2}Ga_{0.8}As/GaAs$ 80Å QW. The dashed curves include spontaneous emission from the 100Å thick GaAs barrier layers on either side of the well (total of 200Å). The emission is normalized such that peak × bandwidth (in nm) yields the spontaneous current density in A/cm².

however, materials growth issues of InGaAs grown directly on AlGaAs make InGaAs/AlGaAs QWs difficult to grow.

The area under the spontaneous emission spectrum in Fig. 4.22 represents the *total* spontaneous emission rate. For example at 4×10^{12} cm^{-2}, the peak is ~ 2 and the bandwidth is ~ 100 nm. The (peak) × (bandwidth) therefore yields a spontaneous current density of ~ 200 A/cm^2 based on the units of the vertical axis. Performing a more exact integration of the area under the curve via Eq. (4.59), we can determine the spontaneous current density as a function of carrier density. This relationship is plotted in Fig. 4.23 for the 80Å GaAs based materials and the 60Å InP based materials.

For low carrier densities in Fig. 4.23, bulk GaAs roughly follows a BN^2 law

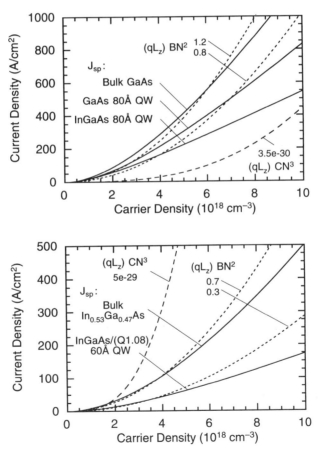

FIGURE 4.23   Spontaneous recombination current density vs. carrier density in GaAs (upper) and InP (lower) based materials (bulk currents per unit area assume an 80Å width for GaAs and a 60Å width for In$_{0.53}$Ga$_{0.47}$As). Indicated values for B have units of 10^{-10} cm^3/s, and values for C have units of cm^6/s.

with $B = 1.2 \times 10^{-10}$ cm^3/s. For bulk $In_{0.53}Ga_{0.47}As$, B is closer to 0.7×10^{-10} cm^3/s. These values are in good agreement with measured values for these two materials. The smaller coefficient in the latter case is the result of a lower optical mode density at longer wavelengths (less modes, less spontaneous emission). For higher carrier densities, the BN^2 law overestimates the recombination rate. This is because the square law is strictly valid only within the Boltzmann approximation. As the injection level becomes highly degenerate, the recombination rate does not increase as rapidly as N^2. To account for this, the bimolecular recombination rate is occasionally written as $B_0 N^2 - B_1 N$. This trend is observed in all cases plotted in Fig. 4.23. Interestingly, the QWs have lower B coefficients than their bulk counterparts.

The Auger recombination current is also plotted in Fig. 4.23 assuming typical values for the two material systems. In GaAs, Auger recombination compared to radiative recombination is small for normal carrier densities ($2-4 \times 10^{18}$ cm^{-3}). Thus, the radiative efficiency in GaAs based materials is quite high and can approach unity at low carrier densities. In the InP system, the Auger current dominates for $N > 2 \times 10^{18}$ cm^{-3}. Thus, in most InP based lasers, the radiative efficiency is below 50% and can often be much lower than this. In this system, where the optical mode density is low enough to yield very low radiative currents, it is unfortunate that another mechanism such as Auger recombination has to turn on and ruin everything.

4.6.5 Gain vs. Current Density

With the current known, we can now proceed to determine the peak gain as a function of current density as shown in Fig. 4.24. The upper plot shows the GaAs based materials with and without the current contributions from Auger recombination and barrier recombination (assuming $C = 3.5 \times 10^{-30}$ cm^6/s). These latter currents can degrade the active material performance for material gains larger than 2000 cm^{-1}. Plotted vs. current density, it is clear that the InGaAs/GaAs QW provides much better gain performance, particularly at currents below 300–400 A/cm^2. The transparency current density for the strained QW is near 50 A/cm^2, in good agreement with experimentally measured values. The unstrained GaAs QW has a transparency current density slightly higher than 100 A/cm^2. And in general for a given gain, the strained QW requires roughly half the current required by the unstrained QW! Thus, strained InGaAs QWs are the active region of choice in GaAs based materials (as long as the application can tolerate the 0.98 μm wavelength). In comparing bulk GaAs to the GaAs QW, we see somewhat better performance out of the QW, however, the "quantum size" effects are not that significant. In fact, the bulk GaAs 80Å well has a lower transparency current density.

In the InP system, only the radiative currents are included in Fig. 4.24. The trends are generally similar to those seen with the gain vs. carrier density. However in this case, the unstrained QW clearly outperforms the bulk $In_{0.53}Ga_{0.47}As$. For the compressively strained and unstrained materials,

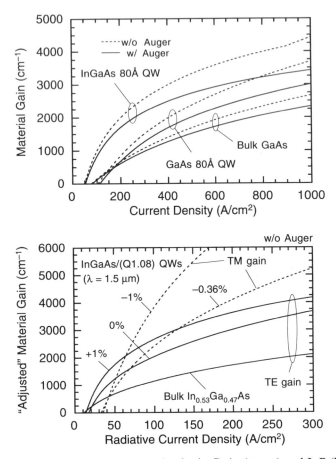

FIGURE 4.24 Peak TE gain vs. current density in GaAs (upper) and InP (lower) based materials (bulk currents per unit area assume an 80Å width for GaAs and a 60Å width for $In_{0.53}Ga_{0.47}As$). The effects of Auger and barrier recombination are included in the upper plot only. Other comments regarding the lower plot can be found in the caption of Fig. 4.20.

transparency radiative current densities are near 10–15 A/cm^2. For the tensile strained QWs, these values are closer to 30–35 A/cm^2. Unfortunately, Auger recombination prevents such low threshold current densities from being realized in lasers.

Curve fits analogous to the gain vs. carrier density fits are listed in Table 4.5. For GaAs based active materials, curve fits are provided for both radiative (spontaneous) current as well as the current including barrier and Auger recombination (assuming 100Å barrier widths on either side of the QW and $C = 3.5 \times 10^{-30}$ cm^6/s). For InP based active materials, only radiative current is fitted (the Auger current can be included by using the carrier density curve

TABLE 4.5   Three- and Two-Parameter Gain vs. Current Density Curve Fits.

Active Material	$g = g_0 \ln\left[\dfrac{J + J_s}{J_{tr} + J_s}\right]$			$g = g_0 \ln[J/J_{tr}]$	
	J_{tr}	J_s	g_0	J_{tr}	g_0
$J_{sp} + J_{bar} + J_{Aug}$					
Bulk GaAs	80	140	1400	80	700
GaAs/Al$_{0.2}$Ga$_{0.8}$As 80Å QW	110	50	1600	110	1300
In$_{0.2}$Ga$_{0.8}$As/GaAs 80Å QW	50	-10	1100	50	1200
J_{sp}					
Bulk GaAs	75	200	1800	75	800
GaAs/Al$_{0.2}$Ga$_{0.8}$As 80Å QW	105	70	2000	105	1500
In$_{0.2}$Ga$_{0.8}$As/GaAs 80Å QW	50	0	1440	50	1440
J_{sp}					
Bulk In$_{0.53}$Ga$_{0.47}$As	11	30	1000	11	500
InGaAs 30Å QW (+1%)	13	2	2800	13	2600
InGaAs 60Å QW (0%)	17	11	1500	17	1200
InGaAs 120Å QW (-0.37%)	32	18	1400	32	1100
InGaAs 150Å QW (-1%)	35	10	1700	35	1500

Inverse Relation	*Differential Gain*
$J = (J_{tr} + J_s)e^{g/g_0} - J_s$	$\dfrac{dg}{dJ}L_z = \dfrac{g_0}{J + J_s}L_z$

$[J] = $ A/cm^2, $[g] = $ cm^{-1}

fits along with an appropriate value for C). Again, all gains are not "adjusted" material gains, but actual material gains that can be used directly in optical mode overlap integrals to determine the modal gain.

In Chapter 2, the current was empirically related to the carrier density through a polynomial fit (i.e., $J \propto AN + BN^2 + CN^3$). Combining Tables 4.4 and 4.5, we have an alternative description of this relationship:

$$J = (J_{tr} + J_s)\left[\frac{N + N_s}{N_{tr} + N_s}\right]^{g_{0N}/g_{0J}} - J_s. \tag{4.90}$$

Here g_{0N} and g_{0J} refer to the fitting parameter g_0 used with N and J, respectively. While this is a more complex six-parameter fit, it does also provide the gain vs. carrier and current density relations. The fit is most useful in converting theoretical gain calculations into practical working models. We will make use of it in the next chapter. It should however be noted that in using Eq. (4.90) below transparency, it is better to use the two-parameter curve fits (i.e., $N_s = J_s = 0$) in Tables 4.4 and 4.5 to insure that J goes to zero when N does. Note also that in the two-parameter limit, $J = J_{tr}[N/N_{tr}]^{g_{0N}/g_{0J}}$, and hence the power relation is governed by the ratio g_{0N}/g_{0J}. Thus for a simple BN^2 relation we would expect g_{0N} to be twice as large as g_{0J}. In comparing the two-parameter fits in Tables 4.4 and 4.5, we see that this is true to an extent.

4.6.6   Experimental Gain Curves

With the gain vs. current density known, we can examine how well the theory works by comparing it to measurements made on in-plane lasers. By measuring the threshold and differential efficiency variations with cavity length, it is possible to extract the internal loss, the internal injection efficiency, and the gain as a function of injected current density. Results of this type of measurement are shown in Fig. 4.25.

The theoretical curves in Fig. 4.25 are found to match quite well in both material systems. The value for C which best matches the InP data also agrees well with other measured values (Eqs. (4.87) and (4.88)). While it is true that in the InP system, the choice of C makes the theory more of a curve-fitting procedure, the theory for the InGaAs/GaAs QW involves very few fitting parameters, and hence represents very good agreement with experiment within the uncertainties involved with the Auger and barrier recombination rates. Figure 2.14 also reveals excellent agreement between the InGaAs/GaAs QW theory and experiment.

4.6.7   Dependence on Well Width, Doping, and Temperature

In addition to picking the best active material, there are other choices to make in designing a laser. For a QW active region, an obvious decision is the well width. Aside from having a preference for a particular lasing wavelength, it is useful to have some guidelines in deciding on the optimum well width. Because each subband of a QW is inherently two-dimensional in nature, we should expect that the performance of the $n = 1$ transition has little dependence on the width of the well. However, in practice it is the spacing between subbands which interferes and causes problems.

For example, the current required to reach transparency and 30 cm^{-1} modal gain in an $In_{0.2}Ga_{0.8}As/GaAs$ QW is plotted as a function of well width on the left in Fig. 4.26. For midrange values between 50–100Å, there is indeed little change in the QW performance. However, at wider well widths the subband spacing becomes smaller (see inset). As a result, other subbands become populated at typical injection levels, degrading the QW performance. In fact, the current starts to become linear with well width as you would expect in bulk material. At narrower well widths, the quantized state is squeezed to the top of the QW, very close to the barrier states (see inset). As a result, the barriers begin to populate significantly at typical injection levels, and again the QW performance is compromised. Thus, there is a window of optimum

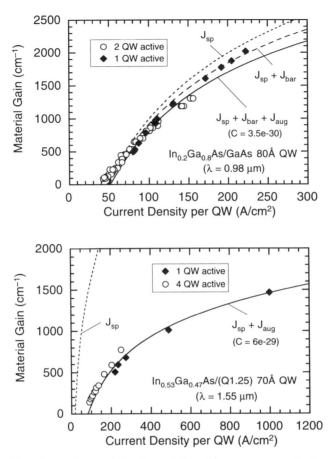

FIGURE 4.25 Experimental material gain vs. injected current curves for $In_{0.2}Ga_{0.8}As/$ GaAs 1 and 2 80Å QW active region 0.98 μm lasers [21] (upper), and $In_{0.53}Ga_{0.47}As/$ (Q1.25) 1 and 4 70Å QW active region 1.55 μm lasers [35] (lower). Theoretical curves are superimposed on the plots. The gain in the lower plot is well represented by the two-parameter expression, $g = 583 \ln(J/81)$ cm^{-1}, where J is given in A/cm^2.

performance, however, for the current density, the window is fairly broad. For the differential gain on the other hand, the functionality is more sharply peaked near 60Å. Away from this peak, the wider well widths reduce the subband spacing, while the narrower widths bring the QW state too close to the barrier states.

Another design consideration involves the doping of the active region. Now doping within the active material itself can alter the band structure near the band edges, degrading the performance. However, by placing the doping in the barrier regions (modulation doping), the charge neutrality condition can be

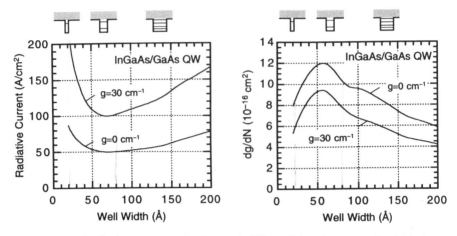

FIGURE 4.26 Radiative current density and differential gain vs. well width in an $In_{0.2}Ga_{0.8}As/GaAs$ QW (the GaAs barrier on either side of the QW is 100Å thick). The curves are for the values at transparency and 30 cm^{-1} *modal* gain (assuming an optical mode width of 2500Å as given by some SCH structure).

altered. In this case, the quasi-Fermi levels are connected through the following equation:

$$N + N_A^- = P + N_D^+, \qquad (4.91)$$

where in quantum wells, the acceptor and donor densities are most easily defined as sheet densities.

The basic effects of p-type doping were discussed in Section 4.6.1. Figure 4.27 provides a more quantitative description by plotting the gain and

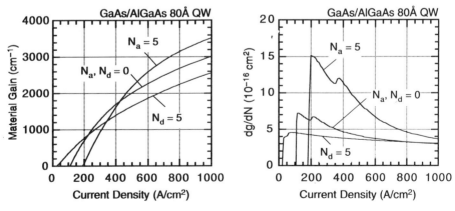

FIGURE 4.27 Peak TE gain and differential gain vs. current density for three different doping levels in a $GaAs/Al_{0.2}Ga_{0.8}As$ 80Å QW. The donor and acceptor sheet densities are in units of 10^{12} cm^{-2}.

differential gain vs. current density for a GaAs/AlGaAs 80Å QW for *n*-type, *p*-type, and undoped material. The downward shift of the quasi-Fermi levels for *p*-type doping increases the *NP* product (since *N* is degenerate and *P* is nondegenerate at transparency), while the upward shift of the quasi-Fermi levels for *n*-type doping reduces the *NP* product. Thus, J_{tr} is increased for *p*-type doping and is reduced for *n*-type doping. However, due to the alignment of the quasi-Fermi levels with respect to the band edges, *p*-type doping increases the differential gain dramatically, while *n*-type doping reduces the differential gain. Thus, depending on the application (ultrahigh-speed laser, or ultralow threshold current), *p*-type and *n*-type doping can add interesting variations to the gain curve. Although neither is as interesting as strain, which can simultaneously reduce J_{tr} and increase dg/dN.

Another parameter which affects the active material performance is an external parameter, the operating temperature. Figure 4.28 illustrates the effects temperature has on the gain vs. current density in both unstrained GaAs QWs and unstrained InGaAs/InP QWs including Auger recombination. In general, the current required to reach a given gain increases with increasing temperature. The dominant cause of this is the broadening of the Fermi occupation probability function which spreads the carriers over a larger energy range for a given overall carrier density. The result is a lower spectral concentration of inverted carriers, which leads to a broadening and flattening of the gain spectrum.

In in-plane semiconductor lasers, the threshold current generally increases exponentially with temperature. As discussed in Chapter 2, this dependence is commonly characterized by

$$I_{th} \sim e^{T/T_0} \rightarrow T_0 = \frac{T_2 - T_1}{\ln(I_2/I_1)}, \tag{4.92}$$

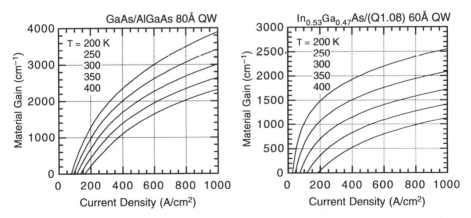

FIGURE 4.28 Peak TE gain vs. current density for different temperatures in a GaAs/Al$_{0.2}$Ga$_{0.8}$As 80Å QW (left) and an In$_{0.53}$Ga$_{0.47}$As/(Q1.08) 60Å QW (right, with $C = 5 \times 10^{-29}$ cm^6/s).

where T_0 is a parameter which characterizes the thermal behavior of the threshold current of the laser. The second equation allows one to estimate T_0 from two threshold measurements at two different temperatures. By comparing the two sets of gain curves in Fig. 4.28, it is evident that InP based materials are affected by temperature more than GaAs based materials. Measurements yield a value of $T_0 \sim 50–100$ K for InP based in-plane lasers and a value of $T_0 \sim 100–150$ K for GaAs based in-plane lasers. These values are smaller than the theoretical curves in Fig. 4.28 would suggest. However, there are other temperature-dependent effects. For example, the Auger coefficient itself is temperature-dependent, and as the latter part of Appendix 2 reveals, carrier leakage out of the active region is also very sensitive to temperature.

REFERENCES

[1] C. Hermann and C. Weisbuch, *Phys. Rev. B*, **15**, 823 (1977).

[2] C. Hermann and C. Weisbuch, *Modern Problems in Condensed Matter Sciences Volume 8*; *Optical Orientation*, ed. V.M. Agranovich and A.A. Maradudin, North-Holland Co., pp. 463–508 (1984).

[3] B. Jani, P. Gilbart, J.C. Portal, and R.L. Aulombard, *J. Appl. Phys.*, **58**, 3481 (1985).

[4] R.J. Nicholas, J.C. Portal, C. Houlbert, P. Perrier, and T.P. Pearsall, *Appl. Phys. Lett.*, **34**, 492 (1979).

[5] H.C. Casey, Jr. and M.B. Panish, *Heterostructure Lasers, Part A: Fundamental Principles*, Academic Press, Orlando (1978). For bandgap shift, also see J. Camassel, D. Auvergne, and H. Mathieu, *J. Appl. Phys.*, **46**, 2683 (1975).

[6] M. Yamada, H. Ishiguro, and H. Nagato, *Jap. J. Appl. Phys.*, **19**, 135 (1980).

[7] M. Yamanishi and Y. Lee, *IEEE J. Quantum Electron.*, **QE-23**, 367 (1987).

[8] M. Asada, Intraband relaxation effect on optical spectra, in *Quantum Well Lasers*, ed. P.S. Zory, Jr., Ch. 2, Academic Press, San Diego (1993).

[9] A.I. Kucharska and D.J. Robbins, *IEEE J. Quantum Electron.*, **QE-26**, 443 (1990).

[10] S.R. Chinn, P. Zory, and A.R. Reisinger, *IEEE J. Quantum Electron.*, **QE-24**, 2191 (1988).

[11] S. Tarucha, H. Kobayashi, Y. Horikoshi, and H. Okamoto, *Jap. J. Appl. Phys.*, **23**, 874 (1984). Also see A. Tomita and A. Suzuki, *IEEE J. Quantum Electron.* **QE-23**, 1155 (1987).

[12] D.A. Kleinman and R.C. Miller, *Phys. Rev. B*, **32**, 2266 (1985).

[13] T. Baba, T. Hamano, F. Koyama, and K. Iga, *IEEE J. Quantum Electron.* **QE-27**, 1347 (1991); *IEEE J. Quantum Electron.*, **QE-28**, 1310 (1992).

[14] S.D. Brorson, H. Yokoyama, and E.P. Ippen, *IEEE J. Quantum Electron.*, **QE-26**, 1492 (1990).

[15] W. Shockley and W.T. Read, Jr., *Phys. Rev.*, **87**, 835 (1952).

[16] R.N. Hall, *Phys. Rev.*, **87**, 387 (1952).

[17] W.T. Tsang, *Appl. Phys. Lett.*, **36**, 11 (1980).

[18] T. Otaredian, *Solid-State Electron.*, **36**, 905 (1993).

[19] V. Swaminathan, J.M. Freund, L.M.F. Chirovsky, T.D. Harris, N.A. Kuebler, and L.A. D'Asaro, *J. Appl. Phys.*, **68**, 4116 (1990).

[20] L. Jastrzebski, J. Lagowski, and H.C. Gatos, *Appl. Phys. Lett.*, **27**, 537 (1975).

[21] S.Y. Hu, S.W. Corzine, K.K. Law, D.B. Young, A.C. Gossard, L.A. Coldren, and J.L. Merz, *J. Appl. Phys.*, **76**, 4479 (1994).

[22] D.D. Nolte, *Solid-State Electron.*, **33**, 295 (1990).

[23] E. Yablonivitch, C.J. Sandroff, R. Bhat, and T. Gmitter, *Appl. Phys. Lett.*, **51**, 439 (1987).

[24] A.R. Beattie and P.T. Landsberg, *Proc. R. Soc. London*, **249**, 16 (1959).

[25] N.K. Dutta and R.J. Nelson, *J. Appl. Phys.*, **53**, 74 (1982).

[26] M. Takeshima, *J. Appl. Phys.*, **43**, 4114 (1972).

[27] R.I. Taylor, R.A. Abram, M.G. Burt, and C. Smith, *Semicond. Sci. Technol.*, **5**, 90 (1990).

[28] M. Takeshima, *Phys. Rev. B*, **29**, 1993 (1984).

[29] B. Sermage, H.J. Eichler, and J.P. Heritage, *Appl. Phys. Lett.*, **42**, 259 (1983).

[30] E. Wintner and E.P. Ippen, *Appl. Phys. Lett.*, **44**, 999 (1984).

[31] M. Takeshima, *J. Appl. Phys.*, **58**, 3846 (1985).

[32] C. Smith, R.A. Abram, and M.G. Burt, *Electron. Lett.*, **20**, 893 (1984).

[33] R.I. Taylor, R.W. Taylor, and R.A. Abram, *Surface Sci.*, **174**, 169 (1986).

[34] S. Hausser, G. Fuchs, A. Hangleiter, and K. Streubel, *Appl. Phys. Lett.*, **56**, 913 (1990).

[35] M.E. Heimbuch, A.L. Holmes, Jr., C.M. Reaves, M.P. Mack, S.P. DenBaars, and L.A. Coldren, *J. Electron. Mat.*, **23**, 87 (1994).

READING LIST

G.P. Agrawal and N.K. Dutta, *Semiconductor Lasers*, 2d ed. Ch. 3, Van Nostrand Reinhold, New York (1993).

S.W. Corzine, R.-H. Yan, and L.A. Coldren, Optical gain in III–V bulk and quantum well semiconductors, in *Quantum Well Lasers*, ed. P. S. Zory, Jr., Ch. 1, Academic Press, San Diego (1993).

K.J. Ebeling, *Integrated Opto-electronics*, Ch. 8, Springer-Verlag, Berlin (1993).

PROBLEMS

4.1 The reduced density of states function determines the total density of state pairs available as a function of photon energy, and is therefore critical in determining the shape of the gain spectrum. Thus, it is worthwhile getting a feel for this function:

(a) Plot $\rho_r(E)$ for bulk GaAs from the GaAs bandgap energy to the

$Al_{0.2}Ga_{0.8}As$ bandgap energy. Be sure to include both C–HH and C–LH transitions.

(b) Plot $\rho_r(E)$ for a fictitious 100Å GaAs QW with infinitely high barriers, assuming parabolic subbands and including all *allowed* C–HH and C–LH subband pairs between the GaAs and $Al_{0.2}Ga_{0.8}As$ bulk bandgaps. Overlay the bulk GaAs $\rho_r(E)$ on top of this curve. Discuss any correspondence you observe between these curves.

(c) Now plot the QW $\rho_r(E)$ in (b) for C–HH transitions *only* and overlay the C–HH component of the bulk GaAs $\rho_r(E)$ on top of this curve. Make a similar comparison for C–LH transitions. Discuss any correspondence you observe between these curves.

(d) Mathematically derive the reasons why the comparisons in (c) work the way they do. *Hint*: Begin by expressing the step edges of the QW $\rho_r(E)$ in terms of the quantum numbers, and then (after some substitutions) compare with the bulk $\rho_r(E)$).

(e) Now plot a realistic QW $\rho_r(E)$ assuming a 100Å GaAs QW with $Al_{0.2}Ga_{0.8}As$ barriers (assume that 60% of the band discontinuity occurs in the conduction band). In your plot, assume parabolic subbands and include all allowed C–HH and C–LH transitions between the GaAs and $Al_{0.2}Ga_{0.8}As$ bulk bandgaps. Also, add to this plot the contributions to $\rho_r(E)$ from the $Al_{0.2}Ga_{0.8}As$ bulk barriers by extending the energy range another 50%. Overlay the bulk GaAs $\rho_r(E)$ on top of this curve. In comparing the bulk GaAs to the GaAs QW, do you qualitatively find much difference in the *total* density of state pairs existing within the energy range plotted?

4.2 The measured absorption curve of a GaAs/$Al_{0.2}Ga_{0.8}As$ 100Å QW (with excitonic effects removed) reveals that the C1–HH1 absorption step has a magnitude of 10^4 cm^{-1} for light with polarization such that the electric field lies in the plane of the QW (TE polarization).

(a) Determine the differential lifetime, τ_{sp}^{21}, for this transition.

(b) With the TE absorption known, we can estimate the TE gain spectrum for different pumping levels. Plot the expected gain spectrum from the QW band edge to the bulk $Al_{0.2}Ga_{0.8}As$ bandgap energy, for two different injection levels: (1) $E_{Fc} \gg E_{c1}$ (i.e., $E_{Fc} \rightarrow \infty$) and $E_{Fv} = E_{hh1}$, and (2) $E_{Fc} \gg E_{c1}$ and $E_{Fv} = E_{hh2}$. Neglect forbidden transitions, C–LH transitions, and bulk barrier transitions in your plot. Also assume that the density of state pairs and the overlap integral is the same for all subband transition pairs. Further assume that the transition matrix element is not a function of energy. Neglect lineshape broadening in your calculation.

(c) What is the *maximum* possible spontaneous emission rate per unit energy at the C1–HH1 subband edge of this QW, assuming the rate is the same for light polarized in any direction?

(d) Considering only C–HH subband transition pairs, what are the radiative currents required to support the two gain spectra plotted in (b), assuming a single QW active region with lateral dimensions of 2 μm × 200 μm? In your estimation, assume that the optical mode density and g_{max} are constants evaluated at the C1–HH1 subband edge. Also assume the emission rate is the same for light polarized in any direction. Be sure to include spontaneous transitions over *all* energies by letting $E_{Fc} \to \infty$.

(e) Can you explain why the quasi-Fermi level positions given in (b) are unrealistic?

4.3 The previous problem was simplified by allowing $E_{Fc} \to \infty$, while E_{Fv} remained finite. For this situation, the electron density would be much higher than the hole density. In reality, charge neutrality in the active region under high injection conditions requires that the electron and hole densities be equal, making the previous problem unrealistic (but useful for academic purposes). A different, more realistic simplification which includes charge neutrality can be made by neglecting *all* but the lowest C1–HH1 subband transition pair in our analysis.

In this problem, we will make use of this approximation to explore the effect of the effective mass asymmetry between the conduction and heavy-hole band. Defining $D = m_{hh}/m_c$, show that charge neutrality leads to the following relationship between the Fermi functions at the C1 and HH1 subband edges (neglecting all other subband and bulk transitions pairs):

$$f_1(E_{hh1}) = (1 - f_2(E_{c1}))^{1/D}. \tag{4.93}$$

Using Eq. (4.93), plot the following quantities as a function of the fraction of filled states at the C1 subband edge (from 0 to 1), for symmetric ($D = 1$) and asymmetric ($D = 5$) bands (assume m_c is the same in both cases):

(a) The fraction of *empty* states at the HH1 subband edge.

(b) The stimulated absorption and stimulated emission Fermi factors.

(c) The subband edge gain normalized to the maximum possible gain in each case.

(d) The carrier density normalized to the density at $f_2(E_{c1}) = 0.5$.

These plots reveal fundamental differences between the symmetric and asymmetric band structures in the way they behave under nonequilibrium conditions.

(e) Which of the two band structures reaches transparency first as a function of the fraction of filled C1 subband edge states? What is the main cause of this?

(f) Does the symmetric or the asymmetric band structure have the lowest transparency carrier density (the carrier density at which $g = 0$)? What are the values for the normalized carrier density in each case?

Assuming the radiative current density roughly follows a BN^2 dependence, how different do you expect the transparency current densities to be?

(g) Derive an expression for the differential gain, dg/dN, explicitly including the dependence of g_{max} on D and m_c (neglect lineshape broadening) (it may be helpful to begin by setting $dg/dN = (dg/df_2(E_{c1}))/(dN/df_2(E_{c1}))$). Does the symmetric or asymmetric band structure provide a higher differential gain at transparency? By how much? Can you explain this behavior qualitatively?

(h) The application of strain in the plane of the QW is known to reduce the in-plane heavy-hole effective mass dramatically. From the above considerations, and what you know about the effective mass asymmetry of unstrained GaAs QWs, can you predict qualitatively what changes in performance in strained QWs can be expected? Should we be using strained QWs in the active regions of lasers? What benefits do you foresee?

4.4 Assume we have a strained QW which has a completely symmetric subband structure in the conduction and valence bands including subband curvatures and subband spacings (also assume that only heavy-hole subbands exist). Now, the modal absorption loss (w/o excitonic effects) of the lowest subband transition is measured to be 400 cm^{-1} in an unpumped laser. If we want to insure that the $n = 1$ gain is always larger than the $n = 2$ gain for modal gains < 200 cm^{-1}, what is the minimum subband spacing between the two lowest conduction band states that we can tolerate? Assume g_{max} is constant and of equal magnitude for each subband transition, and also neglect lineshape broadening effects.

4.5 Comparing Appendix 6 to Chapter 4, what is τ_{sp}^{21} in terms of $|M_T|^2$?

4.6 The surface recombination velocity can be estimated using the simple "broad-area" (i.e., infinite stripe width) threshold carrier density, however, in reality the carrier density profile will vary over the cross section of the active region, particularly when the active width is narrow. In this problem, the effects of a finite diffusion constant for carriers in the active region will be examined.

Assume that the carrier densities in the active region are high enough that any differences in the diffusion profiles of electrons and holes will set up an electric field which will pull the two densities to nearly the same profile. In this *ambipolar* diffusion limit, the hole diffusion rate is enhanced by a factor of ~ 2 by the forward pull of the electrons, and the electron diffusion rate is limited to approximately twice the normal hole diffusion rate by the backward pull of the holes. The overall effect is that we can assume the electron and hole densities are equal everywhere in the active region and are characterized by a single ambipolar diffusion constant, D_{np}. The lateral profile of carriers is then governed by the simple diffusion equation:

$$D_{np} \frac{d^2 N(x)}{dx^2} = -\frac{I(x)}{qV} + \frac{N(x)}{\tau_{np}}. \tag{4.94}$$

The carrier lifetime is in general a function of N, however, to obtain analytic solutions, we can evaluate the lifetime at the broad-area threshold value, $\tau_{np}|_{th} = qL_z N_{th}/J_{th}$.

The problem we wish to solve is the carrier density profile across the width of the active region in the in-plane laser depicted in Fig. 4.13. For this case, we can define two distinct regions: one beneath the contact within w where we assume a uniform current injection profile, and the region outside of w where there is no current injection. Mathematically, with $x = 0$ defined as the center of the stripe, we have $I(x) = I_0$ for $x < w/2$, and $I(x) = 0$ for $x > w/2$. In fabricating the laser we can either leave the active region in place outside of the stripe, or we can remove it by etching through the active region outside of the contact area. The first case leads to carrier outdiffusion, while the second case leads to surface recombination. We would like to compare these two cases.

(i) With the active region in place away from the contact, carriers are free to diffuse outside the stripe width. Draw a sketch of this configuration and solve Eq. (4.94) for $N(x)$ in and out of the stripe assuming the carrier density and its derivative (i.e., the diffusion current) are constant across the $x = w/2$ boundary. Qualitatively sketch $N(x)$.

(ii) With the active region etched away, the carriers recombine at the surface. Draw a sketch of this configuration and solve Eq. (4.94) for $N(x)$ under the stripe assuming the diffusion current (defined by the slope of the carrier density) is equal to the surface recombination current, $D_{np} dN/dx = -v_s N$, at the $x = w/2$ boundary. Place your result in terms of the diffusion equivalent surface recombination velocity, $v_{sD} = \sqrt{D_{np}/\tau_{np}}$. Qualitatively sketch $N(x)$. In comparing $N(x)$ in (ii) to $N(x)$ in (i), what is the significance of v_{sD}? Show that Eq. (4.65) is recovered in the limit of $D_{np} \to \infty$.

Assuming an injection current of 166.4 A/cm^2 (corresponding to the broad-area threshold in Fig. 4.14) and a stripe width of 2 μm, plot $N(x)$ on the same graph for four different cases: (1) $N(x)$ found in (i), (2) $N(x)$ found in (ii) with $v_s = 1 \times 10^4$ cm/s, (3) $v_s = 1 \times 10^5$ cm/s, and (4) $v_s = 5 \times 10^5$ cm/s. Assume $D_{np} = 20$ cm^2/s for all cases. Produce similar plots for stripe widths of 10 μm and 20 μm.

The single quantum-well in-plane laser defined in Fig. 4.14 is expected to lase with the above injected current density. However, the finite stripe width reduces N, which reduces the gain, preventing the onset of lasing at 166.4 A/cm^2. Assuming that the injection current must establish a carrier density *in the center of the stripe* which is equal to the broad-area threshold carrier density in order to lase, calculate and tabulate the new threshold current density for all twelve cases considered above. Compare the threshold current densities for (2) through (4) to the values found using the simple techniques discussed in relation to Eq. (4.67). Answer the following questions:

(a) How does the inclusion of carrier diffusion affect the estimated magnitude of the surface recombination current?

(b) When is carrier diffusion important to consider?

(c) How does the threshold current density with carrier outdiffusion (profile (1)) compare to the three different recombination velocity profiles? Can you use v_{sD} to understand this comparison?

(d) For InGaAs QW active regions, is it better to etch through the active region?

(e) For GaAs QW active regions, is it better to etch through the active region?

In reality, the threshold lasing condition will be more complex than simply setting $N(0) = N_{th}$ because the material gain will be relaxed near the stripe edges, reducing the overall *modal* gain.

(f) Qualitatively discuss how the above-threshold current densities would be modified by the inclusion of this effect.

4.7 Discuss the effects of strain on the threshold energy for the CCCH Auger process by constructing a diagram analogous to Fig. 4.16. Assume $D = 5$ in one case and $D = 1$ in the other (see Problem 4.3 for the definition of D). You may find Appendix 12 helpful in quantifying the exact positions of the four states in each case (use bulk GaAs values for m_c and E_g). How would you expect this difference in the threshold energy to affect the Auger transition rate for a given electron and hole density?

CHAPTER FIVE

Dynamic Effects

5.1 INTRODUCTION

In Chapter 2, the rate equations for both the carrier density in the active region and the photon density of a given optical mode were developed from simple intuitive arguments. The below-threshold and above-threshold limits to the steady-state solutions of the rate equations were considered in order to give a feel for the operating characteristics of the laser. We also took a first look at the small-signal intensity modulation response of the laser. In this chapter we wish to expand upon these simplified discussions considerably, drawing upon the results of Chapter 4 for the evaluation of the various generation and recombination rates.

We will first summarize many of the results obtained in Chapter 2. This review will serve as a convenient reference to the relevant formulas. We will then move on to dynamic effects, starting with a differential analysis of the rate equations. We will derive the small-signal intensity and frequency modulation response of the laser as well as the small-signal transient response. We then consider large-signal solutions to the rate equations. The turn-on delay of the laser and the general relationship between frequency chirping and the modulated output power will also be treated here. Next we consider laser noise. Using the Langevin method, we will determine the laser's relative intensity noise (RIN) and the frequency fluctuations of the laser's output from which we can estimate the spectral linewidth of the laser. We will also discuss the role of the injection current noise in determining the RIN at high power and examine the conditions necessary for noise-free laser operation.

In the final two sections we consider dynamic effects in specific types of lasers. The first section considers carrier transport limitations in separate-confinement heterostructure (SCH) quantum-well lasers. The second deals with the effects of weak external feedback in extended cavity lasers.

185

5.2   REVIEW OF CHAPTER 2

Figure 5.1 summarizes the reservoir model used to develop the rate equations in Chapter 2. To insure particle conservation, each arrow in the flowchart represents the *number of particles* flowing per unit time. This is why the rates per unit volume and the densities are multiplied by the active-region volume V of the carrier reservoir or the mode volume V_p of the photon reservoir. Before deriving the rate equations, we will briefly follow the flowchart from input to output.

Starting with the carrier reservoir, we have the rate of carrier injection into the laser I/q. Of these carriers only $\eta_i I/q$ reach the active region, where η_i is the *injection* or *internal* efficiency of the laser introduced in Chapter 2. The rest recombine elsewhere in the device (this includes all carriers recombining at rates not directly tied to the carrier density in the active region). Once in the carrier reservoir, the carriers have a number of options. Some recombine via non-radiative recombination at the rate $R_{nr}V$. Others recombine spontaneously at the rate $R_{sp}V$, of which a certain fraction emit photons into the mode of interest at the rate $R'_{sp}V$ (or $\beta_{sp}R_{sp}V$ where β_{sp} is the spontaneous emission factor). Other carriers recombine via stimulated emission into the mode of interest at the rate $R_{21}V$. While photons in other modes can also induce stimulated recombination of carriers, we limit ourselves initially to a single mode. Finally, photons in the photon reservoir can be absorbed, serving as an additional source of carriers which are generated at the rate $R_{12}V$. In Chapter 2, we also considered carriers which might leak out of the active region via lateral diffusion

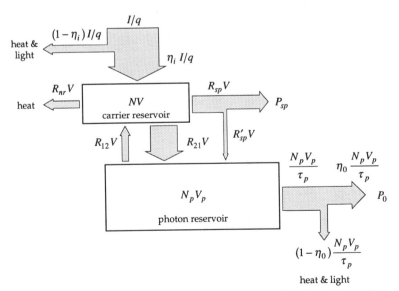

FIGURE 5.1   Model used in the rate equation analysis of semiconductor lasers.

and/or thermionic emission at the rate $R_l V$. However, since $R_l V$ is often negligible and furthermore plays a role identical to $R_{nr} V$ in the rate equations, we will not include it explicitly in this chapter (if we wish to include carrier leakage at any time, we can set $R_{nr} \rightarrow R_{nr} + R_l$).

In the photon reservoir, the stimulated and spontaneous emission rates into the mode provide the necessary generation of photons at the rate $R_{21} V + R'_{sp} V$. Stimulated absorption in the active region depletes photons at the rate $R_{12} V$. All other photons leave the cavity at the rate $N_p V_p / \tau_p$. Of those leaving the cavity, only $\eta_0 N_p V_p / \tau_p$ leave through the desired mirror to be collected as useful output power, P_0, where η_0 is the *optical* efficiency of the laser to be defined below. The rest of the photons exit the cavity through a different mirror or disappear through: (1) free carrier absorption in the active region (which does *not* increase the carrier density), (2) absorption in materials outside the active region, and/or (3) scattering at rough surfaces.

5.2.1   The Rate Equations

By setting the time rate of change of the carriers and photons equal to the sum of rates into, minus the sum of rates out of the respective reservoirs, we immediately arrive at the carrier and photon *number* rate equations:

$$V \frac{dN}{dt} = \frac{\eta_i I}{q} - (R_{sp} + R_{nr})V - (R_{21} - R_{12})V, \qquad (5.1)$$

$$V_p \frac{dN_p}{dt} = (R_{21} - R_{12})V - \frac{N_p V_p}{\tau_p} + R'_{sp} V. \qquad (5.2)$$

Setting $R_{21} - R_{12} = v_g g N_p$ using Eq. (4.32), dividing out the volumes, and using $\Gamma = V/V_p$ (by definition of V_p), we obtain the *density* rate equations derived in Chapter 2 in slightly more general form:

$$\frac{dN}{dt} = \frac{\eta_i I}{qV} - (R_{sp} + R_{nr}) - v_g g N_p, \qquad (5.3)$$

$$\frac{dN_p}{dt} = \left[\Gamma v_g g - \frac{1}{\tau_p} \right] N_p + \Gamma R'_{sp}. \qquad (5.4)$$

It is a matter of preference to use either the density or the number rate equations in the analysis of lasers.[1] Throughout this book we choose to use the density versions. This choice forces us to do a little more book-

---

[1] It is important to note that the gain term in the density versions has a Γ in the photon density rate equation, but not in the carrier density rate equation. While in the number rate equations, the gain term is symmetric. This asymmetry in the density rate equations is often overlooked in the literature.

keeping by explicitly including V and V_p in various places (as we will see in Section 5.5) but in the end places results in terms of the more familiar carrier *density* and related terms (e.g., the differential gain, the A, B, and C coefficients, etc.).

To complete the description, the output power of the mode and total spontaneous power from all modes are given by

$$P_0 = \eta_0 h\nu \frac{N_p V_p}{\tau_p} \quad \text{and} \quad P_{sp} = h\nu R_{sp} V, \tag{5.5}$$

where

$$\eta_0 = F \frac{\alpha_m}{\alpha_m + \langle \alpha_i \rangle}. \tag{5.6}$$

The optical efficiency defined here times the injection efficiency yields the differential quantum efficiency defined in Chapters 2 and 3, $\eta_d = \eta_i \eta_0$. The prefactor F in (5.6) is the fraction of power not reflected back into the cavity which escapes as useful output from the output coupling mirror as derived in Chapter 3. The mirror loss can usually be defined as $\alpha_m = (1/L) \ln(1/r_1 r_2)$, and $\langle \alpha_i \rangle$ is the spatial average over any internal losses present in the cavity. The photon lifetime is given by

$$\frac{1}{\tau_p} = v_g(\alpha_m + \langle \alpha_i \rangle) = \frac{\omega}{Q}, \tag{5.7}$$

where v_g is the group velocity of the mode of interest including both material and waveguide dispersion. The more general definition of τ_p in terms of the cavity Q is useful in complex multisection lasers where α_m can prove difficult to define (see the latter part of Appendix 5). In such cases, we can use the definition: $Q \equiv \omega$ (energy stored in cavity)/(total power lost).

We can use Eq. (4.49) to set $R'_{sp} = \Gamma v_g g n_{sp}/V$ in Eq. (5.4). However, a useful approximation to R'_{sp} is found by setting $R'_{sp} \to \beta_{sp} R_{sp}$ and evaluating β_{sp} at its threshold value:

$$\beta_{sp} \approx \beta_{sp}\bigg|_{th} = \frac{\Gamma v_g g n_{sp}}{\eta_i \eta_r I/q}\bigg|_{th} = \frac{q}{I_{th}\tau_p}\left[\frac{n_{sp}}{\eta_i \eta_r}\right]_{th}, \tag{5.8}$$

where Eq. (4.60) was used to define β_{sp} and $\eta_r = R_{sp}/(R_{sp} + R_{nr})$. The latter equality makes use of Eq. (5.11). The term in brackets at threshold is typically between 1.5 and 2 in short-wavelength materials, while it can be as large as 10 in long-wavelength materials due to low radiative efficiencies (high Auger recombination).

5.2.2   Steady-State Solutions

The steady-state solutions of Eqs. (5.3) and (5.4) are found by setting the time derivatives to zero. Solving Eq. (5.4) for the steady-state photon density and Eq. (5.3) for the dc current, we have

$$N_p(N) = \frac{\Gamma R'_{sp}(N)}{1/\tau_p - \Gamma v_g g(N)}, \tag{5.9}$$

$$I(N) = \frac{qV}{\eta_i}(R_{sp}(N) + R_{nr}(N) + v_g g(N)N_p(N)). \tag{5.10}$$

As the functionality implies, it is useful to think of N as the independent parameter of the system which we can adjust to determine different values of both current and photon density. Solving for $v_g g N_p$ in Eq. (5.9), and using Eq. (5.11) below to define g_{th}, Eq. (5.10) can alternatively be written as

$$I(N) = \frac{qV}{\eta_i}(R_{sp}(N) - R'_{sp}(N) + R_{nr}(N) + v_g g_{th} N_p(N)). \tag{5.10'}$$

In this form, the right-hand side more clearly identifies where the current eventually goes. In fact, (5.10') can be derived by simply drawing a box around the *entire* carrier–photon system in Fig. 5.1 and equating the inward flow with the sum of all outward flows (as opposed to (5.10) which is derived by drawing a box only around the carrier reservoir).

To gain more understanding of how (5.9) and (5.10) behave, we will first consider the limiting forms (below and above threshold) and then we will plot the equations over the entire range. The threshold gain and carrier density as defined in Chapter 2 are

$$\Gamma v_g g_{th} \equiv \frac{1}{\tau_p} \quad \text{and} \quad g(N_{th}) = g_{th}. \tag{5.11}$$

The first two cases below summarize (1) what happens when the gain and carrier density are much smaller than (5.11) and (2) what happens as they approach their threshold values.

5.2.2.1   Case (i): Well Below Threshold   For $N \ll N_{th}$ and $\Gamma v_g g \ll 1/\tau_p$, we can neglect $\Gamma v_g g$ in the denominator of (5.9) and the steady-state solutions become

$$N_p(N) = \Gamma R'_{sp}(N)\tau_p \approx 0, \tag{5.12}$$

$$I(N) = \frac{qV}{\eta_i}(R_{sp}(N) + R_{nr}(N)), \tag{5.13}$$

and

$$P_0 \approx 0 \qquad \text{and} \qquad P_{sp} = \eta_i \eta_r \frac{h\nu}{q} I. \qquad (5.14)$$

With negligible power in the mode, the injection current only needs to resupply carriers lost to spontaneous and nonradiative recombination. The output power is close to zero and the spontaneous power increases approximately linearly with injected current (if Auger recombination is significant, the increase will be sublinear since η_r decreases). These solutions were also derived in Eqs. (2.5) and (2.7). As we approach threshold, the denominator of (5.9) becomes small and the power in the mode starts to build up.

5.2.2.2   Case (ii): Above Threshold   As $N \to N_{th}$ and $\Gamma v_g g \to 1/\tau_p$, we can evaluate all terms in (5.9) and (5.10) at N_{th} *except* for the denominator of (5.9) which contains the difference between the threshold gain and the actual gain:

$$N_p(N) = \frac{R'_{sp}(N_{th})/v_g}{g_{th} - g(N)}, \qquad (5.15)$$

$$I(N) = \frac{qV}{\eta_i}(R_{sp}(N_{th}) + R_{nr}(N_{th})) + \frac{qV}{\eta_i} v_g g_{th} N_p(N), \qquad (5.16)$$

and

$$P_0 = \eta_i \eta_0 \frac{h\nu}{q}(I - I_{th}) \qquad \text{and} \qquad P_{sp} = \eta_i \eta_r \frac{h\nu}{q} I_{th}. \qquad (5.17)$$

The output power in (5.17) was found by rearranging (5.16) and recognizing that $V v_g g_{th} N_p = P_0/(\eta_0 h\nu)$. The *threshold current* of the laser is defined as

$$I_{th} = \frac{qV}{\eta_i}(R_{sp}(N_{th}) + R_{nr}(N_{th})). \qquad (5.18)$$

In this above-threshold limit, the output power increases linearly with current and the spontaneous power saturates at the level found at threshold. From (5.15), it is clear that N and g never actually reach N_{th} and g_{th} for finite output powers (and finite current). They remain ever so slightly below their "threshold" values. These solutions were also derived in Eqs. (2.30) and (2.36).

In lasers where β_{sp} is relatively large, we need to revise our derivation slightly. For example, comparing Eq. (5.16) to Eq. (5.10′) reveals that we have inadvertently dropped a factor, $-R'_{sp}(N_{th})$, in arriving at Eq. (5.16) (it happened when we set $g \to g_{th}$). The more exact threshold current is found by setting $N_p \to 0$ in Eq. (5.10′). Equation (5.18) is then revised to

$$I_{th} = \frac{qV}{\eta_i}((1 - \beta_{sp}(N_{th}))R_{sp}(N_{th}) + R_{nr}(N_{th})). \qquad (5.18')$$

The threshold current as defined here represents the offset in the LI curve at output power levels generated by carrier densities close to N_{th}. If nonradiative recombination is negligible, then as $\beta_{sp}(N_{th}) \to 1$, the threshold current reduces to zero. In such a "thresholdless" laser, *all* injected current is funneled into the lasing mode. For typical lasers, $\beta_{sp}(N_{th}) \ll 1$ and (5.18') reduces to (5.18).

Using Eq. (5.17), the current above threshold can be written as

$$I = I_{th} + \frac{q}{hv} \frac{P_0}{\eta_i \eta_0}. \tag{5.19}$$

Appendix 17 considers methods of minimizing this equation for a given output power.

5.2.2.3 Case (iii): Below and Above Threshold

To observe the transition between below threshold and above, it is useful to plot (5.9) as a function of (5.10). Parameterized light–current (L–I) curves found by varying N from 0 to N_{th} are shown in Fig. 5.2 for a typical in-plane laser and three VCSELs. The details of the VCSEL and in-plane structures are given later in Table 5.1. The plots use Table 4.4 for $g(N)$ and the combination of Tables 4.4 and 4.5 for $R_{sp}(N) + R_{nr}(N)$ (see Eq. (4.90)). We can find $R'_{sp}(N)$ directly from gain calculations[2] (solid curves in upper plot) or we can approximate $R'_{sp}(N)$ using Eq. (5.8) to define the threshold value of β_{sp} (dashed curves in upper plot). For these examples, β_{sp} actually decreases by about 30% as the current increases to threshold. As a result, the approximate dashed curves underestimate the power level by about 30% at very low currents. Above threshold the output power becomes independent of the value of β_{sp}, as Eq. (5.17) verifies.

Each of the "idealized" L–I curves[3] in the upper plot of Fig. 5.2 displays a sudden ~ 30 dB increase in output power (or ~ 20 dB for the smaller VCSELs). This dramatic change of course corresponds to the lasing threshold of the laser and reveals that the transition to lasing is quite sharp. Below the lasing threshold, the power in the mode comes primarily from the spontaneous emission rate, $\Gamma R'_{sp}$. This power is typically well under 1 μW and is larger for smaller devices since the fraction of spontaneous emission into the lasing mode increases with decreasing mode volume (see Eq. (4.49)). As the current increases, the gain approaches the loss and the stimulated emission rate increases dramatically. For the largest VCSEL and the in-plane laser this occurs at

---

[2] Setting $R'_{sp}(N) = \Gamma v_g g n_{sp}/V$, the gain calculations can be used to find both g and n_{sp} as a function of N. For the InGaAs/GaAs 80Å QW considered in Section 4.6, the calculated gain–population inversion factor product can be modeled very well (for $g < 2500$ cm^{-1}) by $g n_{sp} = (850$ cm$^{-1}) \times \ln[1 + N^2/2]$ where N is in units of 10^{18} cm^{-3} (if desired we can combine this with the curve fit for n_{sp} to estimate n_{sp} for $g > 0$).

[3] The L–I curves are ideal in the sense that surface recombination and heating at high injection levels have been neglected. In realistic devices, surface recombination will require additional current to reach threshold (particularly in the smaller devices) and heating will limit the maximum output power of the devices.

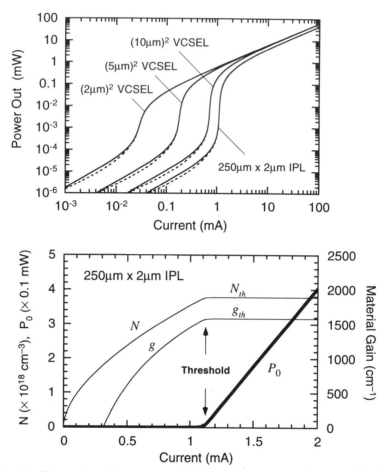

FIGURE 5.2 Upper plot: light vs. current in two different lateral-size 3-QW VCSELs and in a single QW in-plane laser (IPL) (all lasers use InGaAs/GaAs 80Å QWs). Lower plot: Light vs. current on a linear scale for the same in-plane laser. Plot also shows carrier density and material gain vs. current.

~ 0.7 mA and ~ 1.1 mA, while for the smaller VCSELs only ~ 0.2 mA and ~ 30 μA are required due to the smaller active volumes. Above the lasing threshold all curves approach a unity slope which on a log-log plot implies a linear relationship between power and current.

The linear $L–I$ relationship above threshold for the in-plane laser is more clearly illustrated in the lower plot in Fig. 5.2. The $L–I$ curve near threshold which was approximated in Chapter 2 as a discontinuous change in the $L–I$ slope is more accurately a *knee* in the curve. This knee becomes softer for smaller devices due to the higher spontaneous emission rate into the mode. However, when measured experimentally, the light collected into a detector often captures spontaneous emission from more than just the lasing mode, and this can also give the appearance of a softer transition to lasing.

The lower plot in Fig. 5.2 also displays the carrier density and material gain as a function of current. Below threshold, the approximately quadratic carrier density–current relationship and the logarithmic gain–current relationship can be observed. For this example, transparency in the active material occurs at about one-third of the way toward threshold. At threshold, both the carrier density and gain become clamped near their threshold values. Beyond threshold, the carrier density and gain continue to increase, however, the increase is in the fractions of a percent range. Thus for all practical purposes, above threshold we can set $N = N_{th}$ and $g = g_{th}$ as we did in Chapter 2. The only time we cannot use this approximation is when we need the difference $N_{th} - N$ or $g_{th} - g$.

The carrier clamping mechanism illustrated so dramatically in Fig. 5.2 is perhaps best understood by defining a *stimulated* carrier lifetime τ_{st} and writing the total carrier recombination rate as

$$R_{tot}(N) = \frac{N}{\tau_{sp}} + \frac{N}{\tau_{nr}} + \frac{N}{\tau_{st}}, \tag{5.20}$$

where

$$\frac{1}{\tau_{sp}} + \frac{1}{\tau_{nr}} = \frac{R_{sp} + R_{nr}}{N}, \tag{5.21}$$

$$\tau_{st} = \frac{N}{v_g g N_p} \approx \frac{N_{th}}{v_g g_{th}} \cdot \frac{1}{N_p}. \tag{5.22}$$

The inverse dependence of the stimulated carrier lifetime on photon density sets up a negative feedback loop which prevents N from increasing beyond its threshold value. For example, as the current increases and the clamped carrier density increases closer to N_{th}, the gain moves incrementally closer to g_{th}, which increases the output power through Eq. (5.15). This increase in output power reduces τ_{st} just enough to prevent any further increase in N. Carrier clamping is therefore maintained via the photon density's control over the stimulated carrier lifetime. As a result of this interplay, all injected current above threshold $\eta_i(I - I_{th})$ is simply eaten up by the self-adjusting N/τ_{st}. For this reason, $\eta_i(I - I_{th})$ is referred to as the *stimulated emission current*. Note that when $1/\tau_{st} = 1/\tau_{sp} + 1/\tau_{nr}$ (i.e., when $\eta_i(I - I_{th}) = \eta_i I_{th}$), we are at twice threshold.

5.2.3 Steady-State Multimode Solutions

As discussed in Chapter 3, if the laser cavity is not carefully designed, it is inevitable that a number of different resonant modes of the cavity (either axial or lateral) will require similar amounts of gain to reach threshold. To make matters worse, the gain spectrum in semiconductor materials is very broad and can typically provide almost uniform gain across at least a few mode spacings. In such cases, the gain difference $g_{th} - g$ in Eq. (5.15) can be similar for different modes of the cavity at a given injection level. As a result, the photon density can build up in more than one mode.

In such multimode situations, the steady-state solution for the photon density in each mode remains the same as before, however, the steady-state current must include contributions from all m modes:

$$N_{pm} = \frac{\Gamma_m R'_{spm}}{1/\tau_{pm} - \Gamma_m v_{gm} g_m}, \tag{5.23}$$

$$I = \frac{qV}{\eta_i}\left(R_{sp} + R_{nr} + \sum_m v_{gm} g_m N_{pm}\right). \tag{5.24}$$

Figure 5.3 plots the steady-state photon density (5.23) for three modes of the in-plane laser considered earlier, assuming the threshold modal gains for mode 0, 1, and 2 are 50, 60, and 70 cm^{-1}, respectively. From earlier discussions we know the carrier density cannot increase beyond N_{th0} for finite current levels. As a result, the power in modes 1 and 2 cannot increase beyond the levels indicated in the figure. Thus the power in the *side modes* saturates just as the

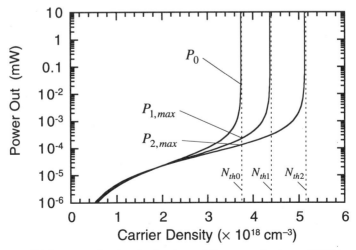

FIGURE 5.3   Light vs. carrier density for three different modes in an SQW in-plane laser. The threshold modal gains for modes 0, 1, and 2 are 50, 60, and 70 cm^{-1}, respectively.

spontaneous power saturates at threshold.[4] However, the closer N_{th1} and N_{th2} are to N_{th0}, the higher this maximum saturated power level will be. The mode suppression ratio (defined as the ratio of power in the main mode to the power in either of the side modes) can be determined from Eq. (5.23) as worked out in Chapter 3. Note that due to the clamping of power in the side modes, the MSR increases linearly with the power in the main mode.

5.3   DIFFERENTIAL ANALYSIS OF THE RATE EQUATIONS

When we want to observe how lasers behave dynamically in response to some perturbation to the system such as a modulation of the current, we must analyze Eqs. (5.3) and (5.4) with the time derivatives included. Unfortunately, exact analytical solutions to the full rate equations cannot be obtained. Therefore if analytical solutions are desired we must make some approximations. In this section we analyze the rate equations by assuming that dynamic changes in the carrier and photon densities away from their steady-state values are small. Such *small-signal* responses of one variable in terms of a perturbation to another can be accommodated by taking the *differential* of both rate equations. Considering I, N, N_p, and g as dynamic variables, the differentials of Eqs. (5.3) and (5.4) become

$$d\left[\frac{dN}{dt}\right] = \frac{\eta_i}{qV}\,dI - \frac{1}{\tau_{\Delta N}}\,dN - v_g g\,dN_p - N_p v_g\,dg, \qquad (5.25)$$

$$d\left[\frac{dN_p}{dt}\right] = \left[\Gamma v_g g - \frac{1}{\tau_p}\right]dN_p + N_p \Gamma v_g\,dg + \frac{\Gamma}{\tau'_{\Delta N}}\,dN, \qquad (5.26)$$

where

$$\frac{1}{\tau_{\Delta N}} = \frac{dR_{sp}}{dN} + \frac{dR_{nr}}{dN} \approx A + 2BN + 3CN^2, \qquad (5.27)$$

$$\frac{1}{\tau'_{\Delta N}} = \frac{dR'_{sp}}{dN} \approx 2\beta_{sp}BN + \frac{d\beta_{sp}}{dN}BN^2. \qquad (5.28)$$

The *differential* carrier lifetime $\tau_{\Delta N}$ depends on the *local* slope dR/dN whereas the total carrier lifetime τ depends on the overall slope R/N. Due to the mixed quadratic and cubic dependence of R on N, $\tau_{\Delta N}$ is typically a factor of 2–3 smaller than τ. The differential lifetime of carriers which radiate photons into

---

[4] Actually, gain compression created by high photon densities can force N to increase in order to maintain $g_0 \sim g_{th0}$. As a result, with increasing power in the main mode, N can increase slightly beyond N_{th0} and the side mode power can to some degree increase beyond the levels indicated in Fig. 5.3.

the lasing mode $\tau'_{\Delta N}$ is typically in the tens of microseconds range and is negligible in most cases.[5]

Equations (5.25) and (5.26) can also be found by performing an expansion about the steady-state of the form $x(t) = x_0 + dx(t)$ for all dynamic variables. Neglecting product terms involving two or more small-signal terms and canceling out the steady-state solutions we would obtain a set of approximate rate equations for $dN(t)$ and $dN_p(t)$ identical to Eqs. (5.25) and (5.26). This procedure was in fact carried out in Chapter 2. The differential approach simply provides a more direct path to the desired equations.

The gain variation dg can be further expanded by assuming it is affected by both carrier and photon density variations:

$$dg = a\, dN - a_p\, dN_p. \tag{5.29}$$

The sign convention reflects the fact that gain increases with increasing carrier density while it decreases or is *compressed* with increasing photon density. In Chapter 4 it was shown that the gain vs. carrier density could be well approximated by a logarithmic function. The gain is also known to be inversely proportional to $(1 + \varepsilon N_p)$, where ε is a constant known as the *gain compression factor*. Thus, we can approximate the gain by

$$g(N, N_p) = \frac{g_0}{1 + \varepsilon N_p} \ln\left(\frac{N + N_s}{N_{tr} + N_s}\right). \tag{5.30}$$

With this expression the gain derivatives become

$$a = \frac{\partial g}{\partial N} = \frac{g_0}{(N + N_s)(1 + \varepsilon N_p)} \equiv \frac{a_0}{(1 + \varepsilon N_p)}, \tag{5.31}$$

$$a_p = -\frac{\partial g}{\partial N_p} = \frac{\varepsilon g}{(1 + \varepsilon N_p)}. \tag{5.32}$$

In Eq. (5.31), a_0 is defined as the *nominal* differential gain—the value of a with zero photon density (with no gain compression). Also note that neither a nor a_p are constants—both tend to get smaller at higher densities.

Replacing dg with Eq. (5.29), collecting like terms and defining a set of *rate*

---

[5] A maximum limit can be placed on $1/\tau'_{\Delta N}$ by defining it as follows:

$$\frac{1}{\tau'_{\Delta N}} = \frac{v_g n_{sp}}{V_p} \frac{dg}{dN} \left[1 + \frac{g}{n_{sp}} \frac{dn_{sp}}{dg}\right] \le \frac{v_g n_{sp}}{V_p} \frac{dg}{dN}.$$

The second term within brackets goes from -1 at transparency to 0 at infinite pumping levels. Thus, the latter inequality defines the largest possible value for $1/\tau'_{\Delta N}$ at positive gains.

coefficients, the differential rate equations become

$$\frac{d}{dt}(dN) = \frac{\eta_i}{qV}dI - \gamma_{NN}\,dN - \gamma_{NP}\,dN_p, \tag{5.33}$$

$$\frac{d}{dt}(dN_p) = \gamma_{PN}\,dN - \gamma_{PP}\,dN_p, \tag{5.34}$$

where

$$\gamma_{NN} = \frac{1}{\tau_{\Delta N}} + v_g a N_p, \qquad \gamma_{NP} = \frac{1}{\Gamma\tau_p} - \frac{R'_{sp}}{N_p} - v_g a_p N_p,$$

$$\gamma_{PN} = \frac{\Gamma}{\tau'_{\Delta N}} + \Gamma v_g a N_p, \qquad \gamma_{PP} = \frac{\Gamma R'_{sp}}{N_p} + \Gamma v_g a_p N_p. \tag{5.35}$$

We have replaced g in these definitions using the steady-state relation: $1/\tau_p - \Gamma v_g g = \Gamma R'_{sp}/N_p$ (see Eq. (5.9)). To interpret the subscripts of the rate coefficients, it is helpful to remember "effect precedes cause." For example, γ_{NP} defines the effect on N caused by changes in P (i.e., N_p). This mnemonic aid allows us to quickly associate γ_{NN} with the differential carrier lifetime, γ_{PP} with the effective photon lifetime, γ_{NP} with the gain, and γ_{PN} with the differential lifetime of carriers which radiate into the mode. The change in gain (5.29) also adds a differential gain term $\propto a$ to the rates caused by N, γ_{NN} and γ_{PN}, and a gain compression term $\propto a_p$ to the rates caused by N_p, γ_{PP} and γ_{NP}. Aside from the small-signal assumptions, these rate coefficients contain no approximations. Well above threshold, N_p is large enough that a few terms can be dropped and the rate coefficients simplify to

$$\gamma_{NN} = 1/\tau_{\Delta N} + v_g a N_p, \qquad \gamma_{NP} = 1/\Gamma\tau_p - v_g a_p N_p, \quad \text{(above threshold)}$$
$$\gamma_{PN} = \Gamma v_g a N_p, \qquad \gamma_{PP} = \Gamma v_g a_p N_p. \tag{5.36}$$

Table 5.1 gives values of these and related parameters for two specific laser structures (the values assume $N = N_{th}|_{\varepsilon=0}$ at and above threshold).

The rate coefficients have been introduced to allow us to conveniently describe the differential rate equations in a compact matrix form:

$$\frac{d}{dt}\begin{bmatrix} dN \\ dN_p \end{bmatrix} = \begin{bmatrix} -\gamma_{NN} & -\gamma_{NP} \\ \gamma_{PN} & -\gamma_{PP} \end{bmatrix}\begin{bmatrix} dN \\ dN_p \end{bmatrix} + \frac{\eta_i}{qV}\begin{bmatrix} dI \\ 0 \end{bmatrix}. \tag{5.37}$$

In this form, the current is seen as the driving term or forcing function. Later when we treat noise in semiconductor lasers, the current forcing function will be replaced by noise sources. If necessary, we could also choose any other parameter to be the forcing function. For example, if we could somehow modulate the mirror loss, then the analysis would proceed by replacing the

TABLE 5.1 List of Common Parameters for Two Laser Structures.

Parameter	In-Plane Laser	VCSEL
d	80 Å	10 μm
w	2 μm	10 μm
L_a	250 μm	3×80 Å
L	250 μm	1.15 μm
Γ_{xy}	0.032	1
Γ_z	1	$1.83 \times 0.0209 = 0.0382$
V	4×10^{-12} cm^3	2.4×10^{-12} cm^3
V_p	1.25×10^{-10} cm^3	0.628×10^{-10} cm^3
$\sqrt{R_1 R_2}$	0.32	0.995
α_m	45.6 cm^{-1}	43.6 cm^{-1}
α_i	5 cm^{-1}	20 cm^{-1}
F_1	0.5	0.9
$\eta_{d1} = \eta_i \eta_0$	$0.8 \times 0.45 = 0.36$	$0.8 \times 0.617 = 0.494$
g_{th}	1580 cm^{-1}	1665 cm^{-1}
τ_p	2.77 ps	2.20 ps
At Threshold:		
N_{th}	3.77×10^{18} cm^{-3}	3.93×10^{18} cm^{-3}
J_{th}	$178.3/0.8 = 223$ A/cm^2	$575.2/0.8 = 719$ A/cm^2
I_{th}	1.11 mA	0.719 mA
η_r	0.840	0.829
a	5.34×10^{-16} cm^2	5.10×10^{-16} cm^2
a_p	2.37×10^{-14} cm^2	2.50×10^{-14} cm^2
τ	2.71 ns	2.63 ns
$\tau_{\Delta N}$	1.57 ns	1.52 ns
$\tau'_{\Delta N}$	44.3 μs	23.0 μs
n_{sp}	1.13	1.11
R'_{sp}	1.02×10^{23} cm^{-3}/s	2.09×10^{23} cm^{-3}/s
$\beta_{sp}|_{th}$	0.869×10^{-4}	1.69×10^{-4}
At $P_{01} = 1$ mW:		
I	3.31 mA	2.32 mA
N_p	2.43×10^{14} cm^{-3}	2.80×10^{14} cm^{-3}
γ_{NP}	1.12×10^{13} s^{-1}	1.18×10^{13} s^{-1}
γ_{PN}	2.95×10^{7} s^{-1}	3.88×10^{7} s^{-1}
γ_{NN}	1.56×10^{9} s^{-1}	1.67×10^{9} s^{-1}
γ_{PP}	1.32×10^{9} s^{-1}	1.93×10^{9} s^{-1}
f_R (Eq. (5.49))	2.907 GHz	3.423 GHz
$\sim f_R$ (Eq. (5.51))	2.904 GHz	3.418 GHz
γ	2.88×10^{9} s^{-1}	3.60×10^{9} s^{-1}
γ_0	0.651×10^{9} s^{-1}	0.686×10^{9} s^{-1}
K	0.265 ns	0.250 ns
$(\Delta v)_{ST}$	1.07 MHz	2.27 MHz
RIN peak	-112 dB/Hz	-111 dB/Hz

Continued

TABLE 5.1   (*Continued*)

Active Material	Material Parameters For Both Lasers ($In_{0.2}Ga_{0.8}As/GaAs$ 80Å QWs @ 980 nm)
Curve fits (from Chapter 4):	
N_{tr}, N_s, g_{0N}	1.8×10^{18} cm^{-3}, -0.4×10^{18} cm^{-3}, 1800 cm^{-1}
J_{tr}, J_s, g_{0J}	50 A/cm^2, -10 A/cm^2, 1100 cm^{-1}
J_{tr}, J_s, g_{0J} (J_{sp} only)	50 A/cm^2, 0, 1440 cm^{-1}
gn_{sp}	850 cm$^{-1} \times \ln[1 + (N/10^{18}$ cm$^{-3})^2/2]$
v_g	$3/4.2 \times 10^{10}$ cm/s
η_i	0.8
ε	1.5×10^{-17} cm^3
A	assumed negligible
B	$\sim 0.8 \times 10^{-10}$ cm^3/s (see Fig. 4.23)
C	3.5×10^{-30} cm^6/s

last term in (5.37) with $v_g N_p \begin{bmatrix} 0 \\ -d\alpha_m \end{bmatrix}$. The rate coefficients would remain unchanged.

For multimode small-signal analysis, we can extend Eq. (5.37) in a natural way:

$$\frac{d}{dt} \begin{bmatrix} dN \\ dN_{p1} \\ \vdots \\ dN_{pm} \end{bmatrix} = \begin{bmatrix} -\gamma_{NN} & -\gamma_{NP1} & \cdots & -\gamma_{NPm} \\ \gamma_{PN1} & -\gamma_{PP1} & 0 & 0 \\ \vdots & 0 & \ddots & 0 \\ \gamma_{PNm} & 0 & 0 & -\gamma_{PPm} \end{bmatrix} \begin{bmatrix} dN \\ dN_{p1} \\ \vdots \\ dN_{pm} \end{bmatrix} + \frac{\eta_i}{qV} \begin{bmatrix} dI \\ 0 \\ \vdots \\ 0 \end{bmatrix}. \quad (5.38)$$

The three rate coefficients γ_{PP}, γ_{PN}, and γ_{NP} for each mode m are still given by Eq. (5.36) with N_p replaced by N_{pm}. However, N_p in the rate coefficient γ_{NN} is replaced by a sum over all photon densities. The solution of Eq. (5.38) for two modes is considered in one of the problems at the end of the chapter. The zeros in the interaction matrix of Eq. (5.38) imply zero interaction or zero coupling between modes. In reality the gain compression of one mode might affect the gain experienced by another mode. Such coupling could be included by replacing the zeros with some *intermodal* gain compression term.

5.3.1 Small-Signal Frequency Response

To obtain the small-signal responses $dN(t)$ and $dN_p(t)$ to a sinusoidal current modulation $dI(t)$, we assume solutions of the form

$$dI(t) = I_1 e^{j\omega t},$$
$$dN(t) = N_1 e^{j\omega t}, \quad (5.39)$$
$$dN_p(t) = N_{p1} e^{j\omega t}.$$

Setting $d/dt \rightarrow j\omega$ and rearranging Eq. (5.37), we obtain

$$\begin{bmatrix} \gamma_{NN} + j\omega & \gamma_{NP} \\ -\gamma_{PN} & \gamma_{PP} + j\omega \end{bmatrix} \begin{bmatrix} N_1 \\ N_{p1} \end{bmatrix} = \frac{\eta_i I_1}{qV} \begin{bmatrix} 1 \\ 0 \end{bmatrix}. \tag{5.40}$$

The determinant of the matrix is given by

$$\begin{aligned} \Delta &\equiv \begin{vmatrix} \gamma_{NN} + j\omega & \gamma_{NP} \\ -\gamma_{PN} & \gamma_{PP} + j\omega \end{vmatrix} \\ &= (\gamma_{NN} + j\omega)(\gamma_{PP} + j\omega) + \gamma_{NP}\gamma_{PN} \\ &= \gamma_{NP}\gamma_{PN} + \gamma_{NN}\gamma_{PP} - \omega^2 + j\omega(\gamma_{NN} + \gamma_{PP}). \end{aligned} \tag{5.41}$$

With this, we can apply Cramer's rule to obtain the small-signal carrier and photon densities in terms of the modulation current:

$$N_1 = \frac{\eta_i I_1}{qV} \cdot \frac{1}{\Delta} \begin{vmatrix} 1 & \gamma_{NP} \\ 0 & \gamma_{PP} + j\omega \end{vmatrix}, \tag{5.42}$$

$$N_{p1} = \frac{\eta_i I_1}{qV} \cdot \frac{1}{\Delta} \begin{vmatrix} \gamma_{NN} + j\omega & 1 \\ -\gamma_{PN} & 0 \end{vmatrix}. \tag{5.43}$$

Expanding the determinants, the small-signal solutions can be written as

$$N_1 = \frac{\eta_i I_1}{qV} \cdot \frac{\gamma_{PP} + j\omega}{\omega_R^2} H(\omega), \tag{5.44}$$

$$N_{p1} = \frac{\eta_i I_1}{qV} \cdot \frac{\gamma_{PN}}{\omega_R^2} H(\omega), \tag{5.45}$$

where the modulation response is conveniently described in terms of the following two-parameter *modulation transfer function*:

$$H(\omega) = \frac{\omega_R^2}{\Delta} \equiv \frac{\omega_R^2}{\omega_R^2 - \omega^2 + j\omega\gamma}. \tag{5.46}$$

We define ω_R as the *relaxation resonance frequency* and γ as the *damping factor*. The physical significance of these parameters was discussed in Chapter 2. Comparing (5.46) to (5.41), the following associations can be made:

$$\omega_R^2 \equiv \gamma_{NP}\gamma_{PN} + \gamma_{NN}\gamma_{PP}, \tag{5.47}$$

$$\gamma \equiv \gamma_{NN} + \gamma_{PP}. \tag{5.48}$$

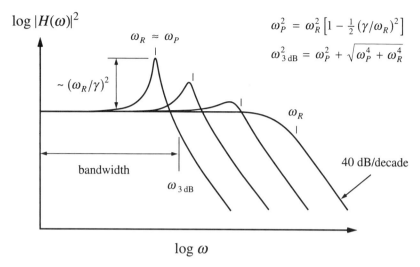

$$\omega_P^2 = \omega_R^2 \left[1 - \tfrac{1}{2}(\gamma/\omega_R)^2\right]$$

$$\omega_{3\,dB}^2 = \omega_P^2 + \sqrt{\omega_P^4 + \omega_R^4}$$

FIGURE 5.4 Sketch of the modulation transfer function for increasing values of relaxation resonance frequency and damping factor, including relationships between the peak frequency, ω_P, the resonance frequency, ω_R, and the 3 dB down cutoff frequency, $\omega_{3\,dB}$.

The densities N_1 and N_{p1} both follow the frequency response of the modulation transfer function, however, the carrier density has an additional zero in the complex frequency plane at $\omega = j\gamma_{PP}$. The general behavior of $H(\omega)$ is shown in Fig. 5.4. It is essentially a second-order low-pass filter with a damped resonance appearing near the cutoff frequency. The intensity modulation can follow the current modulation up to frequencies near ω_R, with an enhancement in the response existing at the relaxation resonance. Beyond the resonance, the response drops off dramatically. The actual peak frequency of the resonance, ω_P, is slightly less than ω_R depending on the damping. The frequency at which the electrical power response drops to half its dc value, $\omega_{3\,dB}$, is somewhat higher than ω_R for small damping. The relations for both ω_P and $\omega_{3\,dB}$ are included in the figure. To understand how we can maximize the modulation bandwidth, or $\omega_{3\,dB}$, we need to evaluate Eqs. (5.47) and (5.48).

The relaxation resonance frequency and the damping factor which characterize $H(\omega)$ can be expanded using Eqs. (5.35):

$$\omega_R^2 = \frac{v_g a N_p}{\tau_p} + \left[\frac{\Gamma v_g a_p N_p}{\tau_{\Delta N}} + \frac{\Gamma R'_{sp}}{N_p \tau_{\Delta N}}\right]\left(1 - \frac{\tau_{\Delta N}}{\tau'_{\Delta N}}\right) + \frac{1}{\tau'_{\Delta N}\tau_p}, \qquad (5.49)$$

$$\gamma = v_g a N_p\left[1 + \frac{\Gamma a_p}{a}\right] + \frac{1}{\tau_{\Delta N}} + \frac{\Gamma R'_{sp}}{N_p}. \qquad (5.50)$$

In practice, the expression for ω_R^2 can be simplified dramatically. For example, the last term is small compared to the first for $N_p V_p > n_{sp}$ (see footnote 5), and hence can be ignored above threshold where $n_{sp} \sim 1$–2 and $N_p V_p \gg 1$. Of the remaining terms, two are $\propto N_p$, and one is $\propto 1/N_p$. The former will dominate at some point above threshold as the photon density increases. For typical numbers (see Table 5.1), this crossover occurs very close to threshold. Hence, we can also neglect the $1/N_p$ term. Comparing the coefficients of the two terms $\propto N_p$, with $\tau_{\Delta N} \gg \tau_p$ and $a \sim \Gamma a_p$, we conclude that $a/\tau_p \gg \Gamma a_p/\tau_{\Delta N}$. Thus, the first term dominates over all other terms and ω_R^2 reduces to

$$\omega_R^2 \approx \frac{v_g a N_p}{\tau_p}. \qquad \text{(above threshold)} \qquad (5.51)$$

This is the same result we found in Chapter 2 with the exception that in this case $a = a_0/(1 + \varepsilon N_p)$. As mentioned in Section 2.7, ω_R can be enhanced by increasing the photon density or output power. This increase continues until the photon density approaches $1/\varepsilon$, at which point the differential gain falls off appreciably due to gain compression. For photon densities well beyond $1/\varepsilon$, ω_R^2 becomes independent of output power, saturating at $v_g a_0/\tau_p \varepsilon$ (however, in practice such high photon densities are not typically encountered, and, furthermore, it is not clear that the form assumed here for gain compression is even valid at such high photon densities).

Since we will be concentrating on the laser performance above threshold, we will specifically define ω_R using (5.51) unless otherwise stated. This simplified definition of ω_R can be used to rewrite the damping factor as

$$\gamma = K f_R^2 + \gamma_0, \qquad (5.52)$$

where

$$K = 4\pi^2 \tau_p \left[1 + \frac{\Gamma a_p}{a} \right] \quad \text{and} \quad \gamma_0 = \frac{1}{\tau_{\Delta N}} + \frac{\Gamma R'_{sp}}{N_p}. \qquad (5.53)$$

For large resonance frequencies, the K-factor describes the damping of the response, and as such is an important parameter in the characterization of high-speed lasers. The damping factor offset γ_0 is important at low powers where the relaxation resonance frequency is small. In practice, K and γ_0 are used as fitting parameters to be extracted from the laser modulation response.

Because the damping increases in proportion to ω_R^2, as we attempt to drive the laser harder to increase ω_R, the response flattens out as illustrated in Fig. 5.4. At some point, the damping becomes large enough that the response drops below the 3 dB cutoff at frequencies less than ω_R. As a result, there is a maximum bandwidth which can be achieved. Using the formulas in Fig. 5.4 combined with the definition of γ (neglecting γ_0), we can determine the modulation bandwidth for low damping as well as the maximum possible bandwidth:

$$f_{3\,\text{dB}} \approx f_R \sqrt{1 + \sqrt{2}} \approx 1.55 f_R, \qquad (\gamma/\omega_R \ll 1) \qquad (5.54)$$

$$f_{3\,\text{dB}}|_{\max} = \sqrt{2}\, \frac{2\pi}{K}. \qquad (\gamma/\omega_R = \sqrt{2}) \qquad (5.55)$$

The modulation bandwidth increases linearly with the relaxation resonance frequency and remains about 50% larger than f_R until damping becomes strong. With strong damping the bandwidth is compromised and eventually decreases with further increases in ω_R and γ. The optimum damping and maximum bandwidth occur when $\omega_P = 0$ and $\omega_R = \omega_{3\,dB}$ as defined in Fig. 5.4. This point is determined by the K-factor, which in turn is determined by the photon lifetime of the cavity. The K-factor therefore defines the intrinsic modulation bandwidth capabilities of the laser.

Returning to the ac photon density modulation response (5.45), we can use the explicit definition of ω_R to simplify the relation. With the above-threshold version of γ_{PN} in Eq. (5.36), we can set $\gamma_{PN} \approx \Gamma v_g a N_p = \Gamma \omega_R^2 \tau_p$. We then obtain

$$\frac{N_{p1}}{I_1} = \frac{\eta_i}{qV} \Gamma \tau_p H(\omega). \tag{5.56}$$

Using Eq. (5.5) to set $\tau_p = \eta_0 h v N_{p1} V_p / P_1$, the ac output power modulation response finally becomes

$$\frac{P_1}{I_1} = \eta_i \eta_0 \frac{hv}{q} H(\omega). \tag{5.57}$$

This result was also found in Eq. (2.54). The electrical power received which is of more fundamental interest is given by the absolute square of Eq. (5.57). In decibels, the frequency response is therefore $\propto 10 \log_{10} |H(\omega)|^2$ or $20 \log_{10} |H(\omega)|$. The phase shift of the modulation is given by $\angle H(\omega)$.

Figure 5.5 gives an example of experimental modulation responses of an InGaAs/GaAs VCSEL at different biases. The resonance peak which occurs at $\sim \omega_R$ is clearly visible in each curve. Note that the resonance shifts to higher frequencies with increasing bias current. This is because the output power and photon density increase with increasing current above threshold which increases ω_R via Eq. (5.51). We can write this relationship more directly by plugging Eqs. (5.5) and(5.17) into Eq. (5.51) to obtain

$$\omega_R^2 = \frac{v_g a}{q V_p} \eta_i (I - I_{th}). \tag{5.58}$$

For high photon densities, it is important to appreciate that $a = a_0/(1 + \varepsilon N_p)$, and hence gain compression can affect the resonance frequency. However, for practical situations its effect is minimal. The curves in Fig. 5.5 also reveal that the resonance peak flattens and broadens out with increasing bias. This is due to the increase in damping present through γ which goes as ω_R^2 from Eq. (5.53). And unlike the resonance frequency, the K-factor is enhanced by ~ 2 due to gain compression, *independent of the magnitude of the photon density*.

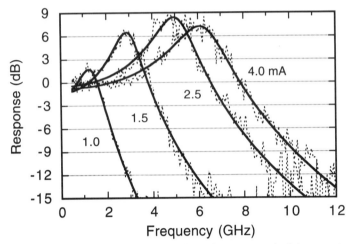

FIGURE 5.5   Small-signal intensity modulation of a mesa-etched intracavity-contacted 7 μm diameter 3-quantum-well InGaAs/GaAs VCSEL at the dc biases indicated. Its threshold current is 0.75 mA, the powers out are 0.2, 0.4, 0.9 and 1.4 mW, respectively, and the MSR is greater than 30 dB over the entire range [1]. Experimental curves are dashed and curve fits are solid.

5.3.2   Small-Signal Transient Response

To find the transient response to the *linearized* system characterized by Eq. (5.37), it is useful to rewrite the double-pole modulation transfer function as the product of two single-pole transfer functions:

$$H(\omega) = \frac{\omega_R^2}{\omega_R^2 - \omega^2 + j\omega\gamma} = \frac{\omega_R^2}{(j\omega + s_1)(j\omega + s_2)}. \tag{5.59}$$

The roots $s_{1,2}$ in the complex frequency plane are given by

$$s_{1,2} = \tfrac{1}{2}\gamma \pm j\omega_{osc}, \qquad \omega_{osc} = \omega_R\sqrt{1 - (\gamma/2\omega_R)^2}. \tag{5.60}$$

Note from Fig. 5.4 that $\omega_{osc}^2 = \tfrac{1}{2}(\omega_P^2 + \omega_R^2)$. From linear systems theory, these complex roots suggest solutions of the following form:

$$e^{-\gamma t/2}(e^{j\omega_{osc}t} \pm e^{-j\omega_{osc}t}), \tag{5.61}$$

plus any additional constants required to satisfy initial and final conditions.

In general we can say that in response to an abrupt change in the system (e.g., a step increase in current, a reduction in the photon lifetime, etc.), the transient carrier and photon densities will oscillate sinusoidally or "ring" at the rate ω_{osc} before eventually decaying to new steady-state values, the decay being characterized by the damping factor (hence the name). When the system is underdamped, i.e., when $\gamma/2\omega_R \ll 1$, the oscillations are characterized by the relaxation resonance frequency ω_R from Eq. (5.60).[6] However, $\gamma \propto \omega_R^2$ so that as ω_R increases with increasing output power, the system approaches critical damping, or $\gamma/2\omega_R \to 1$ and $\omega_{osc} \to 0$. Under these conditions, the densities will simply rise/fall exponentially to their new steady-state values. At even larger output powers the system becomes overdamped, $\gamma/2\omega_R > 1$, and the oscillation frequency becomes imaginary, $\omega_{osc} \to j\gamma_{osc}$, which slows the system down, increasing the rise and fall times of the transient solutions. The maximum modulation bandwidth of the laser is obtained when $\gamma/2\omega_R = 1/\sqrt{2}$ (which is slightly underdamped), yielding $\omega_{osc} = \omega_R/\sqrt{2}$ (and $\omega_P = 0$). Thus, for practical applications the laser is always underdamped and oscillations toward steady-state are expected.

Now let us consider a specific example. We want to know the effect of suddenly increasing the current by dI. If we define the transient responses as $dN(t)$ and $dN_p(t)$, then Eq. (5.61) can be used to obtain their general forms (i.e., damped sines or damped cosines), and Eq. (5.37) can be used to find the initial and final conditions. Initially, the system is at steady-state and we can set $dN(0) = dN_p(0) = 0$. At $t = 0^+$, the current has abruptly increased but no time has elapsed so no change in carrier density is observed. Since the photon density can only increase as a result of an increase in gain, which in turn follows the carrier density increase, it also must still be at its initial value. From this we conclude that $dN(0^+) = dN_p(0^+) = 0$. Equation (5.37) at $t = 0^+$ therefore becomes

$$\frac{d}{dt}\begin{bmatrix} dN(t) \\ dN_p(t) \end{bmatrix} = \frac{\eta_i}{qV}\begin{bmatrix} dI \\ 0 \end{bmatrix}. \qquad (t = 0^+) \qquad (5.62)$$

This gives us our set of initial conditions: the initial slope of $dN(t)$ is finite and proportional to the current step, while the initial slope of $dN_p(t)$ is zero (the carrier density must rise before the gain can begin to make a change in the photon density).

Due to the clamping of the carrier density above threshold, we know that the final value of $dN(t)$ must be approximately zero after the transient has damped out. This fact, together with the initial condition on the slope suggests that $dN(t)$ is a damped *sine* wave of the form:

$$dN(t) = dN_0 e^{-\gamma t/2} \sin \omega_{osc} t. \qquad (5.63)$$

[6] In some treatments, ω_R is *by definition* set equal to ω_{osc}, and the damping factor is set equal to $\gamma/2$ to comply with the natural roots of the transfer function. The problem with setting $\omega_R = \omega_{osc}$ is that ω_{osc} goes to zero at high output powers. The more standard definition of ω_R (as defined in this chapter) allows us to use Eq. (5.51) for all but very small output powers. As for the damping factor, the alternate definition causes no harm, and in fact would perhaps be the better way to define it. However, in the literature, the damping factor is usually defined such that the decay rate is $\gamma/2$, as we have it here. A word of warning: while either set of definitions can be used, one must be careful not to mix and match results found using either set (an error commonly made in the literature).

From Eq. (5.62), we find $dN_0 = \eta_i \, dI/(qV\omega_{osc})$.

For the photon density, we know the initial slope is zero. This suggests a damped *cosine* solution. However, since the current has increased, the steady-state output power must also increase. In other words, the final value of $dN_p(t)$ must have a finite positive value. Hence, the solution is more likely equal to the final value less a damped cosine of the same initial value:

$$dN_p(t) = dN_p(\infty)[1 - e^{-\gamma t/2} \cos \omega_{osc} t]. \tag{5.64}$$

Upon closer examination, we find that the initial slope of the term in brackets is not zero but is equal to $\gamma/2$. For small damping we can neglect this. However for the more general case, we need to add another term which is zero initially but has an initial slope $-\gamma/2$. A damped sine wave with the appropriate prefactor can be used to give

$$dN_p(t) = dN_p(\infty)\left[1 - e^{-\gamma t/2} \cos \omega_{osc} t - \frac{\gamma}{2\omega_{osc}} e^{-\gamma t/2} \sin \omega_{osc} t\right]. \tag{5.65}$$

This function now has zero initial slope and reaches $dN_p(\infty)$ in the steady state.

To determine $dN_p(\infty)$, we know that as $t \to \infty$, the time derivatives must go to zero in Eq. (5.37) allowing us to write

$$\begin{bmatrix} \gamma_{NN} & \gamma_{NP} \\ -\gamma_{PN} & \gamma_{PP} \end{bmatrix} \begin{bmatrix} dN(\infty) \\ dN_p(\infty) \end{bmatrix} = \frac{\eta_i}{qV} \begin{bmatrix} dI \\ 0 \end{bmatrix}. \qquad (t = \infty) \tag{5.66}$$

Using Cramer's rule to solve for the steady-state values, we obtain

$$dN(\infty) = \frac{\eta_i \, dI}{qV} \frac{\gamma_{PP}}{\omega_R^2} \approx \frac{\eta_i \, dI}{qV} \Gamma\tau_p \cdot \frac{a_p}{a}, \tag{5.67}$$

$$dN_p(\infty) = \frac{\eta_i \, dI}{qV} \frac{\gamma_{PN}}{\omega_R^2} \approx \frac{\eta_i \, dI}{qV} \Gamma\tau_p. \tag{5.68}$$

The photon density transient is now completely specified. However, we find that our initial assumption that $dN(t)$ must return to zero in the steady state is not quite right. Two factors affect the final carrier density. First of all, to increase the photon density, the gain must be brought closer to threshold as suggested in Fig. 5.3. As a result, the carrier density must increase slightly. In addition, gain compression reduces the gain at a given carrier density as the photon density is increased. Thus to maintain a given gain, the carrier density must again increase with increasing photon density. Both of these factors are included in γ_{PP}, however, for reasonable photon densities, the gain compression term dominates, leading to the approximation given in Eq. (5.67).

The question now is how to include a finite final value for $dN(t)$. We need a function which has zero slope initially and a finite final value, just like the photon density. In fact, noting that $dN(\infty) = (\gamma_{PP}/\gamma_{PN})\, dN_p(\infty)$ motivates us to write the new improved version of the carrier density transient as

$$dN(t) = dN_0\, e^{-\gamma t/2}\, \sin \omega_{osc} t + \frac{\gamma_{PP}}{\gamma_{PN}}\, dN_p(t), \tag{5.69}$$

where again $dN_0 = \eta_i\, dI/(qV\omega_{osc})$ and $\gamma_{PP}/\gamma_{PN} \approx a_p/a$.

While we have used rather ad hoc methods, one can easily verify that Eqs. (5.65) and (5.69) satisfy the differential rate equations (5.37), which themselves are valid as long as the transients are small compared to the steady-state carrier and photon densities. We could have alternatively derived these solutions through a more rigorous linear systems approach (Laplace transform of $H(\omega) \rightarrow$ impulse response $\rightarrow$ step function response).

Figure 5.6 gives two examples of responses to step function current transients, using both Eqs. (5.65) and (5.69) and an exact numerical solution of the rate equations. The first pair is for a small current step which gives excellent agreement, and the second is for an order of magnitude larger step, both starting from a bias point of twice threshold. For larger bias points proportionally larger transients can be approximated by Eqs. (5.65) and (5.69). However for the best fit, the analytic solutions should use the *final* photon and carrier density values to estimate the rate coefficients (note that both ω_R and γ are larger in Fig. 5.6b).

For small-signal data that involve a square-wave-like current waveform, time-shifted versions of Eqs. (5.65) and (5.69) can be superimposed to give the net time response of the photon and carrier density, respectively. From a linear systems point of view, a square wave input can be described as one positive step function plus a second time-delayed negative step function. The linearity of the system then guarantees that the overall solution is simply the sum of the responses created by each individual step function input.

5.3.3 Small-Signal FM Response or Frequency Chirping

As we saw in the previous sections, current modulation of the active region results in a modulation of both the photon density and the carrier density. The modulation of the carrier density modulates the gain, however, it also *modulates the index* of the active region n_a. As a result, the optical length of the cavity is modulated by the current, causing the resonant mode to shift back and forth in frequency. This frequency modulation (FM) of the laser may be desirable if we wish to dynamically tune the laser. However for intensity modulation (IM) applications, FM or *frequency chirping* broadens the modulated spectrum of the laser, hindering its effectiveness in optical fiber communications.

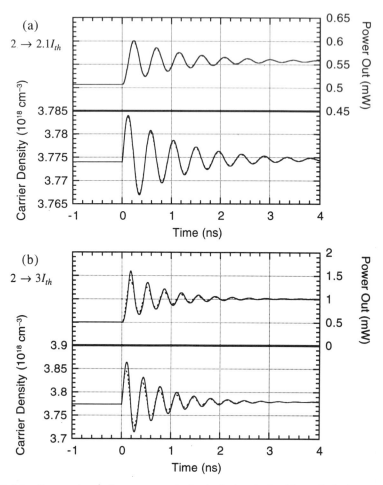

FIGURE 5.6 Comparison of exact numerical calculation (—) with analytical approxima-tions (––) for a small-signal step transient. An in-plane laser with parameters given in Table 5.1 is biased at twice threshold initially. Threshold gain, $\Gamma g_{th} = 50.6$ cm^{-1}; initial carrier density, $N = 3.77 \times 10^{18}$ cm^{-3}; initial photon density, $N_p = 1.23 \times 10^{14}$ cm^{-3}. (a) Transient from 2.0 to 2.1 I_{th}; (b) transient from 2.0 to 3.0 I_{th}.

To derive the general relationship between Δv and Δn_a, we write the frequency of mode m including a passive section as

$$v[\bar{n}_a L_a + \bar{n}_p L_p] = mc/2. \tag{5.70}$$

To find the frequency deviation, we take the differential of this equation and include the possibility of the active index changing with carrier density. The first differential term is

$$\Delta(v\bar{n}_a) = \Delta v\bar{n}_a + v\left[\frac{d\bar{n}_a}{dv}\Delta v + \frac{d\bar{n}_a}{dN}\Delta N\right] = \Delta v\bar{n}_{ga} + v\frac{d\bar{n}_a}{dN}\Delta N. \tag{5.71}$$

For the passive section, we simply have $\Delta(v\bar{n}_p) = \Delta v\bar{n}_{gp}$. In both cases, the g subscript denotes group index: $\bar{n}_g \equiv \bar{n} + v\,d\bar{n}/dv$. Setting the differential of the RHS of Eq. (5.70) to zero and solving for Δv, we obtain

$$\Delta v = -\frac{\Gamma_z v_g}{\lambda}\frac{d\bar{n}_a}{dN}\Delta N, \tag{5.72}$$

where $\Gamma_z = \bar{n}_{ga}L_a/(\bar{n}_{ga}L_a + \bar{n}_{gp}L_p)$ and v_g is the group velocity in the active section. In cases where $L_a \ll \lambda$, it can be shown that standing wave effects as considered in Appendix 5 would also appear in Γ_z (to derive this, one would need to take into account the index discontinuities and resulting reflections between the active and passive sections).

In Chapter 6, we will find that the change in effective propagation index can be related to the change in active material index through $\Delta\bar{n} = \Gamma_{xy}\Delta n$ (see Eqs. (6.12) and (A5.13)), allowing us to set $d\bar{n}/dN = \Gamma_{xy}\,dn/dN$. With this substitution and with $\Gamma_{xy}\Gamma_z = \Gamma\,(=V/V_p)$, we obtain

$$\Delta v = -\frac{\Gamma v_g}{\lambda}\frac{dn_a}{dN}\Delta N. \tag{5.73}$$

The complex index is given by $\tilde{n} = n + jn_i$, and the power gain is related to the imaginary index through: $g = 2k_0 n_i = 4\pi n_i/\lambda$. The relationship between how the real and imaginary indices are affected by the carrier density is described using what is referred to as the *linewidth enhancement factor*:

$$\alpha \equiv -\frac{dn/dN}{dn_i/dN} = -\frac{4\pi}{\lambda}\frac{dn/dN}{dg/dN} = -\frac{4\pi}{\lambda a}\frac{dn}{dN}, \tag{5.74}$$

where a is again used to define the differential gain. With this definition, the frequency shift in response to changes in carrier density becomes

$$\Delta v = \frac{\alpha}{4\pi}\Gamma v_g a\Delta N. \tag{5.75}$$

The amount of frequency modulation is proportional to the linewidth enhancement factor which is typically between 4–6 but can be as low as 2 in some active materials.

Using Eq. (5.44) for the ac carrier density modulation response, we

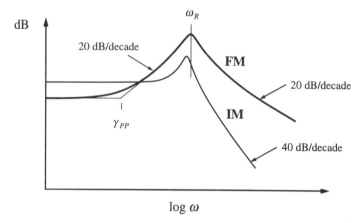

FIGURE 5.7 Qualitative comparison of the FM and IM response of a semiconductor laser.

immediately obtain

$$\frac{v_1}{I_1} = \frac{\alpha}{4\pi} \Gamma v_g a \frac{\eta_i}{qV} \cdot \frac{\gamma_{PP} + j\omega}{\omega_R^2} H(\omega). \tag{5.76}$$

Figure 5.7 compares the qualitative frequency response of this FM and the IM derived in Section 5.3.1. Both FM and IM exhibit a resonance peak, however, the FM response falls off at low frequencies at 20 dB/decade before leveling out below $\gamma_{PP}/2\pi$. Also, the peak of $\omega^2|H(\omega)|^2$ occurs directly at ω_R.

Dividing Eq. (5.76) by Eq. (5.56) and setting $\omega_R^2 = v_g a N_p/\tau_p$, we obtain the simple result:

$$v_1 = \frac{\alpha}{4\pi} (\gamma_{PP} + j\omega) \frac{N_{p1}}{N_p}. \tag{5.77}$$

So we discover that the frequency chirping of the lasing spectrum increases linearly with the intensity modulation depth (since $N_{p1}/N_p = P_1/P_0$). This effect is dramatically illustrated in Fig. 5.8 which displays the lasing spectrum under different degrees of intensity modulation for a constant modulation frequency. The peaks at both extremes of the modulated spectrum result from the time averaging of the sinusoidal modulation signal (i.e., the lasing frequency spends more time on average at the extremes of the sine wave).

If we define the FM modulation index as $M = v_1/f$ and the IM modulation index as $m = P_1/P_0$, then using the absolute magnitude of Eq. (5.77), the ratio of the FM-to-IM modulation index becomes

$$\frac{M}{m} = \frac{\alpha}{2} \sqrt{\left(\frac{\gamma_{PP}}{\omega}\right)^2 + 1}. \tag{5.78}$$

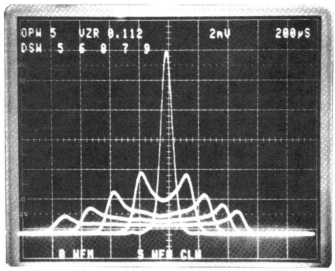

FIGURE 5.8   Time-averaged power spectra of a 1.3 μm InGaAsP laser under sinusoidal modulation at 100 MHz. The spectrum broadens with an increase in the modulation current due to frequency chirping. The horizontal scale is 0.5Å per division. After [2]. (Reprinted, by permission, from *Journal of Applied Physics*).

From Eq. (5.36), $\gamma_{PP} \approx \Gamma v_g a_p N_p$. For typical numbers, this term can be in the hundreds of MHz to few GHz range depending on the output power level. For modulation frequencies beyond this (i.e., $\omega \gg \gamma_{PP}$), $M/m \to \alpha/2$, providing us with a simple and very direct method of measuring α. For lower modulation frequencies, M/m becomes inversely proportional to the modulation frequency. By measuring M/m as a function of ω, one can then determine γ_{PP}, and ultimately determine the gain compression factor ε (via a_p) if curves are taken at different output power levels. Such a measurement is shown in Fig. 5.9. From these curves it is clear that for this laser, γ_{PP} varies from 1–3 GHz, and α is just over 6.

So far we have only considered changes in the index created by the modulation of the carrier density. However, at low modulation frequencies the temperature of the laser is also modulated when we apply current modulation. Since the index varies with temperature, we should expect the frequency modulation to be affected by thermal effects as well. The total FM of the laser may therefore be written as the sum of two contributions:

$$\frac{v_1}{I_1} = \left(\frac{\Delta v}{\Delta I}\right)_{carrier} + \left(\frac{\Delta v}{\Delta I}\right)_{thermal} \tag{5.79}$$

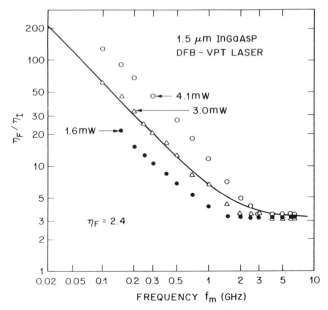

FIGURE 5.9 FM-to-IM modulation index ratio as a function of frequency for different output power levels [3], where $\eta_F \equiv M$ and $\eta_I \equiv m$. (Reproduced, by permission, from *Electronics Letters*).

where

$$\left(\frac{\Delta v}{\Delta I}\right)_{carrier} = \frac{\alpha}{4\pi} \frac{\eta_i}{qV_p} \frac{\varepsilon}{1 + \varepsilon N_p} \cdot (1 + j\omega/\gamma_{PP}) H(\omega),$$

$$\left(\frac{\Delta v}{\Delta I}\right)_{thermal} = \frac{(1 - \eta_{wp}) V_{th} Z_T \, dv/dT}{(1 + j\omega\tau_T)}.$$

In the first equation we have again set $\gamma_{PP} \approx \Gamma v_g a_p N_p$, expanded a_p using Eq. (5.32), and set $1/\tau_p = \Gamma v_g g$ which is a good approximation above threshold. In the second equation, η_{wp} is the wall-plug efficiency of the output power P_{out}/P_{in}, Z_T is the thermal impedance of the laser structure discussed in Chapter 2, and V_{th} is the threshold voltage of the laser which is assumed roughly constant above threshold. Also τ_T is the thermal time constant which is typically in the few microseconds range, yielding thermal cutoff frequencies in the few hundred kilohertz range. Finally, $dv/dT = -(c/\lambda^2) \, d\lambda/dT$ is the shift in the mode frequency with temperature. For 1 μm emission lasers, $d\lambda/dT \sim 0.06–0.08$ nm/K, which translates into $dv/dT \sim -20$ GHz/K.

For a thermal impedance of 0.1 K/mW, the *temperature* tuning below the thermal cutoff frequency is ~ -2 GHz/mA (assuming $(1 - \eta_{wp}) V_{th} \approx 1$ V). For a VCSEL the thermal impedance is closer to 1 K/mW implying an order of magnitude larger temperature tuning than with in-plane lasers. The *carrier*

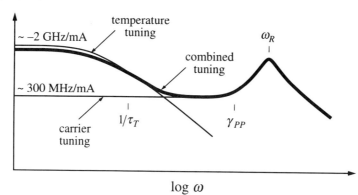

FIGURE 5.10 Sketch of the frequency tuning characteristics as a function of modulation frequency including both temperature and carrier effects (numbers indicated are for a typical in-plane laser). Since the temperature and carrier effects are opposite in sign at low modulation frequencies, one might expect a null in the combined tuning curve at the crossing point. This does not occur because the modulation phase of the temperature tuning shifts by 90° beyond the thermal cutoff frequency.

tuning below $\gamma_{PP}/2\pi$ is closer to ~ 300 MHz/mA for numbers given in Table 5.1, and is opposite in sign to temperature tuning. Figure 5.10 sketches the important characteristics of the combined frequency tuning as a function of modulation frequency. For modulation frequencies between $1/\tau_T < \omega < \gamma_{PP}$ (i.e., ~ 1 MHz $< f < \sim 1$ GHz), the response is flat with a value typically in the range of a few hundred MHz/mA.

5.4 LARGE-SIGNAL ANALYSIS

For deviations from the steady-state which are comparable to the steady-state values themselves, our previous differential analysis of the rate equations fails to provide accurate solutions as evidenced by Fig. 5.6(b). To determine the dynamic response of the laser for large-signal inputs, we must therefore return to the general rate equations (5.3) and (5.4). These equations are valid for large signal modulation provided the nonlinear changes in gain with the carrier and photon densities are included. Furthermore, they hold continuously below threshold and above threshold. Of course if the laser is always kept above threshold, the carrier density does not change by a large amount due to carrier clamping, even during large transients. The problem we face here is that these equations cannot be solved analytically. Therefore, to proceed we must use numerical techniques. Such numerical solutions to the rate equations are in principle found by iterating from one rate equation to the other using a small increment of time Δt in place of dt. As always, the power out can be calculated from N_p, and the frequency chirping can be obtained from the deviations of N from some reference value using Eq. (5.75).

5.4.1 Large-Signal Modulation: Numerical Analysis of the Multimode Rate Equations

In this section we consider numerical solutions to the multimode rate equations to give a feeling for the large-signal dynamic properties of multimode lasers. As mentioned in Section 5.2, the single-mode rate equations as given by Eqs. (5.3) and (5.4) give a good description of the dynamics of operation in single-frequency lasers. Generally, however, several different modes with different resonant wavelengths exist. In well-designed in-plane lasers these are generally due to different axial modes, while in VCSELs these are usually associated with different lateral modes. Each mode will have a different cavity loss and a different gain because of their different wavelength. These differences will be accentuated in lasers with frequency-dependent losses. Since all modes interact with a common reservoir of carriers, they are indirectly coupled even though they are orthogonal solutions to the electromagnetic wave equation.

Writing a separate photon density for each mode indexed by the integer m, Eqs. (5.3) and (5.4) become

$$\frac{dN}{dt} = \frac{\eta_i I}{qV} - (R_{sp} + R_{nr}) - \sum_m v_{gm} g_m N_{pm}, \tag{5.80}$$

$$\frac{dN_{pm}}{dt} = \left[\Gamma_m v_{gm} g_m - \frac{1}{\tau_{pm}} \right] N_{pm} + \Gamma_m R'_{spm}, \tag{5.81}$$

and g_m is given by Eq. (5.30) with a photon density N_{pm}. For numerical analysis, we replace dN, dN_{pm}, and dt by ΔN, ΔN_{pm}, and Δt, respectively, then multiply through by Δt. This time increment is set to some sufficiently small value so that the derivatives are accurately estimated. Since we can expect the response will involve oscillations near ω_R, we should set $\Delta t \ll 1/\omega_R$. The equations are then successively iterated to increment the carrier density and photon density for each mode from the initial values. Thus, after the ith iteration ($t = i\Delta t$), the carrier density is given by $N(i) = N(i-1) + \Delta N$, etc. In practice, numerical techniques such as the fourth-order Runge–Kutta method can be used to reduce the calculation time, since much larger time steps can be used for the same accuracy in the solutions (the simple iteration scheme requires extremely small time steps in order to successfully determine the solution without introducing large errors).

To model the spectral roll-off of the gain near the gain peak, we can often approximate the gain spectrum by a Lorentzian, as illustrated in Fig. 5.11. The gain experienced by each mode m can then be described by

$$g(N, N_p, m) = \frac{1}{(1 + (\Delta m / M^2))} \cdot \frac{g_0}{1 + \sum_n \varepsilon_{nm} N_{pn}} \ln\left(\frac{N + N_s}{N_{tr} + N_s}\right), \tag{5.82}$$

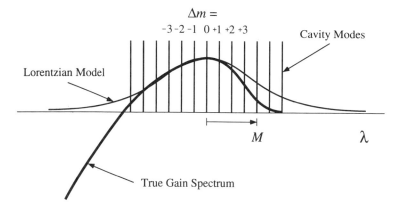

FIGURE 5.11 Simple Lorentzian model of the gain spectrum convenient for analyzing multimode lasers.

where Δm is the mode number measured away from the central mode ($\Delta m = 0$), which is assumed to be aligned with the gain peak, and M is the mode number where the gain has fallen to half of its peak value. In practice, M is adjusted to match the curvature of the gain spectrum near the gain peak, where the details of the spectrum are most relevant. The sum in the denominator accounts for intermodal gain compression.[7]

Figure 5.12 gives numerical plots of the carrier density and photon density for the various modes of an InGaAs/GaAs Fabry–Perot laser. No wavelength-dependent losses are assumed. The current at $t = 0$ increases from zero to $I^+ = 2I_{th}$. As shown, the carrier density initially increases as the active-region reservoir is filled. Little photon density exists until the carrier density reaches its threshold value. At this point, stimulated recombination begins to limit a further increase in carrier density as the photon density increases. This delay before the photon density "turns on" is called the *turn-on delay* of the laser.

The Fabry–Perot laser considered in Fig. 5.12 is obviously not a very good single-mode laser. Initially many modes turn on, and in the steady-state two strong side modes ($\Delta m = \pm 1$) persist. So we discover that the *dynamic* mode suppression ratio (MSR) is quite a bit different (worse) than the steady-state MSR considered at the end of Chapter 3. The challenge for state-of-the-art single-frequency lasers is to maintain a *dynamic* MSR of greater than 30 dB throughout such a turn-on transient.

Figure 5.13 illustrates the predicted turn-on characteristics of a VCSEL in

---

[7] If neighboring modes have no effect on each other's gain then only diagonal terms of ε_{nm} are nonzero ($\varepsilon_{mm} = \varepsilon$, $\varepsilon_{nm} = 0$ for $n \neq m$) and the sum can be replaced by εN_{pm}. On the other extreme, if all modes are affected equally by any photons present, then all terms of ε_{nm} are nonzero ($\varepsilon_{nm} = \varepsilon$) and the sum can be replaced by εN_p. Spectral and spatial hole burning of the carrier population will typically lead to gain compression which is somewhere in between these extremes.

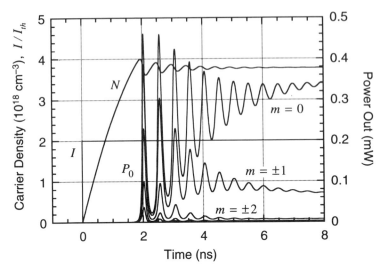

FIGURE 5.12 Large-signal modulation of an SQW InGaAs/GaAs in-plane laser using parameters given in Table 5.1. The relative gain width is set to $M = 25$ as might be appropriate for an in-plane laser. Twenty-five modes ($\Delta m = \pm 12$) are included in the calculation. The sum of gain compression terms is assumed to be εN_{pm}.

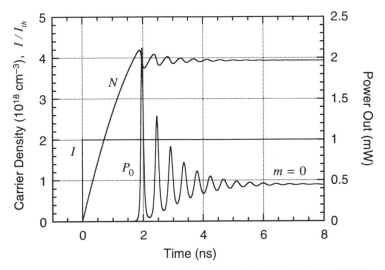

FIGURE 5.13 Large-signal modulation of a short-cavity laser using the VCSEL parameters given in Table 5.1. $M = 3$ is assumed.

which the axial mode spacing is much larger. Everything else is assumed to be the same. Here we observe the desired single-frequency behavior. In practice, such single-*axial* mode operation in VCSELs is compensated for by multi-*lateral* mode operation not considered in this calculation. One challenge in VCSEL design is to eliminate such lateral modes from appearing in the lasing spectrum. Such single axial mode operation can also be achieved with in-plane lasers by using frequency-dependent losses as discussed in Chapter 3 (for example, a DFB or DBR can be employed).

5.4.2   Turn-On Delay

The numerical simulations in Figs. 5.12 and 5.13 clearly illustrate that time is required for the carrier density to build up to the threshold value before light is emitted. This *turn-on delay*, defined as t_d, can be detrimental in high-speed data links, however it can be useful for measuring the carrier lifetime. As a first approximation to estimating t_d, we can simply calculate the initial slope of the carrier density and draw a straight line up to the threshold value N_{th}. To determine the initial slope, we use the carrier density rate equation, assuming no appreciable photon buildup:

$$\frac{dN}{dt} = \frac{\eta_i I}{qV} - (R_{sp} + R_{nr}).$$ (5.83)

If we start with some initial current I_i then the term in parentheses under steady-state conditions can be set equal to $\eta_i I_i/qV$. Now we change the current instantaneously to I_f. The initial slope then becomes

$$\left.\frac{\Delta N}{\Delta t}\right|_{t=0} = \frac{\eta_i}{qV}(I_f - I_i) \approx \frac{N_{th} - N_i}{t_d}.$$ (5.84)

Solving for the turn-on delay, we find

$$t_d \approx \frac{qV}{\eta_i}\frac{N_{th} - N_i}{I_f - I_i}.$$ (5.85)

Replacing N with I using the carrier lifetime or differential carrier lifetime (both evaluated at threshold), the turn-on delay for the two extreme initial bias limits becomes

$$t_d \approx \tau_{th}\frac{I_{th}}{I_f}, \qquad (I_i \approx 0)$$ (5.86)

$$t_d \approx \tau_{\Delta N, th}\frac{I_{th} - I_i}{I_f - I_i}, \qquad (I_i \approx I_{th})$$ (5.87)

where $1/\tau_{th} \approx A + BN_{th} + CN_{th}^2$ and $1/\tau_{\Delta N,th} \approx A + 2BN_{th} + 3CN_{th}^2$. For small initial biases, the turn-on delay is proportional to the total carrier lifetime (~ 3–4 ns), while for initial biases approaching threshold, the turn-on delay becomes proportional to the differential carrier lifetime (~ 1–2 ns). If we know all currents involved, it is a simple matter to estimate both total and differential carrier lifetimes from measured turn-on delays. In practical applications we want to avoid a large turn-on delay. This is accomplished by (1) increasing I_f as much as possible above threshold, or (2) adjusting the bias level I_i close to threshold. In fact for I_i larger than threshold, the turn-on delay becomes very small and can be estimated using the methods discussed in Section 5.3.2.

The first-order estimates of t_d in Eqs. (5.86) and (5.87) neglect the fact that as the carrier density increases, so does the recombination term in Eq. (5.83). This increase in recombination reduces the rate of increase of N from linear to sublinear as illustrated in Fig. 5.14. Thus, while Eqs. (5.86) and (5.87) yield a reasonable first approximation to t_d, they tend to underestimate the value somewhat.

For a constant carrier lifetime (i.e., $R_{sp} + R_{nr} = N/\tau_0$), Eq. (5.83) yields an exponential saturation of the carrier density at N_f:

$$N(t) = N_i + (N_f - N_i)(1 - e^{-t/\tau_0}). \tag{5.88}$$

Of course, N will never reach N_f due to stimulated emission which clamps N at N_{th} as Fig. 5.14 suggests. The time required for N to reach N_{th} defines the turn-on delay and we can set $N(t_d) = N_{th}$. Solving for the turn-on delay we obtain

$$t_d = \tau_0 \ln \frac{I_f - I_i}{I_f - I_{th}}, \qquad (\tau = \tau_0 \neq \tau(N)) \tag{5.89}$$

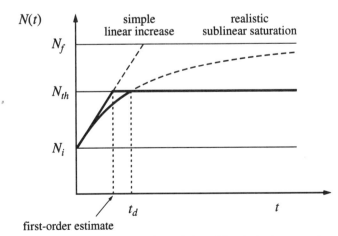

FIGURE 5.14   Carrier density as a function of time. The dashed curves show the carrier density increase without carrier clamping at threshold.

where the constant carrier lifetime has been multiplied through to convert all $N \rightarrow \tau_0 I$. Note from Fig. 5.14 that our initial linear approximation (5.85) is expected to work well when $N_f \gg N_i$ (or equivalently, when $I_f \gg I_i$), as a Taylor expansion of Eq. (5.89) for large $(I_f - I_i)$ verifies.

For short-wavelength lasers, the carrier lifetime is not a constant but is dominated by bimolecular recombination (i.e., $BN^2 \gg AN + CN^3$). For this case, the exponential saturation at N_f becomes a tanh function saturation, and the solution of Eq. (5.83) is given by

$$N(t) = N_f \tanh[BN_f t + \tanh^{-1}(N_i/N_f)]. \tag{5.90}$$

Again setting $N(t_d) = N_{th}$ and solving for t_d, we find

$$t_d = \tau_f \cdot \left[\tanh^{-1} \sqrt{\frac{I_{th}}{I_f}} - \tanh^{-1} \sqrt{\frac{I_i}{I_f}} \right], \quad (BN^2 \gg AN + CN^3) \tag{5.91}$$

where $\tau_f = 1/BN_f = \sqrt{qV/\eta_i I_f B}$, and we have set all $N \rightarrow \sqrt{\eta_i I/qVB}$.

In long-wavelength lasers, Auger recombination dominates and we can assume $CN^3 \gg AN + BN^2$. Unfortunately, no closed-form expression for $N(t)$ can be obtained for this case. However, a solution can still be obtained from the more general definition of t_d, which is found by directly integrating Eq. (5.83) and solving for t_d:

$$t_d = \int_{N_i}^{N_{th}} \frac{dN}{N_f/\tau_f - N/\tau(N)}, \tag{5.92}$$

where $\tau_f = \tau(N_f)$. It is easily shown that both Eqs. (5.89) and (5.91) can be found from (5.92) by setting $\tau(N) = \tau_0$ and $1/\tau(N) = BN$, respectively.

With $1/\tau(N) = CN^2$, the resulting expression for the turn-on delay is rather lengthy, but is important to include for long-wavelength laser applications. Expressing t_d as two components (where the first is typically much larger than the second), we have

$$t_d = t_{d1} + t_{d2}, \quad (CN^3 \gg AN + BN^2) \tag{5.93}$$

where

$$t_{d1} = \tau_f \cdot \tfrac{1}{6} \ln \left[\left(\frac{r_f - r_i}{r_f - 1} \right)^3 \frac{r_f^3 - 1}{r_f^3 - r_i^3} \right],$$

$$t_{d2} = \tau_f \cdot \frac{1}{\sqrt{3}} \left(\tan^{-1} \frac{1 + 2/r_f}{\sqrt{3}} - \tan^{-1} \frac{1 + 2r_i/r_f}{\sqrt{3}} \right).$$

The ratios are $r_f = N_f/N_{th}$ and $r_i = N_i/N_{th}$, or equivalently $r_f^3 = I_f/I_{th}$ and $r_i^3 = I_i/I_{th}$. In addition, $\tau_f = 1/CN_f^2$ where N_f is defined by $CN_f^3 = (\eta_i/qV)I_f$.

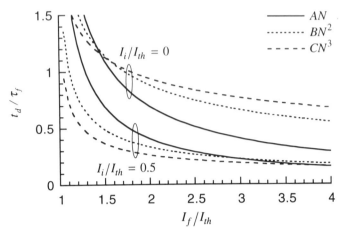

FIGURE 5.15 Normalized turn-on delay for different initial and final currents relative to threshold for the three types of recombination.

The turn-on delay for the case where $AN + BN^2 \gg CN^3$ can also be found.

Figure 5.15 shows t_d/τ_f for various initial-to-threshold and final-to-threshold current ratios using the three different expressions for the turn-on delay (5.89), (5.91), and (5.93). With no prebias ($I_i = 0$), the turn-on delay is generally on the order of the would-be carrier lifetime. In contrast, with a prebias and a final current level that is well above threshold, the turn-on delay can be reduced to just a fraction of the final carrier lifetime.

5.4.3   Large-Signal Frequency Chirping

Here we shall briefly introduce a useful analytic formula to calculate the frequency chirp from the known intensity modulation waveform which holds even under large-signal modulation. Taking the photon density rate equation and solving for the difference between the gain and loss, we obtain

$$\Gamma v_g g - \frac{1}{\tau_p} = \frac{1}{N_p} \frac{dN_p}{dt} - \frac{\Gamma R'_{sp}}{N_p}. \tag{5.94}$$

Because of carrier and gain clamping above threshold, a first-order expansion of the gain around its threshold value remains a good approximation even for large signals, allowing us to set

$$g = g_{th} + a(N - N_{th}) - a_p N_p. \tag{5.95}$$

Substituting this into (5.94), recognizing that $\Gamma v_g g_{th} = 1/\tau_p$, and using Eq. (5.75) to express the carrier density deviation as a frequency deviation, we obtain

$$v(t) - v_{th} = \frac{\alpha}{4\pi}\left[\frac{1}{N_p}\frac{dN_p}{dt} - \frac{\Gamma R'_{sp}}{N_p} + \Gamma v_g a_p N_p\right]. \tag{5.96}$$

With $N_p = N_{p0} + N_{p1}e^{j\omega t}$, the time-varying portion of the frequency chirp formula reduces to the small-signal result (5.77) derived earlier for $N_{p1} \ll N_{p0}$. If the modulated output power varies rapidly (on the order of a nanosecond or less), then the first term in brackets will usually dominate. Under these conditions and using the fact that $N_p \propto P_0$, the frequency chirp $\Delta v(t) = v(t) - v_{th}$ reduces to the simple result

$$\Delta v(t) = \frac{\alpha}{4\pi} \cdot \frac{1}{P_0(t)}\frac{dP_0(t)}{dt}. \tag{5.97}$$

With this equation, the frequency chirp for large-signal modulation can be determined directly from the shape of the modulated signal (e.g., square-wave, sine-wave, Gaussian, etc.). If we are interested in minimizing the frequency chirp, Eq. (5.97) can be used to predict which output waveforms produce the lowest chirp. We can then tailor the current input signal accordingly.

5.5   RELATIVE INTENSITY NOISE AND LINEWIDTH

Up until now we have only considered intensity and frequency responses due to the deliberate modulation of the current or some other cavity parameter. In the steady state, it has been assumed that the carrier and photon densities are constant. However, in reality random carrier and photon recombination and generation events produce instantaneous time variations in the carrier and photon densities, even with no applied current modulation. The variations in photon density lead to variations in the magnitude of the output power, which provides a noise floor, and the variations in carrier density result in variations in the output wavelength, which creates a finite spectral linewidth for the lasing mode. Before launching into the specifics of how to deal with these random noise sources, it is worth considering the implications of intensity and frequency noise in practical laser applications for motivational purposes.

5.5.1   General Definition of RIN and the Spectral Density Function

Figure 5.16 illustrates a noisy laser output for both analog and digital signal transmission. For analog applications, the noise is quantified using the electrical signal-to-noise ratio (SNR). For the laser output defined in Fig. 5.16, the SNR can be written as

$$\text{SNR} = \frac{\langle i_S^2 \rangle}{\langle i_N^2 \rangle} = \frac{\langle (P_1 \sin \omega t)^2 \rangle}{\langle \delta P(t)^2 \rangle} = \frac{m^2}{2}\frac{P_0^2}{\langle \delta P(t)^2 \rangle}, \tag{5.98}$$

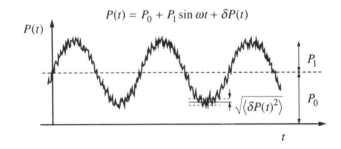

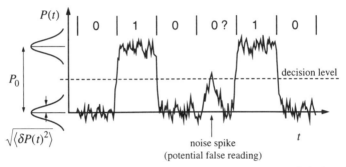

FIGURE 5.16   Noise in modulated laser signals for both analog and digital applications.

where the IM modulation index is given by $m = P_1/P_0$. The $\langle \ \rangle$ denote the time average.

For digital applications, a decision level at the midpoint defines whether a '0' or '1' is recorded. If the noise happens to exceed $P_0/2$ in Fig. 5.16, then a false recording might be made. If the noise has a Gaussian distribution around the mean power level, then in order to reduce the probability of finding $|\delta P(t)| > P_0/2$ to less than 1 in 10^9 (i.e., a *bit-error rate* $< 10^{-9}$), we require that [4]

$$\frac{P_0^2}{\langle \delta P(t)^2 \rangle} > (11.89)^2, \qquad \text{(for BER} < 10^{-9}) \tag{5.99}$$

where $\langle \delta P(t)^2 \rangle$ is the mean-square of the assumed Gaussian noise distribution.

For either of the above two applications, we find it useful to quantify the *relative intensity noise* (RIN) of the laser:

$$\text{RIN} \equiv \frac{\langle \delta P(t)^2 \rangle}{P_0^2}. \tag{5.100}$$

The RIN is often described in decibels, or $10 \log_{10}(\text{RIN})$. For analog applications, if a given electrical SNR is required, then Eq. (5.98) can be used to determine the maximum allowable RIN. For example, if we require the SNR > 50 dB with $m = 1$, then the laser must have a RIN < −53 dB. Alternatively, for a bit-error rate (BER) < 10^{-9} in digital applications, Eq. (5.99) suggests that the laser must have a RIN < −21.5 dB.

To quantify the output power fluctuations (and hence the RIN), it is more convenient to work in the frequency domain, making use of the Fourier transform pairs:

$$\delta P(t) = \frac{1}{2\pi} \int_{-\infty}^{+\infty} \delta P(\omega) e^{j\omega t} \, d\omega, \tag{5.101}$$

$$\delta P(\omega) = \int_{-\infty}^{+\infty} \delta P(t) e^{-j\omega t} \, dt, \tag{5.102}$$

where $\delta P(\omega)$ is the component of the noise which fluctuates at the frequency, ω. Now suppose we were to use a spectrum analyzer to measure the *electrical* power (i.e., the square of the optical power) associated with the noise. If the spectrum analyzer applies a narrowband filter to the signal with a passband described by $F(\omega)$, then the measured mean-square time-averaged signal would be given by

$$\langle \delta P(t)^2 \rangle = \frac{1}{(2\pi)^2} \int_{-\infty}^{+\infty} \int_{-\infty}^{+\infty} \langle \delta P(\omega) \delta P(\omega')^* \rangle F(\omega) F(\omega')^* e^{j(\omega - \omega')t} \, d\omega \, d\omega'. \tag{5.103}$$

For completely random noise, the magnitude of the noise at any given frequency is completely uncorrelated with the magnitude of the noise at any other frequency. As a result, when the product of two frequency components is averaged over time, there is a delta function correlation between them (see Appendix 13). The strength of the delta function correlation is defined as the *spectral density*, $S_{\delta P}(\omega)$, of $\delta P(t)$ at ω, and we can write

$$\langle \delta P(\omega) \delta P(\omega')^* \rangle = S_{\delta P}(\omega) \cdot 2\pi \delta(\omega - \omega'). \tag{5.104}$$

With this substitution, the *measured* mean-square power fluctuation reduces to

$$\langle \delta P(t)^2 \rangle = \frac{1}{2\pi} \int_{-\infty}^{+\infty} S_{\delta P}(\omega) |F(\omega)|^2 \, d\omega. \tag{5.105}$$

If the measurement filter is centered at ω_0, and is narrowband relative to variations in the spectral density, then with $F(\omega_0) = 1$ we obtain

$$\langle \delta P(t)^2 \rangle \approx S_{\delta P}(\omega_0) \int_{-\infty}^{\infty} |F(\omega)|^2 \, df = S_{\delta P}(\omega_0) \cdot 2\Delta f. \tag{5.106}$$

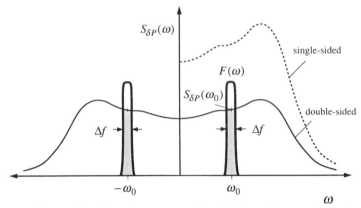

FIGURE 5.17 Measured noise using a narrowband filter. Because the noise spectral density is always an even function of frequency ($S_{\delta P}(-\omega) = S_{\delta P}(\omega)$), we can fold the spectrum in half, if desired, and define a *single-sided* spectrum existing only in the positive frequency domain that is a factor of two larger than the *double-sided* spectrum.

This relation is graphically illustrated in Fig. 5.17 for an arbitrary noise spectrum. Note that the effective measurement bandwidth is $2\Delta f$ since we must include both positive and negative frequencies. We could alternatively define the spectral density as "single-sided" existing only in the positive frequency domain as indicated by the dashed line in Fig. 5.17. In this case, the measurement bandwidth would simply be Δf and the factor of two would be lumped into the single-sided spectral density. The choice of using a single-sided or double-sided spectral density is academic as long as we are consistent. In this chapter, the spectral density will always be defined as double-sided. Finally note that the spectral density has units of (seconds) × (fluctuating variable units)^2.

In terms of the spectral density of the noise accompanying the signal, we can redefine the relative intensity noise as

$$\frac{\text{RIN}}{\Delta f} = \frac{2S_{\delta P}(\omega)}{P_0^2}, \tag{5.107}$$

where Δf is the filter bandwidth of the measurement apparatus (if the spectral density is defined as single-sided, then the factor of 2 should be removed in (5.107)). Because the measurement bandwidth can vary from application to application, it is common to specify the quantity on the left as RIN in dB/Hz or RIN per unit bandwidth. The full RIN is then found by integrating the RIN per unit bandwidth over the (single-sided) detection bandwidth of the system of practical interest. In designing a communications system, the desired SNR or BER sets a maximum limit on the total RIN of the laser. If the RIN spectrum is flat, then the required RIN per unit bandwidth of the laser is found from

$$RIN(dB/Hz) = RIN(dB) - 10 \log_{10}(\Delta f \text{ [in Hz]}). \qquad (5.108)$$

For example, in a digital transmission link with a 2 Gbit/s (1 GHz) system bandwidth and a required BER $< 10^{-9}$ (i.e., RIN < -21.5 dB), the laser must have an average RIN(dB/Hz) < -21.5 dB $- 90 = -111.5$ dB/Hz. If the system bandwidth is increased, the laser RIN per unit bandwidth must be decreased in order to maintain the same total RIN.

5.5.2 The Schawlow–Townes Linewidth

In addition to intensity noise, the laser also produces frequency noise which can adversely affect the lasing spectrum. In single-mode lasers, such as the DFB and DBR types introduced in Chapter 3, the spectral width of the laser's output is reduced to that of a single mode, however, the linewidth of this single mode is still finite due to laser noise. Unless great care is taken, experiments typically show diode laser linewidths much greater than a megahertz. For many applications, such as sensor or communication systems using coherent detection, it is desirable to have linewidths much less than a megahertz. Thus, an understanding of the inherent linewidth of diode lasers is of great practical importance.

The linewidth of a diode laser results from phase fluctuations in its output. These arise from two basic sources: (1) spontaneous emission and (2) carrier density fluctuations. The first is inherent in all lasers, resulting simply from the random addition of spontaneously emitted photons to the quasi-coherent resonant cavity mode. The second is of significance only in diode lasers, and it results from the proportionality between ΔN and Δv characterized by Eq. (5.75) [5]; the constant of proportionality contains the *linewidth enhancement factor*, α. This factor exists because both the gain and the index of refraction depend directly upon the carrier density.

We will develop the full expression for the laser linewidth a little later, after the treatment of frequency noise. For now, we wish to consider a simplified derivation of the spontaneous emission component of the laser linewidth which, while not entirely correct for lasers above threshold, does provide the reader with an intuitive feel for the origin of a finite laser linewidth.

In the derivation of the rate equations in Chapter 2, the cavity lifetime was introduced as the natural decay rate of photons in the resonant cavity in the absence of stimulated or spontaneous emission sources. Therefore, in the absence of sources, Eq. (5.4) has the solution

$$N_p(t) = N_{p0} e^{-t/\tau_p}. \qquad (5.109)$$

The corresponding time dependence of the field is

$$\mathcal{E}(t) = E_0 e^{j\omega_0 t} e^{-t/2\tau_p} u(t), \tag{5.110}$$

where $u(t)$ is a unit step function which turns on at time zero to indicate that the field is instantaneously created at $t = 0$ by a stimulated emission event, for example. The Fourier transform of this time domain response gives the frequency domain response of the laser cavity. The Fourier transform of an exponential is a Lorentzian. This undriven result, sometimes called the cold cavity response, is given by

$$|\mathcal{E}(\omega)|^2 = \frac{|\mathcal{E}(\omega_0)|^2}{1 + (\omega - \omega_0)^2 (2\tau_p)^2}. \tag{5.111}$$

From this we can see that the full-width half-maximum (FWHM) linewidth of the cold cavity is $\Delta\omega = 1/\tau_p$. This spectral width corresponds to the filter bandwidth of the Fabry–Perot resonator mode with no active material present (hence cold cavity response). The key point here is that the resonance width is linked with the photon decay rate. Now, if we add back the stimulated term which is responsible for gain in the cavity, we see from Eq. (5.4) that the same exponential solution in time is obtained, but it is now characterized by a new *effective* cavity lifetime:

$$\frac{1}{\tau'_p} = \frac{1}{\tau_p} - \Gamma v_g g. \tag{5.112}$$

The effective cavity lifetime *increases* as the gain in the cavity compensates for cavity losses. Thus with gain, the FWHM linewidth becomes $\Delta\omega = 1/\tau'_p$, and so as τ'_p increases, the resonance width decreases. As illustrated in Section 5.2.2, Eq. (5.4) can be solved in the steady state for N_p:

$$N_p = \frac{\Gamma R'_{sp}}{1/\tau_p - \Gamma v_g g}. \tag{5.113}$$

Using (5.113) to replace $[1/\tau_p - \Gamma v_g g]$ in (5.112), we can express the driven FWHM linewidth as

$$\Delta\nu_{spon} = \frac{1}{2\pi\tau'_p} = \frac{\Gamma R'_{sp}}{2\pi N_p}. \qquad \text{(spontaneous only)} \tag{5.114}$$

Equation (5.114) is equivalent to the famous Schawlow–Townes linewidth formula [6]. One central conclusion of this formula is that the linewidth varies inversely with photon density (or output power). And because the photon density in a laser can grow very large, the linewidth can collapse into a very narrow spectral line—one of the defining characteristics of lasers.

Unfortunately, this intuitive derivation has some shortcomings. Equation (5.114) does correctly give the below-threshold linewidth and is therefore accurate for amplified spontaneous emission problems. However, above threshold, the nonlinear coupling between the rate equations suppresses one of the two quadrature components of the noise (the field amplitude fluctuations are stabilized above threshold), resulting in a factor of 2 reduction in the linewidth predicted here [7, 8]. With a correction factor of 1/2, Eq. (5.114) becomes the *modified* Schawlow–Townes linewidth formula:

$$(\Delta v)_{ST} = \frac{\Gamma R'_{sp}}{4\pi N_p}. \tag{5.115}$$

The modified Schawlow–Townes linewidth, however, still only considers spontaneous emission noise and does not include carrier noise. To describe carrier noise, we will need to develop the Langevin rate equation approach to laser noise.

5.5.3   The Langevin Approach

To determine the laser RIN and the carrier noise, we must find the spectral density of the output power and carrier noise fluctuations. For this purpose, we introduce *Langevin* noise sources $F_N(t)$ and $F_P(t)$ as the ac driving sources for the carrier and photon densities, respectively. These sources are assumed to be white noise (see Appendix 13), and are assumed small enough that we can make use of the *differential* rate equations. For a constant drive current ($dI = 0$), Eqs. (5.33) and (5.34) become

$$\frac{d}{dt}(dN) = -\gamma_{NN}\, dN - \gamma_{NP}\, dN_p + F_N(t), \tag{5.116}$$

$$\frac{d}{dt}(dN_p) = \gamma_{PN}\, dN - \gamma_{PP}\, dN_p + F_P(t). \tag{5.117}$$

The rate coefficients are defined by Eqs. (5.35) as always. To solve these equations it is again convenient to place them in matrix form:

$$\frac{d}{dt}\begin{bmatrix} dN \\ dN_p \end{bmatrix} = \begin{bmatrix} -\gamma_{NN} & -\gamma_{NP} \\ \gamma_{PN} & -\gamma_{PP} \end{bmatrix}\begin{bmatrix} dN \\ dN_p \end{bmatrix} + \begin{bmatrix} F_N(t) \\ F_P(t) \end{bmatrix}. \tag{5.118}$$

To determine the spectral densities we must first transform to the frequency domain. Replacing all time-dependent variables with equivalent versions of Eq. (5.101), we obtain for each frequency component:

$$\begin{bmatrix} \gamma_{NN} + j\omega & \gamma_{NP} \\ -\gamma_{PN} & \gamma_{PP} + j\omega \end{bmatrix}\begin{bmatrix} N_1(\omega) \\ N_{p1}(\omega) \end{bmatrix} = \begin{bmatrix} F_N(\omega) \\ F_P(\omega) \end{bmatrix}, \tag{5.119}$$

where N_1, N_{p1}, F_N, and F_P represent the components of the noise which fluctuate at frequency ω. This result is analogous to the small-signal result obtained in Section 5.3.1. Using Cramer's rule and the definitions in Section 5.3.1, we immediately obtain

$$N_1(\omega) = \frac{H(\omega)}{\omega_R^2} \begin{vmatrix} F_N(\omega) & \gamma_{NP} \\ F_P(\omega) & \gamma_{PP} + j\omega \end{vmatrix}, \tag{5.120}$$

$$N_{p1}(\omega) = \frac{H(\omega)}{\omega_R^2} \begin{vmatrix} \gamma_{NN} + j\omega & F_N(\omega) \\ -\gamma_{PN} & F_P(\omega) \end{vmatrix}. \tag{5.121}$$

Using Eq. (5.104) we can define the carrier and photon density spectral densities as

$$S_N(\omega) = \frac{1}{2\pi} \int \langle N_1(\omega) N_1(\omega')^* \rangle \, d\omega', \tag{5.122}$$

$$S_{N_p}(\omega) = \frac{1}{2\pi} \int \langle N_{p1}(\omega) N_{p1}(\omega')^* \rangle \, d\omega'. \tag{5.123}$$

Multiplying both sides of Eq. (5.120) by $N_1(\omega')^*$, and both sides of Eq. (5.121) by $N_{p1}(\omega')^*$, taking the time average, and integrating over ω', we finally obtain

$$S_N(\omega) = \frac{|H(\omega)|^2}{\omega_R^4} [\gamma_{NP}^2 \langle F_P F_P \rangle - 2\gamma_{PP}\gamma_{NP} \langle F_P F_N \rangle + (\gamma_{PP}^2 + \omega^2) \langle F_N F_N \rangle], \tag{5.124}$$

$$S_{N_p}(\omega) = \frac{|H(\omega)|^2}{\omega_R^4} [(\gamma_{NN}^2 + \omega^2) \langle F_P F_P \rangle + 2\gamma_{NN}\gamma_{PN} \langle F_P F_N \rangle + \gamma_{PN}^2 \langle F_N F_N \rangle], \tag{5.125}$$

where the Langevin noise source spectral densities are defined by

$$\langle F_i F_j \rangle = \frac{1}{2\pi} \int \langle F_i(\omega) F_j(\omega')^* \rangle \, d\omega'. \tag{5.126}$$

Because F_N and F_P are white noise sources, their noise spectral densities $\langle F_i F_j \rangle$ are uniformly distributed over all frequencies. Hence, the various $\langle F_i F_j \rangle$ can be regarded as constants in the frequency domain (see Appendix 13 for further details). Equations (5.124) and (5.125) therefore reveal that the carrier and photon density fluctuations follow a $(a_1 + a_2\omega^2)|H(\omega)|^2$ spectral dependence, peaking at the relaxation resonance frequency (we will quantify a_1 and a_2 below). This behavior is not surprising since the natural resonance of the carrier–photon system would be expected to accentuate and amplify any noise existing near that resonance. Furthermore, the inverse dependence on ω_R^4 suggests that the fluctuations and corresponding noise decrease with increasing output power. To fully quantify these relationships, we need to expand the terms within square brackets.

5.5.4 Langevin Noise Spectral Densities and RIN

Equations (5.124) and (5.125) reveal that with the Langevin method, the evaluation of the laser noise boils down to evaluating the spectral densities or *noise correlation strengths* $\langle F_i F_j \rangle$ between the various Langevin noise sources. Assuming $N_p V_p \gg 1$ above threshold, Appendix 13 shows that these terms reduce to the following:

$$\langle F_P F_P \rangle = 2\Gamma R'_{sp} N_p, \tag{5.127}$$

$$\langle F_N F_N \rangle = 2R'_{sp} N_p / \Gamma - v_g g N_p / V + \eta_i (I + I_{th}) / qV^2, \tag{5.128}$$

$$\langle F_P F_N \rangle = -2R'_{sp} N_p + v_g g N_p / V_p. \tag{5.129}$$

The various factors appearing in these equations are essentially the summation of the shot noise associated with the random generation and recombination/escape of both photons and carriers in both reservoirs. The shot noise associated with the injected current (which gives rise to the $\eta_i I / qV^2$ term appearing in Eq. (5.128)) can in principle be eliminated or at least reduced by careful design of the current drive circuitry [9]—a point we will return to later in this section.

With Eqs. (5.127) through (5.129) the carrier (5.124) and photon (5.125) spectral density functions can be expanded. The desired output power spectral density function, $S_{\delta P}(\omega)$ needed in Eq. (5.107), can then be calculated. However, this calculation is complicated by an additional noise term in the output power due to the negative correlation between photons reflected and transmitted at the output mirror [10]. Thus, as shown in Appendix 13, $S_{\delta P}(\omega)$ is not simply related to $S_{N_p}(\omega)$ by the expected factor $(h\nu V_p v_g \alpha_m F)^2$ or $(\eta_0 h\nu V_p / \tau_p)^2$. After first deriving $S_{N_p}(\omega)$, Appendix 13 shows that for above-threshold conditions, the output power spectral density can be written as

$$S_{\delta P}(\omega) = h\nu P_0 \cdot \left[\frac{a_1 + a_2 \omega^2}{\omega_R^4} |H(\omega)|^2 + 1 \right], \tag{5.130}$$

where

$$a_1 = \frac{8\pi(\Delta\nu)_{ST} P_0}{h\nu} \frac{1}{\tau_{\Delta N}^2} + \eta_0 \omega_R^4 \left[\frac{\eta_i (I + I_{th})}{I_{st}} - 1 \right],$$

$$a_2 = \frac{8\pi(\Delta\nu)_{ST} P_0}{h\nu} - 2\eta_0 \omega_R^2 \frac{\Gamma a_p}{a},$$

and $(\Delta\nu)_{ST} = \Gamma R'_{sp} / 4\pi N_p$, $I_{st} = qP_0 / \eta_0 h\nu = \eta_i (I - I_{th})$. Also, P_0 defines the power out of the desired facet.

If the emitted field is in a perfectly coherent state, the output power noise (double-sided) spectral density is limited to a minimum value of $h\nu P_0$. This quantum noise floor of the coherent field is often referred to as the *standard quantum limit*, or the shot noise floor. The two contributions within the square

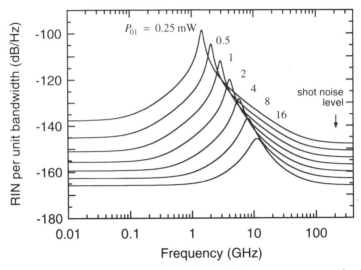

FIGURE 5.18 Calculated relative intensity noise at different output power levels for an InGaAs/GaAs in-plane laser with parameters given in Table 5.1 .

brackets of (5.130) can therefore be thought of as the excess intensity noise and the inherent quantum noise of the laser. At low output powers, the excess noise dominates and is dramatically enhanced near the relaxation resonance frequency of the laser. At high powers, the laser generally quiets down to the standard quantum limit except near the resonance where the excess noise can persist even at very high output powers. Figure 5.18 illustrates these characteristics for the in-plane laser characterized in Table 5.1.

5.5.4.1 Characteristics of the RIN Spectrum The laser RIN plotted in Fig. 5.18 is related to $S_{\delta P}(\omega)$ through Eq. (5.107) and can be written as

$$\frac{\text{RIN}}{\Delta f} = \frac{2h\nu}{P_0}\left[\frac{a_1 + a_2\omega^2}{\omega_R^4}|H(\omega)|^2 + 1\right]. \qquad (5.131)$$

The prefactor $2h\nu/P_0$ is the standard quantum limit for the minimum RIN of the laser. The frequency coefficients, a_1 and a_2, are as defined in Eq. (5.130). The first terms of both a_1 and a_2 are independent of power, while the latter terms depend on P_0^2 and P_0, respectively. At low powers, these latter terms can be neglected and the RIN reduces to

$$\frac{\text{RIN}}{\Delta f} = 16\pi(\Delta\nu)_{ST}\frac{1/\tau_{\Delta N}^2 + \omega^2}{\omega_R^4}|H(\omega)|^2 + \frac{2h\nu}{P_0}. \qquad \text{(low power)} \qquad (5.132)$$

This form of the RIN is commonly used in laser applications, and in fact is a reasonable approximation at higher powers as well, as long as the laser current source is shot noise-limited. With Eq. (5.132), the measured RIN spectrum can be modeled using four fitting parameters: $\tau_{\Delta N}$, γ, ω_R, and $(\Delta v)_{ST}$ [11].

When the excess noise term in Eq. (5.132) dominates, the RIN spectrum levels off to a constant value for $\omega < 1/\tau_{\Delta N}$, which is typically in the range of 100 MHz, as shown in Fig. 5.18 for the 0.25 and 0.5 mW curves. At higher powers, this range of constant RIN increases up to a few GHz, as the low frequency RIN saturates at the shot noise floor. Setting $\omega = 0$ in Eq. (5.131) and using a_1 as defined in Eq. (5.130), the low frequency RIN becomes

$$\frac{\text{RIN}}{\Delta f} = \frac{16\pi(\Delta v)_{ST}}{\omega_R^4 \tau_{\Delta N}^2} + \frac{2hv}{P_0}\left[\eta_0 \frac{\eta_i(I + I_{th})}{I_{st}} + (1 - \eta_0)\right]. \qquad (\omega \ll \omega_R) \qquad (5.133)$$

The first term decreases as $1/P_0^3$ (since $(\Delta v)_{ST} \propto 1/P_0$ and $\omega_R^4 \propto P_0^2$) and hence quickly drops below the shot noise floor with increasing power. The second term converges toward the shot noise $1/P_0$ power dependence at high current levels where $\eta_i(I + I_{th})/I_{st} \to 1$. In Fig. 5.18, we can see this trend by examining the intervals between every doubling of power. At low frequencies, the steps change from 9 dB to 3 dB with increasing power as the power dependence changes from $1/P_0^3$ to $1/P_0$.

At the RIN peak, $a_2\omega^2 \gg a_1$ in Eq. (5.131). Hence, the frequency dependence in this range is characterized by $\omega^2|H(\omega)|^2$. Using Eq. (5.46), it can be shown that the peak of this function occurs at ω_R regardless of the extent of damping, in contrast to $|H(\omega)|^2$ which peaks right at ω_R only when $\gamma/\omega_R \ll 1$ (see Fig. 5.4). Setting $|H(\omega_R)|^2 = (\omega_R/\gamma)^2$ in Eq. (5.131), and neglecting both the noise floor limit and the second term of a_2 in Eq. (5.130), we obtain the simple result

$$\frac{\text{RIN}}{\Delta f} = \frac{16\pi(\Delta v)_{ST}}{\gamma^2}. \qquad (\omega = \omega_R) \qquad (5.134)$$

In general, the damping factor $\gamma = Kf_R^2 + \gamma_0$ from Eq. (5.52). However, at higher powers, $\gamma \approx Kf_R^2 \propto P_0$. Hence, the RIN peak converges toward a $1/P_0^3$ power dependence, since $(\Delta v)_{ST} \propto 1/P_0$. In other words, the RIN peak drops by 9 dB with every doubling of power, as Fig. 5.18 indicates at the higher power levels. Beyond the resonance frequency, the RIN reduces to the standard quantum limit which drops by 3 dB with every doubling of power.

At 1 mW of output power, the laser in Fig. 5.18 roughly has a Schawlow–Townes linewidth of 1 MHz and a damping factor of 3×10^9/s, which gives a peak RIN $= -112$ dB/Hz. This peak level of noise barely satisfies the criterion for a 2 Gbit/s (1 GHz) digital transmission link (see the example following Eq. (5.108)). However, the peak occurs beyond the system bandwidth and will not add to the detected noise. When the 1 mW RIN spectrum in Fig. 5.18 is integrated from 0 to 1 GHz, the average RIN per unit bandwidth is only -140 dB/Hz, easily satisfying the transmission link requirements.

5.5.4.2 Noise-Free Operation The component of the laser noise which derives from the noise of the current source appears in the second term of a_1 in Eq. (5.130). This is one component of the noise we can modify externally, and ideally eliminate. In Eq. (5.131), $a_2\omega^2|H(\omega)|^2$ is responsible for the RIN peak at ω_R, and $2h\nu/P_0$ dominates the spectrum beyond the RIN peak. Thus, $a_1|H(\omega)|^2$ only significantly affects the low frequency RIN away from the RIN peak, or at $\omega \ll \omega_R$. Within the denominator of $a_1|H(\omega)|^2$ then, we can set $(\omega_R^2 - \omega^2)^2 \approx \omega_R^4$ without introducing much error in the total RIN spectrum (this works well as long as $(\gamma/\omega_R)^2 \gg 2/(1 + 2\sqrt{r_p})$ where r_p is the RIN peak relative to the shot noise floor). With this modification, and also concentrating on moderate-to-high powers such that we can neglect the $1/P_0^3$ term of a_1, Eq. (5.131) reduces to

$$\frac{\text{RIN}}{\Delta f} \approx \frac{2h\nu}{P_0}\left[\frac{\eta_0\eta_i(I + I_{th})/I_{st} + (1 - \eta_0) + \omega^2\tau_p''^2}{1 + \omega^2\tau_p''^2}\right] + \frac{2h\nu}{P_0}\frac{a_2\omega^2}{\omega_R^4}|H(\omega)|^2, \quad (5.135)$$

where $\tau_p'' \equiv \gamma/\omega_R^2 \approx \tau_p[1 + \Gamma a_p/a]$. The first term provides the RIN noise floor, while the second term provides the RIN peak which rises above the noise floor near ω_R. The various contributions to the noise floor originate from (in order of appearance in the first term): (1) injected current (I) and carrier recombination (I_{th}), (2) the random selection of output photons (partition noise), and (3) the random delay associated with photons escaping the cavity (only observable at frequencies comparable to or greater than $1/\tau_p''$).

The injected current contribution to the RIN noise floor can be reduced if a current source with sub-shot noise characteristics is used. Yamamoto has suggested and demonstrated along with others that if a high impedance source is used in driving the laser, then sub-shot noise current injection is possible [9, 12]. To accommodate such cases, we can generalize Eq. (5.135) by setting $I \rightarrow S_I/q$, where S_I is the double-sided spectral density of the injection current. For a perfectly quiet current source, $S_I = 0$. For a shot noise-limited current source, $S_I = qI$. If we have a very efficient laser with $\eta_0 \rightarrow 1$, and if we are far above threshold such that $I \gg I_{th}$ and $I_{st} \approx \eta_i I$, then Eq. (5.135) (excluding the RIN peak contribution) reduces to

$$\frac{\text{RIN}}{\Delta f} \approx \frac{2h\nu}{P_0} \cdot \frac{S_I/qI + \omega^2\tau_p''^2}{1 + \omega^2\tau_p''^2}. \qquad \text{(noise floor)} \qquad (5.136)$$

This function is sketched in Fig. 5.19. For a shot noise-limited current source, $S_I/qI = 1$, and the RIN noise floor reduces to a constant at the standard quantum limit, $2h\nu/P_0$, for all frequencies. At frequencies $\ll 1/2\pi\tau_p''$, the noise of the laser comes from the current pumping source. Thus, if a noiseless pumping source is used ($S_I = 0$), the output power noise can be reduced or potentially eliminated below $f \ll 1/2\pi\tau_p''$. However, Eq. (5.135) shows that the current noise contribution is replaced by partition noise as $\eta_0 \rightarrow 0$. Hence it is essential to use an optically efficient laser.

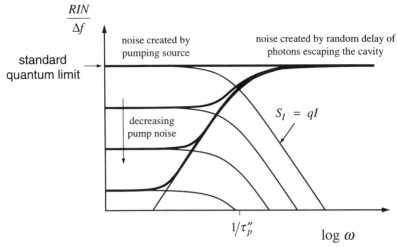

FIGURE 5.19   The two components of the relative intensity noise floor, assuming a high optical efficiency laser. Low frequency noise can be reduced below the standard quantum limit if the pumping source is quiet.

Yamamoto and Machida [9] have shown that the current injection noise into a forward-biased PN junction is dominated by the thermal noise of the series resistor such that the double-sided spectral density is $S_I = 2kT/R_S$. By using a large series resistance the pump noise can be reduced below the shot noise limit of qI. In other words, we require that $R_S \gg 2kT/qI$. For a 1 mA drive current at room temperature, this translates into $R_S \gg 50\ \Omega$. By using a constant current source with very large series resistance, the current input noise can be reduced substantially below the shot noise limit. This is the strategy some researchers have employed to attain noise-free laser operation [9, 10, 12]. To be successful, such lasers must have a very high optical efficiency ($\eta_0 \to 1$) and must be operated well above threshold.

Another caveat we must consider particularly with noiseless lasers is the transmission and detection process. Imagine an ideally quiet laser which emits a perfectly regular stream of photons. In the process of transmission and detection, some of these photons will be lost, randomly selected out of the uniform stream. These random vacancies in the otherwise perfectly regular photon stream produce *partition noise* in the detected signal. For high losses, the magnitude of this partition noise approaches the shot noise limit. Using the Langevin method detailed in Appendix 13 to include this partition noise (see Problem A13.1), we can derive the detected noise spectral density and corresponding RIN [12]:

$$S_{det}(\omega) = \eta_{det}^2(q/h\nu)^2 S_{\delta P}(\omega) + (1 - \eta_{det})qI_{det}, \qquad (5.137a)$$

$$(\text{RIN})_{det} = (\text{RIN})_{laser} + (1 - \eta_{det})(\text{RIN})_{shot}, \qquad (5.137b)$$

where $I_{det} = \eta_{det}(q/h\nu)P_0$ and $(\text{RIN})_{shot} = 2qI_{det}/I_{det}^2$. Here, η_{det} is the photon collection efficiency of the photodetector including coupling and transmission losses in getting from the laser to the detector (current leakage in the detector is not included in η_{det}, since parallel current paths do not necessarily lead to partition noise—the same goes for η_i at the laser end [12]).

In Eq. (5.137a), the coefficient of the first term is the normal transfer function one would expect in converting the power noise to detector current noise. The second term is the standard (double-sided) shot noise term commonly attached to the detector current noise, which more correctly here includes a $(1 - \eta_{det})$ factor. Equation (5.137b) gives the corresponding RIN of the detector current. Both of these equations reveal that to achieve sub-shot noise performance in an optical communications system, it is essential to keep the losses in going from the laser to the detector to a minimum. For example, if half of the light is lost before being converted to current, then the best one can hope to do is drop 3 dB below the standard quantum limit, even if the laser is perfectly quiet.

Note that the argument for high η_{det} is no different from the argument demanding a high η_0 in the laser. Essentially, a perfectly regular stream of electrons injected into the laser must have a one-to-one correspondence with electrons generated in the detector in order to replicate the perfect regularity. Any random division or loss of photons along the way introduces irregularities which show up as shot noise in the receiver current.

5.5.4.3 RIN in Multimode Lasers

Since the derivation leading up to Eq. (5.131) used the single-mode rate equations, it is not surprising that Fig. 5.18 gives a good representation of what is experimentally observed in single-frequency lasers. However, it also is roughly valid for multimode lasers provided all modes are included in the received power. On the other hand, if only one mode is filtered out from a multimode spectrum, it is typically found to contain a much larger noise level, especially at the lower frequencies. This is because of *mode partitioning*. The energy tends to switch back and forth randomly between the various modes observed in the time-averaged spectrum causing large power fluctuations in any one mode. If all modes are included, the net power tends to average out these fluctuations. There are numerous ways in which an optical link can provide the unwanted spectral filtering of a multimode laser's output. For example, in multimode fiber the spatial modes will interfere differently for each frequency in the laser's spectrum, providing a different transmission fraction for each. This is especially accentuated, if there is some incidental spatial filtering in the optical link, so that different spatial modes are coupled differently.

It is also somewhat surprising how large the mode suppression ratio (MSR) for the unwanted modes must be before the laser behaves like a single-frequency laser. Experiments have verified that significant mode partitioning can occur even for MSRs ~ 30 dB, although this is roughly the level at which mode partitioning tends to disappear. Such single-frequency lasers also tend to be very sensitive to spurious feedback from external reflections in the optical path. This sensitivity will be the subject of later discussions in this chapter.

5.5.5 Frequency Noise

For frequency modulation applications, it is useful to determine the frequency jitter or noise. By an extension of Eq. (5.75), we can write

$$\frac{d\phi}{dt} = 2\pi\Delta v(t) = \frac{\alpha}{2} \Gamma v_g a[dN(t)] + F_\phi(t). \tag{5.138}$$

The first equality reminds us that the frequency deviation can be considered a rate equation for the field phase. The latter equation introduces a Langevin noise source for this rate equation which can be associated with the phase noise of the laser. Converting to the frequency domain, we obtain

$$v_1(\omega) = \frac{\alpha}{4\pi} \Gamma v_g a N_1(\omega) + \frac{1}{2\pi} F_\phi(\omega). \tag{5.139}$$

Appendix 13 shows that the correlation strengths for the phase noise source reduce to the following:

$$\langle F_\phi F_\phi \rangle = \frac{\Gamma R'_{sp}}{2N_p}, \qquad \langle F_\phi F_P \rangle = \langle F_\phi F_N \rangle = 0. \tag{5.140}$$

Note that the phase noise source is uncorrelated with photon and carrier density noise sources. To determine the frequency noise spectral density, as before we multiply both sides of Eq. (5.139) by $v_1(\omega')^*$, take the time average, and integrate over ω', to obtain

$$S_v(\omega) = \left(\frac{\alpha}{4\pi} \Gamma v_g a\right)^2 S_N(\omega) + \frac{1}{(2\pi)^2} \langle F_\phi F_\phi \rangle. \tag{5.141}$$

The cross-term does not appear because there is no correlation between the carrier and phase noise. Thus, the frequency noise has two contributions: (1) the carrier noise which induces refractive index changes causing the lasing frequency to fluctuate, and (2) the inherent phase noise of the laser, originating from photons which are spontaneously emitted into the mode. For semiconductor lasers, carrier noise dominates by typically more than an order of magnitude.

We can use Eqs. (5.127) through (5.129) to evaluate the carrier noise spectral density in (5.124). Appendix 13 carries out this task revealing that the carrier density noise spectral density can be well approximated by

$$S_N(\omega) = \frac{8\pi}{(\Gamma v_g a)^2} \left[\frac{\Gamma R'_{sp}}{4\pi N_p}\right] \cdot |H(\omega)|^2. \tag{5.142}$$

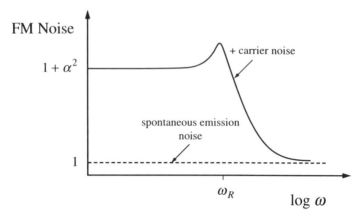

FIGURE 5.20   Frequency noise spectrum illustrating contributions from both carrier and spontaneous emission noise.

With Eq. (5.142) for the carrier noise and (5.140) for the phase noise, the frequency noise spectral density (double-sided) becomes

$$S_v(\omega) = \frac{1}{2\pi} (\Delta v)_{ST} \cdot (1 + \alpha^2 |H(\omega)|^2), \qquad (5.143)$$

where again

$$(\Delta v)_{ST} = \frac{\Gamma R'_{sp}}{4\pi N_p}.$$

The measured FM noise can be modeled using four fitting parameters: α, γ, ω_R, and $(\Delta v)_{ST}$ [11]. A typical FM noise spectrum is sketched in Fig. 5.20. Beyond the relaxation resonance frequency, the carrier noise becomes negligible, reducing the FM noise to the white noise background provided by spontaneous emission phase noise. Additional thermal contributions from the noise on the pumping source can also contribute to the FM noise at low frequencies, as discussed in Section 5.3.3.

5.5.6   Linewidth

To relate the FM noise to the frequency spectrum of the laser, it is useful to introduce the *coherence time* of the laser light. Consider a measurement which mixes the emitted electric field, $\mathscr{E}(t)$, with a time-delayed version of itself, $\mathscr{E}(t - \tau)$. As long as the phases of the two fields are well correlated, the fields will add coherently. We can write the cross-term of the coherent addition, or the *autocorrelation* function of the field as

$$\langle \mathscr{E}(t)\mathscr{E}(t - \tau)^* \rangle \propto e^{j\omega\tau} e^{-|\tau|/\tau_{coh}}. \qquad (5.144)$$

The first factor gives the expected interference fringes created by the coherent mixing. However, because the laser is not emitting a pure single frequency, with increasing time delay, τ, the phases of the two fields become less and less correlated and the interference fringes gradually disappear to the point where the two fields add incoherently. The envelope of the fringe pattern given by the second factor characterizes this coherence decay. As the functionality suggests, the decay can often be described by an exponential with a decay constant defined as the coherence time of the laser, τ_{coh}. However, more generally, τ_{coh} may be a function of the time delay itself, $\tau_{coh}(\tau)$, in which case the autocorrelation does not yield a simple decaying exponential.

Figure 5.21(a) suggests one method of directly measuring the autocorrelation function, allowing us to experimentally extract $\tau_{coh}(\tau)$. Using the measured fringe pattern in Fig. 5.21(a), we can write the envelope of the autocorrelation function as

$$e^{-|\tau|/\tau_{coh}} = \frac{P_1 + P_2}{2\sqrt{P_1 P_2}} \frac{P_{max} - P_{min}}{P_{max} + P_{min}}, \tag{5.145}$$

where $P_1 \propto \langle |\mathcal{E}_1(t)|^2 \rangle$ and $P_2 \propto \langle |\mathcal{E}_2(t - \tau)|^2 \rangle$ are the individual powers in the two legs of the interferometer, and $P_{max} \propto \langle |\mathcal{E}_1(t) + \mathcal{E}_2(t - \tau)|^2 \rangle$ and $P_{min} \propto \langle |\mathcal{E}_1(t) + \mathcal{E}_2(t - \tau)|^2 \rangle$ are the maximum and minimum powers of the interference fringes measured by the detector. The second ratio is often referred to as the *fringe visibility*. The first ratio normalizes the fringe visibility when $P_1 \neq P_2$. Measuring the normalized fringe visibility as a function of the time delay between the two arms yields the envelope of the autocorrelation function.

Once the autocorrelation is known, it can be related to the frequency spectrum of the mode, $P_0(\omega)$, through a Fourier transform (see Appendix 13):

$$P_0(\omega) \propto \mathcal{F}[\langle \mathcal{E}(t)\mathcal{E}(t - \tau)^* \rangle]. \tag{5.146}$$

For example, if τ_{coh} is a constant (i.e., not a function of τ) then the autocorrelation is a decaying exponential and the corresponding lasing spectrum is a Lorentzian.

Experimentally, the lasing spectrum can be measured conveniently using the experimental setup illustrated in Fig. 5.21(b). The laser is coupled through an optical isolator into a fiber (the isolator is necessary to remove unwanted feedback which can affect the laser linewidth (see Section 5.7)). One leg of fiber is fed through an acousto-optic modulator which shifts the light frequency by typically 40 MHz. The other leg of fiber is much longer than the coherence length ($L_{coh} \equiv v\tau_{coh}$) of the laser. When the two incoherent fields are recombined in the detector, they generate a difference-frequency signal current at 40 MHz which contains the combined FM *field* noise of both light sources. The square of this current is therefore equivalent to the combined optical power spectrum. In other words, this technique effectively converts the combined optical power

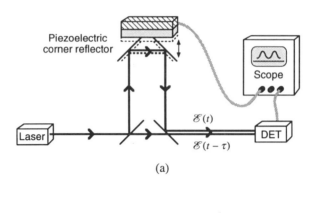

(a)

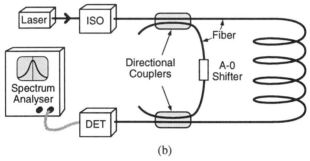

(b)

FIGURE 5.21 Schematics of two techniques to measure linewidth. (a) Coherence length measurement in which one branch of a Mach-Zehnder interferometer is increased in length until fringe visibility is reduced. A piezoelectrically driven mirror provides a small length variation to display interference fringes. (b) Self-heterodyne technique in which a long length of fiber (≫coherence length) is used to mix the laser emission with an incoherent version of itself. The difference frequency signal replicates the combined line-shape of both signals at low frequencies and is easily observed using an RF spectrum analyzer [13].

spectrum to an *electrical power* spectrum centered at 40 MHz which can easily be measured using a radio-frequency spectrum analyzer. However, because it contains the noise of both signals, the electrical power 3 dB-down full-width of the spectrum is *twice* as wide as the original laser linewidth (the factor of 2 is specific to the combination of two *Lorentzian* lineshapes [13]).

Theoretically, the lasing spectrum is found by determining $\tau_{coh}(\tau)$ and applying Eq. (5.146). We can relate the coherence time to the FM noise and the measurement time delay as follows [14]:

$$\frac{1}{\tau_{coh}} = \pi\tau \int_{-\infty}^{+\infty} S_v(\omega) \frac{\sin^2(\omega\tau/2)}{(\omega\tau/2)^2} \, d\omega. \qquad (5.147)$$

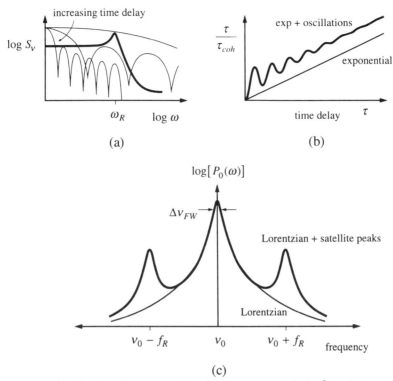

FIGURE 5.22 (a) Overlap between FM noise spectrum and $\text{sinc}^2(\omega\tau/2)$ function for increasing time delays. (b) Resultant functionality of τ/τ_{coh} vs. time delay. (c) Lasing spectrum with and without satellite peaks.

The coherence time is related to the *integrated* frequency noise of the laser times a $\text{sinc}^2(\omega\tau/2)$ function.[8] Because the first zero of $\text{sinc}^2(\omega\tau/2)$ occurs at $\omega = 2\pi/\tau$, the entire function scales toward zero with increasing time delay. This effect is illustrated in Fig. 5.22(a). As the peaks and nulls of $\text{sinc}^2(\omega\tau/2)$ sweep by the resonance peak of the noise with increasing time delay, they produce periodic undulations in the overlap (since the resonance peak is either included or excluded from the overlap). The effect of this on $1/\tau_{coh}$ is shown in Fig. 5.22(b) [14]. The decaying oscillations which appear in the otherwise linear τ/τ_{coh} are the direct result of the resonance peak in the noise. The oscillation

---

[8] This result comes about as follows. Assume the field is $\propto e^{j\omega t}e^{j\phi(t)}$ where $\phi(t)$ is a fluctuating phase created by frequency noise. The autocorrelation is then $\propto \langle e^{j\Delta\phi(\tau)}\rangle = \exp[-\langle\Delta\phi(\tau)^2\rangle/2]$ and we conclude that $\tau/\tau_{coh} \equiv \langle\Delta\phi(\tau)^2\rangle/2$. Using the Fourier transform pair $\Delta\phi(\tau) = \phi(\tau) - \phi(0) \Leftrightarrow \phi(\omega)(e^{j\omega\tau} - 1)$ and Eq. (5.104), we can derive another Fourier transform pair $\langle\Delta\phi(\tau)^2\rangle \Leftrightarrow 2S_\phi(\omega)(1 - \cos\omega\tau)$ where $S_\phi(\omega)$ is the phase noise spectral density. Because phase and frequency are related by a time derivative we can set $S_\phi(\omega) \rightarrow (2\pi)^2 S_v(\omega)/\omega^2$. This leads to $\langle\Delta\phi(\tau)^2\rangle = 2\tau/\tau_{coh} \Leftrightarrow (4\pi)^2 S_v(\omega)\sin^2(\omega\tau/2)/\omega^2$. Expressing the Fourier transform explicitly using Eq. (5.101) and rearranging we obtain Eq. (5.147).

frequency is equal to the peak frequency of the resonance ($\sim \omega_R$), while the magnitude of the oscillations is governed by how sharp and strong the resonance is ($\sim 1/\gamma$).

Eventually, τ/τ_{coh} relaxes back to a linear relationship. In fact, for $\tau \gg 2\pi/\omega_R$ the $\text{sinc}^2(\omega\tau/2)$ function is peaked in the low-frequency portion of the FM noise spectrum, and we can approximate $S_v(\omega) \approx S_v(0)$ inside the integral. This leads to the simple result:

$$\frac{1}{\tau_{coh}} \approx \pi\tau S_v(0) \int_{-\infty}^{+\infty} \frac{\sin^2(\omega\tau/2)}{(\omega\tau/2)^2} d\omega = 2\pi^2 S_v(0). \qquad (\tau \gg 2\pi/\omega_R) \quad (5.148)$$

The integral is equal to $2\pi/\tau$, canceling out the dependence on time delay. Thus, for time delays much greater than $1/f_R$, τ_{coh} is a constant, τ/τ_{coh} increases linearly, and the autocorrelation reduces to a decaying exponential. For shorter time delays, the decaying exponential contains some ripples.

From a physical point of view, it is the relaxation oscillations occurring in response to random absorption/emission events which enhance the FM noise near the resonance frequency. And since these oscillations are created by noise events which occur earlier in time, we could argue that some component of the noise at any given time mirrors the noise which existed $\sim 1/f_R$ seconds ago. Thus, we might expect a higher correlation or coherence between fields which are delayed by time intervals of $\sim 1/f_R$. This is exactly what Fig. 5.22(b) reveals.

Taking the Fourier transform of $e^{-|\tau|/\tau_{coh}}$ produces the power spectrum of the laser as shown in Fig. 5.22(c). Without ripples in τ/τ_{coh}, the spectrum is a Lorentzian. With ripples, the Lorentzian acquires a series of satellite peaks spaced by the resonance frequency. The magnitude of the satellite peaks depends on the damping factor of the laser (less damping $\Rightarrow$ stronger peaks). However, the FWHM linewidth of the spectrum is essentially the same with or without these satellite peaks.

Using the long time delay expression for τ_{coh} (5.148) to isolate the exponential component of the autocorrelation and then Fourier transforming, we obtain the dominant Lorentzian lineshape. In general, if the exponential decay constant is τ_0, then the transformed Lorentzian FWHM is $\Delta\omega = 2/\tau_0$. In the simpler treatment of linewidth leading to Eq. (5.114), τ_0 corresponds to the field decay or $2\tau_p$, yielding $\Delta\omega = 1/\tau_p$. Here, τ_0 corresponds to the coherence time such that

$$\Delta v_{FW} = \frac{1}{\pi\tau_{coh}} = 2\pi S_v(0). \quad (5.149)$$

Inserting Eq. (5.143) with $|H(0)|^2 = 1$, we obtain

$$\Delta v_{FW} = (\Delta v)_{ST}(1 + \alpha^2) = \frac{\Gamma R'_{sp}}{4\pi N_p}(1 + \alpha^2). \quad (5.150)$$

Thus, the modified Schawlow–Townes expression for linewidth is *enhanced* by $1 + \alpha^2$ in semiconductor lasers (this is where α gets its name as the linewidth enhancement factor) [5, 14]. The 1 represents the spontaneous emission noise contribution, and the α^2 represents the carrier noise contribution.

The above-threshold linewidth can be rewritten in terms of external parameters by setting $R'_{sp} = \Gamma v_g g_{th} n_{sp}/V$, $N_p = P_0/h v V_p v_g \alpha_m F$, and $\alpha_m F = \eta_0 \Gamma g_{th}$:

$$\Delta v_{FW} = \frac{(\Gamma v_g g_{th})^2 \eta_0}{4\pi P_0} n_{sp} h v (1 + \alpha^2)$$

$$= 38 \text{ MHz} \times \left(\frac{4.2}{n_g}\right)^2 \left(\frac{\Gamma g_{th}}{50 \text{ cm}^{-1}}\right)^2 \left(\frac{\eta_0}{0.4}\right)\left(\frac{n_{sp}}{1.5}\right)\left(\frac{h v}{1.5 \text{ eV}}\right)\left(\frac{1 \text{ mW}}{P_0}\right)\left(\frac{1 + \alpha^2}{26}\right).$$

$$(5.151)$$

Typical numbers have been used to evaluate the linewidth (the value for η_0 refers to the single-facet efficiency, and hence P_0 refers to single-facet power). Generally speaking, powers in the milliwatts range in a typical semiconductor laser produce linewidths in the tens-of-megahertz range. Thus, to achieve sub-megahertz linewidths it is necessary to increase the power well above 10 mW. Another approach to reducing the linewidth involves the use of external cavities, which we explore in Section 5.7.

5.6 CARRIER TRANSPORT EFFECTS

In the sections so far we have neglected any transport time for carriers to reach the active region. For cases where the intrinsic region of the *pin* diode and the active region are one and the same, such as for simple bulk DH structures, this assumption is generally good. However, for separate-confinement hetero-structures (SCHs) used with quantum-well active regions, it has been found that transport effects must be considered [15]. This structure was introduced in Fig. 1.5. Figure 5.23 gives a slightly more detailed SCH schematic for the present discussion.

For a number of years during the 1980s there was a dilemma as to why quantum-well lasers were not providing the modulation bandwidths predicted by the calculations given above. From Eq. (5.51) and earlier versions, it was clear that the resonant frequency was directly proportional to the differential gain, which is much higher with quantum-well active regions. Thus, it was surprising when experiments showed quantum-well laser bandwidths about the same as for bulk DH structures. Once it was realized that carrier injection delays due to transport effects were significant, laser designs were modified, and bandwidths increased dramatically. In this section we shall review the relevant theory.

With reference to Fig. 5.23 we construct a set of three rate equations for the

SCH design. The change is that the carrier equation has been replaced by two new equations. The first is for the carrier density in the barrier regions, N_B, and the second is for the carrier density in the active region, N. Multiple quantum-well active regions can also be treated by lumping all the barrier regions together. The new rate equations are

$$\frac{dN_B}{dt} = \frac{\Gamma_q \eta_i I}{qV} - \frac{N_B}{\tau_s} + \frac{\Gamma_q N}{\tau_e}, \tag{5.152}$$

$$\frac{dN}{dt} = \frac{N_B}{\Gamma_q \tau_s} - N\left[\frac{1}{\tau} + \frac{1}{\tau_e}\right] - v_g g N_p, \tag{5.153}$$

$$\frac{dN_p}{dt} = \left[\Gamma v_g g - \frac{1}{\tau_p}\right] N_p + \Gamma R'_{sp}, \tag{5.154}$$

where $\Gamma_q = V/V_{SCH}$ is the fraction of the SCH region filled by the quantum-well active region, N_B/τ_s is the loss rate of carriers from the SCH region to the quantum-well active region, N/τ_e is the loss rate of carriers from the quantum wells to the SCH region, and the other symbols are as defined before. In this simplified treatment, we ignore the fact that the holes come from one side of the junction and the electrons from the other. Fortunately, other more in-depth treatments which include holes and electrons separately yield essentially the same results as this simplified analysis.

Limiting our attention to small-signal perturbations, we can linearize the rate equations as we did in Section 5.3 to obtain the differential rate equations. In matrix form, these become

$$\frac{d}{dt}\begin{bmatrix} dN_B \\ dN \\ dN_p \end{bmatrix} = \begin{bmatrix} -\gamma_{BB} & \gamma_{BN} & 0 \\ \gamma_{NB} & -\gamma_{tNN} & -\gamma_{NP} \\ 0 & \gamma_{PN} & -\gamma_{PP} \end{bmatrix}\begin{bmatrix} dN_B \\ dN \\ dN_p \end{bmatrix} + \frac{\Gamma_q \eta_i}{qV}\begin{bmatrix} dI \\ 0 \\ 0 \end{bmatrix}. \tag{5.155}$$

The rate coefficients of the upper left 2×2 submatrix are given by

$$\begin{aligned} \gamma_{BB} &= 1/\tau_s, & \gamma_{BN} &= \Gamma_q/\tau_e, \\ \gamma_{NB} &= 1/\Gamma_q \tau_s, & \gamma_{tNN} &= \gamma_{NN} + 1/\tau_e. \end{aligned} \tag{5.156}$$

The other rate coefficients are as defined before in Eq. (5.35). Converting to the frequency domain and applying Cramer's rule, we obtain

$$N_{p1}(\omega) = \frac{\Gamma_q \eta_i I_1(\omega)}{qV} \cdot \frac{1}{\Delta_t}\begin{vmatrix} \gamma_{BB} + j\omega & -\gamma_{BN} & 1 \\ -\gamma_{NB} & \gamma_{tNN} + j\omega & 0 \\ 0 & -\gamma_{PN} & 0 \end{vmatrix}, \tag{5.157}$$

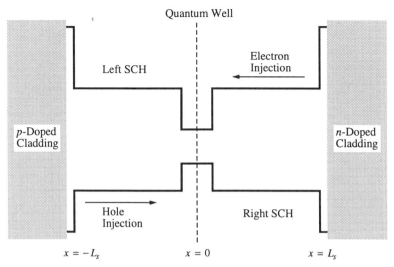

FIGURE 5.23 Schematic diagram of a single quantum-well laser with a separate-confinement heterostructure (SCH) used in the carrier transport model.

where

$$\Delta_t \equiv \begin{vmatrix} \gamma_{BB} + j\omega & -\gamma_{BN} & 0 \\ -\gamma_{NB} & \gamma_{tNN} + j\omega & \gamma_{NP} \\ 0 & -\gamma_{PN} & \gamma_{PP} + j\omega \end{vmatrix}. \tag{5.158}$$

The only nonzero portions of Δ_t are (1) the product of the upper-left element with the determinant of the lower right 2×2 submatrix and (2) the product $\gamma_{BN}\gamma_{NB}(\gamma_{PP} + j\omega)$. Pulling out the factor $(\gamma_{BB} + j\omega)$ in the denominator, setting $\gamma_{BN}\gamma_{NB}/(\gamma_{BB} + j\omega) = \gamma_{BN}\gamma_{NB}/\gamma_{BB} - j\omega(\chi - 1)$, and recognizing that $\gamma_{NN} = \gamma_{tNN} - \gamma_{BN}\gamma_{NB}/\gamma_{BB}$, we obtain

$$N_{p1}(\omega) = \frac{\Gamma_q \eta_i I_1(\omega)}{qV} \cdot \frac{\gamma_{NB}\gamma_{PN}}{\gamma_{BB} + j\omega} \cdot \frac{1/\chi}{\omega_{tR}^2 - \omega^2 + j\omega\gamma_t}, \tag{5.159}$$

where

$$\omega_{tR}^2 = (\gamma_{NP}\gamma_{PN} + \gamma_{NN}\gamma_{PP})/\chi, \tag{5.160}$$

$$\gamma_t = \gamma_{NN}/\chi + \gamma_{PP}, \tag{5.161}$$

$$\chi \equiv 1 + \frac{\gamma_{BN}\gamma_{NB}/\gamma_{BB}}{\gamma_{BB} + j\omega}. \tag{5.162}$$

The new term χ is referred to as the *transport* factor. Using the definitions of the rate coefficients (5.35) and (5.156), and additionally setting $\gamma_{PN} \approx \Gamma v_g a N_p \approx$

$\Gamma \chi \omega_{tR}^2 \tau_p$ and $\tau_p = \eta_0 h \nu N_{p1} V_p / P_1$, we obtain the small-signal output power modulation response:

$$\frac{P_1(\omega)}{I_1(\omega)} = \eta_i \eta_0 \frac{h\nu}{q} \cdot \frac{1}{1 + j\omega\tau_s} \cdot \frac{\omega_{tR}^2}{\omega_{tR}^2 - \omega^2 + j\omega\gamma_t}, \tag{5.163}$$

with

$$\omega_{tR}^2 = \omega_R^2 / \chi \approx v_g(a/\chi) N_p / \tau_p, \tag{5.164}$$

$$\gamma_t = v_g(a/\chi) N_p \left[1 + \frac{\Gamma a_p}{a/\chi} \right] + \frac{1}{\tau_{\Delta N} \chi} + \frac{\Gamma R'_{sp}}{N_p}, \tag{5.165}$$

$$\chi = 1 + \frac{\tau_s/\tau_e}{(1 + j\omega\tau_s)} \approx 1 + \frac{\tau_s}{\tau_e}. \tag{5.166}$$

The approximate expression for the relaxation resonance frequency ignores the last two terms in (5.49), and makes use of the fact that in most lasers, $a/\tau_p \gg \Gamma a_p/\tau_{\Delta N}$. The approximate expression for the transport factor neglects the frequency roll-off of the second term. For typical SCH-SQW lasers at room temperature, $\tau_e \sim 100$–500 ps, $\tau_s \sim 20$–100 ps, and the transport factor $\chi \sim 1.2$.

Using the approximate expression for ω_{tR}^2, the damping factor can be rewritten in terms of the K-factor:

$$\gamma_t = K_t f_{tR}^2 + \gamma_{t0}, \tag{5.167}$$

where

$$K_t = 4\pi^2 \tau_p \left[1 + \frac{\Gamma a_p}{a/\chi} \right] \quad \text{and} \quad \gamma_{t0} = \frac{1}{\tau_{\Delta N} \chi} + \frac{\Gamma R'_{sp}}{N_p}. \tag{5.168}$$

In examining Eqs. (5.164) and (5.165), we conclude that transport across the SCH region *effectively* reduces the differential gain from a to a/χ, and *effectively* increases the differential carrier lifetime from $\tau_{\Delta N}$ to $\tau_{\Delta N}\chi$. The gain compression factor remains unmodified in this model. The reduced differential gain decreases the relaxation resonance frequency and increases the K-factor. The increase in differential carrier lifetime reduces the damping factor offset, γ_{t0}. In addition (and perhaps most significantly), there is a new prefactor which provides a low-pass filtering effect with a cutoff at $\omega = 1/\tau_s$. This prefactor provides an insidious parasitic-like roll-off that is indistinguishable from an RC roll-off.

Figure 5.24 gives examples of the effects of transport in SCH structures. Note that with the narrower SCH region, the low-pass filter effect is eliminated allowing for a much larger modulation bandwidth.

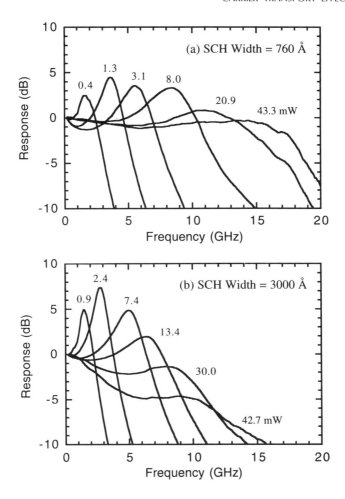

FIGURE 5.24   Modulation response for (a) a narrow SCH and (b) a wide SCH-SQW laser (the lasers are otherwise identical). The modulation response for the wide SCH laser, at comparable power levels, shows the detrimental effects of the low frequency roll-off due to carrier transport. After [15]. (© 1992 IEEE).

5.7   FEEDBACK EFFECTS

In typical applications of diode lasers some light is unintentionally reflected back into the laser cavity. For example, the reflection from the front surface of a fiber is about 4%, and perhaps one-tenth of this may be coupled back into the laser cavity in a pigtailing arrangement. Surprisingly, even such small amounts of feedback can have dramatic effects on the laser's linewidth and noise properties. This is especially true for single-frequency lasers. In fact, 0.4% (-24 dB) is considered to be a large feedback. Experiments have shown that even -60 dB of feedback can be unacceptable with single-frequency lasers.

   To analyze the effects of feedback, we consider the three-mirror cavity results of Chapter 3. Figure 3.7 and Eq. (3.34) are our starting point. They describe an effective mirror concept which folds the effects of the external cavity back into the active laser diode section. From this starting point we explore the effects of feedback on the laser's output power and spectrum. After the static characteristics are quantified, we consider the dynamic effects on the laser's linewidth and noise spectrum.

5.7.1   Static Characteristics

Equation (3.34) describes the vector addition of the primary reflection at the laser facet, r_2, with the feedback term. This is shown schematically in Fig. 5.25. For relatively weak feedback, the external cavity resonance can be neglected (i.e., we can ignore the denominator in Eq. (3.34)), so that the in-phase (ip) and quadrature (q) components of the feedback term, $r_{eff} - r_2 = \Delta r$, become

$$\Delta r_{ip} = t_2^2 \sqrt{f_{ext}} \cos(2\beta L_p),$$
$$\Delta r_q = -t_2^2 \sqrt{f_{ext}} \sin(2\beta L_p),$$

(5.169)

where $-2\beta L_p$ is the round-trip phase of the external cavity. Also, the external feedback level, f_{ext} represents the fraction of emitted *power* coupled back to the laser. It replaces r_3 in Eq. (3.34) to account for additional coupling and propagation losses encountered through one round-trip path of the external cavity.

   The in-phase component of the feedback, Δr_{ip}, affects the laser by changing the magnitude of the effective reflectivity at mirror #2. This modifies the photon lifetime of the cavity, which in turn modifies the threshold gain, threshold carrier density, and threshold current of the laser. The perturbed photon lifetime is given by

$$\frac{1}{\tau_p'} = \frac{1}{\tau_p} + v_g \Delta \alpha_m = \frac{1}{\tau_p} + \frac{v_g}{L}\left[\frac{-\Delta R}{R}\right],$$

(5.170)

where $R = r_1 r_2$ and $\Delta R = r_1 \Delta r_{ip}$. Expanding these terms and defining the

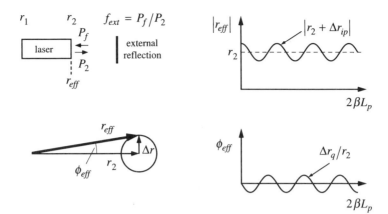

FIGURE 5.25 Polar plot (left) of effective reflection from output side of laser; Δr represents feedback which can change both the amplitude and phase of r_{eff}. Plots of the magnitude and phase of r_{eff} illustrate the quadrature relationship.

round-trip time of the laser cavity as $\tau_L = 2L/v_g$, we can express the change in mirror loss as

$$v_g \Delta\alpha_m = -2 \frac{t_2^2}{r_2} \frac{\sqrt{f_{ext}}}{\tau_L} \cos(2\beta L_p) = -2\kappa_f \cos(2\beta L_p). \tag{5.171}$$

The feedback rate $\kappa_f = (t_2^2\sqrt{f_{ext}}/r_2)/\tau_L$ introduced here represents the fractional increase in the field reflected at mirror #2 per round-trip time of the cavity (i.e., per bounce).

The change in mirror loss gives rise to a shift in the threshold modal gain, or $\Delta(\Gamma g_{th}) = \Delta\alpha_m$. For small changes, we can use the differential gain $a = \Delta g/\Delta N$ to determine the corresponding shift in threshold carrier density, and we can write

$$\Delta g_{th} = -\frac{2\kappa_f}{\Gamma v_g} \cos(2\beta L_p), \tag{5.172}$$

$$\Delta N_{th} = -\frac{2\kappa_f}{\Gamma v_g a} \cos(2\beta L_p). \tag{5.173}$$

The perturbed threshold current can also be described as follows:

$$I'_{th} = I_{th} - |\Delta I_{th}| \cos(2\beta L_p), \tag{5.174}$$

where ΔI_{th} can be found by defining a differential gain per unit current density, a_J, similar to Eq. (5.173). When the external reflection is in phase with the reflection at mirror #2 (i.e. when $2\beta L_p = 2\pi m$), the threshold level decreases due to an increase in r_{eff}. When the external reflection is out of phase (i.e. when $2\beta L_p = 2\pi m + \pi$), the threshold level increases due to a decrease in r_{eff}.

The change in mirror loss also affects the differential efficiency and output power. Expanding Eqs. (3.30) through (3.32) to first-order in $\Delta r_{ip}/r_2$, we can obtain new perturbed values for the total differential efficiency, η_d', the fraction of light emitted through each mirror, $F_{1,2}'$, and the corresponding power out, $P_{01,02}'$. Using Eq. (5.174) and assuming both laser facets are lossless (i.e. $t^2 = 1 - r^2$), the perturbed power out of mirrors #1 and #2 becomes

$$P_{01,02}' = P_{01,02}[1 - \mu_{1,2} \cos(2\beta L_p)], \tag{5.175}$$

where

$$\mu_{1,2} = \left[2\kappa_f \tau_p \left(\frac{\eta_i}{\eta_d} - 1 \right) \right] \pm \left[\kappa_f \tau_L (1 - F_{1,2}) \left(\frac{1 + r_2^2}{t_2^2} \right) \right] - \left[\frac{|\Delta I_{th}|}{I - I_{th}} \right].$$

The upper sign is for μ_2 while the lower sign is for μ_1. As a reminder, $|\Delta I_{th}|$ represents *half* of the peak-to-peak change in threshold current.

Equation (5.175) shows that if we vary the optical path length between the laser and the reflection causing the feedback, the output power will oscillate sinusoidally as the external reflection goes in and out of phase with the reflection at mirror #2. The oscillation is the combined result of (in order of appearance in $\mu_{1,2}$): (1) the change in overall differential efficiency, (2) the change in the fraction of power emitted out of either facet, and (3) the shift in threshold current. The first two effects reinforce each other (upper sign) for light emitted out mirror #2, but work against each other (lower sign) for light emitted out mirror #1. Also, the first effect is enhanced for small η_d, while the second effect is enhanced when the fraction of light emitted from the measured facet is small. The third effect is enhanced near threshold. At certain operating points all three effects can cancel each other out. However, for bias currents well above threshold, the third effect is negligible. By recording the peak-to-peak power variation as the feedback phase is scanned, we can directly measure $\mu_{1,2}$. This provides a method of estimating both κ_f and the feedback level f_{ext}. However, the technique is limited to relatively large feedback levels, since the change in output power for very weak feedback is hard to detect.

A good example of how feedback can affect the *LI* curve of a laser is shown in Fig. 5.26. As shown in the inset, the substrate–air interface of the substrate-emitting VCSEL provides a strong external reflection. The oscillations in output power in this case are not caused by changes in the external path length, but rather by a steady increase in the lasing wavelength with increasing current (which is equally effective at scanning the feedback phase, $-4\pi n_p L_p/\lambda$, through multiples of 2π). We can observe a cancellation between the differential efficiency change and the threshold shift effects at about 3 mA. We can also observe that the oscillations are not entirely sinusoidal as Eq. (5.175) might suggest. This is due to *mode pulling*, where the feedback itself modifies β in often unpredictable ways (we will consider these effects next). To prevent such

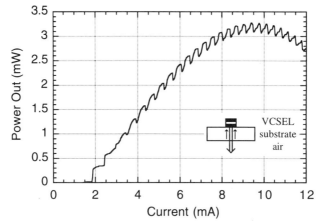

FIGURE 5.26 Continuous-wave LI curve of a substrate-emitting VCSEL with strong feedback from the substrate–air interface. As the current increases, the temperature of the VCSEL increases. This increases both the index and optical path of the resonator cavity, leading to a steady increase in lasing wavelength (and shift in feedback phase) with increasing current. The peaks (valleys) in the output power occur when the feedback field subtracts from (adds to) the recirculating field in the VCSEL cavity (the opposite of what one might think). The rollover of the LI curve beyond 8 mA is due to a variety of thermal effects.

oscillations in output power, an antireflection coating is usually applied to the substrate of the VCSEL.

While Δr_{ip} affects the threshold level and power out, the quadrature component of the feedback, Δr_q modifies the phase angle of the net reflection at mirror #2 from zero to $\phi_{eff} = \Delta r_q / r_2$, affecting the cavity's resonant wavelength. With no feedback, the resonance condition is met when $-2\beta_0 L = 2\pi m$. With feedback, the accumulated round-trip phase must now satisfy: $-2\beta L + \phi_{eff} = 2\pi m \equiv -2\beta_0 L$. In other words, β must adjust itself such that $2(\beta - \beta_0)L = \phi_{eff}$. In a dispersive medium, $\Delta\beta = \Delta\omega/v_g$ allowing us to write the frequency shift due to the external feedback as

$$\Delta\omega_\phi = \frac{\phi_{eff}}{\tau_L} = -\kappa_f \sin(2\beta L_p). \tag{5.176}$$

If we specify the round-trip phase change as $\Delta\phi_{r-t} = -\Delta\omega\tau_L$, we find that this equation is equivalent to $-\Delta\phi_{r-t}/\tau_L = \phi_{eff}/\tau_L$, which reduces to $\Delta\phi_{r-t} + \phi_{eff} = 0$. In other words, the round-trip phase change compensates for the altered phase of mirror #2.

An additional round-trip phase change is created by the shift in threshold carrier density via the linewidth enhancement factor, α. This leads to an additional shift in the lasing frequency. Placing the threshold carrier density shift (5.173) into Eq. (5.75), the resulting carrier-induced frequency shift is given by

$$\Delta\omega_N = -\alpha\kappa_f \cos(2\beta L_p). \tag{5.177}$$

Adding Eqs. (5.176) and (5.177), and using the trigonometric identity: $a \sin\phi + b \cos\phi = \sqrt{a^2 + b^2} \sin[\phi + \tan^{-1}(b/a)]$, the total frequency shift becomes

$$\Delta\omega = -\kappa_f\sqrt{1 + \alpha^2} \sin(2\beta L_p + \phi_\alpha) \equiv \frac{\phi'_{eff}}{\tau_L}, \tag{5.178}$$

where $\phi_\alpha = \tan^{-1}\alpha$ (which varies from 0 to $\pi/2$ as α increases). By defining ϕ'_{eff} as the effective phase of mirror #2 factoring in the carrier-induced round-trip phase shift, Eq. (5.178) becomes equivalent to $-\Delta\phi_{r-t}/\tau_L = \phi'_{eff}/\tau_L$, which reduces to $\Delta\phi_{r-t} + \phi'_{eff} = 0$. Note that α enhances ϕ'_{eff} and pulls the quadrature $\sin x$ dependence toward a $\cos x$ dependence more in sync with the threshold gain variation (5.172).

Our goal is to determine the mode frequency shift, $\Delta\omega$. We start by writing the feedback phase directly in terms of the frequency shift and the external cavity round-trip time $\tau_{ext} = 2L_p/v_{gp}$:

$$2\beta L_p = \Delta\omega\tau_{ext} + 2\beta_0 L_p. \tag{5.179}$$

$2\beta_0 L_p$ is the feedback phase that would exist at the laser's unperturbed frequency. Substituting this into Eq. (5.178) leads to a transcendental equation that cannot be solved for $\Delta\omega$ analytically. However, we can get a feel for the solutions by plotting the left- and right-hand sides individually and looking for frequencies which satisfy the equation. Such a graphical solution is shown in Fig. 5.27. The negative of the round-trip phase shift, $-\Delta\phi_{r-t} = \Delta\omega\tau_L$ for three internal cavity modes and the effective phase of mirror #2, ϕ'_{eff} are all plotted as a function of frequency. A cavity mode appears wherever the phases cancel, $-\Delta\phi_{r-t} = \phi'_{eff}$, which means wherever the two curves intersect.

For weak feedback, only one mode per original internal cavity mode exists. However, for strong feedback we find that multiple modes clustered near each original internal cavity mode become possible. The spacing between these *external cavity modes* is somewhat uneven for the case shown in the figure. However, as the feedback and oscillation amplitude increase, the external mode spacing evens out approaching $\Delta\omega_m = \pi/\tau_{ext}$ (one for every zero crossing). This is half the mode spacing one might expect. For example, the unperturbed internal cavity modes are spaced by $\Delta\omega_m = 2\pi/\tau_L$. The reason for the π phase interval instead of the normal 2π phase interval is that both in-phase and out-of-phase feedback reflection cases (which are spaced by π) represent valid solutions (for $\alpha = 0$, $2\beta_0 L_p = \pi m$), since both cases leave the reflection phase of mirror #2 unperturbed. The catch is that the out-of-phase solutions have a higher threshold gain (see Eq. (5.172)), and hence are not as prominent in a multimode lasing spectrum.

In general, the solutions do not usually yield either pure in-phase or out-of-phase feedback modes (identified by alignment with the extremes in Δg_{th}). However, modes do show tendencies toward one or the other. For example, in Fig. 5.27, the first, third, and perhaps fifth external cavity modes originating from mode 0 might be classified as relatively in phase, while the second and clearly the fourth are closer to out-of-phase modes. Of the three relatively in-phase feedback modes, the one with the lowest threshold gain (i.e., the one with the most in-phase reflection) becomes the dominant mode in the spectrum, replacing what was the internal mode of the laser with no feedback. Note from the figure that the dominant cavity mode is not necessarily the external mode closest to the original internal mode.

The transition from single to multiple mode solutions signifies a distinct change in the mode spectrum of the laser and is useful to quantify. From Fig. 5.27, it is evident that multiple solutions cannot exist when the slope of the straight curve $(-\Delta\phi_{r-t})$ exceeds the maximum slope of the oscillatory curve (ϕ'_{eff}). Using Eq. (5.178), the maximum slope of ϕ'_{eff}/τ_L is $\kappa_f\tau_{ext}\sqrt{1+\alpha^2}$, while the slope of $-\Delta\phi_{r-t}/\tau_L$ is 1. Thus, multiple solutions cannot exist for

$$C = \kappa_f\tau_{ext}\sqrt{1+\alpha^2} < 1. \qquad (5.180)$$

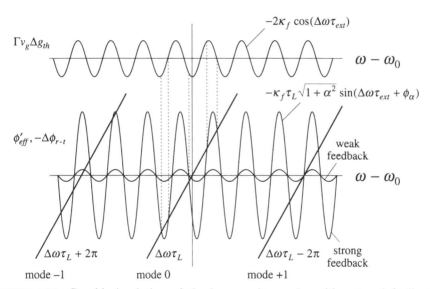

FIGURE 5.27 Graphical solution of the laser cavity modes with external feedback. Cavity modes for both weak and strong feedback are found at the intersection of the straight and sinusoidal phase curves. The relative threshold gains of the external cavity modes clustered near mode 0 in the strong feedback case are indicated by the dashed lines (the curves assume $2\beta_0 L_p = 2\pi m$).

The feedback coefficient, C, characterizes the level of feedback in relation to how it affects the mode structure of the laser ($C = |d\phi'_{eff}/d(\omega\tau_L)|_{max}$). From Fig. 5.27, it is obvious that rapid and/or large oscillations in ϕ'_{eff} can lead to multiple solutions. Equation (5.180) reflects this observation showing that both long external cavities and/or strong feedback can produce multiple external cavity modes.

5.7.2 Dynamic Characteristics and Linewidth

The static behavior of the laser discussed in the previous section considered the basic changes to the threshold, output power, and mode spectrum induced by feedback into the laser. However, such steady-state solutions may not always be stable against random fluctuations in the carrier and photon density. In other words, the inherent noise of the laser can induce dynamic instabilities which can literally run wild under certain feedback conditions. Two dominant factors are responsible for these instabilities.

The first factor is related to the mode solutions depicted in Fig. 5.27. Under certain conditions, the two intersecting curves can run tangent to each other as demonstrated by mode $+1$ with weak feedback (at the zero crossing), or mode -1 with strong feedback (at the extreme of the sine wave). In such cases, the resonance condition is not well defined making the exact position of the mode extremely sensitive to carrier-induced fluctuations in the round-trip phase. The random positioning of the mode also affects the threshold gain, which can lead to sporadic mode hopping if other solutions have similar threshold gains. The second factor involves the external round-trip delay of the feedback field relative to the recirculating mode. This time-delayed reinjection of power and phase fluctuations can accentuate the instability of the laser considerably. The net result is that external feedback can affect the noise, linewidth, and dynamic properties of single-frequency lasers very dramatically. Certain levels of feedback and delay can result in self-pulsations, very large linewidths, or large enhancements in the low-frequency noise.

Experimentally, five distinct regimes of laser performance under feedback have been identified [16]. These are indicated in Fig. 5.28. Let's begin at the bottom of the plot in regime I. For weak feedback and relatively short external cavities, $C < 1$ and only one mode for each original internal laser mode exists. In this regime it has been observed that the laser linewidth is narrowed for certain feedback phases and broadened for others. A similar decrease and increase occurs in the RIN level. Figure 5.27 can be used to explain this behavior. As mentioned earlier, when the two intersecting curves run tangent to each other, the mode position can fluctuate leading to an increase in linewidth. For mode $+1$ with weak feedback, the phase slopes of $-\Delta\phi_{r-t}$ and ϕ'_{eff} are parallel and an instability is expected. However, when the phase slopes are crossed like at mode 0 with weak feedback, the mode position is actually stabilized by the feedback, narrowing the linewidth and quieting

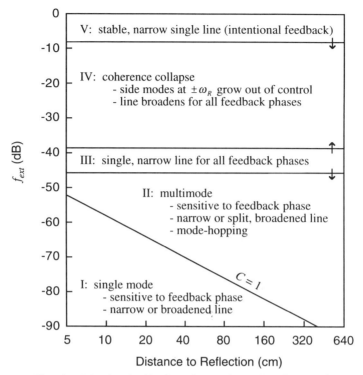

FIGURE 5.28   Sketch of the five feedback regimes. The regime boundaries plotted here were measured on one particular laser and will vary somewhat, depending on the laser structure and bias level. For example, the arrows indicate the movement of the boundaries with increasing output power. After Ref. [16].

the laser. A more rigorous analysis [17] reveals that the laser linewidth varies inversely with the difference between the phase slopes squared,[9] or $\propto [d(\Delta\phi_{r-t} + \phi'_{eff})/d(\omega\tau_L)]^{-2}$. Using Eq. (5.178) to evaluate the derivatives, we can write the linewidth with feedback relative to the unperturbed linewidth as

$$\Delta v = \frac{\Delta v_0}{[1 + C\cos(2\beta L_p + \phi_\alpha)]^2}, \qquad (5.181)$$

where again $\phi_\alpha = \tan^{-1}\alpha$ and C is given by Eq. (5.180).

When the feedback phase is adjusted to $2\beta L_p = 2\pi m - \phi_\alpha$, the phase slopes are crossed and the linewidth is narrowed to $\Delta v_0/(1 + C)^2$. When $2\beta L_p = 2\pi m + \pi - \phi_\alpha$, the phase slopes are parallel and the linewidth is broadened to $\Delta v_0/(1 - C)^2$. In between these extremes (for example at mode -1), the feedback phase approaches zero slope and the linewidth $\rightarrow \Delta v_0$. The crossed

---

[9] This relation holds as long as $\tau_{ext} < \tau_{coh}$, such that the feedback field interferes coherently with the recirculating field in the cavity.

phase slopes configuration of mode 0 is often referred to as having "in-phase" feedback even though this is only true for $\phi_\alpha = 0$. Likewise, the parallel phase slopes configuration of mode $+1$ is referred to as having "out-of-phase" feedback. For large α, both configurations are closer to having a quadrature feedback phase. We can tune the dominant mode of the laser (e.g., mode 0) to either of these configurations by adjusting $2\beta_0 L_p$, which has the effect of shifting all sinusoidal curves in Fig. 5.27 to the left or right by as much as we desire. Every π change in $2\beta_0 L_p$ alternates the phase slope alignment from crossed to parallel. In any case, regardless of the feedback phase, when $C < 0.05$, the feedback modifies the linewidth by $< 10\%$ and the laser can be considered sufficiently isolated for many applications. As $C \to 1$, the narrowing and broadening increase until eventually Eq. (5.181) predicts an infinite linewidth for the out-of-phase feedback configuration (i.e., when the two phase slopes are perfectly tangent to each other). In reality, the laser mode splits into two modes [16, 18] and this defines the boundary between regimes I and II.

For long external cavities, even relatively weak feedback can lead to $C > 1$ and we enter regime II as illustrated in Fig. 5.28. Regime II is characterized by the same feedback phase sensitivity described for regime I. However, in this regime multiple solutions are allowed. Experimentally, the in-phase feedback configuration of mode 0 continues to provide a narrowing single line as C increases. Apparently when the linewidth is stabilized by the crossed phase slopes, the laser favors this mode even in the presence of other mode solutions. For the out-of-phase feedback configuration of mode $+1$, the one mode splits into three mode solutions as the extremes of the sine wave penetrate across the straight phase curve with increasing feedback. The three modes of mode $+1$ can be observed in Fig. 5.27 for strong feedback. Of the three modes, the zero-crossing mode generally has a higher threshold gain and hence does not appear in the spectrum. The outer two mode solutions however can be tuned slightly via $2\beta_0 L_p$ until they have identical threshold gains. For this feedback phase, the laser line splits into two closely spaced modes. Thus, the out-of-phase feedback configuration in regime II is characterized by a double peaked lineshape.

As the feedback increases, the separation between the two modes increases and eventually reaches $2\pi/\tau_{ext}$ (one full cycle of the sine wave). Thus, the overall effective laser linewidth can increase to $\Delta v \sim 1/\tau_{ext}$ [18]. Furthermore, the laser does not exist in both modes at the same time. Rather, it jumps between the two modes at a rate in the few MHz range. This mode hopping creates low-frequency noise that can be undesirable in some applications. However, for both regimes I and II, one might argue that intentional feedback with the proper feedback phase can improve the laser performance. In practice, maintaining the proper feedback phase can prove challenging and it is generally best to either keep $C < 0.05$ or stay away from regimes I and II entirely.

As the two modes of the out-of-phase feedback configuration (mode $+1$) reach a separation close to $2\pi/\tau_{ext}$ with increasing feedback, the rate of jumping decreases and eventually stops altogether marking the transition to regime III.

The laser then settles down to one of the two modes. It is interesting to point out that the two outer modes originating from mode $+1$ have now shifted on the sine wave over to a crossed phase slopes, and hence stable, narrow linewidth configuration. In fact they are approaching this level in Fig. 5.27. As a result, the linewidth is narrowed for both in-phase and out-of-phase feedback configurations, making regime III independent of feedback phase. The boundary between regimes II and III has been found experimentally to be independent of external cavity length (as have all other higher feedback regime boundaries indicated in Fig. 5.28). While regime III offers stable, low noise, narrow single-line laser operation, it unfortunately spans only a small range of feedback levels. The feedback regime opens up for higher output powers [16], but is still relatively small. Designing an intentional feedback system with tight tolerances to maintain the feedback level in regime III is possible but not very practical.

As the feedback level is increased still further, the laser enters a new regime which is characterized by complex dynamic interactions between the many relatively strong multiple external mode solutions. Rate equation models of the field and phase including the time delay of the feedback field reveal unstable solutions and chaotic behavior at these high feedback levels [19]. Indeed, experiments reveal linewidths so large that the coherence length of the laser drops below 1 cm (as opposed to tens of meters). For this reason, regime IV is referred to as the coherence collapse regime. The collapse begins at the transition to regime IV, where the normally small satellite peaks created by noise-induced relaxation oscillations (depicted in Fig. 5.22) grow larger and larger, eventually becoming comparable to the central peak and broadening the linewidth dramatically. It has been suggested that the transition between regimes III and IV occurs when the feedback rate $\kappa_f \sim \omega_R$ [16]. A more refined theoretical estimate of the critical level of feedback is given by [20]

$$f_{ext}|_{crit} = \frac{\tau_L^2}{16C_e^2} (Kf_R^2 + \gamma_0)^2 \left[\frac{1 + \alpha^2}{\alpha^4} \right], \tag{5.182}$$

where we define a laser coupling factor, $C_e = t_2^2/2r_2$ (for Fabry–Perot cavities). The K-factor and γ_0 are given by Eq. (5.53). The critical feedback level is increased for larger output power (via f_R^2), smaller α, longer cavity lengths, and smaller laser coupling factors. The maximum linewidth and RIN level within the coherence collapse regime have also been theoretically estimated and experimentally verified. They are given by [19]

$$\Delta v|_{max} = f_R \sqrt{1 + \alpha^2} \sqrt{\ln 4}, \tag{5.183}$$

$$\frac{RIN}{\Delta f}\bigg|_{max} = \frac{1}{\gamma} = \frac{1}{Kf_r^2 + \gamma_0}. \tag{5.184}$$

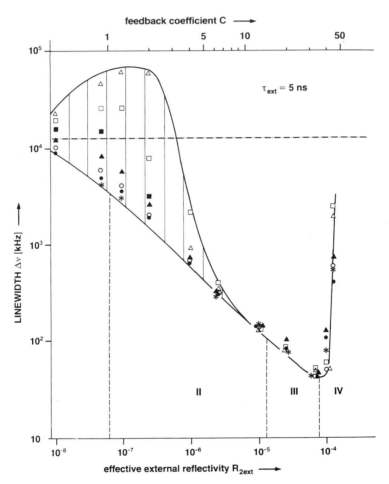

FIGURE 5.29 Spectral linewidth versus the amount of external optical feedback [18]. The roman numerals I–IV denote the regimes given in Fig. 5.28. In this example the assumed parameters are: $\tau_{ext} = 5$ ns, $\tau_L = 9$ ps, $R_2 = 0.32$, $\alpha = 6$, and 5 mW/facet out for an InGaAsP/InP laser ($R_{2ext} \equiv t_2^4 f_{ext}$). (Reprinted, by permission, from *IEEE Journal of Quantum Electronics* © IEEE 1990).

For a resonance frequency of 5 GHz and $\alpha = 5$, we obtain a maximum linewidth of 30 GHz! However, for a typical damping factor of 10^{10} s^{-1} at a few milliwatts of output power, we obtain a RIN of -100 dB/Hz. This level of RIN is large but not extreme. Thus, simple systems may be able to function adequately in regime IV as long as the linewidth is not of critical importance. Figure 5.29 summarizes the impact on laser linewidth due to feedback as it is scanned through regimes I through IV [21]. The envelope superimposed on the measurements reveals the linewidth variation created by the feedback phase.

Finally, at very high feedback levels when the external mirror becomes comparable to the laser's mirror feedback, a stable regime can be observed. Regime V is usually achieved only when the feedback is deliberate and well designed. For example, monolithic complex cavities considered in Chapter 3 fall into this category. An external mirror with good coupling to the laser cavity via an antireflection coating on mirror #2 can also reach feedback levels in regime V. The mode solutions in this feedback regime are stable with very narrow linewidths. If the mirror #2 reflection is small, it is sufficient to treat the entire laser cavity as a solitary laser with mode spacing given by $2\pi/(\tau_L + \tau_{ext})$. The effect on linewidth is then simply calculated by including a fill factor in the modal gain. The expression for linewidth derived earlier (5.151) depends on Γ^2. Thus, all else being equal, we can describe the line narrowing in regime V by

$$\Delta v = \Delta v_0 \left[\frac{\bar{n}_{ga} L_a}{\bar{n}_{ga} L_a + \bar{n}_{gp} L_p} \right]^2. \tag{5.185}$$

This result is not entirely different from Eq. (5.181) evaluated at the in-phase condition, since $C \propto \bar{n}_{gp} L_p / \bar{n}_{ga} L_a$. With typical numbers such as $L_a = 300$ μm, $L_p = 5$ cm, $\bar{n}_{ga} = 4.2$, and $\bar{n}_{gp} = 1$, the line narrowing is $\sim (1/40)^2$. The external cavity can therefore reduce a 100 MHz linewidth down to 62.5 kHz! Combined with a frequency-selective filter to attain single mode operation, such deliberate external cavity feedback can be quite useful for applications demanding highly coherent light. As for unintentional feedback, it is unfortunately very easy to fall into regime IV with small reflections such as given by fiber connectors from moderate distances. Optical isolators that provide greater than 60 dB of isolation are therefore often required in optical communications systems.

REFERENCES

[1] J.W. Scott, B.J. Thibeault, C.J. Mahon, L.A. Coldren, and F.H. Peters, *Applied Phys. Lett.*, **65**, 1483 (1994).

[2] N.K. Dutta, N.A. Olsson, L.A. Koszi, P. Besomi, and R.B. Wilson, *J. Appl. Phys.*, **56**, 2167 (1984).

[3]   R. Schimpe, J.E. Bowers, and T.L. Koch, *Electron. Lett.*, **22**, 453 (1986).

[4]   A. Yariv, *Optical Electronics*, 4th ed., Ch. 10, Saunders College Publishing, Philadelphia (1991).

[5]   C.H. Henry, *IEEE J. Quantum Electron.*, **QE-18**, 259 (1982).

[6]   A.L. Schawlow and C.H. Townes, *Phys. Rev.*, **112**, 1940 (1958).

[7]   M. Lax, *Phys. Rev.*, **160**, 290 (1967).

[8]   H. Gerhardt, H. Welling, and A. Guttner, *Z. Physik*, **253**, 113 (1972).

[9]   Y. Yamamoto and S. Machida, *Phys. Rev. A*, **35**, 5114 (1987).

[10]   Y. Yamamoto and N. Imoto, *IEEE J. Quantum Electron.*, **QE-22**, 2032 (1986).

[11]   K. Kikuchi and T. Okoshi, *IEEE J. Quantum Electron.*, **QE-21**, 1814 (1985).

[12]   E. Goobar, A. Karlsson, G. Björk, and P-J. Rigole, *Phys. Rev. Lett.*, **70**, 437 (1993); W.H. Richardson, S. Machida, and Y. Yamamoto, *Phys. Rev. Lett.*, **66**, 2867 (1991); H-A. Bachor, P. Rottengarter, and C.M. Savage, *Appl. Phys. B*, **55**, 258 (1992); S. Machida, Y. Yamamoto, and Y. Itaya, *Phys. Rev. Lett.*, **58**, 1000 (1987).

[13]   T. Okoshi, K. Kikuchi, and A. Nakayama, *Electron. Lett.*, **16**, 630 (1980).

[14]   C.H. Henry, *IEEE J. Quantum Electron.*, **QE-19**, 1391 (1983).

[15]   R. Nagaragan, M. Ishikawa, T. Fukushima, R.S. Geels, and J.E. Bowers, *IEEE J. Quantum Electron.*, **28**, 1990 (1992).

[16]   R.W. Tkach and A.R. Chraplyvy, *IEEE J. Lightwave Technol.*, **LT-4**, 1655 (1986).

[17]   G.P. Agrawal, *IEEE J. Quantum Electron.*, **QE-20**, 468 (1984).

[18]   J. Helms, C. Kurtzke, and K. Petermann, *IEEE J. Lightwave Technol.*, **LT-10**, 1137 (1992).

[19]   J. Wang and K. Petermann, *IEEE J. Quantum Electron.*, **QE-27**, 3 (1991).

[20]   J. Helms and K. Petermann, *IEEE J. Quantum Electron.*, **QE-26**, 833 (1990).

[21]   K. Petermann, *Laser Diode Modulation and Noise*, p. 277, Kluwer Academic Publishers (1991).

READING LIST

G.P. Agrawal and N.K. Dutta, 2d ed. *Semiconductor Lasers*, Ch. 6, Van Nostrand Reinhold, New York (1993).

K.J. Ebeling, *Integrated Opto-electronics*, Ch. 10, Springer-Verlag, Berlin (1993).

A. Yariv, *Optical Electronics*, 4th ed. Chs. 10 and 15, Saunders College Publishing, Philadelphia (1991).

K. Petermann, *Laser Diode Modulation and Noise*, Chs. 4–9, Kluwer Academic Publishers (1991).

C.H. Henry, Line broadening of semiconductor lasers, in *Coherence, Amplification, and Quantum Effects in Semiconductor Lasers*, ed. Y. Yamamoto, Ch. 2, Wiley-Interscience, New York (1991).

PROBLEMS

5.1   What is the current relative to threshold in Fig. 5.1?

5.2   Derive N_1 for P_1 for a mirror-loss-modulated laser instread of a current-modulated laser (i.e., assume $dI(t) = 0$ and $d\alpha_m(t) = \alpha_{m1}e^{j\omega t}$). What is the major distinction between the two types of modulation?

5.3   Derive $N_{p1} + N_{p2}$ assuming there are two competing modes in the cavity and that intermodal gain compression can be neglected. Under what conditions does the modulation response of the total photon density resemble the modulation response of either of the two modes separately (in other words, under what conditions can we ignore the division of photons into separate modes)?

5.4   Derive P_1/I_1 for a laser which has a series resistance, R, and a parallel shunt capacitance, C, across the diode such that not all of the terminal current, I_1, makes it to the active region. Express the result in terms of an overall transfer function, $H_{RC}(\omega)H(\omega)$, which includes RC parasitics. For a series resistance of $10\,\Omega$ and a laser with negligible damping and a relaxation resonance frequency of $20\,\text{GHz}$, determine the capacitance which reduces the 3 dB bandwidth of the laser to 90% of its value with no parasitics.

5.5   By solving for the poles of Eq. (5.59), show that Eq. (5.60) is correct.

5.6   Using Table 5.1, evaluate all terms in Eq. (5.49), and thus, verify Eq. (5.51).

5.7   Using the analytic approximations for a step function in current, plot the power out of one end of a $300\,\mu\text{m}$ long, $3\,\mu\text{m}$ wide cleaved-facet InGaAs/GaAs 3-QW buried-heterostructure laser vs. time in response to a current which abruptly rises by 10% from an initial value of twice threshold. The laser emits at $0.98\,\mu\text{m}$ and has an internal loss of $5\,\text{cm}^{-1}$, an internal efficiency of 80%, gain and recombination parameters as in Table 5.1 for an 80Å quantum-well, and a confinement factor of 6%.

5.8   For the transient response given in the example of Fig. 5.6(a), plot the frequency chirping vs. time assuming a linewidth enhancement factor of 5.

5.9   For the in-plane device described in Table 5.1 and Fig. 5.6, a small signal sinusoidal current of varying frequency is applied at a bias current of twice threshold. The frequency response peaks at some frequency and then falls off rapidly. What is this peak frequency? By how much is it different from the relaxation resonance frequency?

5.10 A 1.55 µm InGaAsP/InP laser with a linewidth enhancement factor of 5 is modulated from 1.5 to $4.5I_{th}$, creating an output power variation of 1–7 mW. Assuming a trapezoidal output at 1 Gb/s with 0.5 ns linear transitions and 0.5 ns constant power plateaus at each extrema, plot the chirping for two periods of a 10101... sequence.

5.11 A 1.3 µm InGaAsP/InP laser with a threshold carrier lifetime of 3 ns is biased at $0.9I_{th}$. At time $t = 0$ the drive current is suddenly increased to $2I_{th}$. What is the turn-on delay? Make whatever assumptions are necessary.

5.12 Consider the noise emitted from an InGaAs/GaAs VCSEL parameterized in Table 5.1, at 100 MHz.
(a) What is the RIN level for $I = 2I_{th}$? (Can P_0 be assumed small?)
(b) What is the RIN level at $I = 5I_{th}$? (Can P_0 be assumed large?)

5.13 What is the expected linewidth for the two cases of Problem 5.12, assuming a linewidth enhancement factor of 5?

5.14 Estimate the 3 dB modulation bandwidth for the two cases of Problem 5.13.
(a) Neglect transport effects.
(b) Include transport effects wherein the lifetime of carriers in the SCH region due to relaxation into the quantum wells is 50 ps and the leakage out of the well can be neglected.
(c) Include transport wherein leakage out of the well to the SCH region is additionally characterized by a lifetime of 200 ps.

5.15 Verify Eq. (5.145). If the normalized fringe visibility decays exponentially and is equal to 0.2 at a time delay of 3 ns, what is the linewidth of the laser?

5.16 In Fig. 5.26, estimate μ_2 near 2.5 mW of output power. From this determine the feedback level f_{ext} using the VCSEL parameters in Table 5.1, assuming that all of the light is coupled out of mirror #2, and that we are well above threshold. Use a VCSEL cavity length and group index of 1 µm and 4.2, respectively. Also express the feedback level in decibels.

5.17 Estimate the substrate thickness of the VCSEL in Fig. 5.26, assuming a constant voltage of 3 volts above threshold, a thermal impedance of 3°C/mW, a wavelength shift of 0.08 nm/°C, a substrate–air reflection of 32%, and a substrate group index of 4.2.

5.18 The intensity feedback from the end of a fiber is 4% and about one-quarter of this is coupled back into the lasing mode (excluding transmission through mirror #2) in a 300 μm long cleaved-facet 1.55 μm in-plane laser.

(a) Plot the feedback parameter, C vs. L_p, the fiber length ($n_{fiber} = 1.45$), for $\alpha = 0, 2, 4, 6$, up to $L_p = 1$ km.

(b) Plot the linewidth narrowing and broadening factor applicable in regimes I through III vs. C for $0 < C < 10$. Discuss what really happens near $C \sim 1$.

CHAPTER SIX

Perturbation and Coupled-Mode Theory

6.1 INTRODUCTION

This chapter is the first of three that focus more on the electromagnetic aspects of lightwave propagation, particularly as applied to diode lasers and related photonic integrated circuits. In this chapter we introduce the powerful perturbation and coupled-mode approaches to approximately solve very complex problems, which otherwise might only be addressed numerically.

In order to use these approaches we generally must know at least some of the eigenmodes of a relatively simple waveguide configuration. Then, the trick is to express the solution to some perturbed or more complex configuration in terms of this original basis set of eigenmodes. As we shall see, by using the orthogonality relationships amongst the original basis set, it is possible to derive some general formulas that are helpful in solving specific problems. The usefulness of these formulas derives from the fact that they only involve physical dimensions and some of the original basis functions.

To get started, we refer back to Chapter 2 to recall a convenient form for the electric field, $\mathscr{E}$, of some waveguide eigenmode. For mode m, then, the electric field as a function of space and time, Eq. (2.18), can be written as

$$\mathscr{E}_m(x, y, z, t) = \hat{\mathbf{e}}_i E_{0m} U_m(x, y)e^{j(\omega t - \beta_m z)}, \tag{6.1}$$

where $\hat{\mathbf{e}}_i$ is the unit vector along the ith coordinate (giving the polarization direction), E_{0m} gives the magnitude of the field, and U_m is the normalized transverse mode shape for mode m. For convenience, we can also combine the polarization into U_m, so that $\mathbf{U}_m = \hat{\mathbf{e}}_i U_m$. Thus, making use of the orthogonality between eigenmodes, we have

$$\int \mathbf{U}_m^* \cdot \mathbf{U}_n \, dA = \delta_{mn}, \tag{6.2}$$

where δ_{mn} is the Kronecker delta function, which equals unity for $m = n$ and zero otherwise. (For orthogonal polarizations the dot product also gives zero, even for $m = n$.) Orthogonal modes of a uniform waveguide do not interact. It is also worth mentioning that in the normal course of solving a waveguide problem, one finds that the eigenfunctions U_m provide a complete set.

Given that the U_m form a complete set, we can express an arbitrary field in the vicinity of a waveguide by a normal-mode expansion of all of the waveguide eigenmodes (including unguided radiation modes). Thus,

$$\mathscr{E}(x, y, z, t) = \sum_m \mathscr{E}_m(x, y, z, t), \tag{6.3}$$

where the amplitudes of the various terms in the summation are given by E_{0m} in Eq. (6.1). Both Eqs. (6.1) and (6.3) are solutions to the wave equation,

$$\nabla^2 \mathscr{E} + \varepsilon(x, y, z)k_0^2 \mathscr{E} = 0, \tag{6.4}$$

where $\varepsilon(x, y, z)$ is the relative dielectric constant, and k_0 is the free-space propagation constant for the medium of interest. For a single mode of a waveguide, we can also use Eq. (6.1) in (6.4) to obtain a wave equation for the transverse mode profile, U:

$$\nabla_T^2 U + [\varepsilon(x, y, z)k_0^2 - \beta^2]U = 0, \tag{6.5}$$

where we have used $\nabla^2 \mathscr{E} = \partial^2 \mathscr{E}/\partial z^2 + \nabla_T^2 \mathscr{E}$.

Although the higher order modes of even a simple waveguide may be complicated, it is fortunate that we need only know the details of at most two modes for all of the discussion to follow in this chapter. In fact, only one mode must be characterized for many problems.

6.2 PERTURBATION THEORY

6.2.1 Uniform Dielectric Perturbations

Many real waveguide structures involve a slight perturbation from a mathematically more simple structure, for which the eigenmode shapes, U_m, and propagation constants, β_m, are known. The perturbation can usually be expressed in terms of a change in relative dielectric constant, $\Delta\varepsilon$, which for generality can be complex and/or periodic along z.

If we replace ε by $\varepsilon + \Delta\varepsilon$ in Eq. (6.4), we most generally must assume that Eq. (6.3) must be used to represent the perturbed field. However, first we consider sufficiently weak perturbations where scattering to other modes can be neglected. This might occur most easily in a single-mode guide. Also, we assume $\Delta\varepsilon$ is uniform along z for this first example. Thus, in response to $\varepsilon \to \varepsilon + \Delta\varepsilon$, let $\beta \to \beta + \Delta\beta$, and $U \to U + \Delta U$ in Eq. (6.5). That is,

$$\nabla_T^2(U + \Delta U) + [(\varepsilon + \Delta\varepsilon)k_0^2 - (\beta + \Delta\beta)^2](U + \Delta U) = 0. \qquad (6.6)$$

Multiplying this out, and dropping the unperturbed transverse wave equation, which equals zero, and the second-order perturbation terms, we are left with

$$\nabla_T^2 \Delta U + \varepsilon k_0^2 \Delta U + \Delta\varepsilon k_0^2 U - 2\beta\Delta\beta U - \beta^2 \Delta U = 0. \qquad (6.7)$$

Now, we introduce a technique that we shall use again and again in this chapter to simplify complex expressions:

Multiply by the complex conjugate of the transverse mode and integrate over the cross section.

In later cases, modal orthogonality will help to remove many of the unwanted terms. In this case, we obtain

$$2\beta\Delta\beta \int |U|^2 \, dA = \int \Delta\varepsilon k_0^2 |U|^2 \, dA$$

$$+ \int [(\nabla_T^2 \Delta U)U^* + \varepsilon k_0^2 \Delta U U^* - \beta^2 \Delta U U^*] \, dA. \qquad (6.8)$$

As shown in Appendix 14, the second term on the right is negligible, provided that both ΔU and U vanish at infinity and that the unperturbed ε and β are mostly real. This indicates that the small change in the transverse mode shape, ΔU, due to $\Delta\varepsilon$ will have *no effect* on the propagation constant to first order. Thus, we fortunately do not have to worry about how the changing mode shape will change the averaging, at least for this first-order approximation, and we only need know the original unperturbed transverse mode.

Then, solving for $\Delta\beta$, we have the desired perturbation formula

$$\Delta\beta = \frac{\int \Delta\varepsilon k_0^2 |U|^2 \, dA}{2\beta \int |U|^2 \, dA}. \qquad (6.9)$$

If U is normalized according to Eq. (6.2), the denominator integral is just unity. Usually, the index perturbation is limited in lateral extent, and it may even be constant over some range. In these cases the integration is easily performed. It is important to keep in mind that all quantities, except for the actual perturbation, $\Delta\varepsilon$, are for the original unperturbed problem. Below we consider the example of adding a quantum-well active region to a simple three-layer slab waveguide to form a more complex SCH-QW laser.

6.2.2 Quantum-Well Laser Modal Gain and Index Perturbation: an Example

The addition of a quantum well to a simple three-layer slab guide is an example of a relatively easy problem to solve using this perturbation theory. This approach gives a quick and familiar result for a relatively complex, and certainly technologically important, problem.

Figure 6.1 schematically shows the problem at hand. Again, Eq. (6.9) requires only the unperturbed transverse eigenmode, $U(x, y)$, and propagation constant, β, in addition to the actual perturbation, $\Delta\tilde{\varepsilon} = \Delta\varepsilon_r + j\Delta\varepsilon_i$, which in this case is complex. The integration is limited to the area of the active region, A, since $\Delta\varepsilon$ is zero elsewhere. Expanding $\Delta\beta$ in terms of the real effective index and modal gain perturbations, and expanding $\Delta\tilde{\varepsilon}$ in terms of its real index and gain components in the active region,

$$\Delta\beta = \frac{2\pi\Delta\bar{n}}{\lambda} + j\frac{\langle g\rangle_{xy}}{2} = \frac{k_0^2}{2\beta}\frac{2n_A\int_A\left(\Delta n_A + j\frac{\lambda g_A}{4\pi}\right)|U|^2\,dA}{\int|U|^2\,dA}. \qquad (6.10)$$

Solving for the modal gain and effective index perturbations separately, then

$$\langle g\rangle_{xy} = \frac{\dfrac{n_A g_A}{\bar{n}}\int_A|U|^2\,dA}{\int|U|^2\,dA} = \Gamma_{xy}g_A, \qquad (6.11)$$

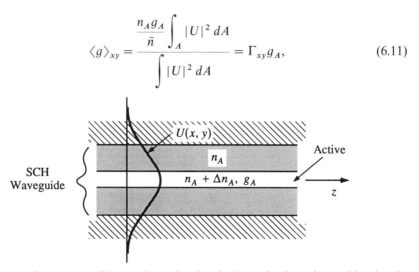

FIGURE 6.1 Quantum-well laser schematic wherein the active layer is considered to be the perturbation.

and

$$\Delta \bar{n} = \frac{\dfrac{n_A \Delta n_A}{\bar{n}} \displaystyle\int_A |U|^2 \, dA}{\displaystyle\int |U|^2 \, dA} = \Gamma_{xy} \Delta n_A, \tag{6.12}$$

where we have assumed that the gain and index perturbations are constant over the active region at g_A and Δn_A, respectively. Again, if U is properly normalized, the integral in the denominator equals one.

With reference to Appendix 5 and Chapter 2, we see that Eqs. (6.11) and (6.12) are the same expressions as derived there for the modal gain and effective index increase for such a waveguide configuration.

6.3   COUPLED-MODE THEORY: TWO-MODE COUPLING

By introducing index and loss perturbations along a waveguide it is possible to facilitate a coupling between basis modes, which, of course, are orthogonal without any such perturbation. Considering the basis set of an isolated uniform waveguide, the perturbation usually has to have some periodicity along the propagation direction to cause such a coupling, since the propagation constants of the two uncoupled modes are different. An important practical example is a grating which has a period such that scattering from one waveguide mode couples coherently to another counterpropagating mode.

On the other hand, if two identical uniform guides are brought together so that their unperturbed degenerate modes overlap, the existence of the index of one guide will perturb the mode of the other such that coherent coupling results. This is called a directional coupler, since only modes traveling in the same direction with about the same velocity can couple. More generally, periodic index perturbations can be added to coupled waveguides to provide more complex couplers. In what follows, we shall first consider simple gratings in a single waveguide, then the coupling of uniform waveguides, and finally, the so-called grating-assisted coupling between two waveguides.

6.3.1   Contradirectional Coupling: Gratings

6.3.1.1 General Theory   Initially, we consider the coupling of two identical modes of the same waveguide propagating in opposite directions. Such coupling is obviously important in reflecting a particular mode within a laser cavity, for example. With this assumption and limiting the discussion to index-guided structures where the phase fronts are perpendicular to the axis of the guide, we can express the net field as the sum of the two identical counterpropagating modes. That is, using Eqs. (6.1) and (6.3)

$$\mathscr{E}(x, y, z, t) = \mathbf{U}(x, y)[E_f e^{j(\omega t - \beta z)} + E_b e^{j(\omega t + \beta z)}], \tag{6.13}$$

where E_f and E_b give the amplitudes of the two counterpropagating modes.

Now, we plug Eq. (6.13) into the wave equation (6.4) with an assumed perturbation, $\varepsilon \to \varepsilon + \Delta\varepsilon(z)$. We let $\Delta\varepsilon$ be z-dependent, since we suspect that this will be necessary to provide the desired coupling. If there is coupling, then the amplitudes E_f and E_b will vary with z, so the full wave equation (6.4) is needed. However, if we assume that the coupling will be weak, the amplitudes should vary relatively slowly. That is, we will be able to drop terms containing their second derivatives in the expansion of the wave equation if the coefficients are not large. Also, for generality we allow a small ΔU, although we suspect that it will not show up in the final result, in analogy with the perturbation result of Section 6.2.2.

Observing that the unperturbed ε and U will generate terms which satisfy the unperturbed wave equation and dropping the second-order terms in the perturbations as well as the second derivatives, we are left with

$$-2j\beta(\mathbf{U} + \Delta\mathbf{U})\frac{dE_f}{dz}e^{-j\beta z} + 2j\beta(\mathbf{U} + \Delta\mathbf{U})\frac{dE_b}{dz}e^{j\beta z}$$
$$= -[\nabla_T^2\Delta\mathbf{U} + (\varepsilon k_0^2 - \beta^2)\Delta\mathbf{U}][E_f e^{-j\beta z} + E_b e^{j\beta z}]$$
$$- \Delta\varepsilon(z)k_0^2\mathbf{U}[E_f e^{-j\beta z} + E_b e^{j\beta z}]. \tag{6.14}$$

We now dot multiply by $\mathbf{U}^*$ and integrate over the cross section. Then, we see that the first term on the right is zero according to Appendix 14, as was the case in Eq. (6.8) above. Next, we recognize that the terms containing $\Delta U(dE_i/dz)$ are really second-order terms and are negligible in comparison to the others. Finally, we divide by $-2j\beta \int |U|^2 \, dA$, to get

$$\frac{dE_f}{dz}e^{-j\beta z} - \frac{dE_b}{dz}e^{j\beta z} = -j\frac{k_0^2}{2\beta}[E_f e^{-j\beta z} + E_b e^{j\beta z}]\frac{\int \Delta\varepsilon(z)|U|^2 \, dA}{\int |U|^2 \, dA}. \tag{6.15}$$

Note that if $\Delta\varepsilon(z)$ were uniform along z, we would be left simply with the perturbation formulas for both forward and backward travelling waves independent of each other, since $dE_f/dz = -j\Delta\beta E_f$, and $dE_b/dz = j\Delta\beta E_b$, in that case. Any dc component in $\Delta\varepsilon(z)$ has exactly the predicted effect of uniformly changing the propagation constants by $\Delta\beta$. However, to couple forward to backward waves and vice versa, it can be seen that $\Delta\varepsilon(z)$ must contain factors $\exp(\pm 2j\beta z)$.

More generally, we allow $\Delta\varepsilon(z)$ be a complex spatially periodic function with possibly a nonuniform cross section. Thus, Fourier analysis can be applied to give

$$\Delta\varepsilon(x, y, z) = \sum_{l \neq 0} \Delta\varepsilon_l(x, y)e^{-jl(2\pi/\Lambda)z}, \tag{6.16}$$

where $2\pi l/\Lambda$ are the various space harmonics of the arbitrary periodic perturbation of fundamental period Λ. Using this in Eq. (6.15), and looking for possible solutions that couple forward to backward waves and vice versa, we identify terms which could provide similar exponential factors on each side of the equation for all z. That is,

$$\frac{dE_f}{dz}e^{-j\beta z} = -j\frac{k_0^2}{2\beta}E_b e^{j\beta z}e^{-jl(2\pi/\Lambda)z}\frac{\displaystyle\int \Delta\varepsilon_{+l}(x, y)|U|^2\, dA}{\displaystyle\int |U|^2\, dA}, \qquad (6.17)$$

and the comparable equation for dE_b/dz from the remaining terms,

$$\frac{dE_b}{dz}e^{j\beta z} = j\frac{k_0^2}{2\beta}E_f e^{-j\beta z}e^{jl(2\pi/\Lambda)z}\frac{\displaystyle\int \Delta\varepsilon_{-l}(x, y)|U|^2\, dA}{\displaystyle\int |U|^2\, dA}. \qquad (6.18)$$

From both Eqs. (6.17) and (6.18) we see that the condition,

$$l\frac{2\pi}{\Lambda} \approx 2\beta = \frac{4\pi\bar{n}}{\lambda}, \qquad (6.19)$$

must be met for dE_f/dz or dE_b/dz to have the same sign over some distance, or put another way, for coherent phasing of the coupling. All other terms can be neglected, since we are only interested in the net effect on E_f and E_b over a distance of several wavelengths or more. That is, the differential equations (6.17) and (6.18) would not provide for any net change in E_f and E_b vs. z unless Eq. (6.19) is satisfied. Equation (6.19) indicates that for a good reflective grating, the period of the index perturbation, Λ, should be some multiple, l, of $\lambda/2\bar{n}$. This is called the Bragg condition. Figure 6.2 illustrates a waveguide containing an index grating for forward to backward wave coupling.

Before continuing to find solutions to Eqs. (6.17) and (6.18), it is convenient to make some substitutions. As already suggested, the propagation constant at the Bragg condition is given by Eq. (6.19) with $\equiv$ replacing $\approx$. More specifically, we let $\beta_0 \equiv l\pi/\Lambda$ which also defines the Bragg wavelength, $\lambda_0/\bar{n} \equiv 2\Lambda/l$ (note that higher orders of l allow us to use a longer grating period for the same λ_0). In addition, we let

$$\kappa_{\pm l} = \frac{k_0^2}{2\beta}\frac{\displaystyle\int \Delta\varepsilon_{\pm l}(x, y)|U|^2\, dA}{\displaystyle\int |U|^2\, dA}. \qquad (6.20)$$

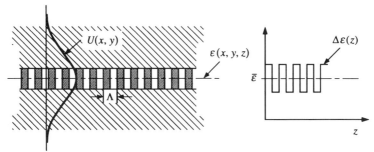

FIGURE 6.2 Waveguide containing index grating for contradirectional coupling. The relative dielectric constant (square of index of refraction) varies periodically along z about some average value, $\bar{\varepsilon}$, so that reflected components add in phase.

As before, if U is normalized according to Eq. (6.2), the denominator drops out. It is also important to remember that $\Delta\varepsilon_{\pm l}$ are Fourier coefficients as given by Eq. (6.16), not the actual amplitude of the index perturbation. For example, with a perturbation of $\delta\varepsilon \cos(2\pi z/\Lambda - \phi)$ two terms would result from Eq. (6.16) for $l = \pm 1$, each with a magnitude, $\Delta\varepsilon_{l=\pm 1} = \delta\varepsilon/2$. That is, we can expand the cosine to find

$$\Delta\varepsilon(x, y, z) = \frac{\delta\varepsilon(x, y)}{2} e^{-j\phi} e^{j(2\pi/\Lambda)z} + \frac{\delta\varepsilon(x, y)}{2} e^{j\phi} e^{-j(2\pi/\Lambda)z}$$

$$= \Delta\varepsilon_{-1}(x, y)e^{j(2\pi/\Lambda)z} + \Delta\varepsilon_{+1}(x, y)e^{-j(2\pi/\Lambda)z}. \qquad (6.21)$$

Thus, we can use Eq. (6.21) and $\delta\varepsilon(x, y) = 2n_g\delta n_g(x, y)$ in Eq. (6.20) to write the coupling coefficient for the cosinusoidal dielectric perturbation with an arbitrary phase relative to $z = 0$. That is,

$$\kappa_{\pm 1} = \kappa e^{\pm j\phi} = \frac{n_g}{\bar{n}} \frac{\pi}{\lambda} \Gamma_{xy_g}(\delta n_g)e^{\pm j\phi}, \qquad (6.22)$$

where for the second equality, we have assumed a uniform grating perturbation over some lateral portion of the waveguide, for which Γ_{xy_g} is the lateral grating region confinement factor, δn_g is the magnitude of the assumed cosinusoidal index perturbation (half the peak-to-peak variation), and n_g is the average index in the grating region.

Finally, we let the *detuning parameter* $\delta \equiv \beta - \beta_0$, so that Eqs. (6.17) and (6.18) become

$$\frac{dE_f(z)}{dz} = -j\kappa_l E_b(z)e^{2j\delta z}, \qquad (6.23)$$

and

$$\frac{dE_b(z)}{dz} = j\kappa_{-l}E_f(z)e^{-2j\delta z}. \qquad (6.24)$$

Equations (6.23) and (6.24) are relatively clear statements of the coupling between the forward- and backward-going lightwaves. They both indicate that the change in amplitude of one wave is directly proportional to the amplitude of the other, and that the proportionality involves a coupling constant, $\kappa_{\pm l}$, which depends upon the overlap of a periodic index perturbation and the modes' energy density, as well as a dephasing factor which deviates from unity if the Bragg condition is not met. It is also worth noting from Eq. (6.20) or Eq. (6.22) that $\kappa_{-l} = \kappa_l^*$ for a real $\Delta\varepsilon(z)$ and β, and that this condition is still approximately true for waveguides with small gains or losses, i.e., β predominately real, as is usually the case.

So now all we need to do is solve the coupled equations (6.23) and (6.24). Unfortunately, these equations are still not in a very convenient form for mathematical solution. The solution for the net electric field along the guide is given by Eq. (6.13) as a linear superposition of the forward and backward fields. We also can express this net field in terms of two new z-varying factors, $A(z) = E_f(z)e^{-j\delta z}$, and $B(z) = E_b(z)e^{j\delta z}$, so that we have,

$$\mathcal{E}(x, y, z) = \mathbf{U}[E_f(z)e^{-j\beta z} + E_b(z)e^{j\beta z}] = \mathbf{U}[A(z)e^{-j\beta_0 z} + B(z)e^{j\beta_0 z}]. \quad (6.25)$$

The latter form combines both coupling and dephasing effects within the z-varying coefficients, A and B, since $\beta_0 = \pi l/\Lambda$ is a constant for a given grating.

Expressing $E_f(z)$ and $E_b(z)$ in Eqs. (6.23) and (6.24) in terms of $A(z)$ and $B(z)$, we obtain

$$\frac{dA}{dz} = -j\kappa_l B - j\delta A, \quad (6.26)$$

and

$$\frac{dB}{dz} = j\kappa_{-l} A + j\delta B. \quad (6.27)$$

These coupled differential equations can now be straightforwardly solved in the usual way by letting

$$A = A_1 e^{-\sigma z} + A_2 e^{\sigma z}, \quad (6.28)$$

and

$$B = B_1 e^{-\sigma z} + B_2 e^{\sigma z}. \quad (6.29)$$

Plugging Eqs. (6.28) and (6.29) into Eqs. (6.26) and (6.27), and separately equating the coefficients of $e^{\sigma z}$ and $e^{-\sigma z}$ (since the intermediate expressions must be true for all z) leads to relationships between A_1 and B_1, and A_2 and B_2, e.g.,

$$B_1 = \frac{j\kappa_{-l}A_1}{-\sigma - j\delta} \quad \text{and} \quad B_2 = \frac{j\kappa_{-l}A_2}{\sigma - j\delta}, \quad (6.30)$$

as well as a solution for σ given by

$$\sigma^2 = \kappa_l \kappa_{-l} - \delta^2 = \kappa^2 - \delta^2, \tag{6.31}$$

where for the last equality we have made use of Eq. (6.22).

For waveguides which contain gain or loss, β is complex, and therefore, so is the detuning parameter, δ. Thus, following previous nomenclature, we also have

$$\tilde{\sigma}^2 = \kappa^2 - \tilde{\delta}^2, \tag{6.32}$$

where

$$\tilde{\delta} = \tilde{\beta} - \beta_0 = \frac{2\pi\bar{n}}{\lambda} + j\frac{\langle g \rangle_{xy} - \alpha_i}{2} - \frac{\pi l}{\Lambda}. \tag{6.33}$$

In cases where $\Delta\varepsilon(z)$ is complex, i.e., where there is a significant periodic gain or loss, κ is complex, and $\kappa_{-l} \neq \kappa_l^*$. However, Eq. (6.32) is still valid since this condition has not been used in its derivation. For simultaneous cosinusoidal index and gain perturbations over some region of the waveguide cross section, the derivation of Eq. (6.22) yields

$$\kappa = \frac{n_g}{\bar{n}} \frac{\pi}{\lambda} \Gamma_{xy_g} \left(\delta n_g + j \frac{\delta g_g}{2k_0} \right), \tag{6.34}$$

where δg_g is the magnitude of the cosinusoidal gain perturbation in the grating region. Thus, Eq. (6.34) can be used in Eq. (6.32) and elsewhere for this case.

Using Eqs. (6.28) and (6.29) with $\sigma = j\sqrt{\delta^2 - \kappa^2}$ in the expression for the propagating electric field, Eq. (6.25), we observe four combinations of phase factors, one for each of the A and B coefficients:

$$j\beta_g = \pm j(\beta_0 \pm \sqrt{\delta^2 - \kappa^2}). \tag{6.35}$$

The propagation constant of the grating, β_g, therefore has four possible solutions at any given frequency, ω, or equivalently, $\delta = \beta - \beta_0 = (\omega - \omega_0)/v_g$. Far away from the Bragg frequency ($|\delta| \gg \kappa$), these solutions are $\pm\beta$ and $\pm(\beta - 2\beta_0)$, which as shown in Fig. 6.3, are (a) the unperturbed forward and backward wave propagation constants, $\pm\beta$, and (b) grating-induced replicas of these propagation constants displaced by $\mp 2\pi/\Lambda$ (for $l = 1$). Away from the Bragg condition, the two sets of solutions do not interact and the latter solutions never contain much energy ($A_2, B_1 \approx 0$).

At the Bragg frequency ($\delta = 0$, $\omega_0 = (c/\bar{n})(\pi/\Lambda)$), the solutions are $\pm(\beta_0 \pm j\kappa)$. That is, all field solutions propagate at $\pm\beta_0$, but grow or decay at the rate κ as they travel due to the coherent transfer of energy from one solution

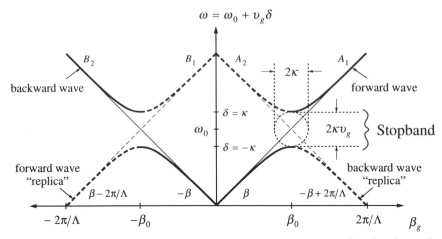

FIGURE 6.3 An ω–β diagram for the coupled β_g-solutions to contradirectional coupling in a grating, with the uncoupled solutions denoted by thinner straight lines. Each of the grating-generated replica solutions (dashed) and ordinary forward and backward wave solutions (solid) correspond to one of the A or B coefficients indicated. The extent of the stopband in both ω and β directions is also shown (where v_g is the group velocity of the unperturbed mode). However, the complex $\tilde{\beta}_g$-solutions which exist throughout the stopband are not shown. Finally, the scale of the stopband has been exaggerated somewhat, considering that $\kappa \ll \pi/\Lambda$ in order to satisfy the weak coupling criterion.

to the other. In this case, the solutions interact heavily and all four A and B coefficients can have significant amplitude. Detuning to $|\delta| = \kappa$, or $\omega = \omega_0 \pm v_g \kappa$, the propagation constants become purely real and equal to $\pm\beta_0$, or $\pm\pi/\Lambda$. Detuning further, the solutions split and approach their uncoupled values asymptotically. Thus, as illustrated in the figure, the coupling between the solutions creates a "stopband" for $|\delta| < \kappa$. Within this stopband, the solutions have complex propagation constants, the real parts remaining fixed at $\pm\beta_0$. It is in this regime where an incident field can decay in the grating and eventually be reflected back out (hence the name *stop*band).

6.3.1.2 Finite-Length Gratings with no Back Reflections

Now, we are finally ready to look at some practical characteristics of a real physical problem. For example, consider the grating of finite length, L_g, in Fig. 6.4. We assume a forward ($+z$-traveling) wave incident from the left that enters the grating at $z = 0$, and divides into a net reflected backward ($-z$-traveling) wave emerging at $z = 0$, as well as a transmitted forward wave at $z = L_g$. Of course, if the grating contains gain or loss the sum of the transmitted and reflected powers will not equal the incident power. Within the grating the forward wave decays as more and more energy is coupled to the backward wave. The backward wave grows as it propagates to the left. We also assume

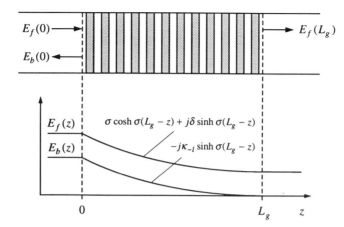

FIGURE 6.4   Grating of finite length. The forward wave decays as the backward wave grows. The backward wave is zero at $z = L_g$, and a portion of the forward wave is transmitted. The functionality of $A(z)$ and $B(z)$ within the grating is also shown. The shaded areas indicate regions of lower index.

no reflections beyond the grating, so the backward wave has no amplitude at $z = L_g$.

For this case of no reflection from the right side of the grating, we have the boundary condition, $E_b(z = L_g) = 0$, which requires that $B(z = L_g) = 0$. The ratio of the amplitudes of the net reflected wave to the incident wave is defined as the grating reflection coefficient,

$$r_g = \frac{E_b(0)}{E_f(0)} = \frac{B(0)}{A(0)} = \frac{B_1 + B_2}{A_1 + A_2}. \tag{6.36}$$

From the right-hand boundary condition at $z = L_g$ and Eq. (6.29), $B_2 = -B_1 e^{-2\sigma L_g}$. Using this and Eq. (6.30), we obtain

$$\tilde{r}_g = -j \frac{\tilde{\kappa}_{-l} \tanh \tilde{\sigma} L_g}{\tilde{\sigma} + j\tilde{\delta} \tanh \tilde{\sigma} L_g}. \tag{6.37}$$

The tildas have been added to explicitly indicate that the coupling constant, κ, detuning parameter, δ, and decay constant, σ, can be complex. The reflectivity from the other side of the grating is given by Eq. (6.37) times $e^{-2j\beta_0 L_g}$, with $\tilde{\kappa}_l$ replacing $\tilde{\kappa}_{-l}$ (the phase factor arises from the definitions of A and B in Eq. (6.25)).

Implicit in the derivation is the assumption of a reference plane at $z = 0$. If the grating is unshifted (i.e., $\phi = 0$ in Eq. (6.22)) and the index perturbation is real then, κ_{-l} is real, and we have a maximum positive deviation at $z = 0$, i.e., a cosinusoidal space harmonic in $\Delta \varepsilon$. (In this case, the grating has

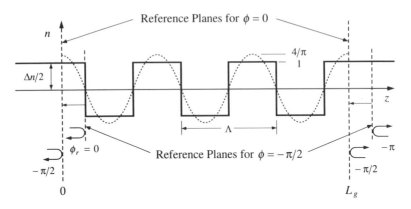

FIGURE 6.5   Illustration of reference planes used to define $z = 0$ and $z = L_g$ in a finite-length grating. The coupled-mode equations assume the index profile has a Fourier component that varies as $\cos(2\beta_0 z - \phi)$. The first set of reference planes are drawn for $\phi = -\pi/2$ which aligns the $z = 0$ plane with an index down step of a square wave grating (these are the reference planes assumed in Chapter 3). The second set shifts the planes to the left for a more symmetric placement by setting $\phi = 0$.

real, positive, and symmetrical Fourier coefficients.) Equation (6.37) shows that at the Bragg frequency, the reflection coefficient has a phase of $-\pi/2$ referenced to this plane. Figure 6.5 illustrates this choice of reference planes and its correspondence to a square wave index grating, for which the reflection phase is more obvious. That is, at the Bragg frequency, where all reflection components add in phase, we know that the reflection phase from a square wave grating must be zero referenced to an index down step. The first Fourier component of this grating has a zero crossing at this point and is a maximum one quarter-wave back. Thus, for a reference plane placed at this maximum of the first Fourier component, the grating phase would be $-\beta\Lambda/2 = -\pi/2$.

More generally, the reflection phase, ϕ_r, of waves incident from the left is determined by $-j\kappa_{-1} \propto -je^{-j\phi}$ at the $z = 0$ plane, for a lossless grating at the Bragg frequency. For waves incident from the right, ϕ_r is determined by $-j\kappa_1 e^{-2j\beta_0 L_g} \propto -je^{j\phi}e^{-2j\beta_0 L_g}$ at the $z = L_g$ plane. These reflection phases are shown in Fig. 6.5 for two different choices of the grating phase, ϕ, and a grating with $L_g = m\Lambda$. When combining a grating with other elements, the symmetric reference planes are a convenient choice. To use them, we adjust the input plane to a cosine peak by setting $\phi = 0$. However if $L_g \neq m\Lambda$, the output plane will not coincide with a cosine peak if it is placed at L_g. To retain the symmetry, we must shift the output plane slightly to create a reference plane separation of $L_r = m\Lambda$. This is accomplished by adding a phase delay, $e^{-2j\beta(L_r - L_g)}$, to the reflection phase from the right. The total reflection phase from the right then becomes $-je^{j\phi}e^{-2j\beta_0 L_r}e^{-2j\delta(L_r - L_g)} \approx -je^{j\phi}$, where the approximation holds for $L_r \sim L_g$, or $\delta = 0$. In practice then, for any grating length, the reflection phase at the Bragg frequency can be set to $-\pi/2$ from both sides of the grating. Away from the Bragg frequency and/or with gain or loss, the reflection phases remain symmetric (for $L_r \sim L_g$), but are no longer exactly equal to $-\pi/2$.

While our choice of reference planes is arbitrary, it is important to realize that our choice of L_g is not, since it determines the reflectivity magnitude. Typically with in-plane lasers, L_g corresponds to the physical length of the grating (i.e. where the uniform index stops to where the uniform index begins again). However, this may not always be the case. For example, in Figs. 6.4 and 6.5 (and the gratings analyzed in Chapter 3), L_g is $\Lambda/2$ longer than the physical length of the grating. The reason for this has to do with how the index outside the grating compares with the *average* index inside the grating. If there is a mismatch, additional reflections are created at the boundaries that can either enhance or reduce the overall grating reflectivity. In coupled-mode theory, this is approximately taken into account by modifying L_g. In Fig. 6.5, each $\Lambda/4$ segment contributes a $\Delta n/2$ step. Thus, adding a mismatch of $\Delta n/2$ at a physical boundary should be roughly equivalent to increasing L_g by $\Lambda/4$, as long as the mismatch reflection adds in phase with the grating reflections. In Figs. 6.4 and 6.5, the index on either side of the grating is $\Delta n/2$ larger than the average grating index. The two $\Delta n/2$ mismatches therefore effectively increase L_g by $\Lambda/2$. In the case of long gratings and small mismatches typically encountered with in-plane lasers, such subtle modifications to L_g are not important to consider.

The power reflection spectrum of the grating, given by the absolute square of Eq. (6.37), is plotted on the left in Fig. 6.6 for different values of κL_g. The curves have similar characteristics to those of Fig. 3.12, with a sin x/x spectrum for small κL_g and the development of a well-defined transmission stopband for large κL_g. The dashed curve tracks the $|\delta| = \kappa$ transition. Outside this transition point, (i.e., for $|\delta| > \kappa$), the reflectivity drops off dramatically. Thus, the transmission stopband (or reflection bandwidth) of the grating can be characterized by $\sim 2\kappa$. From the definition of the grating penetration depth given in Chapter 3, we observe that the reflection bandwidth is roughly the reciprocal of the penetration depth ($L_{eff} \approx 1/2\kappa$).

It may seem surprising that the coupled-mode formula given by Eq. (6.37), which has little resemblence to Eq. (3.52) derived from transmission matrix theory, leads to a nearly identical reflection spectrum (as revealed by the right plot in Fig. 6.6). It turns out that there is a correspondence. It is just obscured by the math. For the interested reader, Appendix 7 reveals this hidden correspondence between the two formulas.

As in Chapter 3, the reflection parameter of interest is κL_g. This can be realized in the present case by evaluating Eq. (6.37) at the Bragg frequency by using Eq. (6.32),

$$\tilde{r}_g = -j \tanh \kappa L_g, \qquad (\delta = 0) \qquad (6.38)$$

and noting that for small κL_g, $|r_g| \to \kappa L_g$. That is, κ can be interpreted as the reflection per unit length, just as in Chapter 3. However, unlike the result of

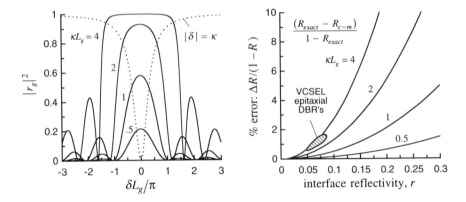

FIGURE 6.6 The left plot shows the power reflection spectrum of the grating for four different values of κL_g vs. the detuning parameter, $\delta \equiv \beta - \beta_0 = (\omega - \omega_0)/\omega_0 \cdot (\pi/\Lambda)(\bar{n}_g/\bar{n})$, where $\Lambda = \lambda_0/2\bar{n}$. Exact and coupled-mode results are indistinguishable on this scale. The right plot shows the percent error in transmission at resonance calculated using the coupled-mode formula. For example, if the exact result predicts $R_g = 99\%$, then a 10% transmission error would correspond to a coupled-mode prediction of $R_g = 98.9\%$.

Chapter 3 discussed in the caption of Fig. 3.12, the coupled-mode prediction is characterized only by the product κL_g, independent of whether κ is small or large. In fact, for large perturbations (large κ, and more significantly, large reflection per half-period, $2r$, using the jargon of Chapter 3), the coupled-mode result begins to break down even for small κL_g. However, even for interface field reflectivities (i.e., $\Delta n/2n$) as high as 18% and $\kappa L_g = 4$, Fig. 6.6 reveals that the error in transmission at the center of the stopband is still less than 10%. Figure 6.6 also shows that the range of interface reflectivities and κL_g used in typical VCSEL semiconductor DBRs would lead to transmission errors of $\approx 1\%$ if the coupled-mode formula (6.37) were used. Thus, coupled-mode theory still works surprisingly well even in this regime. However, in practice the *terminations* used with VCSEL mirrors are far from the average index of the DBRs, and this ultimately makes the coupled-mode formula impractical for such applications. In such cases where the terminations have a very different index (a DBR terminating in air, for example), the analysis at the end of Appendix 7 can be used to determine the peak reflectivity; however, the reflectivity *spectrum* must be calculated numerically using the transmission matrix techniques introduced in Chapter 3.

Other important grating relationships useful for DBR and DFB lasers may also be found in Chapter 3. For example, the reflection for small κL_g is given by Eq. (3.55), and the grating effective reflection plane spacing (or penetration depth), L_{eff}, is given by Eq. (3.59).

The transmission through the grating in Fig. 6.4 is given by

$$t_g = \frac{E_f(L_g)}{E_f(0)} e^{-j\beta L_g} = \frac{A(L_g)}{A(0)} e^{-j\beta_0 L_g}. \tag{6.39}$$

The phase factors arise from how E_f and A are defined relative to the full electric field in Eq. (6.25). Using Eq. (6.28) for $A(L_g)$, the boundary condition $B(z = L_g) = 0$, and Eq. (6.30) to relate the A_i's to the B_i's, we obtain

$$\tilde{t}_g = \frac{\tilde{\sigma} \operatorname{sech} \tilde{\sigma} L_g}{\tilde{\sigma} + j\tilde{\delta} \tanh \tilde{\sigma} L_g} e^{-j\beta_0 L_g}. \tag{6.40}$$

The tildas have again been added to indicate the possibility of complex factors. Note that the prefactor for transmission is $\tilde{\sigma}$ instead of $\tilde{\kappa}_{-l}$ as it was for reflection. This has to be the case because the transmission should not go to zero when the coupling goes to zero. In fact, for a lossless grating with $\kappa = 0$, we see that $\sigma \to j\delta$ and $t_g \to e^{-j\beta L_g}$, which is the simple expected transmission delay between the two reference planes. Also note that because the transmission only depends on the coupling through $\tilde{\kappa}_l \tilde{\kappa}_{-l}$, the phase of the grating relative to the reference planes, ϕ, drops out (from Eq. (6.22)). So unlike the reflection phase, the transmission phase does not depend on the specific placement of the reference planes, only on the separation between. For example, at the Bragg frequency, the transmission phase is $-\pi L_g/\Lambda$, which for $L_g = m\Lambda$, reduces to $-m\pi$, regardless of where we place the reference planes relative to the grating. However, if we separate the reference planes by $L_r \neq L_g$, then the phase factor in Eq. (6.40) should be replaced with $e^{-j\beta_0 L_r} e^{-j\delta(L_r - L_g)} \approx e^{-j\beta_0 L_r}$ (the approximation is good for $L_r \sim L_g$). Finally, it should be noted that the transmission from the other side of the grating is identical to Eq. (6.40).

6.3.2 DFB Lasers

6.3.2.1 No Facet Reflections A DFB laser without any end reflections consists of a finite-length grating with gain. At threshold, the gain is sufficient to overcome internal and transmission losses and provide an output with no input. In terms of the previous section, the characteristic equation for such a laser can be obtained from the poles of r_g or t_g, and these can be obtained by setting the denominator of the grating reflection or transmission coefficient, Eqs. (6.37) or (6.40), respectively, to zero. Thus, the characteristic equation, which determines the threshold gain and wavelength, is

$$\tilde{\sigma}_{th} = -j\tilde{\delta}_{th} \tanh \tilde{\sigma}_{th} L_g. \tag{6.41}$$

The symbols have been already defined, but for clarity we repeat at threshold,

$$\tilde{\sigma}_{th}^2 = \kappa^2 - \tilde{\delta}_{th}^2,$$

$$\tilde{\delta}_{th} = \tilde{\beta}_{th} - \beta_0,$$

$$\tilde{\beta}_{th} = \beta_{th} + j\frac{\langle g \rangle_{xy_{th}} - \langle \alpha_i \rangle_{xy}}{2}, \tag{6.42}$$

$$\beta_{th} = 2\pi\bar{n}/\lambda_{th} \quad \text{and} \quad \beta_0 = \pi l/\Lambda.$$

Again, the fundamental quantities we are usually after are the threshold modal gain, $\langle g \rangle_{xy_{th}}$, and wavelength, λ_{th}. Thus, solutions such as those plotted in Fig. 3.18 can be obtained numerically by plugging these definitions into Eq. (6.41).

The coupling constant κ is calculated from Eq. (6.20) for the specific dielectric grating perturbation involved. In practical DFB lasers a number of different configurations are used. Figure 6.7 shows a popular waveguide cross section, along with calculated values for κ over a range of corrugation depths. Since the thickness of some layer is usually varied in typical semiconductor waveguide gratings rather than the index over the layer as assumed in the above calculations, it is generally most convenient to first calculate the effective index as a function of z over a period of the perturbation. This is accomplished by using Eq. (6.9) repeatedly over the perturbation period. Then, this effective index variation is used as if the index perturbation were uniform across the waveguide cross section. Finally, Fourier analysis is used to obtain the space harmonic of interest. Having obtained the magnitude of this space harmonic, $\delta\bar{n}$, we can replace δn_g in Eq. (6.22) by $\delta\bar{n}$ with $\Gamma_{xy_g} = 1$ and $n_g = \bar{n}$ to obtain

$$\kappa_{\pm 1} = \frac{\pi}{\lambda}\delta\bar{n} \cdot e^{\pm j\phi}. \tag{6.43}$$

For a square wave corrugation pattern, the effective index alternates abruptly between $\bar{n}_1$ and $\bar{n}_2$, implying that $\bar{n}(z)$ is also a square wave with peak-to-peak variation $\Delta\bar{n} = \bar{n}_2 - \bar{n}_1$. The fundamental cosine component of this square wave varies from peak-to-peak by $(4/\pi)\Delta\bar{n}$ which we can set equal to $2\delta\bar{n}$, yielding $\delta\bar{n} = (4/\pi)(\Delta\bar{n}/2)$ (see Fig. 6.5). Thus, for a square wave corrugation, $\kappa = 2\Delta\bar{n}/\lambda$, which is the same result we found in Chapter 3 (see discussion following Eq. (3.56)). For other corrugation patterns, $\bar{n}(z)$ and $\delta\bar{n}$ must generally be calculated numerically, making this approach to determining the coupling constant less than ideal.

The coupling constant has also been derived from a different standpoint. Instead of developing a coupling of waves, the mode is defined in terms of rays bouncing down the guide at some angle and with some effective width which takes into account the penetration into the cladding layers (see Chapter 7). The number of bounces per unit length can easily be determined from this. Next the grating perturbation is considered to be a diffraction grating with a plane wave incident at some angle, which at every bounce diffracts some energy backward at the same angle as the incident ray. The amount of energy diffracted per bounce is characterized by the diffraction efficiency of the grating. The advantage to this approach is that a closed-form expression for the coupling constant can be derived which is in complete agreement with the numerical procedure outlined above. The expression for a three-layer guide is given by [1]

$$\kappa = Gk_0 a \frac{n_g^2 - \bar{n}^2}{2\bar{n}d_{eff}}, \qquad \text{where} \qquad d_{eff} = d + \frac{1}{\gamma_{x1}} + \frac{1}{\gamma_{x3}}. \tag{6.44}$$

The decay rates into cladding regions $i = 1$ and 3 are defined as $\gamma_{xi} = k_0\sqrt{\bar{n}^2 - n_{ci}^2}$, and d is the width of the guiding region. The guide and cladding indices are, n_g and n_{c1}, n_{c3}. The total height of the grating is $2a$, and $k_0 = 2\pi/\lambda$. Finally, G is related to the spatial Fourier coefficient of the grating pattern: (1) $G = 1$ for a sinusoidal variation, (2) $G = 4/\pi$ for a square wave pattern, (3) $G = 8/\pi^2$ for a triangular sawtooth pattern, and (4) $G = (4/\pi) \sin x/x$ for a graded square wave, where $x = 0$ for no grading (perfect square wave) and $x = \pi/2$ for complete grading (perfect sawtooth pattern), and in general, $x = (\pi/2) \times$ (fraction of graded material). For $k_0 a$ approaching unity, G generally becomes more complex than listed here. If a higher-order component of the Fourier spectrum of the grating pattern is used for coupling modes, then G would have to be modified accordingly.

The coupling constant is shown in Fig. 6.7 as a function of the guide width, for a total grating height of 50 nm. We see that there is an optimum guide width which becomes more peaked as the index step between the guide and cladding increases. By converting from a sinusoidal grating to a square wave

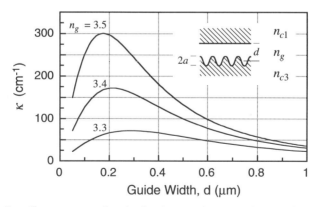

FIGURE 6.7 Coupling constant for the fundamental TE mode as a function of guide width in a three-layer waveguide for three different values of the guiding index. Parameters assumed are: $\lambda = 1.55\,\mu m$, $n_c = 3.17$ (InP at 1.55 μm), $2a = 50$ nm, and $G = 1$ (sinusoidal first-order grating).

grating, the coupling constant can be increased by $4/\pi \approx 1.27$. Also, there is a linear dependence on grating depth, implying that very deep grooves can be used to obtain large κ. In general, the curves in Fig. 6.7 can be scaled for different G and a.

The closed-form expression for the coupling constant Eq. (6.44) can only be used if the effective index of the guide is known, which in general must be found using the techniques discussed in Appendix 3. However, if the guide is symmetric $(n_c = n_{c1} = n_{c3})$, the effective index is given exactly by [2]

$$\bar{n}^2 = n_g^2 b + n_c^2(1 - b), \qquad \text{where} \qquad b \approx 1 - \frac{\ln(1 + V^2/2)}{V^2/2}. \qquad (6.45)$$

In this expression, b is defined as the *field* confinement factor (i.e., the field instead of the field squared is used as the weighting function). As with Γ, we can approximate b using the normalized frequency, $V = k_0 d\sqrt{n_g^2 - n_c^2}$. The approximation in Eq. (6.45) produces less than 1.2% error in the effective index over all ranges of V. In fact, using Eq. (6.45) in Eq. (6.44), we would obtain curves almost indistinguishable from those plotted in Fig. 6.7.

6.3.2.2   With Facet Reflections   As indicated in Fig. 6.8, it is not uncommon to have facet reflections at the end of the grating in practical DFB lasers.

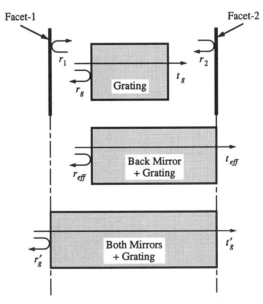

FIGURE 6.8   Progression of calculations to get net transmission and/or reflection from a grating with facet reflections. Any distances to the facet reflections are included in the phases of r_1 and r_2 for convenience. That is, reference planes are those of the grating.

As also indicated, this situation can be modeled using the scattering theory developed in Chapter 3. That is, since we have the reflection and transmission parameters of a symmetrical grating, we can describe the grating by a scattering matrix and include it within a Fabry–Perot resonator formed by the facet reflections.

The grating and back reflector can be combined to get an effective reflector which together with the front reflector form a simple two-mirror resonator for which we can write a characteristic equation. First, we combine the grating and the back reflector assuming any propagation phase is included in r_2. Following Fig. 6.8,

$$r_{eff} = r_g + \frac{t_g^2 r_2}{1 - r_g r_2},$$

(6.46)

and

$$t_{eff} = \frac{t_g t_2}{1 - r_g r_2},$$

(6.47)

where t_2 is the transmission coefficient of facet 2. Equations (6.46) and (6.47), of course, are the same as Eqs. (3.20) and (3.21), although here we have included the phase of the reflections and transmissions relative to the grating reference planes in r_g and r_2, and t_g and t_2, respectively. Note also that with the reference planes of the previous section, the reflection from either side of the grating is r_g, unlike the case in Chapter 3, where we used physical reference planes and the reflection from opposite sides of the grating had opposite signs (see Fig. 6.5).

Now, we can use Eqs. (6.46) and (6.47) as the properties of a new effective back mirror, and write the net transmission through the entire structure. That is,

$$t_g' = \frac{t_1 t_{eff}}{1 - r_1 r_{eff}} = \frac{t_1 t_g t_2}{\left[1 - r_1 r_g - \dfrac{r_1 r_2 t_g^2}{1 - r_g r_2} \right](1 - r_g r_2)}.$$

(6.48)

The characteristic equation is again given by the poles, or,

$$\frac{r_1 r_2 t_g^2}{(1 - r_1 r_g)(1 - r_2 r_g)} = 1.$$

(6.49)

As before, this can be solved for the threshold gain and wavelength using the constituent relationships. For the simple case of no grating, we can set $r_g = 0$, and Eq. (6.49) reduces to the Fabry-Perot threshold condition, $r_1 r_2 t_g^2 = 1$. Alternatively, for $r_1 = r_2 = 0$, it reduces to the standard DFB threshold condition, $t_g = \infty$. And for $r_1 = 0$ ($r_2 = 0$), it reduces to $r_2 r_g = 1$ ($r_1 r_g = 1$), which is the threshold condition for a laser with active mirrors. When solving for threshold, keep in mind that the phase delays implicit in r_1 and r_2 include the round-trip distance between facet-1 or facet-2 and the nearest cosine maximum of the grating index variation (since we have chosen $\phi = 0$ and $L_r = m\Lambda$ for our grating reference planes).

6.3.3   Codirectional Coupling: Directional Couplers

In this section, two different modes propagating in the same direction are considered. These may be (a) orthogonal modes of the same waveguide, or (b) modes of two, initially separate, waveguides. A dielectric perturbation can couple these two modes in either case. In Case (a), the index of the single guide is perturbed, usually in a periodic fashion, and in Case (b) each initially separate guide acts as a dielectric perturbation for the other. As indicated in Fig. 6.9, an example of Case (a) is TE to TM coupling in one guide; and an example of Case (b) is a directional coupler for modes of separate, but closely spaced waveguides. In either case, we can express the net field as

$$\mathscr{E}(x, y, z) = E_1(z)\mathbf{U}_1(x, y)e^{-j\beta_1 z} + E_2(z)\mathbf{U}_2(x, y)e^{-j\beta_2 z}, \qquad (6.50)$$

where $\mathbf{U}_1$ and $\mathbf{U}_2$, and β_1 and β_2 represent the unperturbed solutions for the two uncoupled modes, and the E_i's give the amplitude of each component in the net field.

For coupling between the modes, we need to add a dielectric perturbation. For Case (a), we assume $\varepsilon(x, y, z) = \varepsilon_g(x, y) + \Delta\varepsilon(x, y, z)$, where $\varepsilon_g(x, y)$ is the unperturbed waveguide, and $\Delta\varepsilon(x, y, z)$ is the (possibly periodic) dielectric perturbation seen by both $\mathbf{U}_1$ and $\mathbf{U}_2$. Referring to Fig. 6.10 for Case (b), we observe that $\Delta\varepsilon_2$ acts as the dielectric perturbation for $\mathbf{U}_1$, while $\Delta\varepsilon_1$ acts as the perturbation for $\mathbf{U}_2$. Hence, for this case, $\varepsilon(x, y) = \varepsilon_c + \Delta\varepsilon_1(x, y) + \Delta\varepsilon_2(x, y)$, where the latter two terms act as both waveguides and perturbations simultaneously. The equations to follow are derived for Case (b). However, by setting $\Delta\varepsilon_1 = \Delta\varepsilon_2 = \Delta\varepsilon$ in these equations, we obtain the results for Case (a) as well.

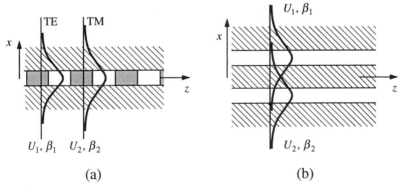

FIGURE 6.9   Codirectional coupling schemes. Uncoupled eigenmodes (a) in the same waveguide and (b) in different waveguides.

Proceeding as before, we add the dielectric perturbation, write the new perturbed mode profiles as $U_i + \Delta U_i$, and plug the assumed net field into the wave equation (6.4). After dropping out second-order terms (assuming $\beta_1 \sim \beta_2$) and the unperturbed solutions, this leads to

$$2j\beta_1 U_1 \frac{dE_1}{dz} e^{-j\beta_1 z} + 2j\beta_2 U_2 \frac{dE_2}{dz} e^{-j\beta_2 z}$$

$$= \Delta\varepsilon_2(x, y) k_0^2 U_1 E_1 e^{-j\beta_1 z} + \Delta\varepsilon_1(x, y) k_0^2 U_2 E_2 e^{-j\beta_2 z}. \qquad (6.51)$$

Additional terms containing ΔU_1 and ΔU_2 analogous to those appearing on the right side of Eq. (6.14) also exist. However, we have left them out of Eq. (6.51) anticipating that we will be able to set them to zero if we dot multiply by U_1^* or U_2^* and integrate over the transverse cross section, as we have done in every other perturbation problem (we go to the trouble of first defining perturbed mode profiles $U_i' = U_i + \Delta U_i$ and then setting the ΔU_i terms to zero to prove that to first order, we only need to know the unperturbed mode profiles, U_i, to estimate the coupling between modes).

Performing the integration of Eq. (6.51) after dot multiplying by U_1^* (the parallel case with U_2^* will be the same except for permuted subscripts), we obtain

$$2j\beta_1 \frac{dE_1}{dz} e^{-j\beta_1 z} \int |U_1|^2 \, dA + 2j\beta_2 \frac{dE_2}{dz} e^{-j\beta_2 z} \int U_1^* \cdot U_2 \, dA$$

$$= k_0^2 E_1 e^{-j\beta_1 z} \int \Delta\varepsilon_2 |U_1|^2 \, dA + k_0^2 E_2 e^{-j\beta_2 z} \int \Delta\varepsilon_1 U_1^* \cdot U_2 \, dA. \, (6.52)$$

To go further, we must consider Cases (a) and (b) separately. However, in both cases we argue that the second term on the left side of the equation and the first term on the right side are negligible or unimportant for coupling. Then, we can proceed to calculate a simple expression for dE_1/dz. Following this, we can write the corresponding expression for dE_2/dz, which follows by multiplying Eq. (6.51) by U_2^* and integrating. The justification for dropping the unwanted terms goes as follows for the two cases.

Case (a): U_1 and U_2 are Eigenmodes of the Same Guide For this case, the second term on the left side of Eq. (6.52) is identically zero, since the unperturbed eigenmodes are orthogonal. The first term on the right side of Eq. (6.52) is not necessarily negligible, but it gives only the single-mode β-perturbation calculated earlier from the dc part of $\Delta\varepsilon$, and no mode-coupling effect. In fact, we already know that $\Delta\varepsilon$ must be periodic along the z-axis to couple the two different β's in this case, so this periodic portion will integrate to zero in this first right-hand term.

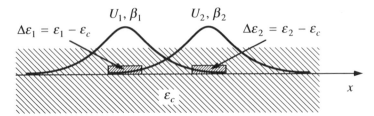

FIGURE 6.10 Cross-sectional schematic of directional coupler showing added index regions for waveguides which also serve to perturb the other waveguide. Eigenmodes and propagation constants refer to unperturbed values.

Case (b): $\mathbf{U}_1$ *and* $\mathbf{U}_2$ *are Eigenmodes of Two Separate Guides*   Figure 6.10 illustrates the overlapping fields in a directional coupler and the fact that the placement of guide #2 near guide #1 is the perturbation to be analyzed for guide #1. On the left side of Eq. (6.52), we argue that the first term is much larger than the second for weakly coupled guides, since there is no place in the cross section of integration where both modes have coincidental large amplitudes. The coefficients of the integrals should be comparable. Put another way, we could say that since we choose to neglect the second term on the left side, the theory is only valid for very small overlap of the two modes quantified by $\int |U_1|^2 \, dA \gg \int \mathbf{U}_1^* \cdot \mathbf{U}_2 \, dA$. (When this condition is not met, a superposition of the exact "supermodes" of the five-layer system is required to determine the net energy transfer from the vicinity of one dielectric stripe to the other. This will be discussed later.)

To neglect the first term on the right side for case (b), we realize that the integration is limited to the cross section of guide #2, where according to the weak overlap assumption, U_1 must be very small, so the first integral on the right is indeed small. The second term survives since $U_1 \gg U_2$ over guide #1 where the integration exists. Also, this term is still much larger than the second term on the left side, since $k_0^2 E_2 \gg \beta_2 \, dE_2/dz$.

With these arguments then, we proceed with the analysis of the directional coupler using only the leftmost and rightmost terms in Eq. (6.52). Thus, for guide #2 perturbing guide #1:

$$2j\beta_1 \frac{dE_1}{dz} e^{-j\beta_1 z} \int |U_1|^2 \, dA = k_0^2 E_2 e^{-j\beta_2 z} \int \Delta\varepsilon_1 \mathbf{U}_1^* \cdot \mathbf{U}_2 \, dA. \qquad (6.53)$$

For guide #1 perturbing guide #2, an equation similar to Eq. (6.52) can be derived by dot multiplying Eq. (6.51) by $\mathbf{U}_2^*$ and integrating. After dropping the small terms,

$$2j\beta_2 \frac{dE_2}{dz} e^{-j\beta_2 z} \int |U_2|^2 \, dA = k_0^2 E_1 e^{-j\beta_1 z} \int \Delta\varepsilon_2 \mathbf{U}_2^* \cdot \mathbf{U}_1 \, dA. \qquad (6.54)$$

Or, after rearranging both,

$$\frac{dE_1}{dz} = -j\frac{k_0^2}{2\beta_1} E_2 e^{-j(\beta_2-\beta_1)z} \frac{\int \Delta\varepsilon_1 \mathbf{U}_1^* \cdot \mathbf{U}_2 \, dA}{\int |U_1|^2 \, dA}, \tag{6.55}$$

and

$$\frac{dE_2}{dz} = -j\frac{k_0^2}{2\beta_2} E_1 e^{-j(\beta_1-\beta_2)z} \frac{\int \Delta\varepsilon_2 \mathbf{U}_2^* \cdot \mathbf{U}_1 \, dA}{\int |U_2|^2 \, dA}. \tag{6.56}$$

These are the basic coupled mode expressions for codirectional coupling. They show that unless $\Delta\varepsilon_i$ contains some spatially periodic factor, β_1 must nearly equal β_2 for coherent coupling (monotonic growth of E_i) over some distance. This is the case for uniform directional couplers. However, if $\beta_1 \neq \beta_2$, then $\Delta\varepsilon_i$ must contribute the difference by including a periodicity. Such is generally the case in our Case (a) where U_1 and U_2 are unperturbed eigenmodes of the same guide, and it can be the situation in our Case (b) if the two coupled waveguides are different.

In the case of $\beta_1 \neq \beta_2$, $\Delta\varepsilon_i$ can be Fourier analyzed, as in Eq. (6.16), and Eqs. (6.55) and (6.56) can be modified to explicity show the z-dependence of $\Delta\varepsilon_i$, analogous to Eqs. (6.17) and (6.18). As in the case of contradirectional coupling, these modified equations give a requirement on the spatial period of the perturbation for coherent addition of the coupling. For the present codirectional coupling case,

$$l\frac{2\pi}{\Lambda} = |\beta_2 - \beta_1| = \frac{2\pi}{\lambda}|\bar{n}_2 - \bar{n}_1|. \tag{6.57}$$

As can be seen, this condition leads to a somewhat coarser period than in the reflective grating case. In fact, the period is generally tens of wavelengths long, since the difference between the mode effective indicies, $\bar{n}_2 - \bar{n}_1$, is usually not so large.

Also, in Case (a) there is no distinction between $\Delta\varepsilon_1$ and $\Delta\varepsilon_2$, since there is only one perturbation of a single waveguide. Thus, the subscripts can be dropped in Eqs. (6.55) and (6.56) in future codirectional coupling results for analyzing this case. In what follows, however, we shall explicitly treat the slighltly more complex Case (b), which generally requires the distinction of $\Delta\varepsilon_1$ from $\Delta\varepsilon_2$. Of course, it also leads to the technologically important four-port directional coupler.

Before moving on to the four-port directional coupler, we choose to rewrite the coupled-mode equations (6.55) and (6.56) in terms of the "normalized amplitudes," a_i's, introduced in Chapter 3. Using Eq. (3.2) then, we make the substitutions,

$$a_1(z)\sqrt{2\eta_1} = E_1(z)e^{-j\beta_1 z},$$

$$a_2(z)\sqrt{2\eta_2} = E_2(z)e^{-j\beta_2 z}. \tag{6.58}$$

In this case, the assumed total electric field may be written as,

$$\mathscr{E} = \sqrt{2\eta_1}\,a_1\mathbf{U}_1 + \sqrt{2\eta_2}\,a_2\mathbf{U}_2. \tag{6.59}$$

From Eqs. (6.59) it can be seen that the power flow in the positive-z direction is just $|a_1|^2 + |a_2|^2$, as should be the case for the normalized amplitudes. Note that the cross terms are negligible for the same reasons as given earlier.

Plugging Eqs. (6.58) into Eqs. (6.56) and (6.57) and neglecting the difference between η_1 and η_2, we find

$$\frac{da_1}{dz} = -j\beta_1 a_1 - j\kappa_{12}a_2,$$

$$\frac{da_2}{dz} = -j\beta_2 a_2 - j\kappa_{21}a_1, \tag{6.60}$$

where

$$\kappa_{12} = \frac{k_0^2}{2\beta_1} \frac{\displaystyle\int_{G1} (\varepsilon_1 - \varepsilon_c)\mathbf{U}_1^* \cdot \mathbf{U}_2 \, dA}{\displaystyle\int |U_1|^2 \, dA},$$

$$\kappa_{21} = \frac{k_0^2}{2\beta_2} \frac{\displaystyle\int_{G2} (\varepsilon_2 - \varepsilon_c)\mathbf{U}_2^* \cdot \mathbf{U}_1 \, dA}{\displaystyle\int |U_2|^2 \, dA}, \tag{6.61}$$

and the integration is shown explicitly as only over the region of perturbation where $\Delta\varepsilon_i = (\varepsilon_i - \varepsilon_c)$ is added to form the additional guide as indicated in Fig. 6.10. That is, $G2$ represents the cross sectional area of added index material to form guide #2, etc. Again, recall that the ε's are always the relative dielectric constants. Also, note that although the coefficient of κ_{ij}, $k_0^2/(2\beta_i) = \omega/(2c\bar{n}_i)$, is linearly proportional to the optical frequency, the overlap integrals can decrease rapidly as the wavelength decreases; thus, κ_{ij} actually tends to decrease slightly with increasing optical frequency in a directional coupler.

In lossless waveguides, $\Delta\varepsilon_1$ and $\Delta\varepsilon_2$ are real and the phase fronts of the modes are flat and perpendicular to the propagation direction, implying that both $\mathbf{U}_1$ and $\mathbf{U}_2$ are real functions. As a result, both κ_{12} and κ_{21} are real. This remains approximately true for waveguides with small gains or losses, as long as $\Delta\varepsilon_1$ and $\Delta\varepsilon_2$ are predominantly real, as is usually the case. In such cases, we can use the real part of the perturbation in each guide to estimate κ_{12} and κ_{21}. In addition, if the guides are symmetric, then $\kappa_{12} = \kappa_{21} = \kappa$ (for asymmetric guides we can still define an average $\kappa = \sqrt{\kappa_{12}\kappa_{21}}$).

Equations (6.60) are already in a convenient form for solution, so we proceed toward a general solution. To this end, we assume trial solutions, $a_1 = a_{10}e^{-j\beta_c z}$, and $a_2 = a_{20}e^{-j\beta_c z}$, where a_{10} and a_{20} are real constants independent of z and β_c is some unknown propagation constant, presumably with multiple solutions, for the coupled waveguides. The general solutions for the $a_i(z)$'s will then be a linear superposition of all of the particular solutions. Plugging the trial solutions into Eqs. (6.60), and solving for β_c, we obtain

$$\beta_c = \frac{\beta_1 + \beta_2}{2} \pm \sqrt{\left(\frac{\beta_1 - \beta_2}{2}\right)^2 + \kappa_{12}\kappa_{21}}, \tag{6.62}$$

or

$$\beta_c = \bar{\beta} \pm s, \tag{6.63}$$

where

$$\bar{\beta} = \frac{\beta_1 + \beta_2}{2} \quad \text{and} \quad s = \sqrt{\left(\frac{\beta_1 - \beta_2}{2}\right)^2 + \kappa_{12}\kappa_{21}}.$$

We can also define a codirectional detuning parameter, $\delta \equiv (\beta_2 - \beta_1)/2$, such that $s^2 = \kappa_{12}\kappa_{21} + \delta^2$, making it the codirectional analog of Eq. (6.31).

With Eq. (6.63), the general solutions are of the form:

$$\begin{aligned} a_1(z) &= e^{-j\bar{\beta}z}[A_1 e^{jsz} + A_2 e^{-jsz}], \\ a_2(z) &= e^{-j\bar{\beta}z}[B_1 e^{jsz} + B_2 e^{-jsz}], \end{aligned} \tag{6.64}$$

and we are now ready for a real problem with real boundary conditions.

6.3.4   The Four-Port Directional Coupler

Figure 6.11 schematically shows a four-port directional coupler formed by two coupled waveguides of finite length, L. In fact, we can derive most of the desired directional coupler expressions without being specific about the length, but we will assume that the outputs are matched so that no reflections exist. This allows us to derive the general scattering parameters for the directional coupler, and from these we can add a wide variety of different boundary conditions that can be handled from the scattering theory.

To get started, we let $a_1(0) = a_1(0)$, and $a_2(0) = 0$, and plug into Eq. (6.64)

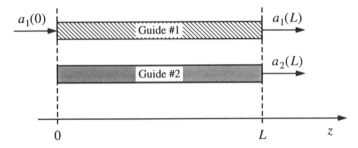

FIGURE 6.11   Four-port directional coupler showing a single input excited.

to find that $B_1 = -B_2$, and $a_1(0) = A_1 + A_2$. Next, we evaluate Eq. (6.60) at $z = 0$, and using Eqs. (6.63) and (6.64), solve for B_2. We find that

$$B_2 = \frac{\kappa_{21}}{2s} a_1(0). \tag{6.65}$$

Again using Eq. (6.64) this gives

$$a_2(z) = -j \frac{\kappa_{21}}{s} a_1(0) e^{-j\bar{\beta}z} \sin sz. \tag{6.66}$$

Next, we solve for A_1 and A_2 using these same equations. This leads to

$$a_1(z) = a_1(0) e^{-j\bar{\beta}z} \left[\cos sz + j \frac{\beta_2 - \beta_1}{2s} \sin sz \right]. \tag{6.67}$$

Therefore, generally, by linear superposition letting the input at port #2, $a_2(0) = a_2(0)$, we have

$$a_1(z) = \left[a_1(0) \left(\cos sz + j \frac{\beta_2 - \beta_1}{2s} \sin sz \right) - j \frac{\kappa_{12}}{s} a_2(0) \sin sz \right] e^{-j\bar{\beta}z}, \tag{6.68}$$

and, likewise by symmetry, we can interchange subscripts to solve for $a_2(z)$,

$$a_2(z) = \left[-j \frac{\kappa_{21}}{s} a_1(0) \sin sz + a_2(0) \left(\cos sz - j \frac{\beta_2 - \beta_1}{2s} \sin sz \right) \right] e^{-j\bar{\beta}z}. \tag{6.69}$$

Figure 6.12 illustrates these normalized amplitudes vs. z for $\beta_1 = \beta_2$ and $\beta_1 \neq \beta_2$ under the initial assumption that $a_2(0) = 0$.

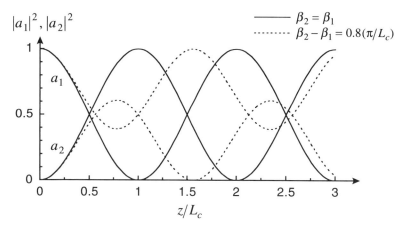

FIGURE 6.12 Energy exchange between two coupled waveguides as a function of propagation distance for matched and mismatched propagation constants.

In the simple case of identical coupled guides, $\beta_1 = \beta_2$ and $s = \sqrt{\kappa_{12}\kappa_{21}} \equiv \kappa$. For a coupler of length L, Eq. (6.66) gives

$$\left| \frac{a_2(L)}{a_1(0)} \right|^2 = \sin^2 \kappa L, \tag{6.70}$$

from which the length, L_c, for full coupling from guide #1 to guide #2 is found to be

$$L_c \equiv \frac{\pi}{2\kappa}. \tag{6.71}$$

This is generally referred to as the *coupling length*. Odd multiples of the length given by Eq. (6.71) also provide full coupling, as indicated by Fig. 6.12.

Viewing the directional coupler as a four-port network, as suggested by Fig. 6.11, it is desirable to develop the four-port scattering matrix, so that the techniques of Chapter 3 can be applied. For the labeling given in Fig. 6.13, Eqs. (6.66) through (6.69) can be used to identify these coefficients by inspection. That is,

$$\mathbf{S} = \begin{bmatrix} 0 & \sqrt{1-c^2} & 0 & -jc \\ \sqrt{1-c^2} & 0 & -jc & 0 \\ 0 & -jc & 0 & \sqrt{1-c^2} \\ -jc & 0 & \sqrt{1-c^2} & 0 \end{bmatrix} e^{-j\bar{\beta}L}, \tag{6.72}$$

where for $\kappa_{12} \approx \kappa_{21} \approx \kappa$,

$$c = \frac{\kappa}{s} \sin sL.$$

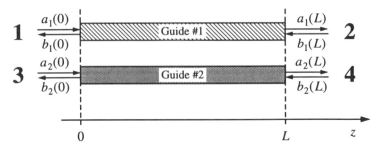

FIGURE 6.13   Schematic of four-port network formed by two coupled waveguides.

Note that we have retained the use of $a_i(z)$ and $b_i(z)$ for the forward and backward waves in guide i, rather than going to the standard "input" and "output" definitions in Fig. 6.13. To conform to the standard scattering theory jaron used in Chapter 3, we must use $a_1 \equiv a_1(0)$, $b_1 \equiv b_1(0)$, $a_2 \equiv b_1(L)$, $b_2 \equiv a_1(L)$, $a_3 \equiv a_2(0)$, $b_3 \equiv b_2(0)$, $a_4 \equiv b_2(L)$, and $b_4 \equiv a_2(L)$.

The general scattering matrix given by Eq. (6.72) can be combined with various boundary conditions to develop scattering matrices of more complex photonic integrated circuits. For example, consider the configuration shown in Fig. 6.14 where guide #2 has reflectors r_3 and r_4 at $z = 0$ and $z = L$, respectively. These provide boundary conditions that create backward traveling waves, $b_1(z)$ and $b_2(z)$. In the present case, it can be seen that $b_2(L) = r_4 a_2(L)$ and $a_2(0) = r_3 b_2(0)$. After a little algebra, the scattering parameters for the resulting two-port network can be obtained. That is,

$$\mathbf{S} =$$

$$\begin{bmatrix} -\dfrac{c^2 r_4 e^{-j\bar{\beta}L}}{1 - r_3 r_4 (1 - c^2) e^{-2j\bar{\beta}L}} & \sqrt{1 - c^2} - \dfrac{c^2 r_3 r_4 \sqrt{1 - c^2}\, e^{-2j\bar{\beta}L}}{1 - r_3 r_4 (1 - c^2) e^{-2j\bar{\beta}L}} \\[4ex] \sqrt{1 - c^2} - \dfrac{c^2 r_3 r_4 \sqrt{1 - c^2}\, e^{-2j\bar{\beta}L}}{1 - r_3 r_4 (1 - c^2) e^{-2j\bar{\beta}L}} & -\dfrac{c^2 r_3 e^{-j\bar{\beta}L}}{1 - r_3 r_4 (1 - c^2) e^{-2j\bar{\beta}L}} \end{bmatrix} e^{-j\bar{\beta}L}.$$

$$(6.73)$$

6.3.5   Codirectional Coupler Filters and Electro-optic Switches

6.3.5.1 $\beta_1 \approx \beta_2$ If two identical waveguides are brought into close proximity to form a directional coupler, the device tends to function over a broad bandwidth of input wavelengths. This is because the unperturbed eigenmodes will be the same and the dispersion properties of each component guide will be the same. Thus, phase-matched coupling will always occur. This kind of dispersion is illustrated in Fig. 6.15, where we have plotted the two β solutions of Eq. (6.62) as a function of optical frequency. The overlap between

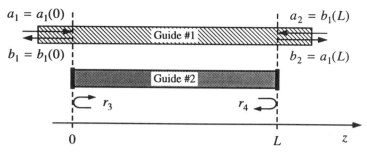

$a_1 = a_1(0)$ $a_2 = b_1(L)$

Guide #1

$b_1 = b_1(0)$ $b_2 = a_1(L)$

Guide #2

r_3 r_4

0   L z

FIGURE 6.14 Two-port formed by placing reflectors at ports 3 and 4 of a directional coupler.

the unperturbed eigenmodes will vary with wavelength, and this results in a change in the coupling constant, κ, according to Eq. (6.61). This gives the gradual spreading of the two propagation constant solutions as κ increases for longer wavelengths (lower optical frequencies).

For a given coupling level, a length can be chosen for 100% coupling from one guide to the other at the operating wavelength. This is given by the coupling length, L_c, in Eq. (6.71) or odd multiples thereof. If the wavelength is now changed, the net transfer would be less for wavelengths either shorter or longer than this according to Eq. (6.70). In fact, to make a wavelength-selective filter, it is common to use a device many coupling lengths long, so that as the

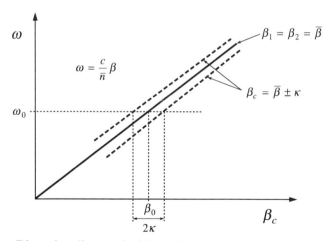

FIGURE 6.15 Dispersion diagram for identical coupled waveguides near the operating wavelength. The operating radial optical frequency, ω_0, results in two propagation constant solutions displaced $\pm\kappa$ from β_0, the value of the uncoupled β for each guide at this point.

wavelength changes and κ changes, the transfer fraction, Eq. (6.71), will vary sinusoidally. Thus, at one wavelength the device can have an odd number of transfer lengths, while at another it can have an even number, yielding either 100% or 0% transfer, respectively. This kind of filter is useful for separating relatively widely spaced wavelengths, such as 1.3 μm from 1.55 μm, however, because κ is a slowly varying function of wavelength, this approach is not effective in providing relatively narrow filter passbands. The quantitative details of the filtering action of such four-port directional couplers are given by Eqs. (6.68) through (6.72). These also allow for slight mismatches in the unperturbed propagation constants as discussed previously.

A voltage-controlled switch or modulator can be constructed of such nominally identical coupled waveguides, if the device is constructed in electro-optic material. The III–V compound semiconductors are electro-optic, and thus, such devices can be compatible with diode lasers. Appendix 15 gives a brief introduction to the electro-optic effect. Without going into a myriad of details, suffice it to say that for certain orientations of electro-optic materials, the application of an electric field with a frequency much lower than that of the optical wave (from dc up to at least 100 GHz) leads to a change in the index of refraction for certain polarizations of the lightwave. Thus, by applying differing electric fields to the waveguides in a directional coupler the propagation constants can be changed slightly, so that $\beta_1 \neq \beta_2$ and the coherence of the coupling can be reduced according to Eqs. (6.66) and (6.67).

As an example of electro-optic modulation, Fig. 6.16 shows a vertical directional-coupler configuration which allows separate control of the applied low-frequency electric fields in each guide, along with a plot of the power transfer fraction $|a_2(L)/a_1(0)|^2$ from Eq. (6.66) for a general directional coupler with $L = L_c$ and $5L_c$. As can be seen, for a directional coupler which

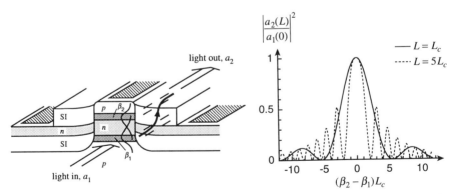

FIGURE 6.16 Schematic of vertical directional-coupler electro-optic switch and example plot of power transfer vs. the normalized deviation of the index of one guide from the other.

is one coupling length long, no transfer occurs between the guides when

$$\frac{\beta_2 - \beta_1}{2} = \sqrt{3}\kappa, \tag{6.74}$$

since s has increased from κ to 2κ at this point. More generally, for a coupler of length $L = mL_c$, the first null in transmission from one guide to the other occurs when the mismatch is

$$(\beta_2 - \beta_1)L_c = \pi\sqrt{[(m + 1)/m]^2 - 1}.$$

Thus, as Fig. 6.16 illustrates, a longer coupler (larger m) requires less of a mismatch to shut off the power transfer.

6.3.5.2 $\beta_1 \neq \beta_2$ If the two waveguide constituents in a directional coupler are not identical, or if two dissimilar modes of the same waveguide are coupled, then a much sharper wavelength filtering action is possible. The conditions for coherent addition of any coupling were discussed in conjunction with Eqs. (6.55) and (6.56). There are several specific possibilities worth mentioning. Two cases have already been outlined in Fig. 6.9. In Case (a) some coarse periodicity in the perturbation was found necessary to couple modes with different propagation constants in the same waveguide. In Case (b), if the β's are different, a modulation of the index will again generally be necessary. In both cases, Eq. (6.57) gives the relationship between the periodicity and the difference in propagation constants. However, in some cases the waveguides can be different, but due to their different dispersive properties, they can have the same propagation constant at some particular wavelength, even with no index modulation. Thus, this becomes a third case, Case (c), where a codirectional waveguide filter is possible. These three cases are summarized by the ω–β diagrams in Fig. 6.17.

To understand the filtering action of these cases more quantitatively, we go back to Eqs. (6.55) and (6.56) and Fourier analyze $\Delta\varepsilon$ as suggested thereafter. This generates a $\pm 2\pi/\Lambda$ term in the argument of the exponential, and replaces $\Delta\varepsilon$ in the integrals by the Fourier coefficients $\Delta\varepsilon_{\pm l}$. Now for the specific case illustrated in the top of Fig. 6.17, $\beta_1 > \beta_2$, and for phase matching we choose $-2\pi/\Lambda$ in Eq. (6.55) and $+2\pi/\Lambda$ in Eq. (6.56). Equivalently, we let

$$\beta_1' = \beta_1 + \frac{2\pi}{\Lambda}, \tag{6.75}$$

and replace β_1 in all of the equations to follow with β_1'. Thus, when we arrive at Eq. (6.66), s and $\bar{\beta}$ are primed, which denotes the use of Eq. (6.75) in place of β_1.

The use of Eq. (6.75) results in a relatively large crossing angle for the two effective propagation constants as shown in Fig. 6.17. Since the proximity of

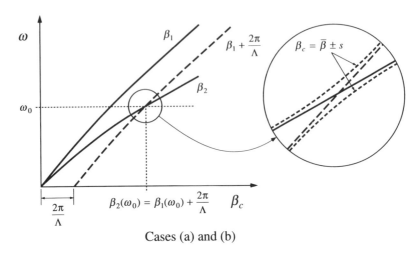

Cases (a) and (b)

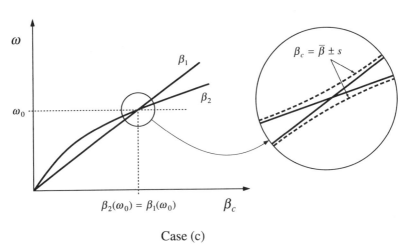

Case (c)

FIGURE 6.17 Dispersion characteristics for two classes of codirectional coupler filters in which two dissimilar modes are coupled. At the top, a grating perturbation is included to provide for phase matching at ω_0. Either two modes of a single guide, Case (a), or coupled waveguides, Case (b), can be used. At the bottom, Case (c), two different waveguide geometries are engineered to have identical β's at ω_0 prior to coupling. The crossing regions are expanded at the right insets to show the pair of new β_c-solutions resulting from the coupling.

the two curves near ω_0 determines the degree of phase coherence for the coupling, the large crossing angle suggests a relatively narrow filter band. In fact, this is the case. Figure 6.18 gives an example of the transfer function, $|a_2(L)/a_1(0)|^2$, vs. the detuning parameter, from Eq. (6.66) using Eq. (6.75).

Switchable or tunable filters result from the combination of the above filters

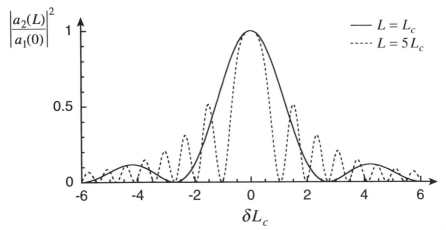

FIGURE 6.18 Power transfer for two different transfer lengths in grating-coupled waveguides vs. the detuning parameter, $\delta \equiv (\beta_2 - \beta_1')/2 = (\omega - \omega_0)/\omega_0 \cdot (\pi/\Lambda)(\Delta \bar{n}_g/\Delta \bar{n})$, where $\Lambda = \lambda_0/\Delta \bar{n}$. The ratio of the *group* effective index difference, $\bar{n}_{g2} - \bar{n}_{g1}$, to the effective index difference, $\bar{n}_2 - \bar{n}_1$, for typical semiconductor coupled waveguides is $\Delta \bar{n}_g/\Delta \bar{n} \approx 3$, which reduces the filter bandwidth accordingly.

with electro-optic material. The application of a field in one guide changes its refractive index which changes the slope of its dispersion curve on the ω–β diagram. Since the slopes of the two dispersion curves are similar and since the offset $2\pi/\Lambda$ does not change, the intersection point, ω_0, changes by a much larger relative amount than the index (much like how the intersection point on a pair of scissors moves more rapidly than the motion of the blades themselves, particularly when the blades are near parallel). That is, the center frequency of the grating-assisted codirectional coupler filter can be tuned by a much larger relative amount than the index. A similar action takes place for Case (c) in Fig. 6.17, since here also, the two uncoupled dispersion curves cross at a small angle. This enhanced tuning rate is in stark contrast to most filters, including the contradirectional grating filter, whose center or Bragg frequency tunes by the same relative amount as the index.

To be more specific, for the grating-assisted codirectional coupler filter shown at the top of Fig. 6.17, where $\beta_2 = \beta_1 + 2\pi/\Lambda$ for phase matching, we find that the center frequency and wavelength vary as

$$\frac{\Delta \omega}{\omega} = -\frac{\Delta \lambda}{\lambda} = \frac{\Delta \bar{n}_2}{\bar{n}_{2g} - \bar{n}_{1g}}, \tag{6.76}$$

in response to an index change in guide #2. The g subscript denotes the group index, which appears due to the frequency dependence of the indices. For Case (c) in Fig. 6.17, we get the same result, but via a different derivation. The tuning enhancement quantified by Eq. (6.76) has been used in tunable lasers, as will be discussed in Chapter 8.

6.4   MODAL EXCITATION

In the process of interconnecting various optical waveguide components together, we must constantly deal with the problem of determining how much power is transmitted and reflected at the junctions. To derive general expressions for this problem we return to the normal mode expansion, Eq. (6.3), which allows us to express an arbitrary excitation field, $\mathscr{E}_e$, in terms of a superposition of the eigenmodes, $\mathscr{E}_m$, of the new waveguide section being excited. Figure 6.19 illustrates the problem schematically. For this exercise it is convenient to work in terms of the normalized amplitudes, since their magnitude squared gives the power flow independent of the impedance of the medium. Thus, we express eigenmode m of the waveguide to be excited as

$$\mathscr{E}_m = E_{0m}\mathbf{U}_m(x, y)e^{-j\beta_m z} = \sqrt{2\eta_m}\, a_m \mathbf{U}_m, \tag{6.77}$$

from which the power flow in the positive z-direction is given by

$$P_{zm} = |a_m|^2 = \int \frac{|\mathscr{E}_m|^2}{2\eta_m}\, dA, \tag{6.78}$$

provided the transverse mode function $\mathbf{U}_m$ is properly normalized according to Eq. (6.2).

According to Fig. 6.19, we assume the given arbitrary field is incident on our waveguide at $z = 0$. Using Eq. (6.3) then,

$$\mathscr{E}_e(0) = \sum_m \mathscr{E}_m(0) = \sum_m \sqrt{2\eta_m}\, a_m(0)\mathbf{U}_m. \tag{6.79}$$

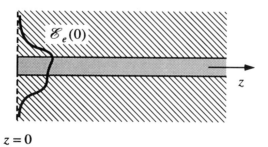

$$z = 0$$

FIGURE 6.19   Illustration of waveguide excitation with some arbitrary field at $z = 0$. This can be expressed in terms of a linear superposition of all the eigenmodes of the waveguide (including radiation modes).

Now, we dot multiply by $\mathbf{U}_l^*$ and integrate over the cross section. That is,

$$\int \mathbf{U}_l^* \cdot \mathscr{E}_e(0) \, dA = \int \sum_m \sqrt{2\eta_m} a_m(0) \mathbf{U}_l^* \cdot \mathbf{U}_m \, dA. \qquad (6.80)$$

Next, we recognize that all terms in the summation except the lth term are zero, and solve for the desired eigenmode amplitude at the entrance to our waveguide, $a_l(0)$.

$$a_l(0) = \frac{1}{\sqrt{2\eta_l}} \int \mathbf{U}_l^* \cdot \mathscr{E}_e(0) \, dA. \qquad (6.81)$$

We can apply Eq. (6.81) for whatever eigenmode number, l, we desire. Fortunately, we only need to know the transverse mode function of the mode(s) in which we are interested.

To illustrate how to use Eq. (6.81) consider the common situation shown in Fig. 6.20, which depicts the joining of two dissimilar waveguides. As shown, power P_{z1} is incident from guide 1, and some fraction P_{z2} is transferred to the fundamental mode of guide 2. Other modes may also be excited, but we are not interested in them initially. (If the guide only supports a single guided mode, then the other modes of the summation are radiation modes, and these would generally be of little interest some distance away.)

The most difficult problem in this and many similar problems is to determine the excitation field on the *right side* of the boundary (region 2). Once it is determined, we can just plug into Eq. (6.81) for the desired a_{20}. Unfortunately, it is a rather complex problem to determine $\mathscr{E}_e(0)$ rigorously. Fortunately, for weak dielectric waveguides, it has been found that the field on the right side of the boundary is similar in shape to that on the left, but it is reduced in amplitude by some transmission coefficient. Also, in this case the transmission coefficient can be well approximated by using the waveguide effective indices in a plane

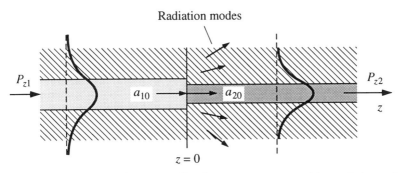

FIGURE 6.20   Illustration of butt-coupling between two waveguides and the excitation of the fundamental eigenmode of the right-hand guide by the left-hand guide.

wave formula. Taking all of this on faith, then,

$$P_{z2}(0^+) \approx P_{z1}(0^-)t^2, \tag{6.82}$$

where

$$t^2 \approx \frac{4\bar{n}_1\bar{n}_2}{(\bar{n}_1 + \bar{n}_2)^2}.$$

Thus,

$$\mathscr{E}_e(0^+) \approx \sqrt{2\eta'_2}\, t a_{10}(0^-)\mathbf{U}_{10}, \tag{6.83}$$

where η'_2 is some effective impedance for the excitation field in region 2. It would be some weighted average of the impedances of the modes to be excited. Applying Eq. (6.81), we find

$$\frac{P_{20}(0^+)}{P_{10}(0^-)} = \left|\frac{a_{20}(0^+)}{a_{10}(0^-)}\right|^2 = \left|t\int \mathbf{U}_{20}^* \cdot \mathbf{U}_{10}\, dA\right|^2, \tag{6.84}$$

where we have neglected the difference between η_{20} and η'_2, which is a good assumption if most of the energy goes into mode 0 or if the impedances of the primary modes excited are not so different. More generally, we could solve for P_{20}/P_{1m} to obtain the excitation of the mth mode.

6.5 CONCLUSIONS

In this chapter we have introduced perturbation and coupled-mode techniques to obtain closed-form analytic solutions to relatively complex problems. These problems are limited only by the requirement that the dielectric perturbation, which creates a change in propagation constant and/or coupling between different modes, is small. Thus, the techniques are very powerful in analyzing many important practical problems. Even in cases where the perturbations are sizable, the techniques are still useful in obtaining approximate results as well as in determining the dependencies on the various parameters of the problem.

For larger perturbations, where the validity of the perturbation and coupled-mode approaches are not good, it is generally necessary to attack the entire problem, usually using some numerical technique. In many cases, however, exact solutions are possible using purely analytical techniques. For example, the case of the quantum-well placed within a separate-confinement waveguide discussed above, is really a five-layer waveguide problem, for which analytic solutions exist. In fact, using the techniques to be introduced in the next chapter, waveguides of many layers can be analyzed.

Another important case that can be solved exactly is the directional coupler, which can be viewed as a different kind of five-layer waveguide. As indicated

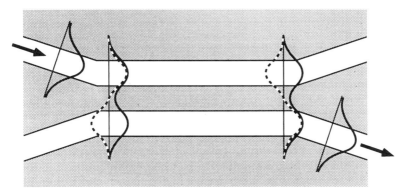

FIGURE 6.21 Directional coupler viewed in terms of the modal interference model. Input excites superposition of first even (solid) and odd (dotted) eigenmodes of the five-layer system. Over the coupling length the even mode travels some number of guided wavelengths, but the high phase-velocity odd mode travels half a wavelength less.

in Fig. 6.21, the first even and odd eigenmodes of the five-layer region can be superimposed to approximate the eigenmode of the individual three-layer guide which excites the coupled region. The phase velocities of the odd and even five-layer eigenmodes are different, so that after some propagation distance, L_c, their superposition can result in most of the energy being aligned with the second high-index region, which in turn, can excite the second exit guide. Thus, most of the problem reduces to calculating the excitation of the different sets of eigenmodes at the boundaries between the uncoupled and coupled regions. Section 6.4 can be used to address this issue.

REFERENCES

[1] H.J. Luo and P.S. Zory, *IEEE J. Quantum Electron.*, **QE-30**, 10 (1994).
[2] K.-L. Chen and S. Wang, *IEEE J. Quantum Electron.*, **QE-19**, 1354 (1983).

READING LIST

K.J. Ebeling, *Integrated Opto-electronics*, Chs. 5 and 6, Springer-Verlag, Berlin (1993).

D.L. Lee, *Electromagnetic Principles of Integrated Optics*, Ch. 8, Wiley, New York (1986).

H.A. Haus, *Waves and Fields in Optoelectronics*, Chs. 7 and 8, Prentice Hall, Englewood Cliffs, NJ (1984).

A. Yariv and P. Yeh, *Optical Waves in Crystals*, Chs. 6 and 11, Wiley, New York (1984).

PROBLEMS

6.1 A symmetric graded-index slab waveguide has a Gaussian transverse mode and an effective index of 3.5. Half of its energy is contained in the central 300 nm of the guide at 1.55 μm.

(a) What is $\Delta\beta$, if a 150 nm wide region from the center to one side of the mode is changed so that the index changes by 0.02?

(b) If the changed region of (a) is only inserted periodically along the waveguide length with a period of $\lambda/2n$ and a 50% duty cycle, what is κ for the resulting grating?

(c) How long must the periodically perturbed waveguide section be for a power reflection coefficient of 0.5?

6.2 Derive an expression for the effective reflection plane separation from a grating's start, L_{eff}, using coupled-mode analysis.

6.3 Using the coupled-mode analysis, show that L_{eff} is also related to the energy decay length L_p in a long grating.

6.4 Using coupled-mode theory for the grating mirror, derive and plot the threshold modal gain vs. the detuning parameter, δ, for a DBR laser which has one cleaved and one grating mirror. The grating mirror also is terminated in a cleave with a relative reflection phase of ϕ. Assume $\lambda = 1.55$ μm, no passive cavity section, an active length of approximately 250 μm, an internal loss of 15 cm^{-1} throughout, a grating $\kappa = 50$ cm^{-1}, and that a mode is aligned with the mirror Bragg condition at a wavelength 1.55 μm. Plot curves for $\phi = 0$, 90°, and 180° with grating κL's of both 0.5 and 1.

6.5 Show how a small propagation loss modifies the grating reflection and transmission coefficients. Express the results as the lossless expressions times multiplicative factors.

6.6 Give an expression for the threshold condition (characteristic equation) for a quarter-wave-shifted DFB laser using coupled-mode results. Neglect any facet reflections.

6.7 From the coupled-mode characteristic equation of a standard DFB laser with no facet reflections, calculate for the first two modes the threshold modal gain $\langle g \rangle_{xy}$ and deviation from the Bragg condition, δL, for

(a) A 1.55 μm InGaAsP/InP DH laser with a $\kappa L = 1$ and a length $L = 400$ μm.

(b) An analogous quarter-wave-shifted DFB, as considered in Problem 6.6, with the same grating and overall length.

Assume the internal modal loss is 20 cm^{-1}. Give the MSR for cases (a) and (b). Assume $\eta_r = 1$, $\beta_{sp} = 10^{-4}$, and $I = 2I_{th}$.

6.8   For a passive grating section formed of 1.3 μm bandgap waveguide material clad by InP with a symmetric triangular grating formed on one side of the 1.3 μm material so that the waveguide's width varies from 200 nm to 210 nm, calculate its coupling constant κ using coupled-mode analysis.

6.9   Use Eq. (6.49) to determine the normalized threshold modal gain, $(\Gamma g_{th} - \alpha_i)L_g$, and wavelength deviation from the Bragg condition, δL_g, for DFB lasers with facet reflections. Plot both values versus the phase of facet reflection #2, ϕ_2, for $\kappa L_g = 1$ and mirror reflection values of

(a) $r_1 = 0$ and $|r_2| = 0.566$,

(b) $r_1 = 0.566$ and $|r_2| = 0.566$, and

(c) $r_1 = 0.566 e^{j\pi/2}$ and $|r_2| = 0.566$.

If necessary, assume $\alpha_i = 15$ cm^{-1}, $L_g = 300$ μm, and $\lambda = 1.55$ μm. Use the analytic equations of this chapter.

6.10   Two identical 3 μm wide channel waveguides with center indexes 0.04 higher than the surrounding cladding material ($n = 3.5$) are formed with their center axes 20 μm apart. The lateral confinement factors are 40%. It is found that the lateral field of the first guide has decayed to 10% of its peak value at the center of the second guide.

(a) What is $\kappa_{21} = \kappa_{12}$? Approximate integrals, do not calculate exact mode shapes.

(b) What is the coupling length for 100% energy transfer?

6.11   What is the 3 dB bandwidth of a directional-coupler filter formed by different width and different index difference guides on InP such that the effective indexes of both are 3.5 at 1.55 μm, but at 1.50 μm the effective indexes of the two guides are 3.51 and 3.52, respectively? At 1.55 μm, $\kappa_{21} \approx \kappa_{12} = 0.01$ μm^{-1}.

6.12   Two three-layer slab waveguides formed of 1.3 μm bandgap InGaAsP/InP are butt-coupled together such that their center axes are aligned. The left guide is 200 nm thick and the right guide is 400 nm thick. Calculate the power loss in coupling across the boundary for the fundamental transverse modes at 1.55 μm. Do the calculation twice. Once for coupling left-to-right and once for coupling right-to-left.

CHAPTER SEVEN

Dielectric Waveguides

7.1 INTRODUCTION

Thus far, we have managed to introduce quite a bit of material which made use of the transverse mode function, U, of some dielectric waveguide without knowing much about its actual form, aside from the brief introduction in Appendix 3. We have chosen this approach in part to emphasize that for many cases one does not need to know the details of all the possible transverse modes that might be supported by some dielectric layer structure. The primary reason has been to maintain a focus on the active device theme of this text and avoid distractions. In fact, we still do not intend to give an extremely detailed treatment of dielectric waveguides because that lies outside the theme of this text. Rather, we wish to introduce several different approaches to solving dielectric waveguide problems to both complement the field theory approach in Appendix 3 as well as provide the student with a broader, and perhaps more intuitive, understanding of the nature of waveguiding in these structures.

We begin by reviewing the reflection of plane waves that are incident at an arbitrary angle from a plane dielectric interface as illustrated in Fig. 7.1. The boundary conditions lead to general expressions for the reflection coefficients of both TE and TM polarizations. If the medium containing the incident plane wave has a higher index of refraction than that beyond the boundary, then we find that total internal reflection is possible. On the incident side of the dielectric boundary, the incident and reflected plane waves create a standing wave with a standing wave-ratio that becomes infinite for incident angles beyond the critical angle. Also, in this case we find that although the reflection coefficient is positive real (has a reflection phase of $0°$) for incident angles smaller than the critical angle, it becomes complex beyond this angle (i.e., has a nonzero reflection phase as well as unity magnitude). Thus, for angles less than the critical angle, a standing wave maximum occurs at the boundary, but for incident angles beyond the critical angle the standing wave maximum moves back from the boundary. In what follows, we quantify the above observations.

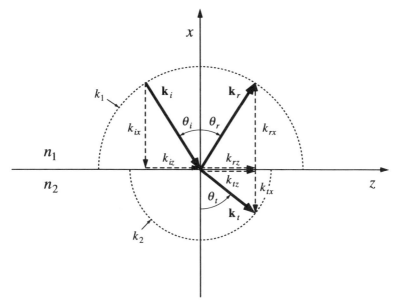

FIGURE 7.1 Illustration of a plane wave incident on a planar boundary. Angle of incidence less than critical angle.

7.2 PLANE WAVES INCIDENT ON A PLANAR DIELECTRIC BOUNDARY

Referring to the nomenclature introduced in Fig. 7.1 and earlier in Chapter 2 and Appendix 3, we express the incident, reflected, and transmitted fields for the TE and TM cases as follows:

TE

$$\mathscr{E}_i(x, z) = \mathscr{E}_i(0, z)\hat{\mathbf{e}}_y e^{jk_{ix}x} e^{-jk_z z}$$

$$\mathscr{E}_r(x, z) = \mathscr{E}_r(0, z)\hat{\mathbf{e}}_y e^{-jk_{ix}x} e^{-jk_z z} \qquad (7.1)$$

$$\mathscr{E}_t(x, z) = \mathscr{E}_t(0, z)\hat{\mathbf{e}}_y e^{jk_{tx}x} e^{-jk_z z}$$

TM

$$\mathscr{H}_i(x, z) = \mathscr{H}_i(0, z)\hat{\mathbf{e}}_y e^{jk_{ix}x} e^{-jk_z z}$$

$$\mathscr{H}_r(x, z) = \mathscr{H}_r(0, z)\hat{\mathbf{e}}_y e^{-jk_{ix}x} e^{-jk_z z} \qquad (7.2)$$

$$\mathscr{H}_t(x, z) = \mathscr{H}_t(0, z)\hat{\mathbf{e}}_y e^{jk_{tx}x} e^{-jk_z z}$$

Because the tangential electric fields, $\mathscr{E}$, and magnetic fields, $\mathscr{H}$, must be equal at $x = 0$, we have already used the fact that $k_{iz} = k_{rz} = k_{tz} = \beta$. Also, from this we have that $\theta_r = \theta_i$, and Snell's law, $k_i \sin \theta_i = k_t \sin \theta_t$. As always the propagation vector components are related by $k_{ix}^2 + k_{iz}^2 = k_i^2 = k_1^2 = k_0^2 n_1^2 = k_0^2 \varepsilon_1$, etc.

In any medium, the magnetic field of a plane wave is related to the electric field by Maxwell's curl equation. For the assumed forms of Eqs. (7.1) and (7.2) we have

$$\mathcal{H} = \frac{1}{\omega\mu} \mathbf{k} \times \mathcal{E}, \tag{7.3}$$

where μ is the magnetic permeability of the medium. Now, using Eqs. (7.1) and (7.2) in Eq. (7.3), and applying the boundary condition that the electric and magnetic fields are continuous at $x = 0$, we can solve for the ratio of the reflected to the incident electric fields [1, 2]. For the TE case with equal permeabilities,

$$\left.\frac{\mathcal{E}_r}{\mathcal{E}_i}\right|_{x=0} = r^{TE} = \frac{k_{ix} - k_{tx}}{k_{ix} + k_{tx}}, \tag{7.4}$$

and for the TM case,

$$\left.\frac{\mathcal{E}_r}{\mathcal{E}_i}\right|_{x=0} = r^{TM} = \frac{k_{ix} - \dfrac{\varepsilon_1}{\varepsilon_2} k_{tx}}{k_{ix} + \dfrac{\varepsilon_1}{\varepsilon_2} k_{tx}}. \tag{7.5}$$

For the case illustrated in Fig. 7.1, the index of refraction in region 1 is larger than that of region 2. Thus, as the incident angle is increased, at some point, k_{iz} equals k_2. This is called the critical angle, θ_c, defined explicitly using Snell's law by $\sin \theta_c = n_2/n_1$. For larger incident angles, k_{tx} must be imaginary to satisfy

$$k_{tx}^2 \equiv k_t^2 - k_{tz}^2 = k_t^2 - k_{iz}^2 = k_2^2 - \beta^2. \tag{7.6}$$

That is, for $k_{iz} > k_2$, $\theta_i > \theta_c$, and

$$k_{tx} = \pm j\sqrt{\beta^2 - k_2^2} = -j\gamma_{tx}, \tag{7.7}$$

where the sign of γ_{tx} is chosen for a decaying solution in region 2. This situation is shown in Fig. 7.2.

Plugging Eq. (7.7) into Eq. (7.4), we see that beyond the critical angle,

$$r^{TE} = \frac{k_{ix} + j\gamma_{tx}}{k_{ix} - j\gamma_{tx}}, \tag{7.8}$$

and $(\theta_i > \theta_c)$

$$r^{TM} = \frac{k_{ix} + j\dfrac{\varepsilon_1}{\varepsilon_2}\gamma_{tx}}{k_{ix} - j\dfrac{\varepsilon_1}{\varepsilon_2}\gamma_{tx}}. \tag{7.9}$$

In both cases, we see that the magnitudes are unity, but the reflected wave

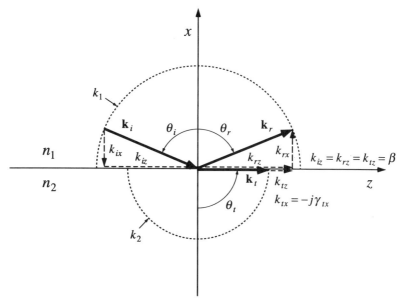

FIGURE 7.2   Plane wave incident at dielectric interface with the angle of incidence larger than the critical angle.

has a phase angle, ϕ. That is, $r = |r|e^{j\phi}$, where from Eq. (7.8) for the TE case,

$$\phi^{TE} = \tan^{-1}\left[\frac{\gamma_{tx}}{k_{ix}}\right] - \tan^{-1}\left[\frac{-\gamma_{tx}}{k_{ix}}\right], \qquad (7.10)$$

or

$$\phi^{TE} = 2\tan^{-1}\left[\frac{\gamma_{tx}}{k_{ix}}\right], \qquad (7.11)$$

where

$$\frac{\gamma_{tx}}{k_{ix}} = \frac{\sqrt{1 - (k_2/\beta)^2}}{\sqrt{(k_1/\beta)^2 - 1}} \qquad \text{and} \qquad \theta_i > \theta_c. \qquad (7.12)$$

For the TM mode using Eq. (7.9) one obtains an equation analogous to Eq. (7.11) in which $\varepsilon_1/\varepsilon_2$ multiplies γ_{tx}.

As indicated in Fig. 7.3, the reflection phase angle is zero for incident angles up to the critical angle, where it begins to increase monotonically toward 180°. This results in standing waves with maxima that move away from the boundary for increasing incident angles, as also indicated. For total reflection there is still energy in region 2, but it decays exponentially away from the boundary and there is no power flow in the x-direction.

The standing waves can be calculated by summing the incident and reflected

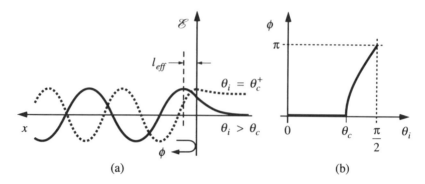

FIGURE 7.3 (a) Illustration of standing waves resulting from plane waves incident at an angle just slightly larger than (dashed) and significantly larger than (solid) the critical angle. (b) Plot of reflection phase angle vs. the angle of incidence.

waves as given by Eq. (7.1). For a TE wave with $\theta_i > \theta_c$, $\mathscr{E}_r(0, z) = \mathscr{E}_i(0, z)e^{j\phi}$. Therefore,

$$\mathscr{E}_i + \mathscr{E}_r = \mathscr{E}_i(0, z)\hat{\mathbf{e}}_y[e^{jk_{ix}x} + e^{j\phi}e^{-jk_{ix}x}]e^{-j\beta z}, \tag{7.13}$$

or

$$\mathscr{E}_i + \mathscr{E}_r = 2\mathscr{E}_i(0, z)\hat{\mathbf{e}}_y e^{j\phi/2}\cos(k_{ix}x - \phi/2)e^{-j\beta z}. \tag{7.14}$$

Figure 7.3 gives two examples for different ϕ, illustrating how the peak of the cosine shifts away from the interface for larger θ_i.

The separation between the maxima and the boundary in Fig. 7.3 is labeled l_{eff}, which is related to the reflection phase by

$$r = |r|e^{j\phi} = e^{-2jk_{ix}(-l_{eff})} = e^{2jk_{ix}l_{eff}},$$

or

$$2l_{eff} = \phi/k_{ix}. \tag{7.15}$$

7.3   DIELECTRIC WAVEGUIDE ANALYSIS TECHNIQUES

7.3.1   Standing Wave Technique

Now, with the above preparation, we can begin to consider the construction of a waveguide that makes use of multiple total internal reflections. In Fig. 7.4 a standing wave resulting from the total internal reflection of a plane wave (such as is shown in Fig. 7.3) is illustrated. The first maxima occurs a distance l_{eff} from the boundary as discussed above. The dashed lines correspond to symmetry planes where for the given angle of incidence another boundary identical to the original one could be inserted without changing the standing wave pattern between the two boundaries.

We can see that with this construction, we have actually formed a waveguide, which traps the plane wave at the original incident angle, forcing it to zigzag

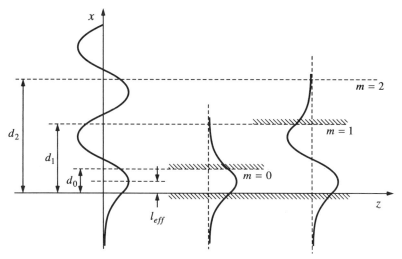

FIGURE 7.4 Construction of waveguides by inserting a second interface (top cross-hatched region) at the symmetry point on a given standing wave (i.e., given ray angle). Given standing wave at left; fundamental ($m = 0$) and first higher-order mode ($m = 1$) illustrated to the right.

indefinitely back and forth so that the net propagation of optical energy is only in the z-direction, as illustrated in Fig. 7.5. The field has the form given by Eq. (7.14) between the boundaries and is evanescent in the two outer region-2s according to $\mathscr{E}_t$ in (7.1) using (7.7) for k_{tx}.

From Fig. 7.4 the constructed waveguide width is seen to be

$$d = 2l_{eff} + m \frac{\lambda_x}{2}, \qquad (7.16)$$

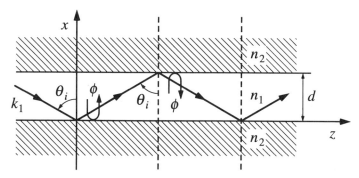

FIGURE 7.5 Illustration of zigzag ray picture of waveguiding. Each ray of the plane wave reflects off alternate boundaries with the same angle of incidence and a reflection phase that provides in-phase addition with other rays propagating in the same direction.

where m is the mode number, equal to zero for the lowest-order symmetric mode, one for the first odd mode, etc., and $\lambda_x = 2\pi/k_{ix}$ is the standing wave wavelength along the x-direction. Now, using Eqs. (7.11), (7.15), and (7.16), we obtain the waveguide dispersion relation for the TE modes,

$$d = \frac{\phi}{k_{ix}} + m\frac{\pi}{k_{ix}} = \frac{2}{k_{ix}}\tan^{-1}\left(\frac{\gamma_{tx}}{k_{ix}}\right) + m\frac{\pi}{k_{ix}}$$

or

$$k_{ix}d = 2\tan^{-1}\left(\frac{\gamma_{tx}}{k_{ix}}\right) + m\pi, \qquad m = 0, 1, 2, \ldots \qquad (7.17)$$

Equation (7.17) is equivalent to Eq. (A3.10), the dispersion relationship derived in Appendix 3 for the symmetric three-layer slab waveguide. Equation (7.12) expands γ_{tx}/k_{ix} in terms of the waveguide propagation constant along z, β. For the TM modes the result is the same, except $\varepsilon_1/\varepsilon_2$ multiplies γ_{tx}.

Using these same techniques the dispersion relationship for a general three-layer asymmetric guide can be derived. As shown in Fig. 7.6 the two cladding layers have different indices of refraction. Thus, the reflection phases at the bottom and top interfaces, ϕ_2 and ϕ_3, and the separations of the standing wave maxima, l_{eff2} and l_{eff3}, respectively, are different also.

Constructing an asymmetric guide with $d = l_{eff2} + l_{eff3}$, and using

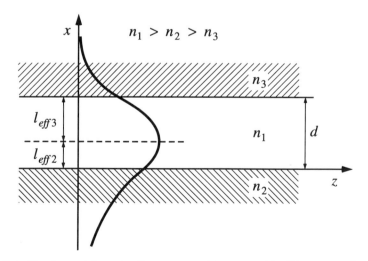

FIGURE 7.6 Fundamental mode of an asymmetric waveguide. The ray angle remains the same throughout, but the phase angle for the top reflection is larger than for the bottom. (Note the change in index definition in this chapter relative to Appendix 3.)

Eqs. (7.15) and (7.11), we find the dispersion relationship for the TE modes to be

$$k_{1x}d = \frac{\phi_2}{2} + \frac{\phi_3}{2} + m\pi = \tan^{-1}\left(\frac{\gamma_{2x}}{k_{1x}}\right) + \tan^{-1}\left(\frac{\gamma_{3x}}{k_{1x}}\right) + m\pi, \qquad (7.18)$$

where as before,

$$k_{1x} = \sqrt{k_1^2 - \beta^2}, \quad \gamma_{2x} = \sqrt{\beta^2 - k_2^2}, \quad \gamma_{3x} = \sqrt{\beta^2 - k_3^2}. \qquad (7.19)$$

The electric fields are given by Eq. (7.14) between the boundaries with $\phi = \phi_2$, and Eq. (7.1) in the evanescent regions using Eqs. (7.7) and (7.19) for k_{tx}.

7.3.2 Transverse Resonance

Another way of analyzing a dielectric waveguide is the transverse resonance technique. This is equivalent to the waveguide construction technique given above and the field-theory technique given in Appendix 3. As indicated in Fig. 7.7, transverse resonance means that the transverse round-trip phase must be a multiple of 2π after a complete cycle of a constituent ray. That is, the fields of an eigenmode must reproduce themselves after the plane wave components have zigzagged up and down one complete transverse cycle as the energy propagates down the guide. This is exactly the same condition as we imposed earlier in determining the modes of a Fabry–Perot resonator, only here we have generalized the situation to the case of nonnormal incidence. In other words, the component of the k-vector normal to the boundaries determines the phase progression along that direction.

For a transverse mode, we must have transverse resonance, or

$$e^{-2jk_{1x}d}e^{j\phi_2}e^{j\phi_3} = e^{-2jm\pi}, \qquad (7.20)$$

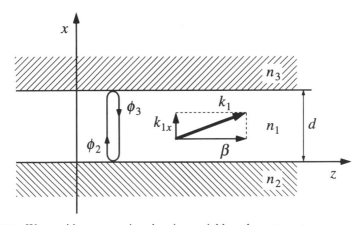

FIGURE 7.7 Waveguide cross section showing variables relevant to a transverse resonance calculation.

where the variables are as defined above in Eqs. (7.19) and (7.11). This implies that

$$2k_{1x}d - \phi_2 - \phi_3 = 2m\pi, \tag{7.21}$$

which is equivalent to Eq. (7.18).

The transverse resonance technique is a relatively simple approach to obtaining a dispersion relationship for a complex dielectric waveguiding structure. That is, regions 2 and 3 can contain a number of dielectric interfaces, and using the techniques of Chapter 3, we can calculate the net reflection coefficient from which we can obtain the reflection phases, ϕ_2 and ϕ_3. Then, Eq. (7.21) can be applied. For a lossless waveguide, the magnitude of these reflection coefficients must be unity. We shall later deal with cases where the reflection is slightly less than unity, but where the optical energy is still relatively well guided aside from a slight propagation loss.

7.3.3 Cutoff and "Leaky" or "Quasi Modes"

The point where the ray angle becomes sufficiently large (measured from the guide axis, z) so that some energy is transmitted at one boundary or the other is generally referred to as *cutoff*. (This is the point where the angle of incidence no longer exceeds the critical angle.) However, the optical energy may still continue to propagate with only modest attenuation since the reflection magnitude at the waveguide walls is still relatively large for angles near cutoff. The light that leaks out of the waveguide also tends to radiate in the forward direction so that it may run parallel to the guide for some distance. These are some of the key differences between dielectric and metal waveguides, in which cutoff tends to result in highly attenuated signals, and/or large reflections.

Put in more mathematical terms, cutoff is where $\phi_j \to 0$, or where $\beta \to k_j$, $j = 2, 3$. In other words, the waveguide propagation constant must always be larger than either of the cladding plane wave propagation constants. This situation is illustrated in Fig. 7.8 in terms of the effective index, $\bar{n} = \beta\lambda/2\pi$.

For the symmetric guide the dispersion relation, Eq. (7.17), becomes

$$k_{1x}d|_{cutoff} \le m\pi, \qquad (n_2 = n_3) \tag{7.22}$$

from which we can see that the fundamental mode, $m = 0$, has no cutoff except at zero frequency, or zero width, d. For the asymmetric guide, the dispersion relation, Eq. (7.18), becomes

$$k_{1x}d|_{cutoff} \le \tan^{-1}\left(\frac{\gamma_{3x}}{k_{1x}}\right) + m\pi. \qquad (n_1 > n_2 > n_3) \tag{7.23}$$

Here we see that the fundamental mode can be cutoff for some finite frequency, or a small enough waveguide width. In Appendix 3 normalized curves for the

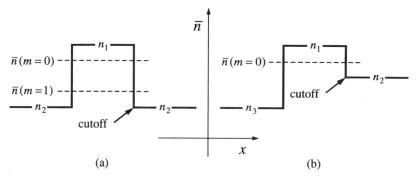

FIGURE 7.8 Illustration of effective index levels for mode numbers, m, in the (a) symmetric and (b) asymmetric cases. For decreasing frequency (increasing wavelength) $\bar{n}$ moves down toward the cutoff levels.

dispersive properties of such waveguides are given. In Fig. 7.9, we plot the effective index vs. normalized frequency for an example asymmetric guide.

Even if the reflection coefficient at one or both waveguide boundaries falls below unity for some ray angle, it is still possible to satisfy the transverse resonance condition. The dashed curves in Fig. 7.9 give the dispersion of these

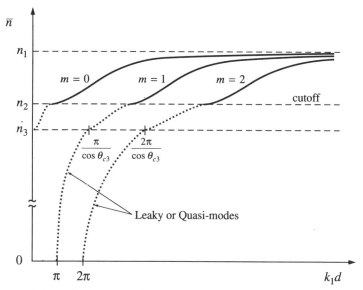

FIGURE 7.9 Schematic example of effective index dispersion curves for three lowest-order modes in an asymmetric slab. Solid curves give effective index for true guided modes. Cutoff is defined where the modes become leaky. Dotted curves indicate locus of points satisfying transverse resonance below cutoff. The angle θ_{c3} is the critical angle at the 1–3 interface. This is not the cutoff angle, which is the critical angle at the 1–2 interface.

"leaky modes" as the ray angle within the waveguide increases toward being normal to the sidewall boundaries, where $\bar{n} \to 0$. Since the reflection phase goes to zero below cutoff, these leaky modes are basically continuations of the guided mode characteristics, and they satisfy Eqs. (7.22) or (7.23) for $k_{1x}d < k_{1x}d|_{cutoff}$, depending upon whether the guide is symmetric or not. For these cases, the optical energy still circulates in the transverse direction and builds up via constructive interference just as in any Fabry–Perot resonator. Thus, optical energy may still be concentrated about the guide axis, and it may be transported along the z-direction with only modest loss. The "modes" of such a structure are usually referred to as *leaky* or *quasi* modes since the propagating energy in the z-direction does change its magnitude along the waveguide. Axial modes of a Fabry–Perot laser cavity are an example of such leaky modes, since there must be mirror transmission for useful output.

7.3.4   Radiation Modes

The leaky modes discussed above are really not true modes but quasi modes as already suggested. Strictly speaking, modes must have uniform profiles and magnitudes along the z-direction. But the guided modes, obtained earlier in this chapter or in Appendix 3, do not form a complete set because a superposition of them can not in general synthesize an arbitrary field. For such a complete set, we need to consider the "radiation modes," which can have other ray angles besides those of the guided and transverse resonant leaky modes.

To construct the radiation modes of some slab waveguide structure as true modes we must insert some additional perfectly reflecting boundaries which can be placed sufficiently far away in the x-direction so that any arbitrary field profile in the vicinity of the central waveguide slab can be completely described. Practically speaking, these inserted boundaries are really more of a mathematically necessary artifact than something that we actually insert in an experiment. Figure 7.10 shows the structure to be analyzed.

As can be seen in Fig. 7.10, perfect reflectors are placed on each side of the waveguide slab a distance $l/2$ away. Now, the transverse resonance technique can be applied to obtain the dispersion relationship for the radiation modes. To make this a bit simpler, we calculate the net reflections looking in the positive and negative x-directions relative to a single reference plane in the center of the waveguide, r_A and r_B, respectively. Then, our transverse resonance condition, Eq. (7.20), reduces to

$$r_A r_B = 1 = e^{-2jm\pi}. \tag{7.24}$$

For the symmetric case in question, $r_A = r_B$, and we refer to the inset in Fig. 7.10 to calculate r_A,

$$r_A = re^{-jk_{1x}d} + \frac{-(1 - r^2)e^{-jk_{2x}l}}{1 - re^{-jk_{2x}l}} e^{-jk_{1x}d}. \tag{7.25}$$

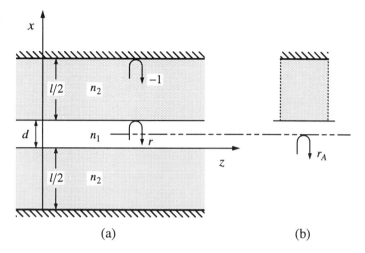

FIGURE 7.10 (a) Waveguide cross section illustrating perfectly reflecting planes inserted at $x = \pm l/2$ to provide for a set of radiation modes. (b) Net reflection from top half, r_A, calculated from the multiple interfaces using k_{ix} as the transverse propagation constant.

Using Eq. (7.4) for r and inserting Eq. (7.25) into the dispersion relation Eq. (7.24), we can derive a dispersion relationship good for both even and odd radiation modes [3],

$$\cot\left(\frac{k_{2x}l}{2}\right) = \frac{k_{1x}}{k_{2x}}\tan\left(\frac{k_{1x}d}{2} - \psi\right), \qquad \psi = \begin{cases} 0 & \text{even} \\ \pi/2 & \text{odd} \end{cases} \tag{7.26}$$

Generally, we choose $l \gg d$, so that the solutions to the dispersion relationship, Eq. (7.26), are closely spaced in frequency (or ray angle). This provides a good basis set that we can use to analyze the propagation of arbitrary optical energy profiles along z. Actually, the set is only complete for describing fields contained completely within l. Thus, l must be at least as large as the extent of the field to be approximated. In practice, setting $l/d > 10$ is usually sufficient to provide the basis functions to synthesize arbitrary field profiles concentrated near the central waveguide.

The field profiles of the radiation modes are standing waves between the outer mirrors with some variation in magnitude and phase over the waveguide slab region. Some of the radiation modes may satisfy transverse resonance for the waveguide interfaces. The fields of this subset will show the expected increase in magnitude over the waveguide due to the coherent addition of multiple reflections there. As set up, we find odd and even modes with respect to the waveguide region. That is, the modes will have either a node or antinode in the center of the waveguide. Figure 7.11 gives examples.

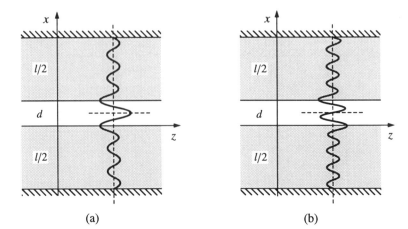

FIGURE 7.11 Schematic examples of even (a) and odd (b) radiation modes constructed by placing perfect reflectors $\pm l/2$ from either side of the waveguide slab. In these examples transverse resonance is approximately satisfied in the waveguide slab as well.

Figure 7.12 summarizes the location of the radiation modes relative to the guided modes on an $\omega-\beta$ plot. Radiation modes that satisfy transverse resonance in the waveguide are shown by the dashed curves. These characteristics are the same as those of the leaky modes discussed earlier. For finite l, the location of the outer mirrors must be chosen properly to satisfy both inner and outer boundary conditions simultaneously.

7.3.5 Multilayer Waveguides

The transverse resonance approach used in the last section for radiation modes can also be applied for the guided modes of multilayer waveguides. That is, Eq. (7.24) is always valid, and it can be applied at the center of a waveguide with any number of dielectric layers. The selection of the center is actually not critical, rather a convenient reference plane that facilitates the subsequent analytical or numerical evaluation of the waveguide dispersion properties is generally used.

Figure 7.13 shows an example of an m-layer waveguide. The evaluation of r_A and r_B can use the techniques developed in Chapter 3 for multilayer reflectors. However, here the axial propagation constant, $k_z \equiv \beta$, is replaced by the transverse propagation constant, k_x, in all of the calculations. For only a few layers, it is possible to derive closed form expressions, as for the radiation modes above. For example, for a five-layer guide, Eq. (3.40) can be applied for both r_A and r_B. If many periods of a pair of layers are used, then the grating formulas can be adapted. That is, Eq. (3.52) can be used for r_A and r_B, provided again that appropriate k_x's replace the β's.

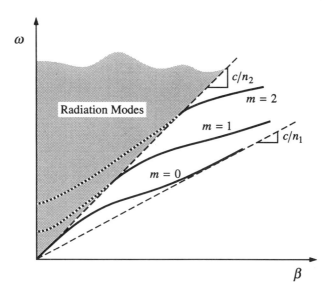

FIGURE 7.12 An ω–β diagram for a symmetric slab waveguide showing relative locations of various modes. Slopes of asymptote lines equal the phase velocities of plane waves in the waveguide slab (c/n_1) and cladding (c/n_2) regions. Local slopes of modal characteristics give the group velocities. Again, dotted curves give dispersion of radiation modes that satisfy transverse resonance in the guide.

7.3.6 WKB Method for Arbitrary Waveguide Profiles

In all of what we have discussed so far in this chapter the index of refraction was constant over some regions, and it jumped abruptly at boundaries there between. Thus, in each region the assumptions involved in the derivation of the transverse wave equation (e.g., uniform dielectric constant) were valid, and exact overall solutions could be found by applying the appropriate boundary conditions.

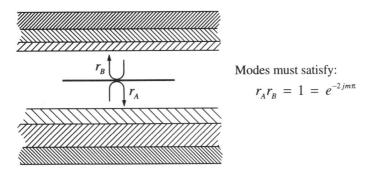

Modes must satisfy:

$$r_A r_B = 1 = e^{-2jm\pi}$$

FIGURE 7.13 Schematic of generalized transverse resonance technique for the determination of modes in multilayer waveguides.

In many practical cases the index varies continuously over some waveguide region rather than discontinuously as above. Thus, strictly speaking the transverse wave equation (6.5) is not valid as discussed in Chapter 6. However, in cases where the dielectric constant varies slowly, so that $\nabla\varepsilon(x, y)$ is small relative to β, we can still apply it with reasonably good results. Otherwise, the wave equation retains terms involving $\nabla\varepsilon(x, y)$ and it becomes very difficult to solve. Figure 7.14 illustrates a waveguide formed by a region in which the index varies continuously.

Wentzel, Kramers, and Brillouin have found that good approximate solutions can be derived in the case of such a slowly varying index profile. Their so-called WKB approximation involves keeping the form of the uniform-medium wave equation, but using a plane wave k-vector that can vary transversely. That is, Eq. (A3.3) becomes

$$\nabla^2 U(x, y) + [k^2(x) - \beta^2]U(x, y) = 0. \tag{7.27}$$

This is the same as Eq. (6.5) where only an x-variation is permitted. Thus, we are still assuming a uniform waveguide along the z-direction.

The WKB approximation also involves neglecting any backscattering due to the slowly varying index. For waveguiding we are only interested in rays traveling roughly perpendicular to the index gradient (or parallel to z). Thus, even for relatively rapid index changes along x, the ray will only experience a

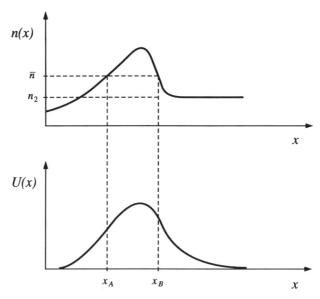

FIGURE 7.14   Transverse index and mode variation for a guide in which the index varies continuously. x_A and x_B are the ray turnaround points where the transverse mode also has points of inflection.

slight gradient. As might be expected in such cases, the ray would bend, and as we shall see, this bending is what provides the equivalent of the zigzagging ray of the analogous three-layer slab guide.

To get started, we express the local plane wave propagation constant, k, in terms of its vector components,

$$k_x^2(x) + k_z^2 = k^2(x), \tag{7.28}$$

where in this case, the index of refraction and the k-vector vary with x. However, for a waveguide mode, $k_z \equiv \beta \equiv k_0 \bar{n}$, independent of x. Thus, we can write

$$k_x(x) = k_0 \sqrt{n^2(x) - \bar{n}^2}. \tag{7.29}$$

The cutoff condition is where $\bar{n} = n_2$, the highest adjacent cladding index.

The waveguide propagation constant can also be expressed in terms of the ray angle, $\theta_z(x)$, from which we can solve for this angle, i.e.,

$$\theta_z(x) = \cos^{-1}\left[\frac{\beta}{k(x)}\right] = \cos^{-1}\left[\frac{\bar{n}}{n(x)}\right]. \tag{7.30}$$

From Eq. (7.30) we see that the ray angle must vary with x, decreasing for decreasing $n(x)$ until $n(x) = \bar{n}$, at which point it is zero. Thus, as indicated in Fig. 7.15, the ray actually turns around at this point, meandering in a sinusoidal-like path as the lightwave propagates along z. At the turnaround points, $x = x_A$, x_B and $k_{1x}(x) = 0$. This is consistent with Eq. (7.29). Thus we have

$$n(x_A) = n(x_B) = \bar{n} \qquad \text{or} \qquad k(x_A) = k(x_B) = \beta. \tag{7.31}$$

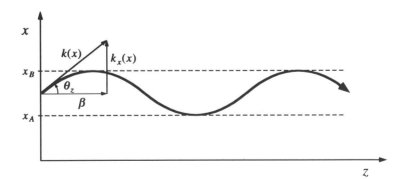

FIGURE 7.15   Sketch of meandering ray path for the case of a smoothly varying index with a maximum between the turnaround points, x_A and x_B. Ray angle with respect to the guide axis, θ_z, shown.

For $x > x_B$, or $x < x_A$, Eq. (7.29) shows that k_x is purely imaginary, indicating that the field decays away to provide the desired trapping of energy along the z-axis. The transverse wave equation (6.5) also shows that the transverse mode has inflection points, i.e., $\partial U^2(x)/\partial x^2 = 0$ at these turnaround points.

Now, we would like to develop the dispersion relationship or characteristic equation. For this, the transverse resonance technique quantified by Eq. (7.21) is used. However, in the present case the round-trip phase across the waveguide is not simply $2k_{1x}d$, but it must be found by integrating $k_x(x)$ from x_A to x_B and back. That is, Eq. (7.21) becomes

$$2 \int_{x_A}^{x_B} k_x(x)\, dx - \phi_2 - \phi_3 = 2m\pi, \tag{7.32}$$

where $k_x(x)$ is given by Eq. (7.29) and the limits of integration can be expressed in terms of the effective index, $\bar{n}$, or propagation constant, β, using the actual index variation in Eq. (7.31).

The remaining problem is to figure out what the ϕ_i's are. For this we use the expressions for the reflection coefficient, either Eqs. (7.4) or (7.5) for the TE or TM modes, respectively. For k_{ix} and k_{tx}, we note that at $x = x_A$ or x_B, the transverse k_x in the present case turns from pure real to pure imaginary according to Eq. (7.29). Thus, the situation is similar to the abrupt index discontinuity case, and the reflection coefficients can be expressed as in Eqs. (7.8) and (7.9) with $k_{ix} = k_x(x_B^-)$ {or $k_x(x_A^+)$} and $\gamma_{tx} = jk_x(x_B^+)$ {or $jk_x(x_A^-)$}. That is, referring to Fig. 7.15, at the top turnaround point assuming a TE mode,

$$r^{TE}(x_B) = \frac{k_x(x_B^-) + j\gamma_x(x_B^+)}{k_x(x_B^-) - j\gamma_x(x_B^+)}. \tag{7.33}$$

Since $k_x(x \to x_B) \to 0$, and $\gamma_x(x \to x_B) \to 0$, we must expand them for $x_B^\pm$, letting the plane wave propagation constant, $k(x_B^-) = \beta + \delta k$, and $k(x_B^+) = \beta - \delta k$. That is, using their definitions given by Eq. (7.19), we find to first order,

$$k_x(x_B^-) = \sqrt{k^2(x_B^-) - \beta^2} = \sqrt{(\beta^2 + 2\beta\delta k) - \beta^2},$$

and

$$\gamma_x(x_B^+) = \sqrt{\beta^2 - k^2(x_B^+)} = \sqrt{\beta^2 - (\beta^2 - 2\beta\delta k)}. \tag{7.34}$$

Thus,

$$r^{TE}(x_B) = \frac{1 + j}{1 - j} = j, \tag{7.35}$$

and likewise for the other turnaround point. For the TM mode, γ_x is multiplied by $\varepsilon_1/\varepsilon_2$, but this approaches unity for a continuous index variation at the turn around point. Therefore, for both TE and TM modes, $\phi_2 = \phi_3 = \pi/2$, and Eq. (7.32) becomes

$$2 \int_{x_A}^{x_B} k_x(x)\, dx = (2m + 1)\pi, \tag{7.36}$$

or, using Eq. (7.29),

$$2k_0 \int_{x_A}^{x_B} \sqrt{n^2(x) - \bar{n}^2}\, dx = (2m + 1)\pi. \tag{7.37}$$

To illustrate the use of the WKB technique, consider the important example of a parabolic variation of the dielectric constant as given by

$$n^2(x) = n_{max}^2 \left[1 - \left(\frac{x}{x_0} \right)^2 \right], \tag{7.38}$$

where we have chosen the origin of the x-axis to be at the maximum of the function. This function, illustrated in Fig. 7.16, is useful only for small index ranges under the WKB approximation, so that the optical energy must be well contained within $|x| \ll x_0$.

Plugging Eq. (7.38) into Eq. (7.37) gives

$$2k_0 \int_{x_A}^{x_B} \sqrt{n_{max}^2 - n_{max}^2 \left(\frac{x}{x_0} \right)^2 - \bar{n}^2}\, dx = (2m + 1)\pi, \tag{7.39}$$

and $x_A = -x_B$ by symmetry. At the integration limits, the argument of the integral is zero, since $n(x_B) = \bar{n}$, or, $k_x = 0$. Therefore, we can solve for x_B from Eq. (7.38) to obtain

$$x_B = \pm x_0 \sqrt{1 - (\bar{n}/n_{max})^2} = \pm x_0 \delta. \tag{7.40}$$

To make the integral more convenient to solve, we make the change of variables, $x = x_0 \delta \sin \varphi$, where φ does not necessarily have any physical meaning. Then,

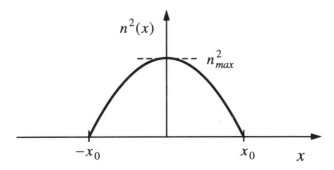

FIGURE 7.16   Parabolic dielectric constant variation given by Eq. (7.38).

performing the integration of Eq. (7.39) using Eq. (7.40) we find the dispersion relationship

$$\bar{n}^2 = n_{max}^2 - \frac{(2m + 1)n_{max}}{k_0 x_0},$$ (7.41)

and plugging back into Eq. (7.40) the maximum ray excursion for mode m is

$$x_B = \pm x_0 \left[\frac{2m + 1}{k_0 n_{max} x_0} \right]^{1/2}.$$ (7.42)

The dielectric constant must continue to vary parabolically according to Eq. (7.38) over a width somewhat larger than the turnaround point separation which is given by twice Eq. (7.42). That is, in most practical examples, it is found that the parabolic medium width must be four or five times x_B for the results to be approximately valid.

Equation (7.41) can be plugged into the transverse wave equation to obtain an expression for $U(x)$. This exercise shows that the eigenmodes of this parabolic medium are Hermite–Gaussian functions of the form,

$$U_m(x) = C_m H_m \left(\frac{\sqrt{2}x}{w_0} \right) e^{-x^2/w_0^2},$$ (7.43)

where $H_m(\xi)$ are the Hermite polynomials and w_0 is the $1/e$ Gaussian spot size. For reference, the first three Hermite polynomials are

$$H_0(\xi) = 1, \qquad H_1(\xi) = 2\xi, \qquad H_2(\xi) = 4\xi^2 - 2.$$ (7.44)

It is also interesting to derive an expression for the ray path illustrated in Fig. 7.15. By definition,

$$\frac{dx(z)}{dz} = \frac{k_x(x)}{\beta} = \frac{k_x(x)}{k_0 \bar{n}}.$$ (7.45)

From Eqs. (7.29), (7.38), and (7.40)

$$k_x(x) = k_0 n_{max} \sqrt{\delta^2 - (x/x_0)^2},$$ (7.46)

and $n_{max}/\bar{n} = (1 - \delta^2)^{-1/2}$. Using Eq. (7.46) in Eq. (7.45),

$$\frac{dx(z)}{dz} = \frac{n_{max}}{\bar{n}} \delta \sqrt{1 - \left(\frac{x}{x_0 \delta} \right)^2},$$ (7.47)

and, again using the convenient change of variables, $x = x_0 \delta \sin \varphi$, we can integrate to obtain

$$x(z) = x_0 \delta \sin \left[\frac{n_{max}}{\bar{n}} \frac{z}{x_0} \right].$$ (7.48)

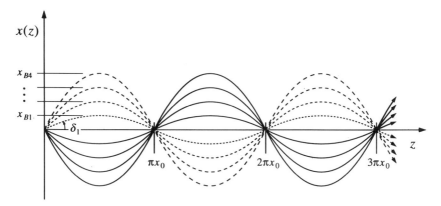

FIGURE 7.17   Illustration of ray paths in a parabolic medium that originate from a point source at the origin. Initial ray angle, δ, results in a sine wave of amplitude, $x_B = x_0\delta$, and period of approximately $2\pi x_0$ (using $n_{max}/\bar{n} \approx 1$).

That is, in a parabolic medium, the rays oscillate around the waveguide axis as sine waves with a period of $2\pi x_0 \bar{n}/n_{max}$. As is the case for the dispersion relationship, Eq. (7.41), this is an exact solution to the approximate wave equation. Since $\bar{n} \approx n_{max}$, note that the ray oscillation period has only a weak dependence upon the initial ray angle, $dx(0)/dz = \delta$. However, the maximum excursion of the ray from the guide axis, $x_B = x_0\delta$, increases in direct proportion. Figure 7.17 illustrates this behavior. The fact that the ray zigzag period along the z-axis does not change much for various angles also suggests that the propagation constant does not change for the various possible modes that satisfy the transverse resonance condition. Equation (7.41) shows this small modal dispersion. Clearly, the parabolically graded waveguide has very different properties from the three-layer slab guide discussed earlier.

Now if we are interested in the eigenmodes of this parabolically graded waveguide, we can substitute the allowed values for $x_0\delta$. That is, using Eqs. (7.40) and (7.42) in Eq. (7.48), we obtain

$$x(z) = x_0 \sqrt{\frac{2m + 1}{k_0 x_0 n_{max}}} \sin\left[\frac{n_{max}}{\bar{n}} \frac{z}{x_0}\right]. \qquad \text{(waveguide eigenmode)} \quad (7.49)$$

Thus, only certain discrete ray angles are possible for the eigenmodes, but they still tend to have nearly the same propagation constant as mentioned above. In many cases, we may be interested in rays that do not satisfy transverse resonance, so Eq. (7.48) should continue to be used in these cases.

In the absence of birefringence, the phase fronts of traveling waves are perpendicular to the ray direction. As in the case of the three-layer slab waveguide, the phase fronts of the constituent rays superimpose to form the net phase fronts or wave crests of the eigenmode. For negligible gain or loss, the result in both cases is an eigenmode with plane phase fronts perpendicular to z. For the parabolic medium, the constituent phase fronts must bend to track the ray path, so they do not appear to be plane waves. For the WKB approximation to be valid, however, these phase fronts must be approximately parallel over distances of a few wavelengths in the x- and z-directions. Put another way, the constituent ray angle can not change appreciably over this distance along the ray.

An example of a practical device that uses a parabolic graded-index waveguide section is the so-called graded-index rod or GRINROD lens. In this case we are not really interested in the eigenmodes of the waveguide, but rather how any given entering rays will propagate and focus after some distance. Although strictly speaking we should always express an entering field as a superposition of the eigenmodes of a waveguide to track its evolution down the guide, it is possible to use a ray-tracing approach if the waveguide is large enough. This makes particular sense, if the propagation distance is small. In the case of the GRINROD then, a pencil-like beam several wavelengths across, but much smaller than the diameter of the GRINROD will tend to propagate like a ray path given by Eq. (7.48). The eigenmode restriction given by Eq. (7.49), which provides for no change in mode cross section as a function of z, is not used because we really would have a large number of interfering eigenmodes superimposed to represent such a pencil-like beam.

GRINRODs are really cylinders with a radial index variation, but we can use our planar analysis to understand how rays will propagate along planes containing the rod axis (so-called meridional rays). For paraxial rays (small angles), Eq. (7.48) shows that a ray entering the GRINROD on axis at $z = 0$ is characterized by a slope $dx(0)/dz \approx \delta \approx \theta_z(0) \equiv \theta_0$. Thus, we write,

$$x(z) = x_0 \theta_0 \sin(z/x_0), \quad \text{(GRINROD)} \quad (7.50)$$

where we already have that the inverse of x_0 is a measure of the curvature of the index-squared profile from Eq. (7.38) and θ_0 is the initial ray angle.

GRINRODS are characterized by a *pitch*, which is really just the ray oscillation period of $2\pi x_0$ for paraxial rays. Thus, a quarter-pitch GRINROD converts any and all rays entering at angles θ_{0i} at its *x–z origin to rays parallel to but spaced a distance* $x_0 \theta_{0i}$ from the z-axis after a quarter oscillation length of $\pi x_0 / 2$. That is, it acts like an ideal lens of focal length $\pi x_0 / 2$. Figure 7.17 illustrates this fact. GRINRODS of any length behave like lenses, but their focusing properties are not as simple as for the quarter- and half-pitch cases.

7.3.7   Review of Effective Index Technique for Channel Waveguides

All of this chapter thus far has dealt only with slab waveguides in which only one-dimensional guiding was considered. For most practical applications two-dimensional guiding, which involves both the transverse (x) and lateral (y) directions, is desired. For most of these structures numerical techniques are

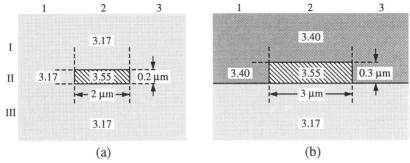

FIGURE 7.18   Example channel waveguide cross sections. (a) Waveguide channel buried in material of equal index on all sides, e.g., InP cladding 1.55 μm InGaAsP guide at 1.55 μm measurement wavelength. (b) Waveguide channel on substrate of one index covered with material of a different index, e.g., 1.3 μm InGaAsP covering 1.55 μm InGaAsP guide on InP substrate at 1.55 μm wavelength.

required to derive accurate results. An exception is a cylindrical geometry for which it is possible to derive results for the circularly symmetric modes in analogy to the slab waveguides discussed above by using cylindrical coordinates in the wave equation. Here again only one variable (radial) is involved.

To analyze two-dimensional or channel waveguide structures involving two independent coordinates the approximate effective index technique is sometimes employed. In Appendix 3, we have already introduced a recipe for obtaining results using this technique. Here we shall briefly review this technique by means of two examples shown in Fig. 7.18.

Example (a) shows a simple buried heterostructure laser in which an InGaAsP active region with a 1.55 μm bandgap is buried in InP by regrowth. For efficiency we shall use the normalized frequency, propagation constant, and asymmetry parameters defined in Eq. (A3.12) of Appendix 3. We explicitly consider the TE mode, but as discussed in the appendix, for small index differences we can neglect the differences between the TE and TM modes in this approximate analysis.

The first step of the effective index technique is to find the transverse effective index of each lateral region as if they were infinitely wide slab guides. In regions 1 and 3 the material is InP for all x. Thus, the effective index in both is clearly $\bar{n}_1 = \bar{n}_3 = 3.17$. In region 2, we use Eq. (A3.12) to obtain

$$V_2 = \frac{2\pi}{\lambda} d [n_{\text{II}}^2 - n_{\text{I}}^2]^{1/2} = \frac{2\pi}{1550 \text{ nm}} 200 \text{ nm} [3.55^2 - 3.17^2]^{1/2} = 1.30$$

and

$$a_2 = 0.$$

From Fig. A3.2, we then read $b_2 = 0.31$. Then, solving for the effective index in the central lateral region, $\bar{n}_2$, we find

$$\bar{n}_2 = [0.31(3.55^2 - 3.17^2) + 3.17^2]^{1/2} = 3.292.$$

Now, we have effective indices for each of the three lateral regions, and we can begin to solve the new lateral effective slab waveguide. Again using Eq. (A3.12),

$$V = \frac{2\pi}{1550 \text{ nm}} \, 2000 \text{ nm} [3.292^2 - 3.17^2]^{1/2} = 6.42,$$

and reading from Fig. A3.2, we have $b = 0.86$. Thus, the net effective index is

$$\bar{n} = [0.86(3.292^2 - 3.17^2) + 3.17^2]^{1/2} = 3.275.$$

As discussed in Appendix 3, we use this net effective index to calculate the lateral k_y in region 2 and the lateral decay constants, γ_y, in regions 1 and 3; however, we use $\bar{n}_2$ to obtain the transverse k_x and decay constants, γ_x, in all regions.

Example (b) in Fig. 7.18 is a little more interesting, since this device is grown with a number of different index regions. Because we have only two different index layers in regions 1 and 3, we have to use a method other than calculating the transverse effective index of a three-layer waveguide there. This would also be true if the guide in these regions were cutoff. Certainly, we can go ahead and calculate the effective index in region 2, then proceed to determine the transverse mode shape, which will be used for all three lateral regions. This latter point is a good clue as to how we should calculate the effective index in the outer regions. That is, we should use the mode shape of the central region to provide the appropriate index weighting for the outer regions. (In fact, this concept even has merit in cases where there is guiding in the outer regions, however, it's generally not used because it's more difficult.)

First, we solve the transverse problem in the central region for this example (b). It should be noted that the cladding layer with the largest index should be chosen as the reference layer—denoted layer III. Otherwise, a will be negative. Evaluating Eqs (A3.12), we find $V_2 = 1.242$, $a_2 = 1.45$, $b_2 = 0.10$, thus $\bar{n}_2 = 3.415$. The small b indicates that we are approaching cutoff in this asymmetric guide.

Now, we can again look at the outer lateral regions. Perhaps the best way to obtain the transverse effective index there is to use our perturbation formula (6.9). We note that regions 1 and 3 can be constructed from region 2 by reducing the index of layer II by 0.15. The perturbation formula naturally uses the eigenmode of region 2, which is exactly what we are looking for. Thus, we proceed with $\Delta \varepsilon = 2n\Delta n = -2(3.55)(0.15) = -1.065$. After some effort in evaluating the perturbation integrals using k_x and γ_x from region 2, we find that $\bar{n}_1 \approx 3.34$. Then, we proceed to solve the lateral problem and obtain

$V = 8.66$, $a = 0$, $b = 0.92$, and $\bar{n} = 3.409$. Thus, we see that the fundamental mode is very near cutoff. Unfortunately, this is also where the effective index technique becomes unreliable, since the evanescent fields are expanding rapidly with slight reductions in the guide's effective index.

7.3.8 Numerical Solutions to the Wave Equation

Many problems cannot be accurately solved by the effective index or WKB techniques. In these cases one generally turns to numerical techniques. Given the availability of computers and the appropriate software in today's world, such numerical techniques are fairly easy to employ. The most straightforward numerical solution to the generalized scalar wave equation (7.27) uses what is called the *finite-difference technique*. As shown in Fig. 7.19, this involves overlaying a grid of some finite period over the lateral waveguide profile to be solved. The lateral normalized field $U(x, y)$ will be found at the nodes of the grid by converting the differential equation (7.27) into a set of finite-difference equations specified at the grid points. The eigenvalue of this set of linear equations will be the effective index of the mode in question. Thus, we wish to convert the partial differential Eq. (7.27) into an eigenvalue problem that can be solved by linear algebra, and linear algebra is easily done on a computer. This is the essence of the finite-difference technique.

The first step in the finite-difference procedure is to appropriately select the computational window. We wish to limit the size of the domain for computational efficiency. On the other hand, the domain should be large enough to contain the field distribution that we want to calculate. The most common approach is to set the field value to zero on the window boundary. Physical

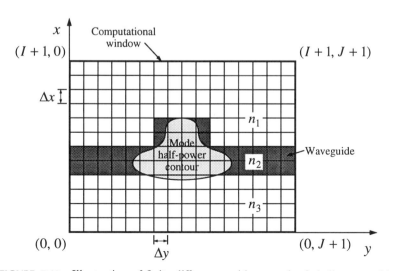

FIGURE 7.19 Illustration of finite-difference grid over a buried-rib waveguide.

considerations should help us in determining the optimum window size. After an initial guess, one can change the window size and check to see if the result is affected. For more accurate calculations on the window boundaries, appropriate boundary conditions to make the window transparent can be used [3].

The second step is to discretize the computational window using a grid as indicated in Fig. 7.19. The grid can be uniform or nonuniform. A nonuniform grid is useful to appropriately sample strongly guided modes which may still have long evanescent tails. Variations in both the transverse and lateral dimensions must be considered. For this introductory treatment, we will assume a uniform grid from this point onward to keep the math as simple as possible. That is,

$$x = i\Delta x, \qquad i = 0, 1, 2, \ldots, I + 1$$
$$y = j\Delta y, \qquad j = 0, 1, 2, \ldots, J + 1 \tag{7.51}$$

Before converting the scalar wave equation (7.27), we expand the operator ∇^2 in rectangular coordinates and factor out the free-space plane wave propagation constant k_0 from the linear term. It then can be written in terms of the lateral index profile $n(x, y)$, and the modal effective index $\bar{n}$, as

$$\frac{\partial^2 U(x, y)}{\partial x^2} + \frac{\partial^2 U(x, y)}{\partial y^2} + k_0^2[n^2(x, y) - \bar{n}^2]U(x, y) = 0. \tag{7.52}$$

Now, we approximate the partial derivatives using the finite differences. This is accomplished by using a second-order Taylor series expansion of the fields at adjacent grid points. That is,

$$U(x + \Delta x, y) = U(x, y) + \Delta x \frac{\partial U(x, y)}{\partial x} + \frac{(\Delta x)^2}{2} \frac{\partial^2 U(x, y)}{\partial x^2} + \cdots$$

and $\hspace{10cm}$ (7.53)

$$U(x - \Delta x, y) = U(x, y) - \Delta x \frac{\partial U(x, y)}{\partial x} + \frac{(\Delta x)^2}{2} \frac{\partial^2 U(x, y)}{\partial x^2} - \cdots$$

Adding the two and solving for the second partial derivative,

$$\frac{\partial^2 U(x, y)}{\partial x^2} \approx \frac{U(x + \Delta x, y) - 2U(x, y) + U(x - \Delta x, y)}{(\Delta x)^2}, \tag{7.54}$$

where the higher-order terms can be neglected if the grid is sufficiently small. Of course, a similar expression can be derived for the second partial with respect to y.

Since we have discretized the space, we only need (or can solve for) the values of the functions at these points. Thus, using Eq. (7.51) we will denote $U(x, y) = U_j^i$, $U(x + \Delta x, y) = U_j^{i+1}$, and $U(x - \Delta x, y) = U_j^{i-1}$. Then,

$$\frac{\partial^2 U(x, y)}{\partial x^2} = \frac{U_j^{i+1} - 2U_j^i + U_j^{i-1}}{(\Delta x)^2},$$

and similarly   (7.55)

$$\frac{\partial^2 U(x, y)}{\partial y^2} = \frac{U_{j+1}^i - 2U_j^i + U_{j-1}^i}{(\Delta y)^2}.$$

The scalar wave equation can now be written in discretized form. That is,

$$\frac{U_j^{i+1} - 2U_j^i + U_j^{i-1}}{(\Delta x)^2} + \frac{U_{j+1}^i - 2U_j^i + U_{j-1}^i}{(\Delta y)^2} + k_0^2(n_j^{i2} - \bar{n}^2)U_j^i = 0. \quad (7.56)$$

For computational ease we divide through by k_0^2 and introduce the dimensionless quantities, $\Delta X^2 = k_0^2 \Delta x^2$ and $\Delta Y^2 = k_0^2 \Delta y^2$. Then, Eq. (7.56) can be written as the matrix eigenvalue equation,

$$\frac{U_j^{i-1}}{\Delta X^2} + \frac{U_{j-1}^i}{\Delta Y^2} - \left(\frac{2}{\Delta X^2} + \frac{2}{\Delta Y^2} - (n_j^i)^2\right)U_j^i + \frac{U_{j+1}^i}{\Delta Y^2} + \frac{U_j^{i+1}}{\Delta X^2} = \bar{n}^2 U_j^i, \quad (7.57)$$

for $i = 0$ to $I + 1$ and $j = 0$ to $J + 1$.

As shown in Appendix 16 this finite-difference matrix equation can be solved numerically for the eigenvalue $\bar{n}$, and the field profile U. This involves some matrix manipulation for compactness and the use of a standard linear matrix-solving algorithm which can be found in most basic numerical analysis software packages. An example problem is also given for clarity. In many cases solutions can be generated very quickly, once the basic finite difference equations are entered.

One can improve the finite difference technique considering the vector nature of the fields [4]. Mode matching techniques that result in equivalent networks have also been developed [5, 6]. Finite-element techniques can also be applied to solve for vectorial fields [7].

7.4 GUIDED-MODE POWER AND EFFECTIVE WIDTH

Many calculations require a normalization of the eigenmode or some integration over a part of it. In this section we calculate the time-averaged Poynting vector power, P, carried by a guided mode along the direction of propagation, which can be useful in such calculations. For the simple three-layer slab guide we find that it can be expressed in terms of the magnitude squared of the electric field in the center of the guide and an effective width, d_{eff}, which accounts for the energy stored in the evanescent fields. Also, for two-dimensional guides in which the lateral width is much wider than the transverse width, we get a similar result.

The time-averaged power propagating in the z-direction is given by

$$P = \frac{1}{2} \operatorname{Re} \iint (\mathscr{E} \times \mathscr{H}^*) \cdot \hat{\mathbf{e}}_z \, dx \, dy. \tag{7.58}$$

The electric field, $\mathscr{E}$, is given by Eq. (6.1) and the magnetic field, $\mathscr{H}$, is obtained from the curl of $\mathscr{E}$. We initially consider the TE modes with the electric fields aligned along the x-direction. Since only $\mathscr{E}_x$ and $\mathscr{H}_y$ survive the vector products, and $\mathscr{H}_y = (j/\omega\mu) \, d\mathscr{E}_x/dz$, we have

$$P = \frac{\beta}{2\omega\mu} \iint |\mathscr{E}_x|^2 \, dx \, dy = \frac{1}{2\eta_g} \iint |\mathscr{E}_x|^2 \, dx \, dy, \tag{7.59}$$

where $\eta_g = \omega\mu/\beta = 377\Omega/\bar{n}$ is the waveguide impedance, and

$$\mathscr{E}_x = E_{0m} U_{xm}(x, y) e^{j(\omega t - \beta z)}. \tag{7.60}$$

The form of the transverse mode shape, $U(x, y)$, for the even (symmetric) modes is given by Eqs. (A3.16), (A3.21), and (A3.22) for a rectangular guide cross section.

For simplicity we wish to look at a guide with a lateral width much wider than the transverse width or $w \gg d$. For the lateral dimension we consider only two extreme cases: (a) a fundamental lateral mode or (b) a uniform field. In both cases since w is large, we neglect any fields for $|y| > w$. Thus, the lateral integration gives in case (a),

$$\int_{-\infty}^{\infty} |U(x, y)|^2 \, dy = |U(x)|^2 \int_{-w/2}^{w/2} \cos^2\left(\frac{\pi}{2w} y\right) dy = |U(x)|^2 \frac{w}{2}, \tag{7.61}$$

(fundamental lateral mode)

and in case (b),

$$\int_{-\infty}^{\infty} |U(x, y)|^2 \, dy = |U(x)|^2 \int_{-w/2}^{w/2} dy = |U(x)|^2 w. \tag{7.62}$$

(uniform lateral field)

Thus, the lateral integration adds only a multiplication by the width in the case of uniform fields or the half-width for the cosinusoidal lateral mode.

Now, we can plug $U(x)$ from Eqs. (A3.16) and (A3.21) into Eq. (7.59) and perform the integration. For the even modes of a symmetric waveguide,

$$P = \frac{w/\rho}{2\eta_g} |E_0 U_0|^2 \left[2 \int_0^{d/2} \cos^2(k_{1x}x) \, dx + 2 \int_{d/2}^{\infty} \cos^2(k_{1x} d/2) e^{-2\gamma_x(x - d/2)} \, dx \right], \tag{7.63}$$

where $\rho = 2$ or 1 for lateral waveguide cases (a) or (b), respectively. Using the waveguide characteristic equation, Eq. (7.17) or (A3.9), to simplify, it can be shown that

$$P = \frac{w/\rho}{2\eta_g}|E_0 U_0|^2 \frac{d_{eff}}{2},\tag{7.64}$$

where

$$d_{eff} = d + \frac{2}{\gamma_x}. \quad \text{(symmetric guide)}\tag{7.65}$$

For the odd modes, it can be shown that the result is the same. Also, if we had considered the more general asymmetric waveguide, we would have obtained

$$d_{eff} = d + \frac{1}{\gamma_{2x}} + \frac{1}{\gamma_{3x}}. \quad \text{(asymmetric guide)}\tag{7.66}$$

In all cases we see that the power propagating down the waveguide is proportional to the magnitude squared of the peak field amplitude in the guide, $|E_0 U_0|^2$, and an effective width, d_{eff}, which equals the slab thickness plus the decay lengths of the evanescent fields in the cladding regions. This is very useful in trying to approximate integrals involving the transverse mode, such as the perturbation integrals of the previous chapter. It is also interesting to note that the integrals in Eq. (7.63) contribute a factor $d_{eff}/2$, just as if the transverse field were a half-period of a cosine that goes to zero at $|x| = d_{eff}/2$. However, such a cosine would have a slightly longer transverse period than the actual cosine of the guide.

If we repeat the above for the TM modes we would find that

$$d_{eff}^{TM} = d + \frac{\vartheta_2}{\gamma_{2x}} + \frac{\vartheta_3}{\gamma_{3x}},\tag{7.67}$$

where

$$\vartheta_i = \left[\frac{k_{1x}^2 + \gamma_{ix}^2}{k_{1x}^2 + (\varepsilon_1/\varepsilon_i)^2\gamma_{ix}^2} \right] \frac{\varepsilon_1}{\varepsilon_i}.$$

In addition to the total power, it is useful to know the relative power contained in the guide and cladding layers, represented by the two integrals in Eq. (7.63). We can write the terms corresponding to each integral conveniently for the fundamental TE mode of a symmetric guide by using the normalized frequency, $V = d\sqrt{k_x^2 + \gamma_x^2}$, introduced in Appendix 3. With Eq. (7.17), we can set $\cos(k_x d/2) = k_x d/V$ and $\sin(k_x d/2) = \gamma_x d/V$ in the evaluated integrals to obtain

$$d_{eff} = \left[d + \frac{2\gamma_x d^2}{V^2} \right] + \left[\frac{2}{\gamma_x} - \frac{2\gamma_x d^2}{V^2} \right].\tag{7.65'}$$

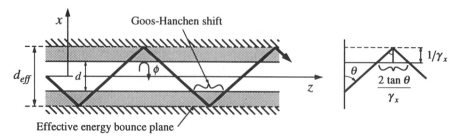

FIGURE 7.20   Zigzag ray picture of energy propagation in a three layer waveguide.

The first set of brackets gives the relative power in the guide and the second set gives the relative power in the cladding layers. Equation (A3.14) gives the resulting fundamental TE mode confinement factor (aside from an $n_1/\bar{n}$ factor).

For weak guiding either through a small index difference or a narrow guide width (characterized by a small V), the tangent function in Eq. (A3.9) reduces to its argument from which we can derive $\gamma_x d \to \sqrt{1 + V^2} - 1 \approx V^2/2$. In this weak-guiding limit, we can use Eq. (7.65') to show that $\Gamma_x \to 2d/d_{eff}$. This result can also be obtained by considering a thin region of width d_A which only samples the peak field of the guide. The power contained in this region is $\propto |E_0 U_0|^2 d_A$. In view of Eq. (7.64), the fraction of power is then $\Gamma_x = d_A/(d_{eff}/2) = 2d_A/d_{eff}$. This result applies equally to a quantum well of width d_A placed at the peak of a mode in an SCH waveguide or to a weakly-guided mode in which the guide itself only samples the peak field, for which we can set $d_A = d$. For the latter case, as the guiding becomes stronger, the prefactor of 2 approaches 1 as both Γ_x and d/d_{eff} converge toward unity. In general, for strong guiding, $d/d_{eff} < \Gamma_x < 1$, and $\Gamma_x \neq d/d_{eff}$ until both are unity. Thus, the d_{eff} defined in Chapter 2 (see Section 2.4) does not refer to this effective width, but the equivalent width of a mode with constant cross-section that is by definition equal to d_A/Γ_x (where d_A is the active region width).

The effective width as defined in Eq. (7.65) is not only useful in relating the peak field to the total power in the mode. It also has a physical interpretation as suggested in Fig. 7.20. Specifically, while the phase fronts of a wave in the zigzag ray picture bounce between effective metal planes spaced apart by π/k_x, the *energy* propagating down the guide effectively bounces off of planes placed at one over the decay length into each cladding region, giving a total width of d_{eff} between bounces. This energy penetration accounts for the evanescent fields in the cladding layers and results in an axial shift and time delay of an incident beam before it is reflected off of the interface. The axial shift is known as the *Goos-Hanchen shift* and represents a very real displacement of energy that is observable in the laboratory.

To rigorously derive the Goos-Hanchen shift, L_{G-H}, we must consider the group characteristics of a spatial *wave packet* reflecting off of the interface beyond the critical angle. Fortunately, we can also guess at the result by noting the familiar relations for phase velocity, ω/β, and energy velocity, $d\omega/d\beta$. For our reflected wave, the phase fronts are advanced along z by ϕ/β, since $\exp[-j\beta z]\exp[j\phi] = \exp[-j\beta(z - \phi/\beta)]$. We therefore predict that the energy is advanced or displaced by $d\phi/d\beta$. A detailed wave packet analysis would in fact confirm this prediction. In other words, the Goos-Hanchen shift is mathematically related to the *dispersion* of the reflection phase angle, ϕ. Using Eq. (7.11) for the reflection phase of a TE mode, we can evaluate the derivative to obtain

$$L_{G-H} = \frac{d\phi}{d\beta} = \frac{2\tan\theta}{\gamma_x}. \qquad \text{(Goos-Hanchen shift)} \qquad (7.68)$$

As shown on the right in Fig. 7.20, this is the same result we would obtain if the energy had to propagate (transversely) an additional $1/\gamma_x$ before reflecting back. With a Goos-Hanchen shift on both sides then, the total effective energy propagation width is just d_{eff} defined in Eq. (7.65) or (7.66), as initially assumed. Also note that the length of the guide required for the energy to make one transverse round-trip is $2d_{eff}\tan\theta$.

7.5   RADIATION LOSSES FOR NOMINALLY GUIDED MODES

There are a few practical situations where lossless guided modes become leaky modes. One occurs when a waveguide is curved to displace the axis of the guide. This is very common in practical devices. Another occurs when some cladding is removed or replaced by higher-index material in a symmetric guide over some device region. This creates an asymmetric guide that may be cutoff, i.e., leaky. A different case is an antiguide which is deliberately constructed to have the largest index in the cladding regions. In these latter cases, we may want to use the energy that leaks out; the axis of the guide remains the z-axis, and the optical leak rate (or propagation loss) can be calculated from the transmission that occurs for a zigzagging ray at one or both waveguide boundaries. The loss rate per unit length is the transmission loss divided by the axial length it takes a ray to cycle back and forth transversely across the guide.

We begin by considering a symmetric antiguide as shown in Fig. 7.21. Even though there is partial transmission at each boundary, the optical energy will be "guided" with relatively low loss provided transverse resonance is satisfied. Since there is π phase shift at each boundary for $n_1 < n_2$, the transverse resonance condition for the antiguide is

$$k_x d = (m + 1)\pi. \qquad \text{(antiguide)} \qquad (7.69)$$

Assuming TE polarization, the transmission at each boundary is given by $(1 - r^2)$, where r is given by Eq. (7.4). A particular ray strikes a boundary

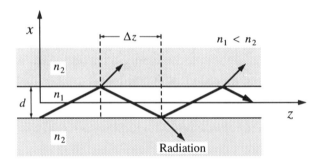

FIGURE 7.21   Illustration of radiation at boundaries in an antiguide for which $n_1 < n_2$.

once in a distance Δz. Setting $k_x \approx k_1 d/\Delta z$ in Eq. (7.69), we obtain

$$\Delta z = \frac{2n_1 d^2}{(m+1)\lambda}, \tag{7.70}$$

for paraxial rays (small angles). Assuming the power decays exponentially according to $P(z) = P(0)e^{-\alpha z}$, we can solve for the attenuation constant as the relative power change per unit length, which can be approximated by

$$\alpha = \frac{1}{\Delta z}\frac{\Delta P}{P} = \frac{(1-r^2)(m+1)\lambda}{2n_1 d^2}. \qquad \text{(antiguide)} \tag{7.71}$$

Equation (7.71) gives the attenuation constant for lightwaves propagating in a symmetric antiguide. Any coherent addition of the radiated fields has been neglected. This is valid in most situations where some absorbing or scattering boundary intercepts the radiated field before it has a chance to add along any appreciable length. The attenuation can be quite high or relatively low depending upon the parameters. For example, consider a typical case where $m = 0$, $\lambda = 1\ \mu m$, $n_1 = 3.5$, $d = 3\ \mu m$ and the index difference gives a ray reflection coefficient, $r^2 = 0.9$. Plugging into Eq. (7.71), we find, $\alpha = 15.9\ \text{cm}^{-1}$, which is not an extremely large loss for semiconductor waveguides. The d used suggests that this would be the lateral waveguide dimension rather than the transverse in the semiconductor case. For small d, e.g., one-tenth of that in the example, the loss would be 100 times larger, and this would be an extremely high loss.

If multiple boundaries are included with reflections that add in phase for some wavelength, the net propagation loss can be significantly reduced for a leaky waveguide. Also, they can operate in a quasi-single mode even for large core widths, since only a particular ray angle experiences low loss. This can lead to desirably low vertical diffraction angles at output facets in semiconductor waveguides. Such "waveguides" have been referred to as antiresonant reflecting

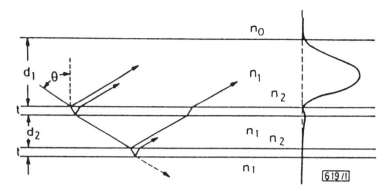

FIGURE 7.22 ARROW slab guide bounded on its upper surface by a low-index medium (n_0) and on its substrate side by one (or more) higher-index (n_2) anti-resonant reflector layers. After [8]. (Reprinted, by permission, from *Electronics Letters*).

optical waveguides or ARROW's [8]. Figure 7.22 shows an example of an ARROW guide.

For low loss we desire a slab mode with a null at the lower interface. This is accomplished with a vertical standing wave that has a node at the interface between the slab and the first high-index reflector layer. If the top surface index difference is large, then the zigzag ray angle in the slab must be, $\theta \approx \cos^{-1}(\lambda/2n_1 d_1)$. For the reflection layer of thickness t and index n_2 to be in antiresonance for high reflection,

$$t \approx \frac{\lambda}{4n_2}(2N+1)\left[1 - \frac{n_1^2}{n_2^2} + \frac{\lambda^2}{4n_2^2 d_1^2}\right]^{-1/2}, \qquad N = 0, 1, 2, \ldots \quad (7.72)$$

The spacing d_2 to the second high-index reflector should also be antiresonant to maximize the net reflection. That is,

$$d_2 = \frac{d_1}{2}(2M+1), \qquad M = 0, 1, 2, \ldots \quad (7.73)$$

Using the transverse resonance recipe of Section 7.3.5 and the technique for calculating radiation loss outlined above, one can rigorously calculate the radiation loss into the substrate. Figure 7.23 gives an example result for the case outlined in Fig. 7.22. The reflector layer thicknesses are varied along with the slab thickness according to the approximate design rules given above.

In addition to having low propagation losses and large fundamental mode size, ARROW modes are also interesting because of the very low overlap of optical energy with the first reflecting layer. Thus, it is possible to have an absorbing reflector layer and still have low waveguide loss. The placement of

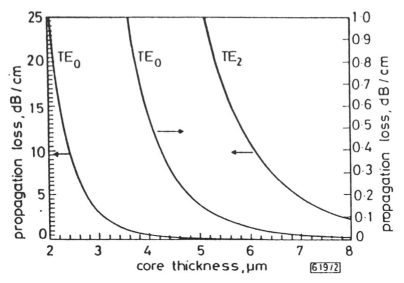

FIGURE 7.23 Computed radiation loss as a function of slab thickness for the two lowest-loss TE modes in a double-reflector-layer ARROW at 1.513 μm [8], with $n_0 = 1.0$, $n_1 = 3.17$, and $n_2 = 3.4$. (Reprinted, by permission, from *Electronics Letters*).

a null at the high-loss layer provides a unique way of placing high-loss layers (e.g., gain layers) immediately adjacent to low-loss guides.

As a third example, consider the radiation loss associated with bending an initially lossless waveguide at a radius of R, as indicated in Fig. 7.24. For this problem we use a first-order perturbation approach in which the unperturbed eigenmode is used in the perturbed problem to calculate the loss. When this mode enters the bend of radius R, it is assumed to stay in the center of the guide with its phase fronts along radii from the center of curvature. From the figure we see that at some radius, x_R, the exponential tail of the unperturbed mode must travel faster than the speed of light in the cladding medium to keep up with the rest of the mode. Clearly, this is not allowed, and we must assume that the light beyond this radius radiates away into the cladding.

To obtain the value of x_R, we note that the local phase velocity is equal to the radial distance times the angular sweep rate, and that it is assumed to equal the unperturbed guide phase velocity, ω/β, at the center of the guide ($r = R$). That is,

$$v_p(r) = r\frac{d\xi}{dt} = r\frac{1}{R}\frac{\omega}{\beta}. \tag{7.74}$$

Now, by the criteria we have adopted, $v_p(R + x_R) = \omega/k_2$. Thus,

$$x_R = R(\beta/k_2 - 1). \tag{7.75}$$

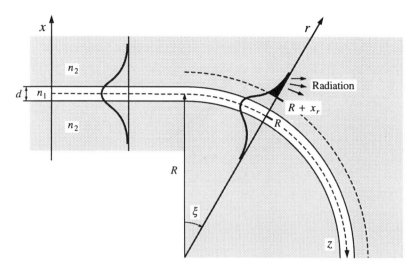

FIGURE 7.24   Illustration of eigenmode entering a waveguide bend of radius R so that the outside tail of the mode must radiate.

As in the antiguide case, the problem is to determine the fraction of power lost, $\Delta P/P$, over some incremental distance, Δz, so that an attenuation constant can be estimated. Again, we shall neglect coherent addition of radiated fields, although this is generally *not* a good assumption. For the relative power lost we use the fraction of power contained in the portion of the mode that radiates (shaded region in Fig. 7.24). This is a very straightforward integration using the unperturbed mode.

Estimating the effective distance it takes this energy to be radiated out of the mode is more complex. For this we use a diffraction distance defined as the distance where the peak power density falls to $1/e$ of its initial value. This is meaningful if the diffracting energy keeps the same functional form as it diffracts, for then, the width of the one-dimensionally diffracting energy in the slab geometry has increased by e-times in this same distance. It turns out that the Hermite–Gaussian functions introduced earlier by Eq. (7.43) as the eigenmodes of a parabolic dielectric constant medium also serve as the eigenmodes of free space in the sense that they maintain the same functional form laterally, even though their characteristic width, w, increases. Thus, they become the natural basis set in which to expand arbitrary excitation field profiles in a uniform medium. This important property comes about because the Fresnel diffraction integral generates the Fourier transform of a function, and the Hermite–Gaussian functions transform back and forth to other Hermite–Gaussian functions, reproducing the original function scaled only in width after some diffraction distance.

Therefore, our procedure for determining the diffraction distance is to expand our radiating excitation field in terms of Hermite–Gaussian basis functions and

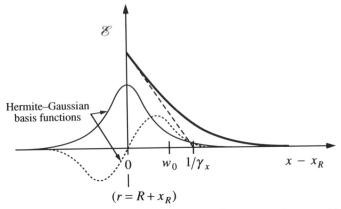

FIGURE 7.25   Radiating part of waveguide eigenmode for $x > x_R$. Superposition of the first two Hermite–Gaussian basis functions can approximate this field. Their characteristic width determines the diffraction distance.

determine the distance, Δz, for their width to increase by e-times. Figure 7.25 illustrates this expansion using only the first two Hermite–Gaussian functions. In the present case the excitation field is a decaying exponential of the form,

$$\mathscr{E}_e(0) = A e^{-\gamma_x(x-x_R)}, \qquad x > x_R, \tag{7.76}$$

and the basis functions are given by Eq. (7.43). By iteration for best fit we find that the Gaussian $1/e$ spot size, w_0, must be approximately

$$w_0 \approx \frac{0.6}{\gamma_x}. \tag{7.77}$$

For a freely diffracting Hermite–Gaussian beam the characteristic width, w, increases from w_0 at $z = 0$ according to [9],

$$w^2 = w_0^2 \left[1 + \left(\frac{\lambda z}{\pi n w_0^2} \right)^2 \right]. \tag{7.78}$$

Interestingly, this is true for all of the higher-order functions as well as the fundamental. Thus, we do not have to keep track of different characteristic widths, and the diffraction problem is completely defined once the excitation problem is solved.

Setting $w/w_0 = e$ in Eq. (7.78) for the desired diffraction length, and using the estimation for the initial width of Eq. (7.77), we obtain

$$\Delta z = \sqrt{e^2 - 1}\, \frac{\pi n w_0^2}{\lambda} \approx \frac{\pi n_2}{\lambda \gamma_x^2}, \tag{7.79}$$

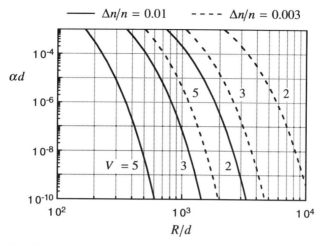

FIGURE 7.26 Bending loss as a function of bend radius for different normalized frequencies and index steps.

where n_2 and γ_x are the refractive index and the unperturbed modal decay constant, respectively, in the cladding region where the radiation is occurring. It is also notable that over this distance, Δz, the wavefronts of the diffracting Hermite–Gaussian functions curve significantly, so that a coherent interaction with the planar phase fronts of the guided mode is inhibited.

Now, we can estimate the attenuation constant, α, from

$$\alpha = \frac{1}{\Delta z}\frac{\Delta P}{P} = \left(\frac{\lambda\gamma_x^2}{\pi n_2}\right)\frac{\dfrac{w}{2\eta_g}\displaystyle\int_{x_R}^{\infty}|\mathscr{E}_y|^2\,dx}{\dfrac{w}{2\eta_g}\displaystyle\int_{-\infty}^{\infty}|\mathscr{E}_y|^2\,dx}. \tag{7.80}$$

Completing the integrations similarly to the work leading up to Eq. (7.64), we obtain

$$\alpha = \frac{\lambda\gamma_x^2}{\pi n_2(\gamma_x d + 2)}\cos^2(k_{1x}d/2)e^{\gamma_x d}e^{-2\gamma_x(\beta/k_2 - 1)R}. \tag{7.81}$$

To uncover important analytical dependences and allow plotting of normalized curves Eq. (7.81) can be approximated by

$$\alpha = C_1 e^{-C_2 R}.$$

From Eq. (7.81), it can be shown that $C_1 \approx \gamma_x^2/k_2$ and $C_2 \approx \gamma_x^3/k_2^2$, assuming weak guiding so that $\beta/k_2 \approx 1 + 0.5(\gamma_x/k_2)^2$. However, more exact calculations have shown that the second quantity is better estimated by $C_2' \approx \frac{2}{3}\gamma_x^3/k_2^2$. Using this and the fact that for weak guiding, $\gamma_x d$ can be expressed solely in terms of the normalized frequency, V, introduced in Appendix 3 (i.e., $\gamma_x d \approx \sqrt{1 + V^2} - 1$), we can generate the normalized plots given in Fig. 7.26.

It is important to note that these calculations do not include the mode mismatch loss between the straight and curved sections.

REFERENCES

[1]   D.L. Lee, *Electromagnetic Principles of Integrated Optics*, Ch. 3, Wiley, New York (1986).

[2]   H.A. Haus, *Waves and Fields in Optoelectronics*, Ch. 2, Prentice Hall, Englewood Cliffs, NJ (1984).

[3]   D.L. Lee, *Electromagnetic Principles of Integrated Optics*, Ch. 4, Wiley, New York (1986).

[4]   M.S. Stern, *IEE Proc., Pt. J.*, **135**, 56 (1988).

[5]   N. Dagli, *IEEE J. Quantum Electron.*, **26**, 90 (1990).

[6]   S.T. Peng and A.A. Oliner, *IEEE Trans. Microwave Theory Tech.*, **MTT-29**, 843 (1981).

[7]   B.M.A. Rahman and J.B. Davies, *IEEE Trans. Microwave Theory Tech.*, **MTT-32**, 20 (1984).

[8]   T.L. Koch, U. Koren, G.D. Boyd, P.J. Corvini, and M.A. Duguay, *Electron. Lett.*, **23**, 244 (1987).

[9]   H.A. Haus, *Waves and Fields in Optoelectronics*, Ch. 5, Prentice Hall, Englewood Cliffs, NJ (1984).

READING LIST

D.L. Lee, *Electromagnetic Principles of Integrated Optics*, Chs. 3–5, Wiley, New York (1986).

H.A. Haus, *Waves and Fields in Optoelectronics*, Ch. 2, Prentice Hall, Englewood Cliffs, NJ (1984).

K.J. Ebeling, *Integrated Opto-electronics*, Chs. 2–4, Springer-Verlag, Berlin (1993).

A. Yariv and P. Yeh, *Optical Waves in Crystals*, Ch. 11, Wiley, New York (1984).

PROBLEMS

7.1   For TM plane waves incident on a planar dielectric boundary, it is possible to get total transmission as well as total reflection. This is called Brewster's angle. Derive an expression for the incident Brewster's angle in terms of the relative dielectric constants of the two media.

7.2   A plane wave with $\lambda = 0.85\,\mu m$ in a medium of index 3.5 is incident on another dielectric medium of index 3.0. The reflection coefficient at the planar interface is determined to be $1/60°$. What is the angle of incidence for TE polarization? Repeat assuming TM polarization.

7.3   A symmetric three-layer slab guide has a core region 0.4 µm thick. Its index is 3.4 and that of the cladding regions is 3.2.

(a) Determine the ray angle for the reflecting plane waves that make up the fundamental TE guided mode.

(b) What guide width will provide the same ray angle for the first higher-order odd TE mode?

(c) If the cladding index were changed to 3.0 on one side, what guide width would be necessary to obtain the same ray angle for the fundamental TE mode? Assume $\lambda = 1.55$ µm.

7.4   Plot the effective index vs. center layer thickness, d, for $0.1 < d < 0.2$ µm with $\lambda = 1.3$ µm for the multilayer slab guide shown in Fig. 7.27. Add dispersion curves for any higher order modes as they become lossless.

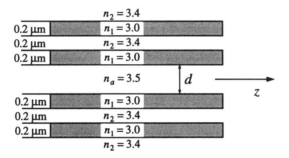

FIGURE 7.27   Schematic of multilayer slab waveguide.

7.5   In exciting a 0.3 µm thick GaAs slab waveguide, it seems that much of the excitation energy goes into radiation modes. In order to expand this excitation in terms of the eigenmodes of the system, perfectly reflecting planes are assumed to exist at a distance ± 30 µm on each side of the slab. The cladding material is $Al_{0.2}Ga_{0.8}As$ and the wavelength is 1.0 µm. Plot the propagation constant vs. mode number for all of the modes.

7.6   Using the WKB method, determine the effective index vs. waveguide base width for the fundamental mode of the triangular index profile shown in Fig. 7.28 with $4.0 < d < 1.0$ µm. Assume $\lambda = 1.0$ µm. Indicate any regions where the WKB approximation may not be accurate. Also, plot the ray path for the mode at $d = 2.0$ µm.

7.7   A quarter-pitch GRINROD is 1 mm in diameter and 2 mm long. The maximum index in the center is 1.6. If an object is placed 1 mm in front of the GRINROD, what is the total distance from the object to the image, and what is the magnification?

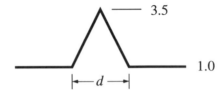

FIGURE 7.28   Triangular index profile waveguide.

7.8 A pencil-like beam of 1.3 μm light, polarized perpendicular to the plane of incidence, strikes an air interface from a medium of index of 3 at an incident angle of θ_i. Plot the shift of the reflected beam center vs. θ_i for $0° < \theta_i < 90°$.

7.9 What is the loss of the fundamental "leaky mode" that satisfies transverse resonance for an antiguide consisting of a slab 1.0 μm thick of index 3.2 sandwiched between media of index 3.4? Assume $\lambda = 1.3$ μm.

7.10 In order to fan out from a directional coupler to adjacent fibers, symmetric S bend waveguide sections are required as shown in Fig. 7.29. Assuming BH waveguides of 1.3 μm Q-material surrounded by InP for the 1.55 μm lightwaves, waveguide thicknesses of 0.4 μm and widths of 2.0 μm and bend radii of R_b to accomplish the total lateral shift of 70 μm, what is the total excess loss due to the four bend sections between input and output ports? $R_b = 100$ μm, 150 μm.

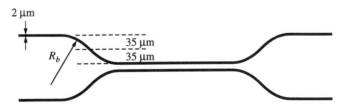

FIGURE 7.29   Directional coupler with input and output S-bends for fan out to fibers.

7.11 Find the effective index, $\bar{n}$, and the field profile, $U(x, y)$, at 1.55 μm of the fundamental mode of the rib waveguide shown in Fig. 7.30 by numerically solving the scalar wave equation using the finite-difference technique. *Hint*: This is a strongly guiding waveguide, so the computational window boundaries on which $U = 0$ can be placed fairly close to the core of the waveguide. In the initial trial, a 4.8 μm wide (*y*-direction) by 2.3 μm high (*x*-direction) window together with a grid size given by $\Delta x = 0.1$ μm and $\Delta y = 0.4$ μm should be sufficient. Place the window to leave 0.25 μm above, 0.65 μm below and 1.4 μm on either side of the high-index rib. For more details see M.J. Robertson et al., Semiconductor waveguides: analysis of optical propagation in single rib structures and directional couplers, *IEE Proceedings, Part J*, **132**(b), (1985).

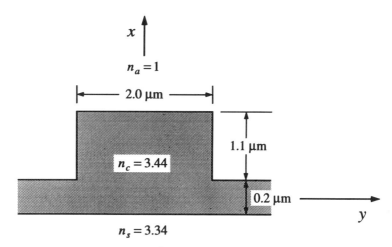

FIGURE 7.30 Rib waveguide cross section.

CHAPTER EIGHT

Photonic Integrated Circuits

8.1 INTRODUCTION

In the last two chapters we introduced techniques to analyze waveguides and complex waveguide junctions. In this chapter we shall apply these techniques to practical examples of photonic integrated circuits (PICs). The purpose is to illustate the importance of these techniques and generate proficiency in their use, rather than to give a complete summary of PIC technology. In keeping with the active device theme of this text we shall emphasize PICs which include diode lasers or amplifiers, or which, in fact, are diode lasers of a relatively complex design. The incorporation of modulators with lasers will also be emphasized.

To qualify as a PIC the device must have at least two sections with an optical waveguide junction there between. In Chapter 3 a few examples have already been introduced. For example, Fig. 3.16 shows a three-section DBR laser with separate gain, modal phase, and Bragg wavelength control electrodes. Each contacts a region with somewhat different optical waveguide properties. In this chapter we shall begin with a brief review of such reflective grating-based waveguide devices. Other than the multiple section laser analysis introduced previously, the main issue in such devices is efficient modal excitation (i.e., low scattering or mode mismatch loss) at the various waveguide junctions. A laser-modulator and a widely tunable laser will be considered explicitly.

The second group of examples deals with the use of waveguide directional couplers for removing energy from a device or for coupling different devices together. A ring laser with a directional coupler output tap and a coherent receiver will give examples of PICs using directional couplers. As will be seen, these are considerably more complex structures than the in-line multiple section devices. The limitations on waveguide bending radius are particularly important in these cases.

Thirdly, widely tunable lasers and filters that use codirectionally coupled filters will be investigated. Here we shall review how grating-assisted couplers are designed given a certain set of constraints. This discussion will also briefly introduce the acousto-optically coupled waveguide filter. The coupling of modes within one waveguide as well as coupling between modes of two nominally separate guides will be included.

Finally, we review important numerical techniques for analyzing wave-guide junctions whose characteristics cannot be accurately understood from analytic techniques. The beam-propagation method (BPM) is the primary technique used. Examples of its utility include the zero-gap directional coupler and Y-branching junctions.

8.2 TUNABLE LASERS AND LASER-MODULATORS WITH IN-LINE GRATING REFLECTORS

As discussed in Chapter 3 grating reflectors can be incorporated within waveguides to provide a frequency-selective reflection. Assuming the waveguide is within the *i*-region of a *pin* heterojunction diode, the index can be changed by either applying a forward current to inject carriers (free-carrier plasma effect) or a reverse bias to increase the electric field (electro-optic effect). Thus, the center (Bragg) frequency of the grating reflector can be tuned in direct proportion to the index change. As shown in Fig. 3.16, tunable DBR lasers have been constructed by incorporating this tunable DBR mirror with gain and phase-shift regions. The addition of the phase-shift region allows the cavity mode to be shifted in wavelength independent of the DBR mirror's center frequency. Also, this region gives the flexibility of shifting the cavity mode together with the mirror center frequency for true continuous tunability of a single mode. Without the phase-shift region a tuning of the grating index moves the center frequency of the mirror across the axial mode spectrum to provide a discrete mode-jump tuning.

8.2.1 Two- and Three-Section DBR Lasers

Figure 8.1, shows a general multisection DBR laser with three separately controlled sections: one each for gain, mode phase shifting, and shifting of the frequency-selective mirror. Currents applied to these sections primarily affect the three aspects of gain, mode tuning, and mode filtering, respectively, as shown in part (c). However, it is generally not possible to have each electrode only control one aspect of the single-frequency laser's operation. For example, a shifting of the grating index will shift the lasing mode slightly as well as the center frequency of the grating.

For a two-section embodiment, the phase-shift section is left out. We can control the gain and filter wavelength directly. The mode locations will move only slightly along with the grating tuning according to the fraction of the mode volume that resides in the grating. As the grating current is

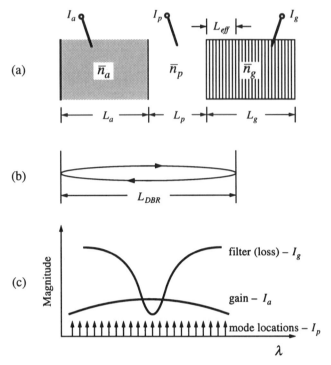

FIGURE 8.1   Three-section DBR. (a) Laser schematic; (b) effective cavity that determines mode spacing and tuning characteristics; (c) sketch of relative spectral characteristics of the cavity filter, gain, and modes along with currents that predominately affect the specified element.

adjusted, the filter will move across the mode comb, alternately selecting axial modes. Its center wavelength, λ_B, tunes according to

$$\frac{\Delta\lambda_B}{\lambda_B} = \frac{\Delta\bar{n}_g}{\bar{n}_g},\tag{8.1}$$

where $\bar{n}_g$ is the effective index in the grating. A given mode will lie at the grating's Bragg frequency only at one point near the middle of its selection during this tuning of the grating filter. The frequency spacing between selected modes at these points is

$$\Delta f = \frac{c}{2[(\bar{n}_a)_g L_a + (\bar{n}_g)_g L_{eff}]},\tag{8.2}$$

where $(\bar{n}_a)_g$ and $(\bar{n}_g)_g$ are the group indexes in the active and mirror sections. Figure 8.2 illustrates an actual experimental example of such a two-section discretely tunable DBR. This device has found considerable use in wave-length-division multiplexing (WDM) systems experiments.

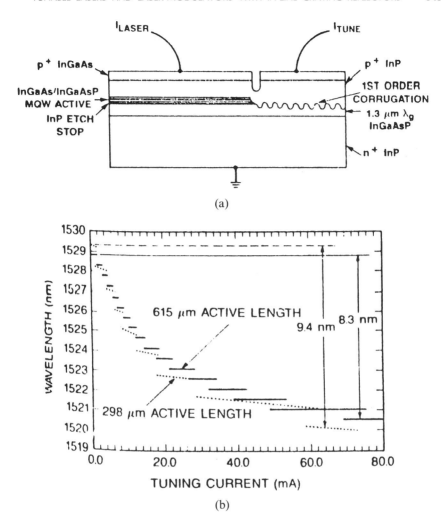

FIGURE 8.2 (a) Schematic of the axial structure of a two-section DBR laser, and (b) output tuning characteristic versus the tuning current to the grating section. After [1]. (Reprinted, by permission, from *Appl. Phys. Lett.*)

Ideal continuous single-mode tuning is possible by including a finite phase-shift section [2]. Here we attempt to move the mode comb along with the grating filter by simultaneously applying a current to the phase-shift section. Because this section only encompasses a portion of the mode volume, we suspect that the index change here will have to be larger than in the grating. Specifically, we desire the wavelength of the mode, λ_m, to be the same as the Bragg wavelength of the DBR mirror, λ_B, for best modal purity (best MSR). For wavelength tuning, the relative change in wavelength of the mode should equal the relative change in the grating's Bragg wavelength, or

$$\frac{\Delta\lambda_m}{\lambda_m} = \frac{\Delta\lambda_B}{\lambda_B} = \frac{\Delta\bar{n}_g}{\bar{n}_g}. \tag{8.3}$$

Using Eq. (3.65) we can then solve for the required change in index of the phase shift region to accomplish this alignment,

$$\frac{\Delta\bar{n}_p}{\bar{n}_p} = \frac{\Delta\lambda_m}{\lambda_m}\left(1 + \frac{\bar{n}_a L_a}{\bar{n}_p L_p}\right) - \frac{\Delta\bar{n}_a}{\bar{n}_a}\frac{\bar{n}_a L_a}{\bar{n}_p L_p}. \tag{8.4}$$

Thus, to obtain continuous single-mode tuning in the three-section DBR, the primary tuning signal is applied to adjust the index of the grating, and then a secondary signal is derived to adjust the phase shift region to satisfy Eq. (8.4). In practice, it is not necessary to measure all of the independent parameters in Eq. (8.4) to derive the desired tracking signal. This is because the output power experiences a local maximum when this ideal alignment is obtained, since the cavity loss is minimized there. Thus, a simple feedback loop can be constructed that senses the output power and adjusts the signal to the phase control electrode to locate a local maximum. Figure 8.3 illustrates one simple design which has been demonstrated. In principle, this feedback loop could be internal to the laser package.

To accomplish the ultimate goal of a tunable laser with only two control inputs—one for power level and another for wavelength—the second independent feedback loop in Fig. 8.3 is added to control the current to the gain

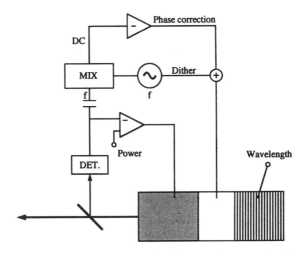

FIGURE 8.3   Circuit that uses first-order feedback to set the power level and a primitive AFC to lock the mode wavelength to the Bragg wavelength of the grating, an alignment where a local maximum in output power exists.

section so that a given reference current from a power-monitoring detector is maintained. This latter feedback loop is already commonly applied to most lasers in lightwave systems environments in order to level the output power. In practice, control curcuits tend to be somewhat more complex than suggested by the figure.

8.2.2   Two-Section Example Problem

We wish to design a tunable two-section 1.55 μm InGaAs/InP DBR (a gain region with a cleave at one end and passive grating at the other) which has output frequencies equally spaced by 50 GHz when the index of the grating is tuned. That is, we seek a design very similar to that illustrated in Fig. 8.2. A 4-quantum-well gain region with characteristics as given by Fig. 4.25 is assumed. This is contained in a quaternary separate-confinement waveguide of bandgap wavelength 1.25 μm, and this waveguide is clad by InP as illustrated in Fig. 8.4. The product of transverse and lateral confinement factors is found to be 6%. The grating region is formed by removing the quantum wells and etching a fundamental-order triangular sawtooth grating with a peak-to-peak depth of 50 nm on the top side of the remaining 0.3 μm quaternary waveguide. It is desired to tune the output over 12 of the axial modes (spaced by 50 GHz). We shall use a BH waveguide width of 3 μm, an internal efficiency of 0.8, an internal loss of 10 cm^{-1} along the entire device length, and a grating length to give a power reflection of 70% in the absence of loss.

For our design specification, we need to determine the grating and gain region lengths; plot the power-out of the cleaved end vs. current into the gain section; and plot the output frequency deviation vs. the current injected into the grating, assuming only radiative recombination in the 1.25 μm bandgap Q-material.

We first determine the necessary grating length. The net power reflection gives us κL_g, but we need to calculate κ from the waveguide geometry. For the passive grating region we first calculate the effective index and width using Appendix 3. We find normalized frequency $V = 1.39$; a normalized propagation parameter, $b = 0.30$, and $\bar{n} = 3.231$, using an interpolated index of

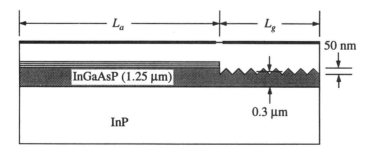

FIGURE 8.4   Two-section DBR laser with separate gain and mirror tuning electrodes.

$n_1 = 3.37$ from Table 1.1 for the waveguide slab. Also, the transverse k-vector and decay constants are $k_x = 3.879 \ \mu m^{-1}$, and $\gamma_x = 2.539 \ \mu m^{-1}$, respectively. Thus, from (7.65), $d_{eff} = 1.088 \ \mu m$.

The grating coupling constant can be found from (6.44), using $G = 8/\pi^2$, $a = 25 \times 10^{-3} \ \mu m$, and $n_1 = 3.37$. We find $\kappa = 0.0107 \ \mu m^{-1}$. From $r_g^2 = 0.70 \approx \tanh^2 \kappa L_g$, we determine a grating length $L_g = 115 \ \mu m$. Also, we can calculate the grating penetration depth, $L_{eff} = 0.5 \ (\tanh \kappa L_g)/\kappa = 39.1 \ \mu m$.

For 50 GHz mode spacing we use Eq. (8.2) to find $L_{DBR} = L_a + L_{eff} = c/(2\langle \bar{n} \rangle_g \Delta f) = 750 \ \mu m$, using an average effective group index, $\langle \bar{n} \rangle_g = 4$. Subtracting the grating penetration, the active length becomes

$$L_a = L_{DBR} - L_{eff} = 750 \ \mu m - 40 \ \mu m = 710 \ \mu m. \tag{8.5}$$

For the output power plot, we need to determine the threshold current and the differential efficiency for the cleaved end. The threshold gain Γg_{th} is

$$\Gamma g_{th} = \alpha_i + \alpha_m = 10 \ cm^{-1} + \frac{1}{0.075 \ cm} \ln \frac{1}{\sqrt{0.32 \times 0.70}} = 20 \ cm^{-1}. \tag{8.6}$$

Therefore, $g_{th} = (750/710)(1/0.06)20 \ cm^{-1} = 352 \ cm^{-1}$. From Fig. 4.25, $J_{th} = 150 \ A/cm^2$ per well $\times 4$ wells $= 600 \ A/cm^2$. Then, $I_{th} = wL_a J_{th}/\eta_i = 16 \ mA$. The differential efficiency for the cleaved end #1 is

$$\eta_{d1} = F_1 \eta_i \frac{\alpha_m}{\Gamma g_{th}} = 0.4 F_1. \tag{8.7}$$

The fraction coupled out end #1,

$$F_1 = \frac{t_1^2}{(1 - r^2) + \dfrac{r_1}{r_g}(1 - r_g^2)} = 0.77. \tag{8.8}$$

Therefore, $\eta_{d1} = 30.8\%$. Thus, the power out the cleaved end is

$$P_{01} = \eta_{d1} \frac{h\nu}{q}(I - I_{th}) = 0.246(I - 16 \ mA) \ mW/mA. \tag{8.9}$$

For the desired 12 output frequencies spaced by 50 GHz, we need a total grating index change of

$$\Delta \bar{n}_g = -\bar{n}_g \left(\frac{\Delta f}{f} \right) = -3.24(12 \times 50 \ GHz)\frac{1.55 \ \mu m}{c} = -0.010. \tag{8.10}$$

Now, $\Delta \bar{n}_g = (\partial \bar{n}/\partial N)N$ where $\partial \bar{n}/\partial N \approx -\Gamma_{xy} 10^{-20} \ cm^3$. Γ_{xy} is the confinement factor for the passive 0.3 μm guide $\approx V^2/(2 + V^2) = 49\%$. For only radiative recombination, N is proportional to the square root of the grating tuning current. Solving for the frequency shift of the grating vs. the tuning current to obtain the desired plot,

$$\Delta f = f \frac{\Gamma_{xy} 10^{-20}}{\bar{n}_g} \left[\frac{\eta_i I_g}{qVB} \right]^{1/2} = 376 \sqrt{I_g} \, \text{GHz/mA}^{1/2}. \qquad (8.11)$$

Figure 8.5 gives the desired plots of output power from the cleaved end vs. gain region current Eq. (8.9) as well as the output frequency vs. mirror tuning current Eq. (8.11) for this example problem. The solid curve represents the center frequency of the grating reflection. The bold horizontal bars represent the current range over which the laser stays in the selected axial mode. In reality the wavelength will change slightly over that range as the mirror reflection phase varies. Also, in practice we find that the assumption of only radiative

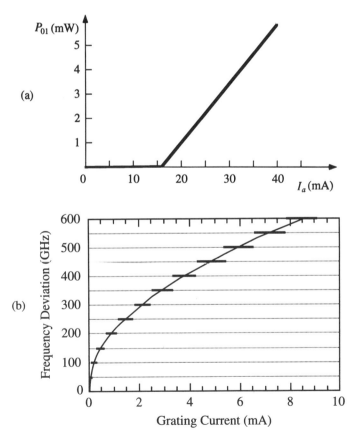

FIGURE 8.5 (a) Optical power from the cleave vs. active-region current, and (b) output frequency deviation vs. tuning current of example two-section DBR.

recombination in this example is inaccurate for currents > 1 mA. At 1 mA the carrier density, given by the square rooted bracket in Eq. (8.11), is 7×10^{17} cm^{-3}. At this point Auger recombination accounts for about 17% of the total recombination (using $C = 3 \times 10^{-29}$ cm^6/s). Thus, we would actually need 1.2 mA at this point. At 8 mA, the Auger current would be about one-third of the total. Thus, we would actually need 12 mA of total current to get the desired carrier density.

8.2.3   Extended Tuning Range Four-Section DBR

To obtain a wider tuning range than the 8–10 nm possible with the three-section DBR, research on other extended tuning range lasers has been carried out. Figure 8.6 gives some examples. One of these—the four-section modulated-grating DBR—builds upon the principles of the basic three-section DBR design.

As shown in Fig. 8.7 this modulated-grating device deviates from the three-section DBR only in that two separately contacted grating reflectors

• Coupled Y-cavity laser

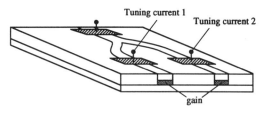

• Grating-assisted co-directional coupler

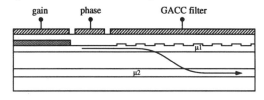

• Sampled grating DBR laser

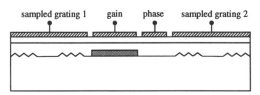

FIGURE 8.6   Examples of extended tuning range lasers. After [3–6].

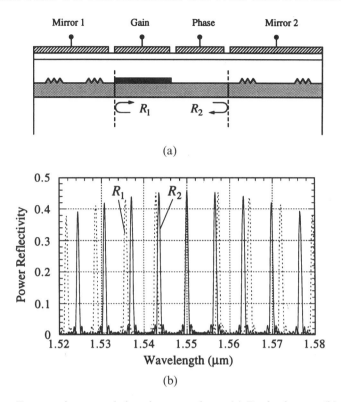

FIGURE 8.7   Four-section extended tuning range laser. (a) Device layout; (b) individual sampled-grating mirror reflectivities, R_1 and R_2. Alignment at 1.55 µm is indicated, so that the net mirror loss will only have a single minimum at 1.55 µm. A slight change in index of one mirror will shift the alignment position to another pair of peaks.

are used for the end mirrors, and each of these contains a periodic modulation of the amplitude or phase of the grating. This periodic spatial modulation, which can be as simple as a periodic blanking of the grating, creates a corresponding reflection spectrum with periodic maxima in the frequency domain. This can be understood most simply in the special case of weak reflections in a very long grating with very short grating bursts. Here the impulse response is seen to be a comb function, and the Fourier transform of a comb function is another comb function in the frequency domain. Figure 8.8 illustrates how the periodic blanking of a continuous grating results in a comb of reflection orders.

As suggested above, the simplest form of modulated grating is the sampled grating in which a uniform grating is periodically blanked, perhaps with a double exposure during the lithographic fabrication step. For a finite-length grating the reflection coefficient for each peak in the comb of reflection maxima has the same form as the unsampled grating, i.e., Eq. (6.37), however, in this

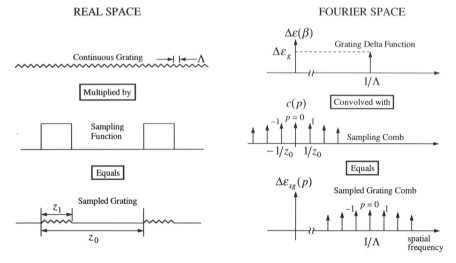

FIGURE 8.8 Sampling of a continuous grating is accomplished by multiplying it by a sampling function. In the Fourier domain the delta function spectrum of the continuous grating is replicated into a comb.

case the total reflection spectrum is composed of a superposition of these reflection components, one for each peak (reflection order) of the comb. Also, the coupling constant, κ_p, and the normalized propagation constant, σ_p, are functions of the duty factor of the sampling function, z_1/z_0, and the peak (or order) number, p, in the comb of reflection peaks. For simple periodic blanking of a grating, the coupling constant is given by

$$\kappa_p = \kappa_g \frac{z_1}{z_0} \frac{\sin(\pi p z_1/z_0)}{(\pi p z_1/z_0)} e^{-j\pi p z_1/z_0}, \tag{8.12}$$

where κ_g is the coupling constant for the continuous, unsampled grating, z_1 is the length of the grating burst, and z_0 is the sampling period. From Eq. (6.37) the reflection coefficient for one of the reflection orders is,

$$r_p = \frac{-j\kappa_p^* \tanh[\sigma_p L_g]}{\sigma_p + j\delta_p \tanh[\sigma_p L_g]}, \tag{8.13}$$

where

$$\sigma_p = \sqrt{|\kappa_p|^2 - \delta_p^2},$$

$$\delta_p = \frac{2\pi n}{\lambda} - j\frac{\alpha}{2} - \frac{\pi}{\Lambda} - \frac{\pi p}{z_0},$$

and α is the net propagation power loss. The net total sampled-grating reflection coefficient is

$$r_g = \sum_p r_p. \tag{8.14}$$

In Eq. (8.14) we should only use the largest r_p at any wavelength in the summation to be consistent with the assumptions of the coupled-mode formalism which only considers one Fourier order at a time.

As indicated in Fig. 8.7, the 4-section sampled-grating DBR laser makes use of two different sampled-grating mirrors. By sampling the gratings at different periods, reflection maxima with different wavelength periods are created in each mirror. Thus, as shown in part (b), if a certain reflection maximum from one mirror is aligned with one in the second mirror, the others will be misaligned, and the product of the two reflectivities which determines the cavity loss, will only have one maximum. That is, the laser will still be a good single-frequency laser with very good MSR. Now, if both mirrors are tuned together, e.g., by connecting their electrodes, we are left with an effective three-section DBR that functions identically to the normal three-section DBR. Thus, ~ 8 nm of continuous single-mode tunability might be expected at 1.55 μm by simultaneous tuning of the phase electrode.

However, if the index of one grating is tuned differently from the other, adjacent maxima will successively line up. We refer to this differential adjustment to obtain alignment of different reflection maxima as *channel changing*, whereas the above joint adjustment of the two mirrors is *fine tuning*. If we wish to have full wavelength coverage, the channel spacing between mirror reflection maxima should be less than the fine-tuning range of ~ 8 nm. Since the periodicities of the maximas may be only slightly different, only a relatively small differential index change is necessary to switch channels. Thus, with a differential effective index change of less than $\sim 0.1\%$, it is possible to switch the alignment point across many different reflection maxima channels. As can be seen, this sliding-scale action is very similar to the function of a vernier scale. Figure 8.9 gives experimental results from ridge-waveguide sampled-grating devices near 1.5 μm. As can be seen a large tuning range with very good side mode suppression is possible.

8.2.4 Laser-Modulator or Amplifier

One of the more important PICs being developed in recent years is the laser-modulator. Usually the laser is also a tunable laser, so as shown in Fig. 8.10, we again have at least three waveguide sections butted together. In this case, however, one of the sections (the modulator) is outside of the laser cavity beyond the DBR mirror. The modulator section is somewhat analogous to the phase-shift section in the continuously tunable laser. However, here we desire an intensity modulator rather than a phase modulator, so that the laser can operate cw and the emitted lightwave can be modulated external to the cavity.

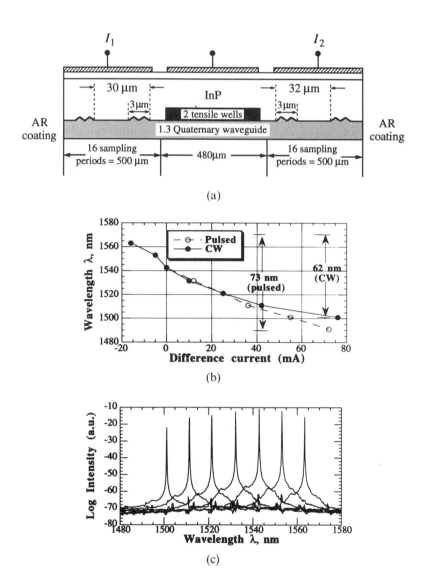

FIGURE 8.9 (a) Schematic of experimental sampled-grating device that uses two tensile-strained InGaAs quantum wells. (b) Pulsed and cw tuning characteristics vs. $(I_1 - I_2)$. (c) Spectra at the 7 bias points indicated in part (b). The cw output level was ~ 10 mW at 1543 nm. After [6].

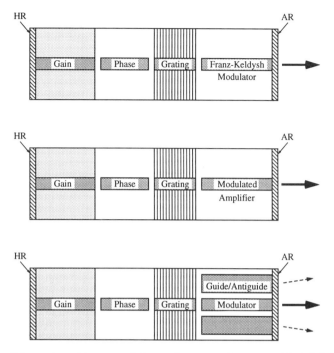

FIGURE 8.10 Schematics of laser-modulators incorporating a tunable DBR laser. The top illustrates a Franz–Keldysh absorption modulator, the center shows a modulated post-amplifier, and the bottom illustrates a guide/antiguide refractive modulator.

The main reasons for interest in the laser-modulator PIC is that the external modulator adds *less wavelength chirp* in the process of modulation, and its *modulation bandwidth* can be higher than that of a laser which is optimized for tunability or some other purpose.

If wavelength tuning is not necessary, a somewhat simpler laser-modulator PIC can be formed by using a simple single-section DFB laser, rather than the three-section DBR shown in Fig. 8.10. Figure 8.11 shows such a configuration along with some experimental results [7]. As shown the DFB in this example utilizes a quarter-wave shifted design, and the modulator uses electroabsorption from the Franz–Keldysh effect. For low-chirp operation, it was found that the isolation resistance between the laser and modulator sections had to be larger than 500 kΩ. These kinds of devices are being developed for use with long optical fiber links such as in undersea cable.

8.2.4.1 Franz–Keldysh Modulator Perhaps the most popular intensity modulator is the bulk Franz–Keldysh modulator. This modulator consists simply of a straight section of waveguide with an active region of slightly larger bandgap energy than the photon energy of the lightwave to be modulated. The

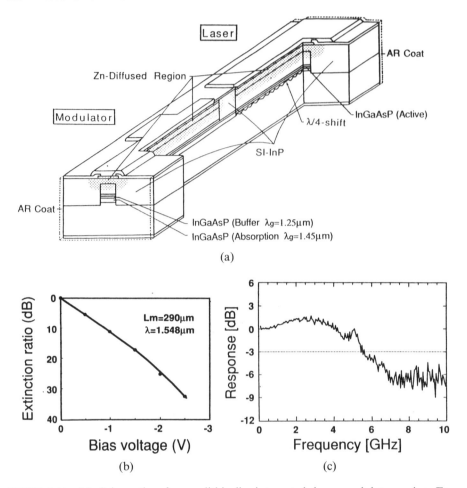

FIGURE 8.11 (a) Schematic of monolithically integrated laser-modulator using Fe doped semi-insulating regrowth to form the lateral BH structure and isolate the sections. The laser is a 300 µm long asymmetric quarter-wave-shifted DFB, and the modulator is a 290 µm long buffer layer loaded electroabsorption structure. The separating isolation region is 50 µm long. (b) Modulation extinction ratio as a function of bias voltage to the modulator. (c) Modulator frequency response. After [7]. (Reprinted, by permission, from H. Tanaka, M. Suzuki, and I. Matsushima, *IEEE J. Quantum Electron.* **QE-29,** 1708 (1993). © IEEE.)

application of a reverse bias field lowers the absorption edge via the Franz–Keldysh effect and reduces the emitted light. As indicated in Fig. 8.12, the Franz–Keldysh effect describes the electron–hole excitation with below-band-gap photons due to the possibility of lateral carrier tunneling with an applied electric field. It can equivalently be understood in terms of the extension of carrier wavefunctions into the forbidden gap when the field is applied.

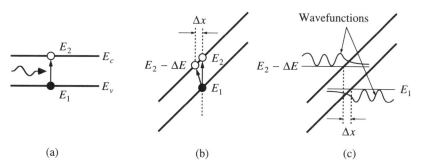

FIGURE 8.12 Schematic of photon absorption to generate an electron–hole pair; (a) with no electric field, and (b) and (c) with a strong applied field so that photons lower in energy by ΔE can be absorbed. (b) illustrates the tunneling model and (c) the equivalent wavefunction overlap model.

The reduction in required photon energy, $\Delta E/q$, for ionization is given by the product of the applied electric field, $\mathscr{E}$, and the effective tunneling distance, Δx. Tunneling is important at distances up to several nanometers, and electric fields of ~ 2–3×10^5 V/cm are easily applied in *pin* structures. Thus, a lowering of the absorption edge by $\mathscr{E}\Delta x \sim 10$ meV (or increasing the absorption wavelength ~ 10 nm near 1.3 µm) is obtainable. As a result, the attenuation constant for a given wavelength within several nanometers of the absorption edge can be increased from a few cm^{-1} to well over 100 cm^{-1} for a few volts applied in practical *pin* DH waveguides. For a desired modulation ratio, P_0/P_{in}, and available change in modal loss, $\langle \Delta\alpha_{FK} \rangle$, the required modulator length is given by

$$L_m = \frac{1}{\langle \Delta\alpha_{FK} \rangle} \ln \frac{P_0}{P_{in}}. \tag{8.15}$$

Modulator lengths ~ 300 µm are typical for practical Franz–Keldysh modulators. Another important design consideration is that the zero field loss, $\langle \alpha_0 \rangle$, must be sufficiently small so that the available output power is not reduced too much. This restricts just how close the operating wavelength can be to the unperturbed absorption edge.

Figure 8.13 gives results of detailed experimental measurements of the absorption near the band edge as electric fields are applied. Also shown for comparison are similar measurements for quantum-well material. In the quantum-well case the absorption is primarily due to excitonic effects. Thus, it has been dubbed the quantum-confined Stark effect to emphasize the different physics involved. As can be seen, the primary difference with the quantum-well absorption edge is that it is much sharper. As a result, larger field effects are possible over a narrow wavelength range. However, because the effect is more confined to the proximity of the absorption edge, devices using it require well-defined wavelengths and are more temperature sensitive as compared to

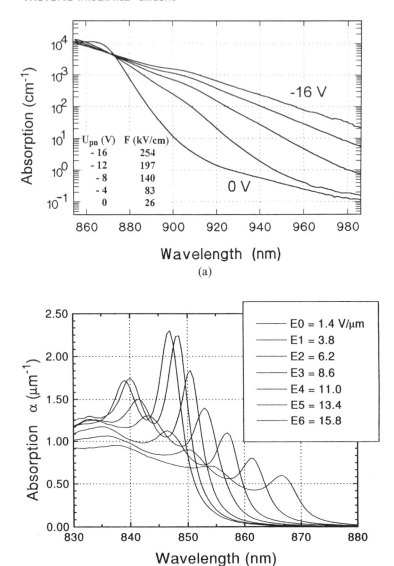

FIGURE 8.13 Absorption edges of (a) bulk and (b) quantum-well material as electric fields are applied. Applied fields are given in the figures. For (b) $d_{QW} = 9.5$ nm and $d_{barrier} = 3.5$ nm of $Al_{0.3}Ga_{0.7}As$. After [8, 9]. (Reprinted, by permission, from *IEEE Photonics Technology Letters* © 1993 *IEEE, and Optical and Quantum Electronics.*)

bulk Franz–Keldysh effect devices. Thus, for use with tunable lasers, the bulk-effect devices may be the better choice.

8.2.4.2 Modulated Amplifier

For widely tunable lasers, even the bulk-effect absorption modulators may not have enough optical bandwidth to serve as effective external modulators. To achieve broader optical bandwidth, different kinds of intensity modulators have been investigated. The second two examples in Fig. 8.10 are capable of providing modulation over a broader optical bandwidth. The modulated amplifier consists simply of a gain section in series with the laser. For fabrication ease, this can be the same gain material as that in the laser. Thus, the optical bandwidth should be as large as the laser's gain material. In fact, depending upon how the amplifier section is biased, it may be possible to have an even larger optical bandwidth than the laser. This fact can be appreciated by reviewing the example gain spectra given in Chapter 4.

Besides chirp, the primary question in the modulated amplifier case regards the magnitude of the modulation bandwidth. For the best on/off ratio it is desirable to swing the modulation current above and below the transparency value. However, as can be seen from the carrier density rate equation (2.15), the carrier decay time is comparable to the normal carrier lifetime when the gain is small (or negative). This is typically ~ 3 ns, which would limit the modulation bandwidth to ~ 100 MHz. However, if the amplifier is biased at a relatively high level, where the stimulated emission term dominates, and then modulated, it may be possible to obtain a modulation bandwidth > 1 GHz with a reasonable (~ 10 dB) on/off ratio.

With stimulated emission, we can use Eq. (2.15) to obtain an effective carrier lifetime, τ', given by

$$\frac{1}{\tau'} = \frac{1}{\tau} + a' v_g N_{p0} = \frac{1}{\tau} + a' \frac{P\Gamma_{xy}}{w\,dh\nu}, \tag{8.16}$$

where $a' = g/N$ at the bias point and P is the optical power flowing in the mode. From this we see that the optimum bias point is at the knee of the gain curve where g/N is a maximum. The modulation bandwidth limited by carrier lifetime is

$$f_c = \frac{1}{2\pi\tau'} = \frac{1}{2\pi}\left[\frac{1}{\tau} + a' \frac{P\Gamma_{xy}}{w\,dh\nu}\right]. \tag{8.17}$$

Plugging in values for a 1 mW/μm^2 input power density at 1.55 μm wavelength, and a gain bias point at the knee of the gain characteristic where $a' \sim 10^{-15}$ cm^2, we obtain $\tau' \sim 1$ ns. Higher input power densities will decrease this, and as the photon density grows along the amplifier's length, τ' would also decrease more. Thus, gigahertz operation appears to be possible. However, we would not expect to be able to operate this amplifier modulator much above the gigahertz level unless very high input power densities were used.

If the input (or output) power level of the amplifier becomes relatively large, we must also consider the power saturation characteristics of the amplifier. When the stimulated emission term becomes comparable to the other carrier recombination terms in the carrier rate equation, the carrier density and, hence, the gain decrease with increasing optical power. Substituting for the photon density as in Eq. (8.16), the carrier rate equation becomes

$$\frac{dN}{dt} = \frac{\eta_i I}{qV} - \frac{N}{\tau} - g\frac{P\Gamma_{xy}}{w\,dh\nu}. \tag{8.18}$$

At low optical powers in the steady state, $N = N_0 = \eta_i I\tau/(qV)$, and over some bias range we can approximate the gain by $g = a(N - N_{tr})$. Therefore, the steady-state gain can be written as $g_0 = a[\eta_i I\tau/(qV) - N_{tr}]$. Then, solving for the steady-state carrier density and gain for higher optical powers where the stimulated emission term must be included, we obtain

$$N = \frac{\eta_i I\tau}{qV} - g\frac{P\Gamma_{xy}\tau}{w\,dh\nu}, \tag{8.19}$$

and

$$g = \frac{g_0}{1 + P/P_s}, \tag{8.20}$$

where

$$P_s = \frac{w\,dh\nu}{a\Gamma_{xy}\tau}. \tag{8.21}$$

For quantum-well separate-confinement waveguide amplifiers typical values lead to $P_s \sim 1 - 10\,\text{mW}$.

Now, to get the net amplifier response, we must integrate over the length of the gain region using

$$\frac{dP}{dz} = gP. \tag{8.22}$$

Inserting Eq. (8.20) into (8.22) and performing the integration over a length L, we obtain an implicit relation for the large-signal gain, $G = P(L)/P(0) = P_o/P_{in}$,

$$G = G_0 \exp\left[-\frac{G-1}{G}\frac{P_o}{P_s}\right], \tag{8.23}$$

where $G_0 = \exp(g_0 L)$ is the unsaturated gain for $P_o \ll P_s$. From this we can derive the output saturation power, P_{os}, where the gain has fallen to half of G_0,

$$P_{os} = \frac{G_0 \ln 2}{G_0 - 2}P_s. \tag{8.24}$$

Thus, we note that P_{o_s} is slightly smaller than P_s.

Another issue with the modulated amplifier approach is the broadband spontaneous emission level. Although it is easy to keep the spectral density level low, the integrated spontaneous energy can be comparable to the energy at the desired output wavelength if the input signal level is not large enough. Careful spatial filtering can be used to eliminate off-axis spontaneous energy, but there is still the inherent spontaneous emission into the mode of interest. It can be shown [10] that the noise figure, $F_A = (\mathrm{SNR})_{in}/(\mathrm{SNR})_{out}$ must be $\geq 2n_{sp}$, the population inversion factor introduced in Chapter 4. In fact, a quantitative measure of this spontaneous emission relative to the gain can be obtained from the details in Chapter 4. The minimum noise figure is increased further by internal loss, α_i, and any facet reflections. Neglecting the facet feedback, we can write the noise figure as [10],

$$F_A = 2n_{sp}\left(\frac{g}{g - \alpha_i}\right) = 2\frac{f_2(1 - f_1)}{f_2 - f_1}\left(\frac{g}{g - \alpha_i}\right).$$

Typically, $F_A \sim 5$ dB in semiconductor amplifiers.

8.2.4.3 Guide/Antiguide Modulator The third example of an integrated modulator is the guide/antiguide modulator [11] shown in Fig. 8.10. Unlike the other two examples, this device uses refractive index effects to achieve intensity modulation. However, unlike other devices of this class, such as Mach–Zehnder or directional-coupler modulators, it does not rely upon length-dependent mode beating or interference effects. Thus, it tends to be less dependent upon wavelength and temperature changes. It consists of a field-induced guide region surrounded by identical field-induced cladding regions. It functions as an intensity modulator by using applied voltages to alternately create either a laterial index guide or antiguide profile as indicated in Fig. 8.14.

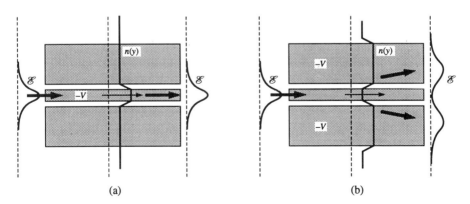

(a)   (b)

FIGURE 8.14 Schematic of guide/antiguide modulator in (a) on and (b) off states.

Thus, like the Franz–Keldysh and amplifier modulators, it basically consists of a straight section of guide that does not require long waveguide bends or other transitions. Radiated energy in the antiguiding state must be spatially filtered in the output guide; this would naturally occur in fiber coupling.

8.2.5 Laser-Modulator Example Problem

Figure 8.15 shows a schematic of a tunable laser-modulator to be designed. The device is to operate at a minimal bias current at 5 mW out, have a ±2 nm continuous tuning range, and a 10:1 on/off ratio with a 2V p–p applied modulation. The characteristics of the active layers in the gain, phase shifting, DBR, and modulator regions are given by Fig. 4.25, the example of Section 8.2.2, and reference 7, respectively. The problem is to determine the lengths of the various sections. A final criterion is that the total injected current density to any section can not exceed 10 kA/cm² due to anticipated problems with device heating and reliability.

We also need to determine the mismatch loss at each waveguide junction, so that we can calculate the total internal loss. Each region will have slightly different modal cross sections and waveguide impedances, and reducing the mismatch loss between these sections is one of the key problems in fabricating a viable device. For this step we use the modal excitation formula (6.81), which leads approximately to (6.84).

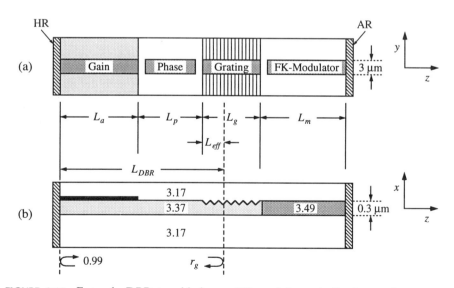

FIGURE 8.15 Example DBR tunable-laser—FK-modulator, indicating various parameters: (a) top view; (b) cross-sectional side view. The internal modal loss is assumed to be 10 cm⁻¹ throughout; the internal quantum efficiency for current injection is assumed to be 0.8 throughout. Separate current sources are attached to each top electrode.

First, we calculate the effective indices of each waveguide section, and then we calculate the confinement factors in the active and passive regions. From Section 8.2.2, $\bar{n}_p = 3.24$. Also, $\bar{n}_g = 3.23$. The addition of the 4-quantum-well active region gives $\bar{n}_a = 3.30$. From perturbation theory, $\Gamma_{xya} = 0.06$, and from Section 8.2.2, $\Gamma_{xyp} = 0.49$. From this information we can plug into Eq. (6.84) for the various interfaces. For the gain-to-phase shift region we find a power transmission of 98%. Between the phase-shift and grating regions or the grating and modulator regions, the transmission is near unity. Most of the energy not transmitted between the gain and phase-shift regions is lost to forward-propagating radiation modes. However, a little ($\sim 0.2\%$) is reflected back. This can show up as a slight modulation on the cavity loss, but since our primary mirrors have much larger reflectivities, we shall neglect it here.

Before we can pursue the issue of optimum mirror reflectivity, we need to determine the minimum lengths for the phase-shift section and modulator. The phase-shift section must be sufficiently long to provide continuous tuning over 4 nm. From Eq. (8.4), neglecting the difference in effective indices and phase shift in the gain region, we find

$$\frac{\Delta \bar{n}_p}{\bar{n}_p} = \frac{4}{1550}\left(1 + \frac{L_a}{L_p}\right). \tag{8.25}$$

Thus, we see that even for $L_p \gg L_a$, we require $\Delta \bar{n}_p/\bar{n}_p > 0.0026$. If we let $L_p = L_a$, then $\Delta \bar{n}_p/\bar{n}_p \le 0.005$, over nearly all of the tuning range. (The required maximum grating effective index change is $4/1550$, or just half this amount.) This is about as large as we can expect to get without adding significant free carrier losses with injected currents, thus, we make the choice, $L_p = L_a$. From Eq. (3.66) we can calculate the maximum tuning current required for either the phase-shift or grating sections. That is,

$$I_j = -\frac{qw\,dL_j}{\Gamma_{xy}\eta_i\tau \times 10^{-20}\text{ cm}^3}\Delta \bar{n}_j, \tag{8.26}$$

where the subscript $j \equiv g$ or p. For $\Delta \bar{n}_g = -0.01$, $\Delta \bar{n}_p = -0.02$, and a carrier lifetime $\tau = 3$ ns, $I_g = 90.9L_g$ mA/mm and $I_p = 181.8L_g$ mA/mm. The maximum current density in the phase shifter is therefore 6.1 kA/cm², a sufficiently low value.

For the external modulator, we use Eq. (8.15) with absorption numbers from [7]. First, we determine a zero-bias built-in voltage of 0.8 V for a relatively heavily doped *pin* structure, and assume modulation of 0–2 V. Thus, the internal modulator field varies between 0.8 V/0.3 μm and 2.8 V/0.3 μm. From ref. [7] for 1.45 μm bandgap material at 1.55 μm, we obtain $\Delta\alpha_{FK} = 140$ cm⁻¹. Using a confinement factor of 0.66, and $P_0/P_{in} = 10$, in Eq. (8.15), we obtain

$$L_m = \frac{0.66}{140\text{ cm}^{-1}}\ln(10) \approx 250\text{ μm.} \tag{8.27}$$

With a 10 cm^{-1} residual loss, the net on-loss in the modulator is 1.1 dB. Thus, we must get 6.4 mW out of the grating mirror to get 5 mW to the output facet.

In order to select the active region and grating lengths for minimum drive current, we refer to Appendix 17. In this appendix, L_p is the total passive length including the grating penetration depth. From the discussion there we also know that we will minimize the drive current for the largest mirror reflectivity and shortest gain length that is consistent with other practical constraints. One practical constraint is the additional loss in propagating through the grating that is not included in Appendix 17. Another is heating that will occur for high current densities. Thus, we know that we need to use a nominal-length grating. As a first design iteration we will use the existing design of Section 8.2.2, since the reflection was relatively high. For this design, $r_g^2 = 0.7$, $L_g = 115$ µm, and $L_{eff} = 40$ µm. Thus, using the effective mirror model, we have an additional propagation factor through the remaining grating length of $\exp[-2(40)0.001] = 0.92$; or we need $P_{01} = 6.9$ mW at the effective mirror output.

Reviewing Appendix 17, we see that our present case doesn't really fit into any case considered due to the grating penetration. Case C (fixed L_p/L_a) would appear to be the closest; however, after plugging in our numbers, we see that it leads to an optimum threshold gain far above the knee of the gain characteristic, where the current density $\gg 10$ kA/cm^2, which would be impractical from the standpoint of device heating, carrier leakage, and accelerated aging. The other extreme would be Case A1 (fixed L), where the optimum threshold gain is at the knee of the gain vs. current characteristic. Working through these two extremes and using Fig. 4.25, we obtain $g_{opt}(C) \sim 1500$ cm^{-1}; $g_{opt}(A1) \sim 600$ cm^{-1}. Since no case exactly fits our example, we choose the average of these two extremes as a first iteration. At this point $J_{th} \sim 1.9$ kA/cm^2 for the 4-quantum-well active region. Then, we can calculate L_a (assuming $L_p = L_a$), using $\Gamma_{xy} g_{th} L_a = \alpha_i(2L_a + L_{eff}) + \ln(1/R)$,

$$L_a|_1 = \frac{\alpha_i L_{eff} + \ln(1/R)}{\Gamma_{xy} g_{th}|_1 - 2\alpha_i} = \frac{10 \text{ cm}^{-1}(0.004 \text{ cm}) + \ln(1/\sqrt{0.7})}{0.06(1050 \text{ cm}^{-1}) - 2(10 \text{ cm}^{-1})} = 51 \text{ µm.} \quad (8.28)$$

For this first trial the required drive current to the active region is

$$
\begin{aligned}
I_{a1} &= \frac{J_{th} w L_a}{\eta_i} + \frac{1}{\eta_i}\left[\frac{\alpha_i L}{\ln(1/R)} + 1\right]\frac{q}{h\nu} P_{01} \\
&= \frac{1900(3 \times 51 \times 10^{-8})}{0.8(10^{-3})} \text{ mA} + \frac{1}{0.8}\left[\frac{0.001(142)}{\ln(1/\sqrt{0.7})} + 1\right]\frac{1.55}{1.24} 6.9 \text{ mW} \quad (8.29) \\
&= 3.64 \text{ mA} + 19.36 \text{ mA} \\
&= 23.0 \text{ mA.}
\end{aligned}
$$

Here we observe that the first term is the threshold current and the second is the current required to obtain 6.9 mW of output. We also note that the device is being operated at about 6.3 times threshold at a net input current density of ~ 15 kA/cm^2. This clearly exceeds the allowed drive current density. In this case, we might expect significant problems with heating and resulting carrier leakage and degradation. Thus, our optimization problem is superseded by the need to reduce the operating current density to 10 kA/cm^2. This will require a lower threshold current density and longer length. Also, we need to look at other mirror reflectivities to find an optimum design.

In order to see the tradeoffs between the selected mirror reflectivity and active region length most clearly, we plot the calculated total drive current from Eq. (8.29) versus the active region length determined from Eq. (8.28) in Fig. 8.16. For the gain curve the analytical expression given in the figure caption of Fig. 4.25 is used. Besides the initial value of $R_g = r_g^2 = 0.7$, output grating power reflectivity values of 0.5 and 0.3 are shown in Fig. 8.16. The vertical arrows indicate the first point that exceeds a total current density of 10 kA/cm^2 on each curve. All points to the left have higher current density and are excluded from consideration. Therefore, even though the higher mirror reflectivity combined with short active regions has a lower mimimum drive current, the current density is too high to be considered.

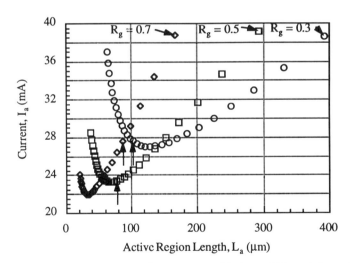

FIGURE 8.16 Required drive current for DBR tunable-laser FK-modulator for 5 mW output power. 6.9 mW is required at the effective mirror plane of the DBR laser section. The points represent threshold gain increments of 50 cm^{-1}. The leftmost point for each reflection value is for a threshold gain of 2000 cm^{-1}. The rightmost points for the $R_g = 0.7$ and 0.5 cases are for 550 cm^{-1}, while for $R_g = 0.3$, the endpoint is at 600 cm^{-1}. The vertical arrows mark the point at which the operating current density reaches 10 kA/cm^2. The solution chooses $L_a = L_p$, assumes zero coupling loss between sections, and uses the parameters given in the example.

From this figure we see that the 50% grating power reflectivity is the best compromise to achieve the lowest drive current while not exceeding 10 kA/cm² in current density. Just to the right of the vertical arrow we have a threshold gain of 1100 cm⁻¹, a threshold current density of 2.1 kA/cm², a total drive current of 23.6 mA, and a drive current density of 9.5 kA/cm² for an active-region length of 83 μm. Thus, this is the best solution for the given criteria. A summary of these and other device parameters follows:

$$
\begin{aligned}
&r_g = 0.707, &&L_g = 84 \ \mu\text{m}, &&L_{eff} = 34 \ \mu\text{m}, \\
&L_a = 83 \ \mu\text{m}, &&L_p = 83 \ \mu\text{m}, \\
&I_{th} = 6.6 \ \text{mA}, &&J_{th} = 2.1 \ \text{kA/cm}^2, \\
&I_a = 23.6 \ \text{mA}, &&J_a = 9.5 \ \text{kA/cm}^2.
\end{aligned}
\tag{8.30}
$$

In finishing this design we should note that even 10 kA/cm² is a high current density for lasers in the InGaAsP materials system. As shown in Fig. 8.16 the total drive current increases relatively slowly for increasing L_a away from the minimum. Thus, it is generally found desirable to design devices with active regions quite a bit longer than the value for minimum current, since a large reduction in current density can be achieved. For example, if we increase the length to 173 μm, the drive current increases to 29.6 mA, but the drive and threshold current densities drop to 5.7 and 1.1 kA/cm², respectively. Also, with a mirror reflectivity of 30% in the above example, a 170 μm active length requires a current of 27.9 mA with drive and threshold current densities of 5.5 and 1.7 kA/cm².

8.3 PICs USING DIRECTIONAL COUPLERS FOR OUTPUT COUPLING AND SIGNAL COMBINING

A different class of PICs uses directional couplers to tap and couple optical energy from one waveguide to another. This lateral or transverse coupling opens new opportunities as compared to PICs which use only axial coupling of waveguides. Mode mismatch and fabrication problems in accurately butt coupling two minute semiconductor waveguides are avoided. As in microwave circuits energy can be tapped without strongly perturbing the original wave-guide. As an example, directional couplers can tap energy from a laser cavity just like a partially transmissive mirror, but the output ends up in a parallel waveguide still within the semiconductor material for further processing. As a second example, lightwaves from two waveguides can be combined into one or both to add or mix two separate signals. This process is central to forming an optical heterodyne receiver. In these examples as well as others, one of the key difficulties with using directional couplers is the waveguide transitions necessary to separate the close spacing required in the coupling region. These usually

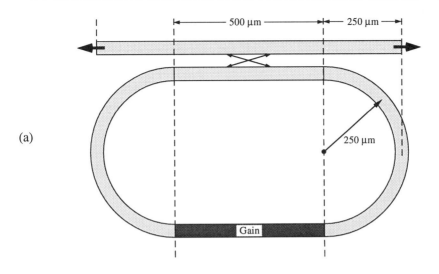

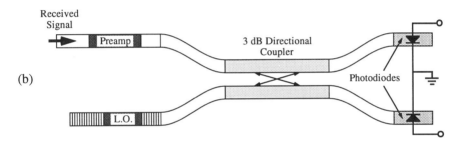

FIGURE 8.17 Example directional-coupler devices: (a) ring laser with directional-coupler output tap; (b) heterodyne receiver with integrated preamp and local oscillator. It is further assumed that the internal modal losses are 10 cm^{-1} throughout both structures (not counting any radiation losses) and the internal efficiency is 0.8. The spacing in the directional coupler is 1 μm.

require gradual waveguide bends that use up a considerable length on the circuit.

Figure 8.17 illustrates the use of a directional coupler as both an output coupler for a ring laser and a signal combiner in a monolithic coherent receiver. In the following two sections we shall explore examples of each. We assume the same passive and active waveguide regions as in the previous examples.

8.3.1 Ring Laser with a Directional Coupler Output Tap

Referring to Fig. 8.17, part (a) shows an example ring laser problem. We wish to analyze the performance of the device assuming a 4-quantum-well gain region as in Fig. 4.25. The passive waveguide cross section in the x-direction is similar to that considered in the previous examples; i.e., it is 0.3 μm thick, has an index of 3.37 and is clad by InP with an index of 3.17 at 1.55 μm. However, we here adjust the lateral width to insure single lateral mode operation. Again, a BH structure is assumed with a lateral cladding of InP also. Primarily, we are interested in the L–I characteristic. Thus, we first need to determine the output coupling loss (analogous to mirror loss). Then, we can proceed to calculate the threshold modal gain, threshold current and differential efficiency.

For the calculation of the coupling ratio, we first need to determine κ, the guide-to-guide coupling constant from Eq. (6.61). Assuming the lowest-order symmetric lateral mode, we can plug Eqs. (7.60), (A3.16), and (A3.19) into Eq. (6.61) and solve. For identical guides coupled laterally we find

$$\kappa = \Gamma_x \frac{2k_0^2(\varepsilon_1 - \varepsilon_c)\gamma_y \cos^2(k_{1y}w/2)}{\beta w_{eff}(\gamma_y^2 + k_{1y}^2)} e^{-\gamma_y s}, \tag{8.31}$$

where Γ_x is the transverse confinement factor, ε_1 and ε_c are the relative dielectric constants (indices squared) of the waveguide slab and surrounding cladding, γ_y is the lateral decay constant outside of the guides, w is the lateral waveguide width, w_{eff} is the effective width (given by Eq. (7.65) with y replacing x), and s is the waveguide separation in the coupler region.

First, we need to determine the guide width, w, and the lateral propagation and decay constants using the effective index technique. For single lateral mode, $V_y \leq 3.25$. Therefore, from Eq. (A3.12), $w \leq 1.2$ μm. Then, using $\bar{n}_2 = 3.24$; $\bar{n} = 3.216$, $k_y = 1.60/\mu m$, and $\gamma_y = 2.20/\mu m$. These numbers give $w_{eff} = 2.51$ μm. Then, we can plug into Eq. (8.31) for the coupling constant between the two guides in the directional coupler,

$$\kappa = 0.49 \frac{2(2\pi/1.55 \ \mu m)^2(3.216^2 - 3.17^2)2.2/\mu m \times \cos^2(1.6 \times 1.2/2)}{[2\pi(3.216)/1.55 \ \mu m]2.51 \ \mu m(2.2^2 + 1.6^2)/\mu m^2} e^{-2.2 \times 1}$$

$$= 15.7 \ cm^{-1}. \tag{8.32}$$

The fractional power coupled from one guide to the other over the 500 μm long coupler length is

$$c^2 = \sin^2 \kappa L = \sin^2(15.7 \times 0.05) = 0.498, \tag{8.33}$$

omitting internal losses. This means that $(1 - c^2) = 0.502$ of the power is transmitted through one of the guides, again omitting losses.

For the ring laser we need to identify the various lengths, so that we can use Eq. (2.22) or (2.23) for the threshold modal gain. We only consider propagation in one direction, since there is no coupling between forward and backward waves. Clearly, the active length $L_a = 500$ μm, and the total passive length, $L_p = (\pi 500 + 500)$ μm $= 2071$ μm. The equivalent of the mean mirror reflectivity, R, in Eqs. (2.22) and (2.23) is just $(1 - c^2)$ as might be verified by reviewing the derivation leading up to Eq. (2.22). Thus, the output coupling loss (equivalent of mirror loss) is

$$\alpha_m = \frac{1}{L_{tot}} \ln \frac{1}{1-c^2} = 2.7 \text{ cm}^{-1}. \tag{8.34}$$

Next, we must determine the additional loss due to radiation at the bends. The internal modal loss is given as 10 cm^{-1} throughout, but this does not include the radiation losses. For this we use Eq. (7.81) or equivalently, Fig. 7.26. For the present example, the normalized frequency, $V = 3.25$, and the ratio of radius to guide width is 313. Therefore, from the figure the radiative attentuation on the bends, α_R, is

$$\alpha_R = 10^{-3}/w = 6.3 \text{ cm}^{-1}. \tag{8.35}$$

Thus, the average internal modal loss is

$$\langle \alpha_i \rangle = \frac{(2 \times 500 \ \mu\text{m})10 \text{ cm}^{-1} + (1571 \ \mu\text{m})16.3 \text{ cm}^{-1}}{2571 \ \mu\text{m}} = 13.8 \text{ cm}^{-1}. \tag{8.36}$$

Now, we can calculate the threshold volume modal gain as

$$\Gamma g_{th} = \langle \alpha_i \rangle + \alpha_m = 16.5 \text{ cm}^{-1}. \tag{8.37}$$

The confinement factors for the transverse, lateral, and axial directions are 0.06, 0.84, and 0.194, respectively. Thus, the threshold gain, $g_{th} = 1688 \text{ cm}^{-1}$, from which we can extrapolate from Fig. (4.25) that $J_{th} = 5.2 \text{ kA/cm}^2$. Then, the threshold current, $I_{th} = J_{th} w L_a / \eta_i = 52 \text{ mA}$. This is a very high threshold gain and current density. In a real case we would try to include more quantum wells to increase the modal gain by increasing the transverse confinement factor. For example, with six quantum wells the threshold gain and current density would be reduced to 1125 cm^{-1} and 3.3 kA/cm^2, more reasonable values for 1.55 μm.

For the complete L–I curve, we also need to determine the differential quantum efficiency. Using the above numbers, but excluding loss in the output waveguide, the differential efficiency is

$$\eta_d = (0.8) \frac{2.7}{16.5} = 13.1\%. \tag{8.38}$$

For the output power, this must be multiplied by the additional attenuation factor from the coupler to the output facet. The coupler loss has been included in the ring cavity, but not for the energy coupled over. A careful review of the theory reveals that the average propagation constant is used in the propagation delay for both the coupled and uncoupled portions of the lightwave. If it is

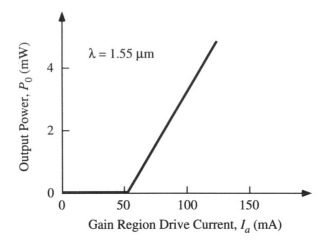

FIGURE 8.18 $L–I$ characteristic of example ring laser.

complex, an attenuation factor will be introduced in both cases. That is, in the present case the coupled fractional power is $c^2 \exp(-\alpha_i L)$, and the uncoupled fractional power is $(1 - c^2) \exp(-\alpha_i L)$. We have already included all of the loss in the uncoupled fraction by including the coupler length in L_p, but this only includes the loss in one guide for the coupler as a whole (i.e., half of the coupler loss for the coupled fraction). Thus, we must include an additional factor of $e^{-(10 \times 0.25)}$ for the energy coupled over. (Put another way, we have modeled the coupler as an effective lumped coupler at the center of the actual coupler.) Since we have an additional 250 μm in the extension guide after the coupler, the net reduced differential efficiency becomes

$$\eta'_d = 0.131 e^{-(10 \times 0.05)} = 7.9\%. \tag{8.39}$$

The $L–I$ characteristic (Fig. 8.18) can then be determined from Eq. (2.36), using the values for I_{th} and η'_d above. This ring laser is obviously not a very efficient device, primarily because of the long sections of lossy passive waveguide.

8.3.2   Integrated Heterodyne Receiver

Part (b) of Fig. 8.17 illustrates a monolithically integrated heterodyne receiver. Such structures have been widely studied as receivers for coherent fiber-optic communication systems. Figure 8.19 shows a structure developed by Koch and Koren [12]. The reasons for using a heterodyne receiver include possible gains in receiver *sensitivity* and wavelength *selectivity* as compared to a direct detection system preceded by some sort of optical filtering. (In a later section we will also consider such a direct detection receiver.) The received photocurrent flowing from the detector diodes in the present case is proportional to the

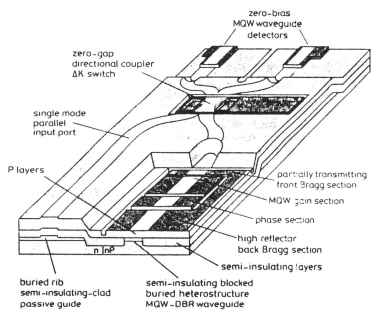

zero-bias
MQW waveguide
detectors

zero-gap
directional coupler
ΔK switch

single mode
parallel
input port

P layers

partially transmitting
front Bragg section

MQW gain section

phase section

high reflector
back Bragg section

semi-insulating layers

n InP

buried rib
semi-insulating-clad
passive guide

semi-insulating blocked
buried heterostructure
MQW-DBR waveguide

FIGURE 8.19 Schematic diagram of MQW balanced heterodyne receiver photonic integrated circuit, containing a continuously tunable LO, a low-loss buried-rib parallel input port, an adjustable 3 dB coupler, and two zero-bias MQW waveguide detectors [12]. (Reprinted, by permission, from *Electronics Letters.*)

product of the amplitudes of the received signal and the local oscillator (LO). Thus, for increased LO powers, the receiver sensitivity increases. Because the signal and LO lightwaves are "mixed" in the detector diodes, the output current also has an IF carrier frequency equal to the difference of the optical frequencies of the two lightwaves. Thus, a relatively simple microwave filter can be used to reject unwanted adjacent wavelength channels in a WDM application. The preamplifier shown in our example is not always used. It is desired only to compensate for the coupling and waveguide losses. Also, it can cause problems with cross talk if a number of wavelength channels are present.

We wish to consider an example problem to solve for the required lengths and currents to make a viable heterodyne receiver as shown in Fig. 8.17(b). Needless to say, there are a variety of elements in this structure that need to be specified. Some of the individual component parameters are specified in the figure caption. We wish to complete a design for such a device with a preamp gain which makes up for any input coupling and propagation losses and still gives 6 dB of net optical gain, a local oscillator with 10 mW of output power, radiation bend losses sufficiently low that the overall propagation loss in the bends is minimized (considering the 10 cm^{-1} of background), and a 3 dB directional coupler to provide a balanced drive to the differentially connected detectors. We assume waveguides, gain sections, and a directional coupler similar to those analyzed for the ring laser in part (a). Furthermore, we require that the guides be spaced by 141 μm for fiber coupling, and we also find that the input coupling loss is 4 dB.

Assuming the directional coupler is identical to that above, for which $\kappa = 15.7$ cm^{-1}, we can calculate its required length to provide a 3 dB coupling. That is, using Eq. (6.70).

$$L_{DC}(3\ \text{dB}) = \frac{1}{\kappa} \sin^{-1} \sqrt{0.5} = \frac{\sqrt{2}}{2(15.7\ \text{cm}^{-1})} = 450\ \mu\text{m}. \qquad (8.40)$$

The additional transmission factor due to the internal modal loss over this distance is $\exp(-0.001 \times 450) = 0.64$, or -2 dB.

Now, for the bend radii we wish to minimize overall loss, knowing that the background propagation loss due to scattering and free-carrier absorption is a relatively high 10 cm^{-1}, or 4.34 dB/mm. Each S-bend must move the guide laterally by 70 μm, or each segment must move the guide by ~ 35 μm in the y-direction. That is, referring to Fig. 8.20, we see that the angle swept by each half S-bend using arcs of radius R is $\theta_b = \cos^{-1}[(R - 35)/R]$.

The total loss in traversing an S-bend is, therefore,

$$\alpha_T L_b = [\langle \alpha_i \rangle + C_1 e^{-C_2 R}](2R) \cos^{-1}\left(\frac{R - S_l/2}{R}\right), \qquad (R > S_l/2) \quad (8.41)$$

where S_l is the lateral waveguide displacement to be achieved (in this case 70 μm), and $C_1 \approx \gamma_x^2/k_2$ and $C_2' \approx \frac{2}{3}\gamma_x^3/k_2^2$ are the radiation bend loss constants given with Eq. (7.81). Equation (8.41) can be minimized with respect to R by taking its derivative and identifying the zero crossing on a plot of the result. However, since a plot is required for the transcendental equation, we might as well just plot Eq. (8.41) directly. The result is plotted in Fig. 8.21 for the present parameters. As can be seen the optimum R is found to be 205 μm. The total loss in each S-bend at this minimum is found to be 1.12 dB.

The total loss for the input signal which must be compensated by the

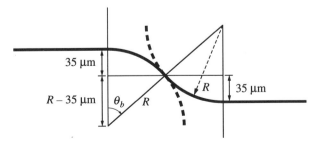

FIGURE 8.20 Single-guide S-bend used at each end of the coupler for both guides.

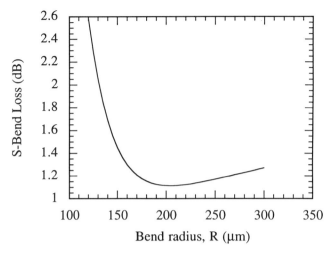

FIGURE 8.21 Plot of S-bend loss vs. bend radius for $\gamma_y = 2.2/\mu m$, $k_2 = 12.85/\mu m$, and $S_l/2 = 35\ \mu m$, and a background propagation loss of $\alpha_i = 10\ cm^{-1}$.

preamp, in addition to the desired 6 dB net circuit gain, is 4 dB + 2(1.15 dB) + 2.0 dB = 8.3 dB. Thus, the net preamp gain must be 14.3 dB. From Fig. 4.25 the point of maximum gain per unit current with four quantum wells is $(600\ cm^{-1}, 900\ A/cm^2)$. Thus, with the transverse lateral confinement factor of $\Gamma_{xy} = 0.05$, the required amplifier length to operate at this point is

$$L_{P-A} = \frac{G}{(\Gamma_{xy}g - \alpha_i)} = \frac{14.3\ dB}{(0.05(600) - 10)\ cm^{-1}(4.34)\ dB} = 1628\ \mu m, \quad (8.42)$$

where we have reduced the net preamp gain by its internal loss of $10\ cm^{-1}$. The drive current required would be $I_{P-A} = JwL_{P-A}/\eta_i = 22.0\ mA$. This very long device results from an internal loss that is comparable to the modal gain. Thus, we might consider operating higher up the gain curve to save real estate. For example, if we use the bias point $(1050\ cm^{-1}, 1.9\ kA/cm^2)$ considered in Section 8.2.5 above, we would obtain a required length of 775 μm, which would need a drive current of 22.1 mA. The minimum current occurs at the point $(900\ cm^{-1}, 1.5\ kA/cm^2)$ with a length of 941 μm, where it is 21.2 mA. Thus, we again see how broad the minimum in drive current tends to be. The current density with the 775 μm length is still only $2.4\ kA/cm^2$, so we will select this design as being more desirable than the point of minimum current. That is, the final preamp design is given by

$$L_{P-A} = 775\ \mu m, \qquad I_{P-A} = 22.1\ mA. \qquad (8.43)$$

Finally, the DBR local oscillator is to put out 10 mW presumably in the most efficient manner. Here we are assuming that the only passive cavity length is the penetration length in the gratings. For the back mirror we wish minimal transmission with a reasonable grating length. Let's look at $\kappa L_g = 2$. Then, $\tanh \kappa L_g = 0.96$, which is a relatively high value. Using $\kappa = 80$ cm^{-1} as in the earlier DBR design, the grating length, $L_{gb} = 250$ μm, and the effective penetration length for the back mirror, $L_{effb} = 60$ μm. Since the preamplifier is so long, this grating should not use up any extra real estate on the substrate. Here, our gain level is higher than in the earlier case, and we wish more power out. Thus, the output mirror should be relatively low in reflectivity. That is, lower than the value selected in Section 8.2.5 as justified in Fig. 8.16. Thus, here we select the 30% mirror considered in that earlier section. There we found that $L_{gf} = 59$ μm and $L_{efff} = 26$ μm.

Since the preamplifier was so long, we still have lots of room for gain without increasing the chip size. That is, $L_a|_{max} = (775 - 250 - 59)$ μm $= 466$ μm. Checking with Appendix 17, we see that a considerably shorter active region would minimize the current, but again the power density is high. Thus, we'll again use the (1150 cm^{-1}, 1.9 kA/cm^2) point and see what length this gives us. Analogous to Eq. (8.10), but without the passive section, we find that $L_a = 148$ μm. Using Eq. (8.11) this leads to a threshold current of 4.22 mA and a current above threshold to get to 10 mW of 21.7 mA. However, the resulting current density is 14.6 kA/cm^2, so we need to use a longer length. Using the conservative knee of the gain curve (600 cm^{-1}, 900 A/cm^2) considered initially for the preamp, we would need a length of 344 μm, which is shorter than the space available, so this is the length we'll use. Here the threshold current is 3.1 mA and the current to get to 10 mW is 26.8 mA. The total drive current of 29.9 mA results in a current density of 7.2 kA/cm^2, so this more conservative design seems okay.

To summarize, we choose the following lengths and drive current for the LO laser,

$$L_{gb} = 250 \text{ μm}, \qquad L_{gf} = 59 \text{ μm},$$
$$L_a = 344 \text{ μm}, \qquad I_{LO} = 29.9 \text{ mA}. \tag{8.44}$$

Its output characteristic is shown in Fig. 8.22.

8.4 PICs USING CODIRECTIONALLY COUPLED FILTERS

As discussed in Chapter 6, directional couplers can become wavelength filters if the two coupled waveguides are dissimilar. Also, the percentage of tuning can be larger than the fractional index change, so as already pointed out in Fig. 8.6, a laser with a tuning range, $\Delta\lambda/\lambda > \Delta n/n$, can be constructed. Such filters are also potentially useful in tunable receivers for WDM systems. One difficulty with such devices is that the bandwidth of the filter also tends to become large as the tuning range is increased. Thus, it can be difficult to construct lasers with both large tuning range and MSR, and even more difficult to construct

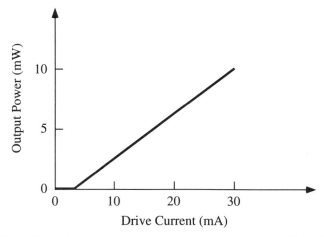

FIGURE 8.22   Light output vs. drive current for the local oscillator laser.

receivers with many resolvable channels having low cross talk. Nevertheless, these codirectionally coupled filters promise to be very important elements in future photonic integrated circuits.

There are two important examples that deserve mention, the acousto-optically tuned filter (AOTF) and the grating-assisted codirectionally coupled (GACC) filter. Both use an additional k-vector in the effective index of one guide to accomplish phase matching between two coupled waveguides. In the case of the AOTF, the effective index modulation is accomplished via the acousto-optic effect with a copropagating ultrasonic wave of wavelength $\Lambda(f_a)$; in the case of the GACC filter the effective index modulation is done directly by modulating the geometry of the waveguide with a periodicity of Λ. The AOTF allows real-time control over the added grating via the acoustic frequency, f_a, whereas tuning in the GACC case is accomplished by changing the background index either electro-optically or by current injection, similar to the cases discussed previously. In both cases maximum coupling occurs when

$$\beta_1 = \beta_2 + \frac{2\pi}{\Lambda} \quad \text{or} \quad \lambda = (\bar{n}_1 - \bar{n}_2)\Lambda, \tag{8.45}$$

as discussed in Section 6.3. Because the difference between the effective indexes of the two coupled modes is generally <0.1, the period of the index perturbation is usually >10 times the optical wavelength. This makes the corrugation fairly coarse in GACC filters, and allows for VHF acoustic frequencies in AOTFs, since the velocity of sound is typically $\sim 10^{-5}\ c$.

If the index difference is tuned by $\Delta\bar{n}$, the filter center frequency tunes according to Eq. (6.76) or

$$\frac{\Delta\lambda}{\lambda} = \frac{\Delta\bar{n}}{\bar{n}_{1g} - \bar{n}_{2g}} = F\frac{\Delta\bar{n}}{\bar{n}_{g1}}, \tag{8.46}$$

where the tuning enhancement factor, $F = \bar{n}_{g1}/(\bar{n}_{g1} - \bar{n}_{g2})$, gives the increase in tuning range as compared to devices which tune by the same relative amount as the index. Also, this tuning enhancement is accompanied by a similar increase in the filter's FWHM optical bandwidth, $\Delta\lambda_{1/2}$. For a uniform interaction region of length L_C, this bandwidth, as outlined in Fig. 6.19, is given by

$$\Delta\lambda_{1/2} = 0.8 \frac{F}{\bar{n}_{g1}L_C} \lambda^2. \tag{8.47}$$

Thus, for a reasonably narrow bandwidth filter, F cannot be too large. This implies that the effective index difference between the two coupled modes must be relatively large.

Because the acoustic wavelength in AOTFs is much larger than the optical wavelength for reasonable tuning enhancement, it is generally difficult to acousto-optically modulate one waveguide of two coupled optical waveguides without modulating both. Thus, AOTFs usually involve TE-to-TM conversion in a single optical waveguide. That is, they are usually an example of Case (a) in Section 6.3.3. Since it is difficult to obtain very much effective index difference by waveguide dispersion alone, the necessary dispersion to fabricate relatively narrowband filters must come from material birefringence. This requirement eliminates the semiconductor materials we are considering in this book. Highly anisotropic materials like lithium niobate (LiNbO$_3$) are required. Thus, we shall not consider specific examples of AOTFs here. However, guided-wave versions are becoming very important in WDM lightwave systems.

8.4.1 The Grating-Assisted Codirectionally Coupled Filter and Related Devices

Figure 8.23 schematically illustrates one technique of constructing a GACC filter. Here we show the introduction of the necessary index modulation by corrugating the side of one of the coupled waveguides.

This tunable filter can be used in both lasers and receivers as mentioned

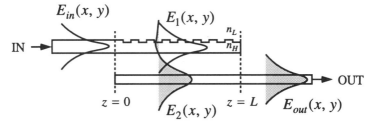

FIGURE 8.23 GACC filter with corrugated-waveguide index modulation to achieve coherent coupling.

earlier. In either case, there are several different possible geometries that can be used. Figure 8.24 shows two different laser structures and a two-stage receiver. Although the AlGaAs/GaAs system is shown, most work has been carried out in the InGaAsP/InP system for operation near 1.5 µm.

The first laser structure incorporates a *pnp* (or *npn*) structure so that the upper and lower guides can both be biased separately. This has been used to provide gain in the lower guide and index tuning in the upper guide along the coupler. The uncoupled section can be short, and it is only used to attenuate uncoupled energy in the lower guide by applying a reverse bias there. The advantage is that the axial confinement factor can be large, since there is gain along the coupler. The disadvantage is the difficulty in fabricating the necessary contacts and electrically isolating the two sections of the lower active waveguide.

The second laser structure, Fig. 8.24 (b), overcomes some of the fabrication difficulties by using an inactive lower waveguide. But now an active–passive

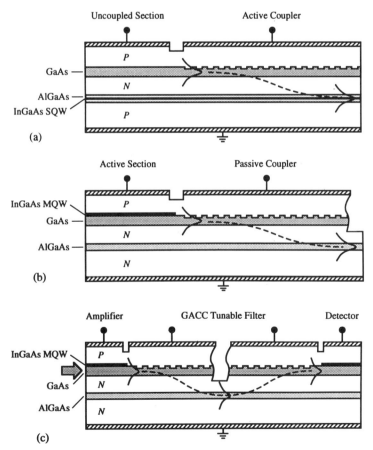

FIGURE 8.24 Schematics GACC tunable devices: (a) active-coupler tunable laser, (b) passive-coupler tunable laser, and (c) photonic integrated-preamp tunable receiver.

junction must be formed in the top guide, and the resulting laser has a relatively small axial confinement factor, since the coupler is all passive. In fact, for good single-mode behavior, we wish to maximize the fraction of the cavity filled with the coupler filter, since any additional length results in a reduced mode spacing and more modes within the filter's passband. Using Eq. (8.47), we can relate the mode spacing, $\Delta\lambda_m$, to the FWHM of the filter, $\Delta\lambda_{1/2}$,

$$\frac{\Delta\lambda_m}{\Delta\lambda_{1/2}} = \frac{1}{1.6F}\frac{L_C}{L_T}, \tag{8.48}$$

where L_T is the total effective cavity length experienced by the laser mode. That is, the right side of Eq. (8.48) is the reciprocal of the number of modes under the filter's passband. As the tuning enhancement, F, is increased, so are the number of included axial modes. Fortunately, only a fraction of these modes near the passband center are important in the spectrum of a laser. Comparing the central mode to the next adjacent mode in the case of Fig. 8.23(b), an approximate analytic expression can be derived [5] for the mode suppression ratio (MSR),

$$\text{MSR} = \frac{2P_{out}}{\hbar\omega v_g n_{sp}\Gamma g_{th}\ln(1/R)}\left(\frac{L_c}{L_T}\frac{1}{F}\right)^2\frac{L_T}{L_T - L_c}. \tag{8.49}$$

Here we observe a quadratic dependence on the coupler length and inverse tuning enhancement. This is plotted in normalized form in Fig. 8.25 along with the normalized filter bandwidth as a function of the tuning enhancement factor, F. As can be observed, for $L_C^2/L_T L_u \sim 1$, it is still possible to have a reasonable

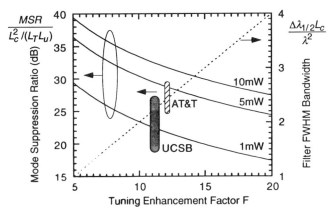

FIGURE 8.25   Calculated mode suppression ratio normalized by $L_C^2/(L_u L_T)$ and plotted against tuning enhancement factor F at 1 mW, 5 mW, and 10 mW power levels. $n_g = 4$, $\hbar\omega = 1$ eV, $\Gamma_{xy}g_{th} = 50$ cm^{-1}, $R = 0.3$, and $n_{sp} = 1.2$ were assumed. $L_u = L_T - L_C$. Theory after [5]; shaded areas refer to early experimental results at UCSB [5] and AT&T [4].

MSR at high output powers. Of course, it is not as good as in normal DFB or DBR lasers.

As indicated in Fig. 8.24(c), this same GACC filter can be incorporated into integrated tunable receivers. The configuration shown uses the same concept as the laser of part (b), but two stages are cascaded, and AR coatings are necessary on the facets to prevent regenerative feedback or lasing. The operation of the various sections in the receiver is very different from the case of the laser. The first section of gain material is used as a preamplifier; the two-stage filter selects one wavelength channel from the incoming spectrum and removes the spontaneous emission out of this band; and the second region of gain material is reverse biased to function as a waveguide detector. Presumably the same gain material can be used for the preamplifier and the waveguide detector, so that the fabrication is basically the same as for the laser of part (b).

Having two or more stages of filtering provides an additional degree of freedom to shape the passband for the received signal. Higher rejection of adjacent channels is possible and a more flat-topped transfer function is also possible to allow for slight deviations in the wavelength of the incoming signal. Effective removal of out-of-band amplified spontaneous emission is very important since the incoming signals may be relatively small. The coupling constant along the length of each filter can also be weighted to further reduce the side-lobe level for the filter.

The design of the amplifier involves an exercise similar to the one carried out in Section 8.3.2 above. That is, it is generally necessary to compensate for input coupling, propagation, and filter insertion losses to avoid a loss of receiver sensitivity. Usually, it would be desirable to have additional preamp gain to improve receiver sensitivity. But since the preamp amplifies all of the incoming channels, a large gain here is not as desirable as adding an additional gain stage after a first channel-dropping filter. This could be accomplished by cascading two of the Fig. 8.24(c) structures. Then, the second active region would become the second amplifier rather than a detector, and a third active section would be the detector. Again, this two-amplifier, four-stage filter device could be fabricated with the same steps as for the receiver shown.

The waveguide detector design involves a tradeoff of its quantum efficiency and capacitance, since both increase with length. Receiver sensitivity is inversely proportional to the capacitance, since the load resistor must be chosen to provide enough bandwidth to receive the highest frequency components of the data. The required detector length for a desired quantum efficiency, η_D, is given by

$$L_D = \frac{1}{\Gamma_{xy}\alpha_D} \ln \frac{1}{1 - \eta_D}, \tag{8.50}$$

where $\Gamma_{xy}\alpha_D$ is the incremental modal absorption constant in the detector. The absorption constant, α_D, is just the negative of the gain at zero injection given in Chapter 4. From Fig. (4.18) for example, we see that $\alpha_D \sim 5000 \text{ cm}^{-1}$ in

InGaAs quantum wells. For an RC-limited receiver bandwidth equal to the bit rate, B (i.e., a receiver bandwidth twice the highest fundamental frequency component in the data), the required load resistor, R_L, is

$$R_L = \frac{1}{2\pi B C_D}. \tag{8.51}$$

Assuming a FET amplifier front end, the received signal is proportional to the signal voltage on the gate. This voltage is equal to the detector photocurrent, i_D, times this load resistance. For an input optical power of P_i, we can use Eqs. (8.50) and (8.51) to obtain

$$V_G = i_D R_L = \frac{P_i q \eta_D}{hv} \frac{d\Gamma_{xy}\alpha_D}{2\pi B \varepsilon_D w \ln(1 - \eta_D)^{-1}}, \tag{8.52}$$

where d, ε_D, and w, are the thickness, dielectric constant, and width of the waveguide detector region, respectively, used to calculate the capacitance. Effective values may be necessary to account for fringing and parasitic capacitance.

Using the waveguide design of Section 8.3, $\alpha_D \sim 5000$ cm^{-1}, and a desired η_D of 90%, we find from Eq. (8.50) that

$$L_D = \frac{1}{0.05(5000 \text{ cm}^{-1})} \ln \frac{1}{1 - 0.9} = 92 \text{ μm.}$$

The capacitance for this length of the example waveguide under reverse bias is ~ 100 fF, allowing for some fringing and parasitic capacitance. Thus, for a data rate of 1 Gb/s, Eq. (8.51) gives a load resistor,

$$R_L = \frac{1}{2\pi B C_D} = \frac{1}{2\pi(10^9/\text{s})(10^{-13} \text{ F})} = 1.6 \text{ k}\Omega.$$

For 1 μW of optical input power to the detector, we then find that the FET gate signal voltage, $V_G = 1.8$ mV.

8.5   NUMERICAL TECHNIQUES FOR ANALYZING PICs

8.5.1   Introduction

Numerous photonic integrated circuits cannot be accurately analyzed by simple analytic techniques. Also, even where analytical techniques are available, numerical techniques are oftentimes employed to verify the designs before proceeding to invest serious effort in making new structures. In fact, numerical techniques are becoming increasingly available and user-friendly. In Section 7.3.8, we introduced the finite-difference technique to analyze dielectric waveguides. In this section we introduce the *beam-propagation method (BPM)* to analyze arbitrary waveguide propagation problems. In particular, we consider a finite-difference BPM as one of the best approaches. As we shall find it builds upon the finite-difference techniques developed in Chapter 7 and Appendix 16.

The most rigorous way of handling electromagnetic wave propagation in integrated optics is to solve Maxwell's equations with appropriate boundary conditions. However, such PIC structures have certain features which makes this approach very difficult to implement. The main reason is the large aspect ratio between the propagation distance and the transverse or lateral dimensions of the propagating energy. The cross section may be contained in a 2 μm × 5 μm window, but the propagation distance could be centimeters long. Therefore, to establish a fine enough grid for a boundary-value approach would pose a great challenge for computer memory and CPU performance. Fortunately, the guided waves in PICs have certain other properties which allow some approximations to be made. For example, in most cases the scalar wave equation is sufficient to describe the wave propagation. Secondly, the phase fronts of guided waves are nearly planar or their plane wave spectra are quite narrow. Therefore, they are paraxial. Thirdly, index changes along the propagation direction tend to be small and gradual in many situations. Hence, as discussed in Chapter 6, the wave amplitudes change slowly and back reflections are negligible in these cases. Under these conditions it is possible to reduce the scalar wave equation (6.4) to the *paraxial wave equation*, which can be written as

$$2jk_0 n_r \frac{\partial \psi}{\partial z} = \frac{\partial^2 \psi}{\partial x^2} + \frac{\partial^2 \psi}{\partial y^2} + k_0^2 [n^2(x, y, z) - n_r^2]\psi, \tag{8.53}$$

where n_r is a reference index that describes the average phase velocity of the wave, and $\psi(x, y, z) = E(z)U(x, y)$. That is, n_r determines the rapidly varying component of the wave, and ψ includes the slowly varying amplitude along the propagation direction. Thus for the polarization of interest, $\mathscr{E}(x, y, z) = \psi(x, y, z) \exp[-jk_0 n_r z]$. The index n_r must be carefully chosen to get accurate answers. In guided-wave problems it is usually chosen as the index of the substrate.

The paraxial wave equation describes an initial value problem as opposed to a boundary value problem. As a result, one can start with an arbitrary initial wave amplitude, $\psi(x, y, 0)$, which could be a Gaussian beam formed by a lens, for example. The resulting amplitude Δz away can be found by integrating the paraxial wave equation over Δz. Repeating this procedure one can find the evolution of the initial field profile over the photonic integrated circuit. Note that one only needs the field values at $z = 0$ in order to calculate the field values at $z = \Delta z$. Therefore, there is no need to store or manipulate the field values at every grid point in the z-direction as required in the solution of the boundary-value problem. Thus, BPM is much more computationally tractable and efficient. Furthermore, all parts of the wave, including the guided and radiation spectrum, are handled together. So there is no need for modal decomposition or to neglect the radiation part of the spectrum. However, since it originates from the paraxial wave equation, back reflections and wide-angle propagation are not handled.

The initial procedure for BPM involved operator techniques and FFTs [13]. But recently it has been shown that algorithms which are much more efficient and robust can be generated by using finite-difference techniques [14]. Due to its significant advantages only the finite-difference BPM (FD-BPM) will be described here. It is possible to generate FD-BPM algorithms using implicit or explicit techniques. In this basic treatment only an implicit algorithm based on the Crank–Nicolson algorithm [15] will be given. The reader is referred to the literature for discussion of the explicit FD-BPM [16].

8.5.2   Implicit Finite-Difference Beam-Propagation Method

The basic idea of the implicit finite-difference beam-propagation method (FD-BPM) is to approximate the paraxial wave equation (8.53) using finite-difference techniques described in Section 7.3.8. This requires choosing a computational window in the transverse dimension (x, y) as well as choosing a grid in the z-direction. The computational window must be large enough to contain the desired field distribution all along the propagation path. However, since radiated fields can always be present, setting the field values to zero at the boundaries of the computational window can create difficulties. Since this boundary condition effectively creates a reflecting boundary, the radiated fields will reflect back and create spurious field distributions. One way to eliminate this difficulty is to use absorbing boundaries. This is achieved by introducing an artificial complex index distribution around the computational window to generate a lossy boundary. The radiated fields are then absorbed before reaching the edge of the window. In practice, less than 10 mesh points around the boundary with a complex index tends to be sufficient. Also, the imaginary part of the index is tapered so as to avoid reflections. Remember to keep the absorber outside of the desired field profile. An alternative approach is to use "transparent boundary conditions" [17] to avoid the absorbers.

With the computational window and z-grid appropriately chosen, we then have $x = p\Delta x$, $y = q\Delta y$, and $z = l\Delta z$. The Crank–Nicolson algorithm is chosen since it is both unconditionally stable and unitary [18]. The reader is warned that a straightforward implementation of the FD-BPM can lead to an unstable algorithm which is not power conserving, thus allowing slight round-off errors to lead to nonphysical results. With the Crank–Nicolson scheme fields will not diverge or diminish without any physical reason regardless of the mesh size, Δx, Δy, or Δz. But accuracy will be lost as the mesh size gets larger.

To convert Eq. (8.53) into the finite-difference form, we let $\psi(x, y, z) = \psi_{p,q}^l$, and approximate the z-partial on the left side by a forward difference to obtain

$$\frac{\partial \psi}{\partial z} = \left[\frac{\psi_{p,q}^{l+1} - \psi_{p,q}^l}{\Delta z} \right]. \tag{8.54}$$

The right-hand side involves only x and y derivatives which are approximated using the regular finite-difference approximations for second-order partial derivatives given by Eq. (7.54). This is refined by taking the average of the discretization at $z = l\Delta z$ and $z = (l + 1)\Delta z$. That is,

$$\frac{\partial^2 \psi}{\partial x^2} = \frac{1}{2} \left[\frac{\psi_{p+1,q}^{l+1} - 2\psi_{p,q}^{l+1} + \psi_{p-1,q}^{l+1}}{\Delta x^2} + \frac{\psi_{p+1,q}^l - 2\psi_{p,q}^l + \psi_{p-1,q}^l}{\Delta x^2} \right], \tag{8.55}$$

$$\frac{\partial^2 \psi}{\partial y^2} = \frac{1}{2} \left[\frac{\psi_{p,q+1}^{l+1} - 2\psi_{p,q}^{l+1} + \psi_{p,q-1}^{l+1}}{\Delta y^2} + \frac{\psi_{p,q+1}^l - 2\psi_{p,q}^l + \psi_{p,q-1}^l}{\Delta y^2} \right]. \tag{8.56}$$

Also, for the final term,

$$k_0^2(n^2 - n_r^2)\psi = k_0^2 \left[\frac{(n_{p,q}^{l+1})^2 + (n_{p,q}^l)^2}{2} - n_r^2 \right] \frac{\psi_{p,q}^{l+1} + \psi_{p,q}^l}{2}. \tag{8.57}$$

For normalization we divide the whole equation by k_0^2 and call $k_0\Delta z = \Delta Z$, $k_0\Delta x = \Delta X$, and $k_0\Delta y = \Delta Y$. Then substituting Eqs. (8.54) through (8.57) in Eq. (8.53) and grouping terms with common q indices, we obtain

$$-\frac{\psi_{p-1,q}^{l+1}}{2\Delta X^2} + [-b\psi_{p,q-1}^{l+1} + a_{p,q}\psi_{p,q}^{l+1} - b\psi_{p,q+1}^{l+1}] - \frac{\psi_{p+1,q}^{l+1}}{2\Delta X^2}$$

$$= \frac{\psi_{p-1,q}^l}{2\Delta X^2} + [b\psi_{p,q-1}^l + c_{p,q}\psi_{p,q}^l + b\psi_{p,q+1}^l] + \frac{\psi_{p+1,q}^l}{2\Delta X^2}, \tag{8.58a}$$

where

$$b = \frac{1}{2\Delta Y^2},$$

$$a_{p,q} = \frac{2jn_r}{\Delta Z} + \frac{1}{\Delta X^2} + \frac{1}{\Delta Y^2} - \frac{1}{2} \left[\frac{(n_{p,q}^{l+1})^2 + (n_{p,q}^l)^2}{2} - n_r^2 \right],$$

$$c_{p,q} = -a_{p,q} + \frac{4jn_r}{\Delta Z}.$$

For $p = 1$ to P and $q = 1$ to Q within the computational window, we obtain $P \times Q$ coupled equations at each z step, one Eq. (8.58a) for each value of p and q. The solution to these equations is best handled with matrices. The following reduction of the equations to matrix form runs parallel to Appendix 16 and assumes the same zero-field boundary conditions discussed in more detail there.

We first compact the y-direction into matrix notation by defining a vector which encompasses y for each x position, p:

$$\boldsymbol{\psi}_p^l = \begin{bmatrix} \psi_{p,1}^l \\ \vdots \\ \psi_{p,Q}^l \end{bmatrix}.$$

Then, by vertically listing all Q equations (8.58a) for a given p, we can group common p indices into a matrix-difference equation along x:

$$-\mathbf{B}\boldsymbol{\psi}_{p-1}^{l+1} + \mathbf{A}_p\boldsymbol{\psi}_p^{l+1} - \mathbf{B}\boldsymbol{\psi}_{p+1}^{l+1} = \mathbf{B}\boldsymbol{\psi}_{p-1}^l + \mathbf{C}_p\boldsymbol{\psi}_p^l + \mathbf{B}\boldsymbol{\psi}_{p+1}^l, \quad (8.58b)$$

where

$$\mathbf{B} = \frac{1}{2\Delta X^2}\mathbf{I},$$

$$\mathbf{A}_p = \begin{bmatrix} a_{p,1} & -b & 0 & \cdots & 0 & 0 \\ -b & a_{p,2} & -b & 0 & & 0 \\ 0 & -b & a_{p,3} & -b & 0 & \vdots \\ \vdots & 0 & & \ddots & & 0 \\ 0 & & 0 & -b & a_{p,Q-1} & -b \\ 0 & 0 & \cdots & 0 & -b & a_{p,Q} \end{bmatrix},$$

$$\mathbf{C}_p = \begin{bmatrix} c_{p,1} & b & 0 & \cdots & 0 & 0 \\ b & c_{p,2} & b & 0 & & 0 \\ 0 & b & c_{p,3} & b & 0 & \vdots \\ \vdots & 0 & & \ddots & & 0 \\ 0 & & 0 & b & c_{p,Q-1} & b \\ 0 & 0 & \cdots & 0 & b & c_{p,Q} \end{bmatrix}.$$

and $\mathbf{I}$ is the $Q \times Q$ identity matrix. If there is no y-direction in the problem of interest, then $\boldsymbol{\psi}_p^l \to \psi_p^l$, $\mathbf{A}_p \to a_p$, $\mathbf{C}_p \to c_p$, $\mathbf{B} \to 1/2\Delta X^2$, and Eq. (8.58b) reduces to a simple difference equation along x.

Now writing all P equations (8.58b) along x in matrix form, we obtain

$$
\begin{bmatrix}
\mathbf{A}_1 & -\mathbf{B} & 0 & \cdots & 0 & 0 \\
-\mathbf{B} & \mathbf{A}_2 & -\mathbf{B} & 0 & & 0 \\
0 & -\mathbf{B} & \mathbf{A}_3 & -\mathbf{B} & 0 & \vdots \\
\vdots & 0 & & \ddots & & 0 \\
0 & & 0 & -\mathbf{B} & \mathbf{A}_{P-1} & -\mathbf{B} \\
0 & 0 & \cdots & 0 & -\mathbf{B} & \mathbf{A}_P
\end{bmatrix}
\begin{bmatrix}
\psi_1^{l+1} \\
\psi_2^{l+1} \\
\psi_3^{l+1} \\
\vdots \\
\psi_{P-1}^{l+1} \\
\psi_P^{l+1}
\end{bmatrix}
$$

$$
=
\begin{bmatrix}
\mathbf{C}_1 & \mathbf{B} & 0 & \cdots & 0 & 0 \\
\mathbf{B} & \mathbf{C}_2 & \mathbf{B} & 0 & & 0 \\
0 & \mathbf{B} & \mathbf{C}_3 & \mathbf{B} & 0 & \vdots \\
\vdots & 0 & & \ddots & & 0 \\
0 & & 0 & \mathbf{B} & \mathbf{C}_{P-1} & \mathbf{B} \\
0 & 0 & \cdots & 0 & \mathbf{B} & \mathbf{C}_P
\end{bmatrix}
\begin{bmatrix}
\psi_1^{l} \\
\psi_2^{l} \\
\psi_3^{l} \\
\vdots \\
\psi_{P-1}^{l} \\
\psi_P^{l}
\end{bmatrix},
\qquad (8.58c)
$$

which we can write symbolically as

$$
\mathbf{A}\psi^{l+1} = \mathbf{C}\psi^l \quad \text{with} \quad \psi^l =
\begin{bmatrix}
\psi_1^l \\
\vdots \\
\psi_P^l
\end{bmatrix}. \qquad (8.59)
$$

The right-hand side is known since both $\mathbf{C}$ and ψ^l (the values of the field at the previous propagation step) are known. To solve for the unknown field values in the next step we have to solve this set of linear equations. As we propagate the beam we need to do this at every propagation step. If the problem is two-dimensional, i.e., if there is only one transverse dimension, $\mathbf{A}$ becomes a tridiagonal matrix. There are very efficient algorithms to solve for a tridiagonal system of equations [18]. Therefore, it is very advantageous to convert a three-dimensional problem into a two-dimensional problem using the effective index approximation, if possible. In the full three-dimensional case it is generally better to use iterative techniques rather than inversion of matrix $\mathbf{A}$ at each step.

8.5.3 Calculation of Propagation Constants in a z-invariant Waveguide from a Beam Propagation Solution

The field distribution $\psi(x, y, z)$ everywhere in a z-invariant waveguide can be calculated by applying the algorithm represented by Eq. (8.59). In the course of the propagating beam calculation, one calculates the correlation function, which is

$$P(z) = \iint \psi^*(x, y, 0)\psi(x, y, z) \, dx \, dy. \tag{8.60}$$

On the other hand, $\psi(x, y, z)$ can be represented by the superposition of othogonal eigenfunctions of the z-invariant waveguide, which is

$$\psi(x, y, z) = \sum_n E_n U_n(x, y) \exp(-j\beta_{pn}z), \tag{8.61}$$

where $U_n(x, y)$ and β_{pn} are the eigenfunction and the propagation constant of the nth mode as obtained from the paraxial wave equation (8.53). In this expansion it is assumed that degeneracy does not exist, which is a good approximation for dielectric waveguides. If (8.61) is substituted into (8.60), one obtains

$$P(z) = \sum_n |E_n|^2 \exp(-j\beta_{pn}z). \tag{8.62}$$

The Fourier transform of (8.62) is

$$P(\beta) = \sum_n |E_n|^2 \delta(\beta - \beta_{pn}). \tag{8.63}$$

Thus, one can find the propagation constant, β_{pn}, by numerically calculating the correlation function, $P(z)$, Fourier transforming it, and locating the peak in the Fourier domain. Ideally the accurate determination of β_{pn} can only be done by infinitely propagating the beam or when the electric field value is known over all z because only when z extends to infinity will the Fourier transform of (8.62) yield (8.63). However, in practice, one can propagate a beam only a finite length, hence field values over a certain z-range, or z-window, are known. In mathematical terms this is equivalent to multiplying (8.60) with a window function $w(z)$, which accounts for the finite length of propagation. Then the Fourier transform of the correlation function, $P_w(z)$, becomes

$$P_w(\beta) = \sum_n |E_n|^2 L(\beta - \beta_{pn}), \tag{8.64}$$

where the lineshape function for the propagation distance, D, is defined by

$$L(\beta - \beta_{pn}) = \frac{1}{D} \int_0^D \exp[j(\beta - \beta_{pn})z] w(z) \, dz. \tag{8.65}$$

Knowing this lineshape function the propagation constant can be quite accurately determined from the spectrum, $P_w(\beta)$, using curve fitting. In the calculation the Hanning window function, $w(z) = 1 - \cos[(2\pi z)/D]$, is used as is typically done in the literature. The eigenfunction of the paraxial wave equation (8.53) is identical to that of the original scalar Helmholtz equation. However, the propagation constant of the Helmholtz equation, β_h, is found from that of the paraxial equation, β_p, using the relation,

$$\beta_h = k_0 n_r (1 + 2\beta_p/k_0 n_r)^{1/2}. \tag{8.66}$$

The details of calculating the peak position from the spectrum are described in [12].

8.5.4 Calculation of Eigenmode Profile from a Beam Propagation Solution

If both sides of (8.61) are multiplied by $D^{-1}w(z)\exp(j\beta z)$ and integrated from 0 to D, we can obtain

$$\psi(x, y, \beta) = \frac{1}{D}\int_0^D \psi(x, y, z)\exp(j\beta z)w(z)\,dz \tag{8.67}$$

$$= \sum_n E_n U_n(x, y)L(\beta - \beta_{pn}). \tag{8.68}$$

Thus, for $\beta = \beta_{pi}$, $\psi(x, y, \beta_{pi})$ can be expressed as

$$\psi(x, y, \beta_{pi}) = E_i U_i(x, y)L(0) + \sum_{n \neq i} E_n U_n(x, y)L(\beta_{pi} - \beta_{pn}). \tag{8.69}$$

Equation (8.69) shows that the eigenmode profile $U_i(x, y)$ can be determined by evaluating the integral (8.67) with $\beta = \beta_{pi}$ provided that most of the excited power belongs to the ith mode which is the mode of interest. In practice, such excitation can be achieved in most cases for dielectric waveguides. The detailed description of this method can be found in [19].

REFERENCES

[1] T.L. Koch, U. Koren, and B.I. Miller, *Appl. Phys. Lett.*, **53**, 1036 (1988).

[2] L.A. Coldren and S.W. Corzine, *IEEE J. Quantum Electron.*, **QE-23**, 903 (1987).

[3] W. Idle, M. Schilling, D. Baums, B. Laube, K. Wunstel, and O. Hildebrand, *Electron. Lett.*, **27**, 2268 (1991).

[4] R.C. Alferness, U. Koren, L.L. Buhl, B.I. Miller, M.G. Young, T.L. Koch, G. Raybon, and C.A. Burrus, *Appl. Phys. Lett.*, **60**, 3209 (1992).

[5] Z.-M. Chuang and L.A. Coldren, *IEEE J. Quantum Electron.*, **QE-29**, 1071 (1993).

[6] V.J. Jayaraman, Z.-M. Chuang, and L.A. Coldren, *IEEE J. Quantum Electron.*, **QE-29**, 1824 (1993).

[7] H. Tanaka, M. Suzuki, and Y. Matsushima, *IEEE J. Quantum Electron.*, **QE-29**, 1708 (1993).

[8] B. Knüpfer, P. Kiesel, M. Kneissl, S. Dankowski, N. Linder, G. Weimann, and G.H. Döhler, *IEEE Photonics Tech. Lett.*, **5**, 1386 (1993).

[9] G.D. Boyd and G. Livescu, *Optical and Quantum Electron.*, **24**, S147 (1992).

[10] T. Saitoh and T. Mukai, Ch. 7 in *Coherence, Amplification and Quantum Effects in Semiconductor Lasers*, ed. Y. Yamamoto, Wiley, New York (1991).

[11] T.C. Huang, Y. Chung, L.A. Coldren, and N. Dagli, *IEEE J. Quantum Electron.*, **QE-29**, 1131 (1993).

[12] T.L. Koch, U. Koren, R.P. Gnall, F.S. Choa, F. Hernandex-Gil, C.A. Burrus, M.G. Young, M. Oron, and B.I. Miller, *Electron. Lett.*, **25**, 1621 (1989).

[13] M.D. Feit and J.A. Fleck, Jr., *Appl. Opt.*, **19**, 1154 (1980).

[14] Y. Chung and N. Dagli, *IEEE J. Quantum Electron.*, **QE-26**, 1335 (1990).

[15] G.D. Smith, Numerical solution of partial differential equations, in *Finite Difference Methods*, Oxford University Press, New York (1985).

[16] Y. Chung and N. Dagli, *IEEE J. Quantum Electron.*, **QE-27**, 2296 (1991).

[17] G.R. Hadley, *IEEE J. Quantum Electron.*, **QE-28**, 363 (1992).

[18] W.H. Press, B.P. Flannery, S.A. Teukolsky, and W.T. Vetterling, *Numerical Recipes, the Art of Scientific Computing*, Cambridge University Press, New York (1988).

[19] W.H. Press, B.P. Flannery, S.A. Teukolsky and W.T. Vetterling, *Appl. Opt.*, **19**, 2240 (1980).

READING LIST

K.J. Ebeling, *Integrated Opto-electronics*, Chs. 12 and 13, Springer-Verlag, Berlin (1993).

T.L. Koch and U. Koren, Photonic integrated circuits, in *Integrated Optoelectronics*, ed. M. Dagenais, J. Crow, R. Leheny, Academic Press, New York (1994).

L.A. Coldren, Lasers and modulators for OEICs, in *Integrated Optoelectronics*, ed. M. Dagenais, J. Crow, R. Leheny, Academic Press, New York (1994).

PROBLEMS

These problems may draw from material in previous chapters and appendices.

8.1 Design a tunable two-section 1.55 µm InGaAsP/InP DBR (a gain region with a cleave at one end and a passive grating at the other) which has output frequencies equally spaced by 100 GHz when the index of the grating is tuned. Assume a 4-quantum-well gain region with characteristics as given by Fig. 4.25 with quaternary barriers of bandgap wavelength 1.25 µm. This active region is placed on top of a 0.25 µm waveguide also of 1.25 µm bandgap material and all clad with InP. (You must determine the confinement factor, effective index, etc.) The grating region is formed by removing the quantum wells and etching a fundamental-order triangular sawtooth grating with a peak-to-peak depth of 30 nm on the top side of the remaining 0.25 µm quaternary waveguide. This again is coated with InP. It is desired to tune the output over eight of the axial modes (spaced by 100 GHz). Assume a BH waveguide width of 1.5 µm, an internal loss of 10 cm^{-1} along the entire device length, an internal carrier injection efficiency of 70%, and provide a grating length to give a power reflection of 50% in the absence of loss.

(a) Determine the grating and gain region lengths.

(b) Plot the power out of the cleaved end vs. current into the gain section.

(c) Plot the output frequency deviation vs. the current injected into the grating assuming both radiative and Auger recombination in the 1.25 µm bandgap Q-material.

8.2 Design a 1.55 µm DFB laser with an integrated amplifier used as a modulator. The DFB laser is a quarter-wave-shifted design with each grating half having $\kappa L = 0.7$. Its total length is 400 µm. The active region consists of four quantum wells in a separate-confinement waveguide as described by the gain curve of Fig. 4.25, and the transverse confinement factor is found to be 6%. This active region is the same throughout, and there is no significant gap between the laser section and the amplifier section. Assume ideal AR coatings on the output facets, an internal efficiency of 70%, and an internal loss of 15 cm^{-1} throughout. The laser is biased to output 3 mW cw into the amplifier section. We desire the amplifier-modulator to operate as fast as possible, to have a 10 dB optical on/off ratio, and to have an on-level output of 10 mW.

(a) Determine the length and width of the amplifier (same width for laser—cannot exceed 5 µm for single lateral mode).

(b) Determine the total laser current and the on/off level currents to the amplifier.

(c) What is the maximum modulation data rate?

(d) Can the modulation rate be improved for different laser and modulator biases?

8.3 It is desired to design a quarter-wave-shifted 1.55 µm DFB with a maximum overall power efficiency (power out/power in) at an output power of 10 mW. (Assume a series resistance that scales inversely with active area, which equals 10 Ω at an area of 100 µm^2. Estimate the jujction voltage by the approximate quasi-Fermi level separation.) At 300 K, we assume a gain as determined by Fig. 4.25, a transverse confinement factor of 6%, an internal efficiency of 70%, an internal loss of 15 cm^{-1}, and a $\kappa = 50$ cm^{-1}. However, we also empirically decrease the gain curve and the internal efficiency by 1% for each 1°C temperature rise. We assume that the thermal impedance can be estimated by Eq. (2.65) with a substrate thickness, $h = 100$ μm and effective thermal conductivity $\xi =$ W/cm-°C. Assume lateral cladding is InP.

(a) What length and width is optimum?

(b) Plot the resulting $L-I$ curve.

8.4 It is desired to design a 1.55 μm GACC laser as illustrated in Fig. 8.24(b) with a tuning enhancement factor of $F = 15$ and a maximum MSR at an output of 5 mW. We assume a gain characteristic as in Fig. 4.25, a transverse gain region confinement factor of 6%, an internal efficiency of 70%, $n_{sp} = 1.2$, $n_{g1} = 4$, an internal loss of 15 cm^{-1} in the active and passive sections of the top guide, an internal loss of 5 cm^{-1} in the lower guide, a design for 100% guide-to-guide coupling, and an active region length of 300 μm. Also, the operating current density in the active region is limited to 8 kA/cm^2.

(a) What is the corrugation period in the coupler?

(b) What is the required κ for 100% coupling?

(c) What is the optimum coupler region length for maximum MSR?

(d) Plot the $L-I$ characteristic for a 3 μm active region width from the left facet when the coupler is tuned for alignment of a mode with its filter peak.

8.5 Consider the slab guide geometry shown below. The free-space wavelength is 1.3 μm.

Choose the computational window as shown in the figure, i.e., 1 μm above the core in the air and 1.5 μm below the core in the substrate. Introduce absorbing regions 0.5 μm thick on the boundaries of the computational window. In the absorbing regions we need to introduce a negative imaginary part to the actual index distribution. For example, the indices of the absorbers in the air and in the substrate can be chosen as $1 - j0.08$ and $3.1 - j0.08$ respectively. Now set up a grid such that $\Delta x = 1/6$ μm. This should result in 18 grid points as indicated in the figure. Excite this geometry at $z = 0$ with a Gaussian beam whose profile is given as,

$$\psi(x, 0) = \exp\left[\frac{-\left(x - \frac{10}{6}\right)^2}{\left(\frac{1}{6}\right)^2}\right],$$

where x is in micrometers.

This is a Gaussian whose center is located on the lower grid point in the core, i.e., on the grid point closer to the substrate interface. Using the FD-BPM propagate this Gaussian 100 μm along the slab. You can choose

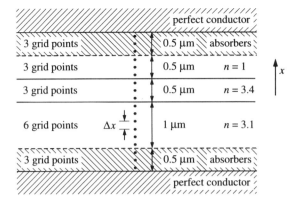

$n_r = 3.1$ and $\Delta z = 5\ \mu m$. At the end of $100\ \mu m$ compare the resulting field distribution with the analytical one discretized on the same mesh. In this case you may find it advantageous to calculate the inverse of the matrix **A** once, since the geometry is invariant along the z-direction. You can improve your accuracy by increasing the number of mesh points and the propagation distance, but this will increase the computational effort.

Review of Elementary Solid-State Physics

A1.1 A QUANTUM MECHANICS PRIMER

A1.1.1 Introduction

In quantum mechanics, the properties and motion of particles are defined in terms of a wave (or state) function, Ψ; its magnitude squared gives the probability density of finding a particle at some point in time in a volume element dV. Or, put another way, the density of particles at some point in space is proportional to $\Psi^*\Psi$. Note that $\int \Psi^*\Psi \, dV = 1$, for properly normalized state functions, since the probability of the particle being somewhere is unity. In our case we are interested in both electrons and photons as particles. For photons this description is roughly equivalent to standard electromagnetic theory where the wavefunction is analogous to a normalized electric field. Maxwell's equations give the description of photon fields. In this appendix we shall focus more specifically on the properties of electrons.

In quantum mechanics, measurements are limited in accuracy by the uncertainty principle, $\Delta x \Delta p_x \geq \hbar/2$, where p_x is the momentum in the x-direction. (In electromagnetic theory the equivalent statement is that $\Delta x \Delta k_x \geq 1/2$.) The expected (or mean) value of some observation is calculated by operating on the wavefunction with the operator, A, corresponding to the observable, a. The operation to obtain the mean value is analogous to a standard weighted average,

$$\langle a \rangle = \int \Psi^* A \Psi \, dV, \tag{A1.1}$$

where a is a possible observation of the operator A. In many cases the operator simply multiplies the observable variable, in others it is more complex, such as momentum, p, where it is $-i\hbar\nabla$.

The motion of particles is governed by Schrödinger's equation,

$$\frac{-\hbar^2}{2m} \nabla^2 \Psi + V\Psi = i\hbar \frac{\partial \Psi}{\partial t}, \tag{A1.2}$$

where m is the particle's (e.g., electron) mass, V is the potential energy operator (same as observable), and $(-\hbar^2/2m)\nabla^2$ is the kinetic energy operator ($=p^2/2m$). Together, these two form the overall energy operator, the so-called *Hamiltonian*. The state function can be expressed as the product of space-dependent and time-dependent factors, $\Psi(r, t) = \psi(r)w(t)$. If we substitute into Eq. (A1.2) and divide by ψw, we obtain a function on the left which only depends on r and a function on the right which only depends on t. Thus, to be valid for all r and t, each side must equal a constant, E:

$$\frac{-\hbar^2}{2m} \frac{\nabla^2 \psi}{\psi} + V = \frac{i\hbar}{w} \frac{\partial w}{\partial t} = E. \tag{A1.3}$$

From this we immediately have

$$w(t) = Ce^{-i(E/\hbar)t}, \tag{A1.4}$$

from which we can identify $E = \hbar\omega$, where ω is the radian frequency of oscillation. For the time-independent part,

$$\frac{-\hbar^2}{2m} \nabla^2 \psi + V\psi = E\psi. \tag{A1.5}$$

The general solution for a uniform potential can be written as the sum of two counterpropagating plane waves,

$$\psi(r) = Ae^{ikz} + Be^{-ikz}, \tag{A1.6}$$

where

$$k^2 = \frac{2m}{\hbar^2}(E - V), \tag{A1.7}$$

is found by substituting back into Eq. (A1.5).

A1.1.2   Potential Wells and Bound Electrons

Electrons are confined by some potential depression in most situations. The most fundamental example is the atom, where electrons are bound by the confining potential of the positively charged nucleus. For the simple case of the hydrogen atom, $V(r) = -q^2/[4\pi\varepsilon_0 r]$, and analytic solutions to Schrödinger's time-independent Eq. (A1.5) can be found. However, for atoms with higher atomic numbers and many electrons, only numerical solutions are possible. Nevertheless, the electron always experiences some sort of confining potential.

When solids are formed from these atoms, the more weakly bound electrons near the exterior of the atom are significantly influenced by the attractive potential of neighboring atoms. In fact, in covalently bonded solids, the outer valence electrons are shared by many atoms, and they develop wavefunctions that extend throughout the crystal. In such cases, the details of the original atomic confining potential are lost. Thus, we shall not dwell on that problem unduly. Rather, we shall investigate the properties of an electron in a simple rectangular potential well to develop the concepts of confined wavefunctions and discrete energy levels common to atoms.

It will later be shown that by coupling together a series of such wells, a periodic potential is formed which leads to electronic properties very similar to those in real crystals. Thus, we can learn much about the properties of electrons in solids by taking this course. As is well known, one of the key results is that electrons in solids can behave much like free electrons with plane wave solutions and a parabolic E–k relationship as illustrated by Eq. (A1.7). However, they appear to have an *effective mass, m^**, that is different from the free electron mass. Also, this effective mass approximation is usually limited to relatively low kinetic energies. Finally, we shall consider quantum-confined structures that include heterostructures to form much larger potential wells than for a single atom. Nevertheless, the mathematics is very similar and we will be able to apply much of what we develop in this section.

First, consider the one-dimensional potential well of width l shown in Fig. A1.1. The simplest method of solution is to recognize that there are three separate regions of uniform potential, where the solution to Schrödinger's equation will have the form of Eq. (A1.6). Then, if we assume that the effective electron mass is the same in all regions, we can develop a complete wavefunction by requiring that the value and slope of the constituent solutions in each of the three regions match at the boundaries. That is, we would not expect any discontinuity in the probability density function. Looking for bound solutions,

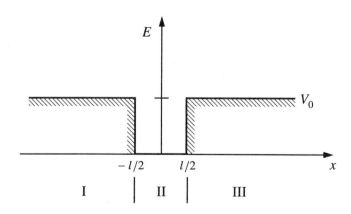

FIGURE A1.1   One-dimensional potential well for electron confinement.

for which $E < V_0$, we can rewrite the general solution Eq. (A1.6) in each region. In the central region II,

$$\psi_{II} = \begin{cases} A \cos kx & \text{(symmetric solutions)} \\ A \sin kx & \text{(antisymmetric solutions),} \end{cases} \tag{A1.8}$$

where $k^2 = 2mE/\hbar^2$. In region III,

$$\psi_{III} = Be^{-\gamma x}, \tag{A1.9}$$

where $\gamma^2 = 2m(V_0 - E)/\hbar^2$. In region I, $\psi_I = Be^{\gamma x}$, but by symmetry, we only need to use the single boundary condition at $x = l/2$ between regions II and III. At $x = l/2$, we have that $\psi_{II} = \psi_{III}$ and $\psi_{II}' = \psi_{III}'$. For the symmetric solutions, this gives

$$A \cos\left(\frac{kl}{2}\right) = Be^{-\gamma l/2}, \tag{A1.10a}$$

and

$$Ak \sin\left(\frac{kl}{2}\right) = B\gamma e^{-\gamma l/2}. \tag{A1.10b}$$

Dividing Eq. (A1.10b) by (A1.10a), we obtain the characteristic equation,

$$k \tan\left(\frac{kl}{2}\right) = \gamma. \tag{A1.11}$$

Similarly, for the antisymmetric solutions, we obtain

$$k \tan\left(\frac{kl}{2} - \frac{\pi}{2}\right) = \gamma, \tag{A1.12}$$

where $\cot x = -\tan(x - \pi/2)$ has been used to illuminate the similarity between the symmetric and antisymmetric characteristic equations.

The electron energy, E, appears on both sides of these charactcristic equations via k and γ, implying that only discrete values of E will satisfy the requirement that the wavefunction and its derivative be continuous across the boundaries. Because the tangent function is periodic, multiple solutions can be found for E, leading to a discrete set of wavefunctions which satisfy the boundary conditions.

Figure A1.2 shows the first few wavefunctions drawn schematically on their respective energy levels over the potential well for reference. These clearly represent bound solutions. There are solutions with $E > V_0$ but they are not bound, and their wavefunctions extend to $\pm\infty$. An interesting property of all of the solutions is that they must be *orthogonal*. That is, if we multiply one wavefunction by the complex conjugate of another and integrate over all space, the integral must be zero. If the wavefunctions are normalized so that the

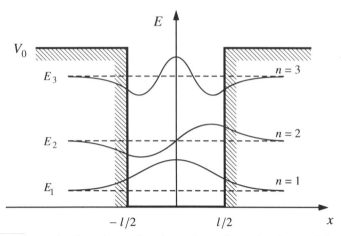

FIGURE A1.2   Energy levels and wavefunctions of one-dimensional potential well. Three bound solutions illustrated.

integral of the product of a wavefunction times its own complex conjugate is unity, then the wavefunctions would be *orthonormal*.

To determine the bound solutions, we need to solve the characteristic equations. For infinitely large V_0, such that the wavefunction goes to zero at the boundaries, i.e., $\psi(l/2) = 0$, the characteristic equation for both symmetric and antisymmetric cases becomes simply

$$\frac{kl}{2} = \frac{n\pi}{2}, \qquad n = 1, 2, 3, \ldots \tag{A1.13}$$

where odd (even) quantum numbers correspond to symmetric (antisymmetric) states. The corresponding discrete energy levels in terms of the quantum numbers are

$$E_n = n^2 E_1^\infty, \tag{A1.14}$$

where

$$E_1^\infty = \frac{\hbar^2 k_1^2}{2m} = \frac{\hbar^2 \pi^2}{2ml^2} = 3.76(m_0/m)(100\ \text{Å}/l)^2 \quad \text{in meV.}$$

When V_0 is reduced from infinity, the discrete energies can still be found using (A1.14), however, the quantum numbers in this case are no longer simple integers, but are real numbers which we will refer to as n_{QW}. For example, if $V_0 = 25E_1^\infty$, the infinite-barrier integer quantum numbers $n = 1, 2, 3, 4, 5$ become $n_{QW} = 0.886, 1.77, 2.65, 3.51, 4.33$.

To calculate n_{QW} for an arbitrary V_0, we need to solve the characteristic equations given in Eqs. (A1.11) and (A1.12). Using Eq. (A1.14), combined with the definitions for k and γ given below Eqs. (A1.8) and (A1.9), the characteristic equations can be conveniently normalized:

$$\tan\left[\frac{\pi}{2}n_{QW}\right] = \frac{1}{n_{QW}}[n_{max}^2 - n_{QW}^2]^{1/2} \quad \text{(symmetric)} \quad \text{(A1.15)}$$

$$\tan\left[\frac{\pi}{2}(n_{QW}-1)\right] = \frac{1}{n_{QW}}[n_{max}^2 - n_{QW}^2]^{1/2} \quad \text{(antisymmetric)} \quad \text{(A1.16)}$$

where

$$n_{QW} \equiv \sqrt{\frac{E_n}{E_1^\infty}} \quad \text{and} \quad n_{max} \equiv \sqrt{\frac{V_0}{E_1^\infty}}. \quad \text{(A1.17)}$$

These equations can be solved graphically by plotting both the left-hand side (LHS) and the right-hand side (RHS) as a function of n_{QW}. Figure A1.3 illustrates this procedure for four different values of V_0.

Note that only a finite set of quantum numbers exist for a given potential barrier, V_0. The normalized variable, n_{max} when rounded up to the nearest integer, yields the largest number of bound states possible for a given V_0. For example, with $V_0 = 3E_1^\infty$, from Eq. (A1.17), we find that $n_{max} = \sqrt{3} \approx 1.73$. Thus, only two bound states are possible under these circumstances. This is perhaps demonstrated more clearly by plotting the possible n_{QW} as a continuous function

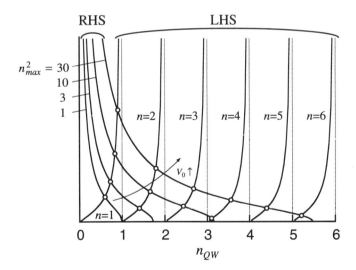

FIGURE A1.3 Graphical solution to Eqs. (A1.15) and (A1.16). The intersections between the LHS and RHS of the equations yield the possible values of n_{QW} for a given n_{max} (or equivalently V_0). The odd (even) quantum numbers displayed next to each tangent curve correspond to the LHS of the symmetric (antisymmetric) characteristic equation.

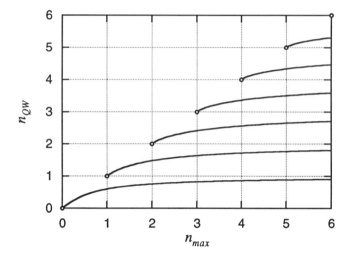

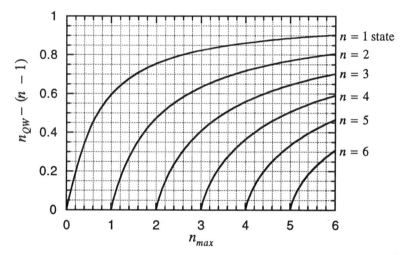

FIGURE A1.4   Plot of quantum numbers as a function of the maximum allowed quantum number which is determined by the potential height, V_0. The quantum numbers are related to V_0 and E through Eq. (A1.17). The lower plot gives a close-up view of the curves (which have been shifted vertically to fit on the same scale).

of n_{max}. Figure A1.4 gives all possible solutions for $n_{max} \leq 6$ (which covers nearly all practical ranges of interest). Note that all quantum numbers approach their integer limit as n_{max} increases toward infinity. In addition, the quantum numbers cease to satisfy the equations (indicated by the open circles) when a given quantum number approaches the integer value of the next lowest state. The lowest quantum number can be approximated to within $\pm 1\%$ using the following formula:

$$n_{QW} \approx \frac{2}{\pi} \tan^{-1}[n_{max}(1 + 0.6^{n_{max}+1})]. \tag{A1.18}$$

A1.2 ELEMENTS OF SOLID-STATE PHYSICS

A1.2.1 Electrons in Crystals and Energy Bands

Electrons in crystals experience a periodic potential originating from the regularly spaced wells at the lattice ions. Figure A1.5 gives a schematic picture along one dimension of such a lattice. As predicted in Chapter 1, when N_A atoms are coupled in such a manner, each atomic energy level of the constituent atoms splits into a *band* of N_A discrete levels. However, this splitting is only significant for the uppermost energy levels where the two atoms interact.

There are several approaches that have been applied to solve this problem. The Kronig–Penney model approximates the actual periodic potential of Fig. A1.5 by a square wave potential, then uses the single rectangular well solution above as a starting point. However, the result is a complex transcendental equation that must be solved numerically. A second approach which provides better closed-form analytic solutions is the coupled-mode approach of Feynman et al. [1]. For accuracy some fairly complex functions need to be evaluated, but by leaving them in general form, we can still get a good picture of the nature of the solutions.

The first step is to go back to Schrödinger's equation and consider a possible general solution for a perturbed system, such as the atom which has been placed into a crystal. The isolated atom had a set of orthonormal wavefunction

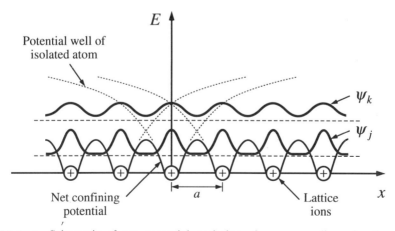

FIGURE A1.5 Schematic of net potential variation along a one-dimensional crystal lattice. Wavefunctions of nominally bound (ψ_j) and free (ψ_k) electron states are illustrated.

solutions just as we obtained for the rectangular potential well. When we perturb the original potential, a new set of orthonormal wavefunctions will exist. But now it may be impossible to solve Schrödinger's equation. It is common to use a superposition of the original set of orthonormal functions to express the new solutions. We shall use this kind of "normal mode expansion" later when we discuss optical solutions.

In the present case we let

$$\Psi = \sum_j w_j(t)\psi_j(r), \tag{A1.19}$$

plug into Eq. (A1.2), multiply by ψ_k^*, and integrate. Then, we have

$$\sum_j w_j(t)\int \psi_k^* H\psi_j \, dV = i\hbar \sum_j \frac{dw_j}{dt}\int \psi_k^*\psi_j \, dV, \tag{A1.20}$$

where the Hamiltonian, $H = [(-\hbar^2/2m)\nabla^2 + V]$, in which V includes the perturbation. Since the original basis functions are orthonormal, the last integral is zero[1] unless $j = k$. Using the shorthand notation,

$$H_{kj} \equiv \int \psi_k^* H\psi_j \, dV, \tag{A1.21}$$

we finally have

$$i\hbar \frac{dw_k}{dt} = \sum_j H_{kj}w_j. \tag{A1.22}$$

This is the desired coupled-mode equation which is independent of the spatial variables. It illustrates that the probability density will flow back and forth among the various original states as a function of time to form the new states. Note that with $k = j$ in Eq. (A1.21) we have the equation to determine the expected value of energy for that original wavefunction—the eigenvalue E that we have been evaluating previously. Thus, the diagonal terms in the H_{kj} matrix are these energy eigenvalues for the respective unperturbed states. The off-diagonal terms represent the coupling strength between the various states. They determine the magnitude of the energy splitting experienced by some original state. It is also important to realize that for most of what we are doing here we do not have to know the actual form of the wavefunctions or even the magnitudes of the matrix elements, H_{kj}. Experimental measurements are usually used to determine the actual values.

Our first example is that of coupling just two identical atoms together. For sufficiently weak coupling, we can approximate the effect on a particular state

---

[1] Actually, in some important cases of interest, ψ_k and ψ_j may include basis functions of laterally displaced atoms to better approximate the perturbed solution. Thus, for some terms in the summation the integral is only small rather than identically zero.

by using only the basis function for that state from each atom in the summation. (Clearly, for vanishingly small coupling, these give the exact solution.) Then, Eq. (A1.22) can be expanded into two coupled-mode equations:

$$i\hbar \frac{dw_1}{dt} = H_{11}w_1 + H_{12}w_2,$$

$$i\hbar \frac{dw_2}{dt} = H_{21}w_1 + H_{22}w_2. \tag{A1.23}$$

Letting the energy of the state in question $H_{11} = H_{22} = E_0$, the coupling energy $H_{12} = H_{21} = \Delta E$, and then, assuming solutions $w_j(t) = C_j \exp(-iEt/\hbar)$ and plugging into Eq. (A1.23), we obtain a characteristic equation from which we must have

$$E = E_0 \pm \Delta E. \tag{A1.24}$$

Thus, the original energy level at E_0 for the isolated atom has split into two levels spaced equally on either side by the magnitude of the off-diagonal matrix element, ΔE. This same process for N_A atoms leads to N_A levels spaced symmetrically about the original level.

Now we are ready to illustrate how *energy bands* are formed when a large number of atoms are coupled together in a crystal. First we consider a simple one-dimensional crystal. Figure A1.6 illustrates a row of atoms spaced by a distance a, similar to the situation of Fig. A1.5.

Our first approximation will be to neglect the perturbation from all atoms except nearest neighbors. Then we can consider a general atom, the kth atom, which can represent every atom in this long chain. From Eq. (A1.22), taking $H_{11} = H_{kk} = E_1$ and $H_{12} = H_{kk \pm 1} = \Delta E$,

$$i\hbar \frac{dw_k}{dt} = \Delta E w_{k-1} + E_1 w_k + \Delta E w_{k+1}. \tag{A1.25}$$

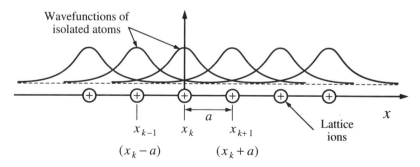

Wavefunctions of
isolated atoms

x_{k-1} x_k x_{k+1}

$(x_k - a)$ $(x_k + a)$

a

Lattice
ions

x

FIGURE A1.6   One-dimensional lattice of coupled atoms to derive energy bands using the coupled-mode approach.

Again letting $w_j(t) = C_j \exp(-iEt/\hbar)$ and plugging into Eq. (A1.25), we obtain a characteristic expression,

$$EC_k = E_1 C_k + \Delta E [C_{k-1} - C_{k+1}]. \tag{A1.26}$$

However, since the subscript k corresponds to the general lattice location, x_k, and the neighbors are at $x_k \pm a$, we can rewrite Eq. (A1.26) letting $C_k \Rightarrow C(x_k)$ and $C_{k\pm1} \Rightarrow C(x_k \pm a)$. Then, we have a difference equation in terms of the spatial variable x. This is solved by letting $C(x_k) = K \exp(ikx_k)$. Finally plugging in this assumed solution,

$$Ee^{ikx_k} = E_1 e^{ikx_k} + \Delta E [e^{ik(x_k - a)} + e^{ik(x_k + a)}],$$

or,

$$E = E_1 + 2\Delta E \cos ka. \tag{A1.27}$$

Equation (A1.27) indicates that in this infinite one-dimensional crystal a continuum of energy values between $E = E_1 \pm 2\Delta E$ is allowed. This is the familiar energy band that solid-state and semiconductor engineers are always referring to. (A later section of this appendix will remind us that for finite crystals, the discrete levels in any real situation really are very closely spaced.)

This same development of bands happens for all of the higher-lying energy levels when atoms are bonded together to form crystals. Thus, the next higher-lying band at energy E_2 also splits into a band due to nearest-neighbor coupling energy $\Delta E'$. Therefore, it provides a new band with $E' = E_2 + 2\Delta E' \cos ka$, where in direct bandgap semiconductors, the sign of $\Delta E'$ is reversed. Also, the overlap of wavefunctions is larger for the higher lying energy levels. Thus, according to Eq. (A1.21), the coupling energy is larger, and the bands become wider. Figure A1.7 illustrates these two bands. As indicated, one period of the plot is sometimes referred to as a Brillouin zone. Since the curves repeat themselves for larger k-values, we usually need concern ourselves only with the first Brillouin zone.

In semiconductors all states of all bands up to the valence band are full, and in the next higher-lying band, called the conduction band, they are empty at $T = 0 \, \text{K}$. We could imagine that Fig. A1.7 represents the conduction and valence bands of a direct bandgap semiconductor such as GaAs or InP. The potentials affecting electrons in such semiconductors are a little more complicated than described by this simple example, in which only nearest-neighbor interactions are considered. So, the E–k plots are not perfect sine waves. Also, in these materials the valence band actually divides into two bands called the *light-hole* and *heavy-hole* bands. These originate because of the asymmetric wavefunctions involved, and the difference in overlap that can occur for different relative orientations when Eq. (A1.21) is evaluated.

For a real three-dimensional crystal with lattice constants a, b, and c, the same procedures can be carried out using a three-dimensional version of Eq. (A1.22) with coupling coefficients ΔE_x, ΔE_y, and ΔE_z, and a three-dimensional

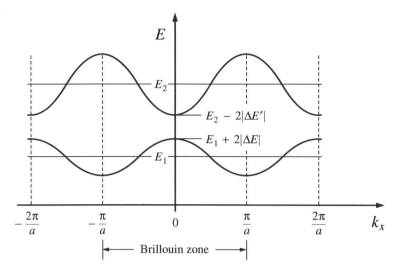

FIGURE A1.7 Energy bands created by a one-dimensional chain of coupled atoms. Two bands resulting from two original discrete states are shown.

envelope wavefunction,

$$w(x, y, z, t) = e^{-iEt/\hbar} e^{i(k_x x + k_y y + k_z z)}, \tag{A1.28}$$

to obtain

$$E = E_1 + 2\Delta E_x \cos k_x a + 2\Delta E_y \cos k_y b + 2\Delta E_z \cos k_z c. \tag{A1.29}$$

The real situation is still more complex than the first-order calculation resulting in Eq. (A1.29). Figure A1.8 illustrates the actual band structure for both GaAs and InP along the $\langle 1\,0\,0 \rangle$ and $\langle 1\,1\,1 \rangle$ directions.

A1.2.2   Effective Mass

Near the top of the valence band and near the bottom of the conduction band it is sometimes possible to approximate the shape of these E–k extrema by parabolas. In these cases the concept of an effective mass is useful, and simple expressions for the density of states are possible. However, the concept of an effective mass has also been extended to limited regions within nonparabolic bands where the parabolic approximation is still valid.

To determine an expression for the effective mass and show that the parabolic band is desired, we follow a semiclassical approach in which we calculate the acceleration of an electron in a solid under the force of an applied electric field. The force $q\mathscr{E}$ on a particle may be classically expressed as the time rate of change of its momentum, p. Quantum mechanically $p = \hbar k$. Thus, the force is

$$F = q\mathscr{E} = \hbar \frac{dk}{dt}. \tag{A1.30}$$

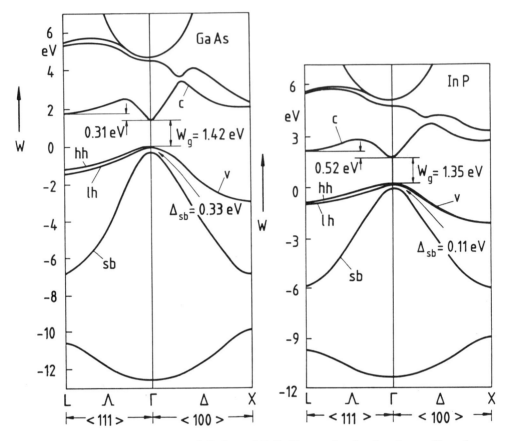

FIGURE A1.8 Band structure of GaAs and InP. The conduction band as well as the heavy-hole, light-hole, and split-off valence bands are labeled by, c, hh, lh, and sb, respectively ($W_g \equiv E_g$). (Reprinted, by permission, from K.J. Ebeling, *Integrated Opto-electronics*, Springer-Verlag, 1993).

The velocity of particles is defined by their group velocity, $v_g = d\omega/dk = (1/\hbar)\, dE/dk$, which shows the proportionality of velocity to the slope of the $E-k$ characteristic. Since the acceleration, *acc.*, is the time derivative of the velocity, we can write

$$acc. = \frac{dv_g}{dt} = \frac{dv_g}{dk}\frac{dk}{dt} = \frac{1}{\hbar}\frac{d^2 E}{dk^2}\frac{dk}{dt}. \qquad (A1.31)$$

Dividing Eq. (A1.30) by (A1.31), and defining an effective mass, $m^* = F/acc.$, we obtain

$$m^* = \frac{\hbar^2}{d^2 E/dk^2}. \qquad (A1.32)$$

Thus, for parabolic bands, as observed for uniform potentials, e.g., Eq. (A1.7), the electron will move much like a free particle, but with an effective mass, m^*, related to the curvature of the band. For nonparabolic bands, m^* is not constant and the local slope and curvature of the $E-k$ relationship must be used to obtain the velocity and acceleration of a particle with energy E.

A1.2.3   Density of States using a Free-Electron (Effective Mass) Theory

We just learned above that an electron in a crystal can behave much like a free electron moving in a region of uniform potential if it is at a point on the $E-k$ diagram that is parabolic. This is a remarkable result, since we know that the potential within a crystal is very nonuniform. Nevertheless, this revelation allows us to treat some very complex problems. For example, if we consider a crystal of finite dimensions, d_x, d_y, d_z, we can more or less ignore the crystal lattice potential which is periodic on the scale of the lattice constant a, provided that $d_j \gg a$ (Fig. A1.9). But, we must use a different effective mass as determined by the curvature of the $E-k$ diagram.

By considering electron energies near band extrema, where the $E-k$ curve tends to be parabolic, we can now consider reusing some of the same physics that we developed in Section A1.1.2 for electrons in very simple potential wells that had uniform potential regions. That is, we can now find the states in finite pieces of crystal or pieces with potential wells created by double heterostructures as described in Chapter 1. The simplest case is when the potential barriers are large so that we can assume that the bound wavefunctions go to zero at the boundaries. Then, from Eqs. (A1.13) and (A1.14), we have

$$E = \frac{\hbar^2 k^2}{2m^*} = \frac{\hbar^2}{2m^*}[k_x^2 + k_y^2 + k_z^2], \tag{A1.33}$$

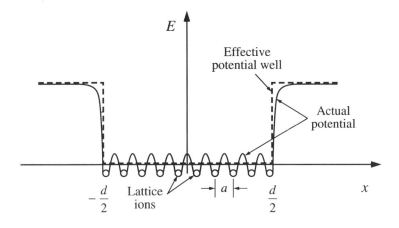

FIGURE A1.9   Potential plot for crystal (or quantum-well) of thickness d. The dashed well is an approximation to the actual potential.

where we have included all three dimensions for completeness and assumed that the effective mass is the same in all directions. Applying the boundary conditions for a large barrier, $k_j d_j = n_j \pi$,

$$E = \frac{\hbar^2 \pi^2}{2m^*} \left[\left(\frac{n_x}{d_x} \right)^2 + \left(\frac{n_y}{d_y} \right)^2 + \left(\frac{n_z}{d_z} \right)^2 \right]. \tag{A1.34}$$

From Eq. (A1.34) we note that we have an energy state for each (n_x, n_y, n_z) set of quantum numbers. To determine the size of the energy spacing between states, we can evaluate the coefficient, $\hbar^2 \pi^2 / 2m_0 = 376$ meV-nm^2, where we have used the free electron mass rather than the effective mass. In GaAs the electron mass, $m^* = 0.067 m_0$, so we should use $15(376) = 5640$ meV-nm^2 for the coefficient in the conduction band. From this we can see that the energy separation between states is quite small. For example, for a cube with $d_j = 1$ µm, the difference in energy between the first two states, $E(211) - E(111) \approx 17 \times 10^{-3}$ meV. Since $kT \approx 26$ meV at room temperature, we see that this energy difference is less than one thousandth of a kT. On the other hand, for a cube with dimensions $d_j \sim 10$ nm, this energy difference is ~ 170 meV, or more than $6kT$ at room temperature.

From the above, we conclude that for dimensions $d_j \gtrsim 1$ µm, quantum effects are not going to be very noticeable at room temperature, and the E–k diagram can be treated as describing a continuum of states. We shall refer to this as the *bulk* regime. On the other hand for $d_j < 100$ nm, the discreteness of the energy levels indicated in Eq. (A1.34) must be considered. We shall refer to this as the *quantum-confined* regime.

Even though the states may be very closely spaced in the so-called bulk regime, we still need to be able to count them to determine the carrier density and the energy to which they would have to be filled for a given carrier density. In the smaller structures, we again need an effective method of counting states. The method commonly used is to define a *density of states*, ρ, which when integrated over some range gives the number of states in that range. The density of states can be expressed in terms of a number of variables (e.g., E, p, or k) in a number of different coordinate systems. If N_s is the number of states up to some point, we can generally state that

$$N_s(u) = V \int_0^u \rho(u) \, du, \tag{A1.35}$$

where u is the desired variable and V is the volume. Once we have this definition, we can then state that

$$\rho(u) \, du = \frac{1}{V} dN_s(u). \tag{A1.36}$$

It should be realized that $\rho(u)\, du$ can be defined and used regardless of the size regime in which we find ourselves. For the smaller structures, we find that it contains discontinuities and even impulse functions, but it is still a good function.

In order to determine $\rho(u)\, du$ for the various cases of interest, we follow a standard procedure: (1) determine the number of states by calculating the volume in state or n-space, $N_s(n)$; (2) substitute for the desired variable, $n = f(u)$, which gives $N_s(u)$; and (3) apply Eq. (A1.36) to get the desired $\rho(u)\, du$. A few examples are useful for future reference.

For the first example, we consider *bulk* dimensions, a *spherical* coordinate system, and *energy* as the variable. Spherical coordinates imply that we are considering a spherical state space. Equation (A1.34) is written in rectangular coordinates, but as stated after it, each set of quantum numbers, or each volume element in n-space, represents a state that can be occupied by an electron. Figure A1.10 illustrates this n-space. The first step is to calculate the volume $N_s(n)$

$$N_s(n) = \tfrac{4}{3}\pi n^3 \cdot 2 \cdot \tfrac{1}{8}. \tag{A1.37}$$

The first factor just gives the standard expression for volume. However, we must multiply by 2, since two states actually exist at each allowed energy because of spin degeneracy. And only positive quantum numbers are allowed, so we have the factor of $\tfrac{1}{8}$. Now, for the second step, we use Eq. (A1.34) in spherical coordinates, (identical to Eq. (A1.14)), solve for n in terms of E, and plug back into Eq. (A1.37):

$$N_s(E) = \frac{\pi}{3}\left(\frac{2m^*Ed^2}{\hbar^2\pi^2}\right)^{3/2}. \tag{A1.38}$$

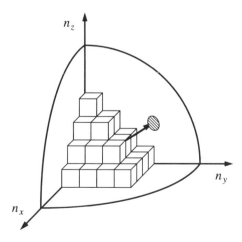

FIGURE A1.10   State space or n-space in spherical coordinates. Each block corresponds to a particular state and has unit dimensions.

For the third step, we now apply Eq. (A1.36), and use $V = d^3$,

$$\rho(E)\, dE = \frac{1}{2\pi^2}\left[\frac{2m^*}{\hbar^2}\right]^{3/2} E^{1/2}\, dE. \tag{A1.39}$$

Equation (A1.39) is our final result. It will be of much use in calculating carrier densities, gain, and other quantities associated with bulk active regions in lasers.

For the second example, we consider *bulk* dimensions, *rectangular* coordinates, and *momentum* as the variable. Then, the volume in n-space is, $N_s(n) = 2n_x n_y n_z$. The momentum in each direction, j, is

$$p_j = \hbar k_j = \frac{\hbar\pi n_j}{d_j} = \frac{hn_j}{2d_j}. \tag{A1.40}$$

Solving for n_x, n_y, and n_z, and plugging into $N_s(n)$, we get

$$N_s(p_x, p_y, p_z) = 2\left(\frac{2}{h}\right)^3 p_x p_y p_z(d_x d_y d_z). \tag{A1.41}$$

Applying Eq. (A1.36) to (A1.41) gives the desired density of states,

$$\rho(p_x, p_y, p_z)\, dp_x\, dp_y\, dp_z = 2\left(\frac{2}{h}\right)^3 dp_x\, dp_y\, dp_z. \tag{A1.42}$$

The density of all states with a given momentum, $\rho(p)$, can be obtained from this result by setting $dp_x\, dp_y\, dp_z = 4\pi p^2\, dp/8$. The factor of 8 is required because (A1.42) defines the density of *standing wave* states, $\rho_{SW}(p_x, p_y, p_z)$, which do not distinguish between positive and negative values of momentum. Hence, the density is limited to the first quadrant. We can also define a density of *plane wave* states, $\rho_{PW}(p_x, p_y, p_z)$, which can travel in any direction. Using periodic boundary conditions, we have $k_j = 2\pi n_j/d_j$ instead of $k_j = \pi n_j/d_j$, but now we consider both positive and negative values of n_j as unique states. In three dimensions then, $\rho_{PW}(p_x, p_y, p_z) = \rho_{SW}(p_x, p_y, p_z)/2^3$, and is distributed over all quadrants of momentum space. So for $\rho_{PW}(p_x, p_y, p_z)$, we can set $dp_x\, dp_y\, dp_z = 4\pi p^2\, dp$. In either case, we obtain $\rho_{PW}(p) = \rho_{SW}(p) = 8\pi p^2/h^3$, from which we also obtain $\rho(k) = (k/\pi)^2$ (since $\rho(k)dk = \rho(p)\, dp$).

For the third example, we consider a *quantum well* (small dimension in one direction), *cylindrical* coordinates, and *energy* as the variable. We shall let d_x be the small dimension. As always the energies are given by Eq. (A1.34), but we need to develop a density of states for the y–z plane, which will be summed for each n_x. Figure A1.11(a) gives a plot of the energy, E, relative to the k_x–k_z

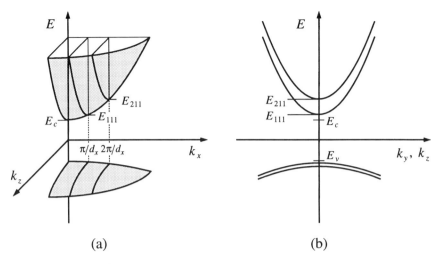

FIGURE A1.11   (a) Three-dimensional $E-k$ plot showing discrete jumps in k_x due to small d_x. (b) Projection perpendicular to the k_z axis.

plane for a quantum-well region using Eq. (A1.34). The lowest-lying states for $n_x = 1$ and 2 are labeled by the quantum numbers (n_x, n_y, n_z). Since d_x is small, there are no states near $k = 0$. Figure A1.11(b) replots the energy vs. k_z (or k_y) in a two-dimensional graph for clarity. In part (a) only positive k_x and k_z are shown.

Now to determine the density of states for this quantum well, we start with the disk of Fig. A1.12 and determine $N_s(n)$. The volume of the unit height disk in the first quadrant, multiplied by 2 for spin, is

$$N_s(n_{yz}) = \frac{\pi}{2}(n_y^2 + n_z^2).$$

(A1.43)

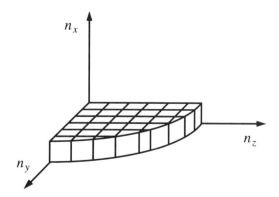

FIGURE A1.12   Two-dimensional state space which occurs for each n_x in a quantum well.

Using the y and z terms from Eq. (A1.34) for E_{yz} and $(n_y^2 + n_z^2)$, letting $d^2 = (d_y^2 + d_z^2)$, and assuming that the effective masses are the same in y and z,

$$N_s(E_{yz}) = \frac{m^*d^2}{\pi\hbar^2} E_{yz}. \qquad (A1.44)$$

Again, we apply Eq. (A1.36) and recognize that this density of states is for $n_x = 1$. Thus, generally

$$\rho(E) = \frac{1}{d_x} \sum_{n_x} \frac{m^*}{\pi\hbar^2} \mathscr{H}(E - E_{n_x}), \qquad \text{(for QW)} \qquad (A1.45)$$

where n_x and E are related by Eq. (A1.14) with $l = d_x$, and $\mathscr{H}(E - E_{n_x})$ is the Heaviside unit step function.

Figure A1.13 compares the densities of states for the bulk and quantum-well active regions, Eqs. (A1.39) and (A1.45) respectively. As can be seen the bulk curve forms an envelope for the steps of the quantum-well case, which correspond to the energies where the quantum numbers are $(n_x, 1, 1)$.

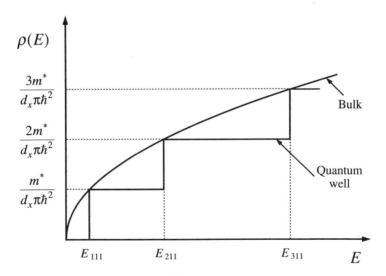

FIGURE A1.13 Density of states for an infinite-barrier quantum well and bulk material. If the barrier is not infinite, the quantum-well energies decrease slightly. If desired, the density of state plateaus can be decreased by using an effective $d_x = d_x^*$ (a different one for each state) so that the extrema continue to intersect the bulk characteristic.

REFERENCES

[1]   R.P. Feynman, R.B. Leighton, and M.L. Sands, *The Feynman Lectures on Physics*, Addison-Wesley, New York (1964).

READING LIST

L. Solymar and D. Walsh, *Lectures on the Electrical Properties of Materials*, Chs. 3–8, Oxford University Press, New York (1984).

C. Kittel, *Introduction to Solid State Physics*, Chs. 7 and 8, Wiley, New York (1986).

H. Kroemer, *Quantum Mechanics*, Ch. 7, Prentice Hall, Englewood Cliffs, NJ (1994).

Relationships between Fermi Energy and Carrier Density and Leakage

A2.1 GENERAL RELATIONSHIPS

In Appendix 1 the groundwork of energy bands and densities of states within these bands was outlined. Here we attempt to understand how carriers fill these states in semiconductors. This will allow us to relate the carrier density to the ranges of energies that must be occupied. They key missing element is the state occupation probability, $f(E)$, at energy E. For a large density of states, this function is equivalent to the fraction of states occupied at energy E. For particles in solids the relevant function is the Fermi–Dirac distribution,

$$f(E) = \frac{1}{e^{(E - E_F)/kT} + 1}, \tag{A2.1}$$

where the energy E_F is called the Fermi level. This probability distribution is one-half at $E = E_F$, and closely approaches unity for E a few kT below E_F, and zero for E a few kT above E_F. In fact, for E a few kT on either side of E_F, the asymptotic approach to these limits is exponential. These regions are sometimes referred to as the Boltzmann tails, since they imitate the classical Maxwell–Boltzmann distribution. At $T = 0$ K, $f(E)$ is essentially a step function, stepping from unity to zero at $E = E_F$.

Now, we can write the carrier density, N, as the integral over energy of the density of filled states in the conduction band,

$$N = \int \rho_c(E) f(E) \, dE, \tag{A2.2}$$

and similarly, we can write the hole density, P, as the integral of the density of unfilled states in the valence band,

$$P = \int \rho_v(E)[1 - f(E)] \, dE, \qquad (A2.3)$$

where $\rho_c(E)$ and $\rho_v(E)$ are the density of states (per unit volume per unit energy) in the conduction and valence bands, respectively. Figure A2.1 gives schematics of the density of filled and unfilled states vs. electron energy in the conduction and valence bands, respectively, for a bulk semiconductor under nonequilibrium conditions at $T = 0\,\text{K}$ and $300\,\text{K}$. Figure A2.2 gives analogous plots for a quantum well.

In both cases we have only shown the heavy-hole band in the valence band. In reality, the light-hole band also is a significant part of $\rho_v(E)$, and some of the holes occupy states there. The light-hole band is a little less important in quantum wells because of the separation of the light-hole band from the heavy-hole band at the finite k_x where the lowest energy is found.

Equations (A2.2) and (A2.3) give a relationship between the carrier densities and the Fermi level, E_F, for a known density of states. As a very simple example consider a quantum well at $T = 0\,\text{K}$ with an electron density of N. In this case

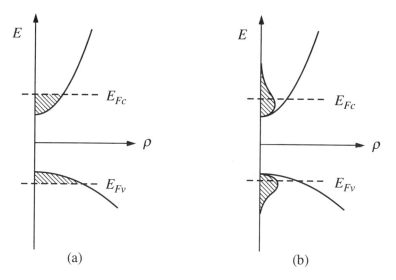

(a) (b)

FIGURE A2.1 Density of filled states in the conduction and valence bands of a bulk semiconductor under nonequilibrium conditions at (a) $T = 0\,\text{K}$ and (b) $T = 300\,\text{K}$. Under nonequilibrium conditions, such as when a current is flowing, separate Fermi functions and quasi-Fermi levels, E_{Fc} and E_{Fv}, are used for the electrons and holes, respectively.

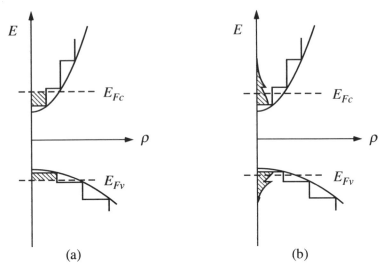

FIGURE A2.2   Density of filled states in the conduction and valence bands of a quantum well under nonequilibrium conditions at (a) $T = 0$ K and (b) $T = 300$ K.

the step-like Fermi function merely sets the limits of integration from the first allowed state, E_{111} to E_F as shown in Fig. A2.2(a). From Eqs. (A1.45) and (A2.2), and assuming we do not approach $n_x = 2$, we have

$$N = \frac{m^*}{\pi\hbar^2 d_x}(E_F - E_{111}).$$ (A2.4)

Thus, we can solve for E_F as

$$E_F = E_{111} + \frac{\pi\hbar^2 d_x}{m^*}N.$$ (QW; $n_x = 1$; $T = 0$ K)   (A2.5)

At finite temperatures, the Fermi function is no longer a simple step function, and we must use the general form (A2.1) inside the integral. This makes the integration more complex, but using a quantum-well density of states function, the integral still has a closed-form solution. Repeating the above calculation at finite temperature, we have

$$N = \frac{1}{d_x}\sum_{n_x}\frac{m^*}{\pi\hbar^2}\int_{E_{n_x 11}}^{\infty}\frac{dE}{e^{(E - E_F)/kT} + 1},$$ (A2.6)

where $\mathscr{H}(E - E_{n_x})$ in Eq. (A1.45) has shifted the lower integration limit. Making the substitution $u = \exp[-(E - E_F)/kT]$, allows the integration to be readily performed. Then, for the situation of Fig. A2.2(b),

$$N = \frac{kTm^*}{\pi\hbar^2 d_x} \sum_{n_x} \ln[1 + e^{(E_F - E_{n_x11})/kT}]. \qquad \text{(QW)} \qquad \text{(A2.7)}$$

For $n_x = 1$ and $(E_F - E_{111}) \gg kT$, note that Eq. (A2.7) reduces to Eq. (A2.4).

A2.2 APPROXIMATIONS FOR BULK MATERIALS

Unfortunately, the finite-temperature calculation for carrier density is not as simple using a bulk-like density of states function, as in Fig. A2.1(b). The general form for the carrier density in bulk material is given by

$$N = \frac{1}{2\pi^2} \left(\frac{2m^*}{\hbar^2}\right)^{3/2} \int_{E_c}^{\infty} \frac{\sqrt{E - E_c}}{1 + e^{(E - E_F)/kT}} \, dE, \qquad \text{(A2.8)}$$

where E_c is the conduction band edge. Defining $v = (E_F - E_c)/kT$ and $y = (E - E_c)/kT$, we can write the integral in a more normalized form:

$$N = N_c \frac{2}{\sqrt{\pi}} \int_0^{\infty} \frac{\sqrt{y}}{1 + e^{y-v}} \, dy \equiv N_c \frac{2}{\sqrt{\pi}} F_{1/2}(v), \qquad \text{(A2.9)}$$

$$N_c \equiv 2\left(\frac{m^*kT}{2\pi\hbar^2}\right)^{3/2} = 2.51 \times 10^{19} \times \left[\frac{m^*}{m_0}\right]^{3/2} \times \left[\frac{T}{300\ K}\right]^{3/2} \text{cm}^{-3}, \quad \text{(A2.10)}$$

where $F_{1/2}(v)$ is known as the Fermi–Dirac integral of order $1/2$ (referring to the $y^{1/2}$ in the numerator), and N_c is the "effective" density of states in the conduction band. For $v \ll 0$ (in other words, for $E_F \ll E_c$), the 1 in the denominator can be neglected reducing the Fermi function to the *Boltzmann* exponential limit, allowing the integral to be evaluated in closed-form. The result reduces to

$$N \approx N_c e^{(E_F - E_c)/kT}. \qquad \text{(bulk, Boltzmann limit)} \qquad \text{(A2.11)}$$

For larger v, other approximation techniques must be used. Tables A2.1 and A2.2 summarize various approximations to the Fermi–Dirac integral (and their range of validity) which have been formulated over the years (see Blakemore [1] for an overview). The first table finds N in terms of v, while the second table finds v in terms of N.

The error associated with the various approximations listed in Table A2.1 are plotted in Fig. A2.3. The Boltzmann limit discussed above is valid for large negative v, but quickly begins to overestimate the carrier density as v approaches zero (that is, as E_F approaches the band edge). At the other extreme, the

TABLE A2.1   Summary of Approximations for Carrier Density $N = N_c \dfrac{2}{\sqrt{\pi}} F_{1/2}(v)$.

Formal Name	Mathematical Expression	Range Over Which Fractional Error* $< 1\%$	
Fermi–Dirac Integral	$F_{1/2}(v) = \displaystyle\int_0^\infty \frac{\sqrt{y}}{1+e^{y-v}}\, dy$ $v \equiv (E_F - E_c)/kT$ $y \equiv (E - E_c)/kT$	exact definition	
Boltzmann approximation	$F_{1/2}(v) \approx \dfrac{\sqrt{\pi}}{2} e^v$	$v < -3.5$	
Unger approximation	$F_{1/2}(v) \approx \dfrac{\sqrt{\pi}}{2} z[a_1 + a_2 z + \cdots]$ $a_1 = 1$ $a_2 = \dfrac{1}{2}\left(1 - \dfrac{1}{\sqrt{2}}\right) \approx 0.146\,45$ $z \equiv \ln(1 + e^v)$	1st order ($a_2 = 0$) 2nd order ($a_2 = 0.146\,45$) 2nd order ($a_2 = 0.15$)	$v < -2.7$ $v < 1.7$ $v < 7.6$
Sommerfeld approximation	$F_{1/2}(v) \approx \tfrac{2}{3} v^{3/2}[a_1 + a_2 v^{-2} + \cdots]$ $a_1 = 1$ $a_2 = \dfrac{\pi^2}{8} \approx 1.233\,7$	1st order (0) 2nd order (1.233 7) 2nd order (1.3)	$v > 11$ $v > 2.9$ $v > 1.7$
Modified Sommerfeld approximation	$F_{1/2}(v) \approx \tfrac{2}{3} z^{3/2}\left[1 + \dfrac{\pi^2}{8} z^{-2} + \cdots\right]$ $z \equiv \ln(1 + e^v)$	1st order (0) 2nd order (1.233 7)	$v > 11$ $v > 3$
Global approximation (one of many)	$F_{1/2}(v) \approx [(1.3 + 0.3x)F_L^p$ $\quad\quad + (0.76 + 0.24x)F_H^p]^{1/p}$ $F_L = \dfrac{\sqrt{\pi}}{2} z$ $F_H = \tfrac{2}{3} z^{3/2}$ $p = 3 + x - 0.19 e^{-(v-3.25)^2/8}$ $x \equiv \tanh(0.8(v - 2.7))$ $z \equiv \ln(1 + e^v)$	error $< 0.06\%$ $-\infty < v < \infty$ note that $z \to e^v$ $x \to -1$ $\quad (v \ll 0)$ $z \to v$ $x \to 1$ $\quad (v \gg 0)$	

* error $\equiv |F_{1/2} - F_{\text{calc}}|/F_{1/2}$

TABLE A2.2   Summary of Approximations for Fermi Level $v = f(N/N_c)$.

Formal Name	Mathematical Expression	Range Over Which Abs Error* < 0.04†	
Inverse Fermi–Dirac integral	$v = F_{1/2}^{-1}[F_{1/2}(v)]$ $$= F_{1/2}^{-1}\left[\frac{\sqrt{\pi}}{2} N/N_c\right] \equiv f(r)$$ $v \equiv (E_F - E_c)/kT,\ r \equiv N/N_c$	exact definition	
Boltzmann approximation	$v \approx \ln r$	$v < -2.1$	
Joyce–Dixon approximation	$v \approx \ln r + A_1 r + A_2 r^2 + \cdots$ $A_1 = 1/\sqrt{8}$ $A_2 \approx -4.950\,09 \times 10^{-3}$ $A_3 \approx \ \ 1.483\,86 \times 10^{-4}$ $A_4 \approx -4.425\,63 \times 10^{-6}$	1st order $\quad v < 2.1$ 2nd order $\quad v < 4.2$ 3rd order $\quad v < 5.6$ 4th order $\quad v < 6.8$	
Padé approximation	$v \approx \ln r + A_1 r$ $\quad + [K_1 \ln(1 + K_2 r) - K_1 K_2 r]$ $K_1 \approx 4.7 \leftrightarrow 5.2$ $K_2 = \sqrt{2|A_2|/K_1}$ A_1, A_2 as defined above	$K_1 = 4.7$ error $< 0.0004\quad v < 9.8$ $K_1 = 4.9$ error $< 0.02\qquad v < 18.8$ $K_1 = 5.2$ error $< 0.1\qquad v < 28$	
Inverse second-order Unger approximation	$v \approx \ln\left\{\exp\left[\dfrac{1}{2a_2}(\sqrt{1 + 4a_2 r} - 1)\right] - 1\right\}$	$(a_2 = 0.146\,45)\quad v < 2.8$ $(a_2 = 0.15)\qquad v < 7.4$	
Inverse first-order Sommerfeld approximation	$v \approx \left(\dfrac{3\sqrt{\pi}}{4} r\right)^{2/3}$	$v > 20$	
Nilsson's global approximation	$v \approx \dfrac{\ln r}{1 - r^2} + \dfrac{f_H}{1 + (0.24 + 1.08 f_H)^{-2}}$ $f_H = \left(\dfrac{3\sqrt{\pi}}{4} r\right)^{2/3}$	error < 0.006 $-\infty < v < \infty$	

* error $\equiv |v - v_{\text{calc}}|$, † $\Delta E_F \approx \pm 1$ meV @ RT.

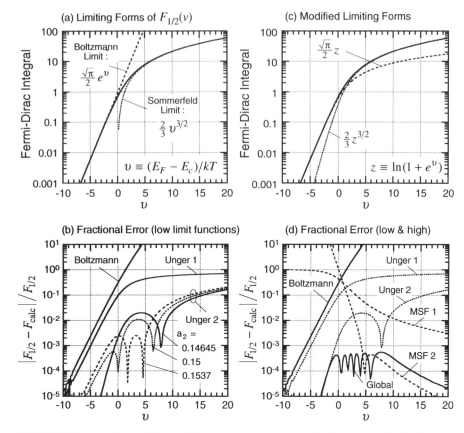

FIGURE A2.3 (a) Plot of Fermi–Dirac integral along with first-order limits for low and high v. (b) Fractional error of Boltzmann and first- and second-order Unger approximations. (c) Modified first-order limits using the z function substitution introduced by Unger. (d) Fractional error of Boltzmann, first- and second-order Unger, first- and second-order modified Sommerfeld (MSF), and global approximations. (There is little difference between the error in the Sommerfeld and modified Sommerfeld expansions.)

Sommerfeld expansion (which essentially determines the zero-temperature carrier density with correction terms added to account for the finite-temperature smoothing of the Fermi function) can be used to determine the Fermi–Dirac integral for large positive v. The first-order terms of the Boltzmann and Sommerfeld limits are plotted in (a). As can be observed, the two limits fail to cover the range $-2 < v < 5$. Unger [2] introduced a function z which led him to a useful expansion of the Fermi–Dirac integral for small v. The fractional error of the Unger expansion is plotted in (b). The multiple curves are plotted using the second-order expansion coefficient as a curve-fitting parameter. Increasing a_2 from its true value of ~ 0.14645 to 0.15 to 0.1537 reveals the trend toward reducing the initial hump in the curve at the expense of increasing the error elsewhere. Note that with $a_2 = 0.15$, the Unger expansion provides an error less than $\sim 1\%$ for v as large as 7 (actually the initial jump dips below 1% when $a_2 = 0.151$, but there is something to be said for round numbers). What this means is that E_F can penetrate $7kT$ into the conduction band and the Unger approximation will still provide an accurate estimate of the carrier density. Higher-order terms in the Unger expansion can also be used but they do not significantly increase or extend the accuracy of the approximation.

For some applications it is desirable to know the Fermi function for values of v even larger than the Unger approximation is capable of handling. For these situations, one could switch to the second-order Sommerfeld expansion, but it would be advantageous to have a single expression to cover the entire range of v. One could think of constructing a global approximation by bridging the two limiting approximations. However, the Sommerfeld expansion is unfortunately invalid for $v \leq 0$. For the purpose of constructing a global approximation, we can substitute z for the two limiting forms given in (a). The resulting expressions are plotted in (c). Note that the "modified" limiting forms are better behaved than the limiting forms in (a). The global approximation given in Table A2.1 bridges the gap between these "modified" limits [3]. The overall expression for the global approximation is a bit complicated, but it does get the job done—it approximates the Fermi–Dirac integral to within 0.06% over the entire range of v, as shown in (d), with excellent convergence on either side of the "trouble area."

Table A2.1 found N in terms of v. However, oftentimes we know the carrier density, and would like to determine the Fermi level. This requires calculating the inverse of the Fermi–Dirac integral. Various approximations used to achieve this are summarized in Table A2.2. Of these, the Joyce–Dixon approximation [4] is perhaps the most well known. Its popularity stems from the fact that it simply adds correction terms to the Boltzmann approximation, which is useful for the analytical description of semiconductor devices. The Unger approximation appears in this list as well. In fact, one very attractive feature of the second-order Unger approximation is that it is invertible [5]! Figure A2.4 plots the various approximations for comparison. In (a) we see that the even-order Joyce–Dixon series expansions diverge rapidly for $v > 10$, while the odd-order terms remain well-behaved. The Unger expansion actually tracks the exact solution better than any of the Joyce–Dixon expansions. However, (b) reveals that the absolute error for the second-order Unger expansion is larger than the higher-order Joyce–Dixon expansions for small v. To extend the range, Nilsson [6] constructed a global inverse approximation by attempting to bridge the inverted Boltzmann and Sommerfeld expansions. The error in his global approximation is shown in (c). While the absolute error remains low, the convergence on either side of the "trouble area" is not particularly strong.

The Padé approximation [7] shown in (d) is a variant of the Joyce–Dixon approximation which avoids the highly divergent properties associated with the even-order Joyce–Dixon series. The Padé approximation is by far the most

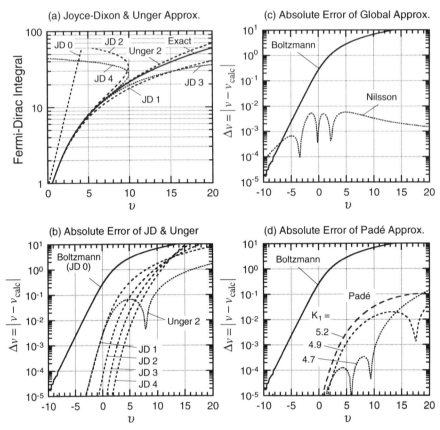

FIGURE A2.4   (a) Plot of Fermi–Dirac integral along with various orders of Joyce–Dixon (JD) approximations and the second-order Unger approximation. (b) Comparison of the absolute error of these approximations. (c) Absolute error of Nilsson's global approximation. (d) Absolute error of the Padé approximation for different values of the adjustable parameter, K_1.

accurate approximation for $v < 10$ with K_1 set to 4.7. One nice feature of this approximation is that it is adjustable. For example, to extend the range one can sacrifice accuracy slightly by increasing K_1 up to a maximum of 5.2 (beyond which the error becomes very large).

At this point, we can see that the number of approximations available for the Fermi–Dirac integral allow us to determine the relationship between carrier density and Fermi level in bulk material to a very good degree of accuracy. In fact, it could be argued that the quantum-well Fermi level is more difficult to calculate than the bulk Fermi level because the summation in the expression for quantum-well carrier density (Eq. (A2.7)) prevents one from inverting the equation.

A2.3   CARRIER LEAKAGE OVER HETEROBARRIERS

In double heterostructures the band diagrams shown in Figs. A2.1 and A2.2 for the bulk and quantum-well active layers are not valid for energies greater than the band offsets between the active and cladding materials. That is, once carriers fill to the top of the barrier they are free to diffuse into the cladding regions. At finite temperatures shown in parts (b) of these figures, the carriers in the high-energy tails would ideally extend to energies above the barriers. Thus, in practice there is a "carrier leakage" out of the active region which results in a leakage current as the carriers diffuse away. With increased temperatures the fraction of carriers in the high-energy tail as well as the carrier leakage current will increase.

Figure A2.5 is a plot of the carrier distribution for an electron density of 2×10^{18} cm^{-3} in a GaAs active region clad by AlGaAs for which the net effective barrier height is 0.3 eV. The horizontal axis is the electron spectral density $\rho_c(E)f(E)$. To calculate the density of electrons above the barrier, N_{lk}, indicated by the shaded region, we integrate the electron spectral density from the top of the effective barrier (which also includes any residual band-bending effects). Thus, as in Eq. (A2.2), but beginning the integration at E_B rather than E_c, we obtain

$$N_{lk} = \int_{E_B}^{\infty} \rho_c(E)f(E)\, dE, \qquad (A2.12)$$

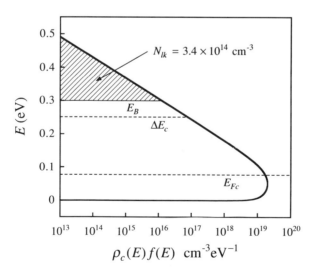

FIGURE A2.5   Room temperature electron energy vs. the electron spectral density in the conduction band of GaAs with a total integrated density of $N = 2 \times 10^{18}$ cm^{-3}. The leakage density, N_{lk}, above the 0.3 eV AlGaAs barrier is shaded [8]. The total effective barrier height ($E_B - E_c$) includes residual band bending under forward bias as well as the conduction band discontinuity, ΔE_c.

or for a bulk-like density of states,

$$N_{lk} = \frac{1}{2\pi^2} \left(\frac{2m^*}{\hbar^2}\right)^{3/2} \int_{E_B}^{\infty} \frac{(E - E_c)^{1/2}}{1 + \exp[(E - E_{Fc})/(kT)]} \, dE. \quad (A2.13)$$

For the assumed barrier height of $E_B = 0.3$ eV, and for a Fermi level to yield a total density $N = 2 \times 10^{18}$ cm^{-3} carriers from Eq. (A2.2), one finds that $N_{lk} = 3.4 \times 10^{14}$ cm^{-3} at room temperature as shown in Fig. A2.5.

Some of these high-energy electrons will escape the active region and diffuse into the p-cladding material as illustrated in Fig. A2.6. The fraction that actually do however involves a number of factors. For example, some electrons approaching the cladding barrier will be reflected even though they have sufficient energy to pass over the barrier (it is much like photons being reflected off of a dielectric interface). For a simple potential step the reflection coefficient is not significant, dropping below 50% for electrons with energies just 3% greater than the barrier. However, specially designed cladding layers can reflect a much higher portion of the high-energy electrons.

Another factor to consider is that in a thermal distribution, the electron velocities are randomly distributed in all directions (with just a slight preference toward the direction of the current flow). Therefore only a fraction of the high energy electrons are even moving toward the p-cladding barrier. Finally, some electrons which make it into the p-cladding layer will diffuse right back into the active region. Thus, we must consider the flow of electrons in both directions. Thermionic emission theory can be used to estimate the maximum supply rate of electrons to the p-cladding layer by calculating the average thermal velocity of high-energy electrons directed toward the barrier. However, the tricky part is determining how many might be flowing back.

An alternative approach to studying the detailed flow of electrons across the barrier is to simply assume that the electron populations on both sides of the barrier are in thermal equilibrium such that we can match the quasi-Fermi levels across the barrier. If we assume this to be the case, then in the Boltzmann

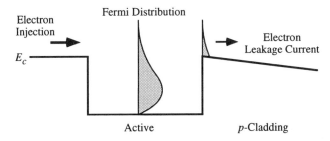

FIGURE A2.6 Schematic band diagram of a double heterostructure showing electron energy distribution in the active layer and leakage current.

limit, the electron population at the edge of the p-cladding layer is given by

$$N_{p0} = N_c e^{(E_{Fc} - E_B)/kT}. \quad \text{(assuming } E_{Fc}|_{act} = E_{Fc}|_{clad}) \quad \text{(A2.14)}$$

This estimate of the actual leakage carrier density which establishes itself in the p-cladding layer is typically 20–30% of the value suggested by Eq. (A2.13). It also reveals the strong exponential dependence of the leakage carrier density, and eventually, the carrier leakage current on temperature.

Once we know the electron density at the edge of the p-cladding layer, we can estimate the electron leakage current by assuming that the electrons diffuse away transversely into the p-cladding material as minority carriers with an assumed initial density N_{p0} and a diffusion length L_n. Taking the $x = 0$ origin to be at the barrier on the p-side (opposite side from which they were injected), the distribution of electrons in the p-type cladding layer can then be written as

$$N(x) = N_{p0} e^{-x/L_n}, \quad \text{(A2.15)}$$

where we can express $L_n = \sqrt{D_n \tau_n}$, with τ_n being the minority carrier lifetime and D_n being the diffusion constant in the p-cladding material. From the Einstein relationship, $kT/q = D/\mu$, we can use the measured mobility to estimate the diffusion constant.

Assuming that any residual electric field would have a negligible effect, and also assuming a sufficiently thick p-layer, the associated diffusion current of the leakage electrons (neglecting the sign) is given by

$$J_n|_{x=0} = qD_n \frac{dN}{dx} = qD_n \frac{N_{p0}}{L_n} = qL_n \frac{N_{p0}}{\tau_n}. \quad \text{(A2.16)}$$

The last form shows that the diffusion current is equivalent to the recombination of a uniform carrier density spread over one diffusion length. For the example of Fig. A2.5 with $N = 2 \times 10^{18}$ cm^{-3} and $E_{Fc} - E_c = 79$ meV, we find that a 0.3 eV barrier yields $N_{p0} = 8.5 \times 10^{13}$ cm^{-3} from Eq. (A2.14) using the N_c of GaAs for simplicity (for AlGaAs, N_c is slightly larger and might need to include other conduction band minima depending on the x value). For $L_n = 5 \ \mu$m and $\tau_n = 5$ ns, we obtain $J_n \sim 1.4$ A/cm^2 at room temperature. For holes a similar leakage current will exist. For $P = 2 \times 10^{18}$, we have $E_v - E_{Fv} = -28$ meV. Assuming a net barrier of 0.25 eV and using $N_{v,hh} + N_{v,lh}$ of GaAs for simplicity, we calculate $P_{n0} = 1.4 \times 10^{14}$ cm^{-3}. For $L_p = 1 \ \mu$m and $\tau_p = 5$ ns, we find that the hole leakage current at the other barrier is $J_p \sim 0.5$ A/cm^2.

Since typical laser threshold current densities are at least a few hundred A/cm^2, we conclude that for this high barrier GaAs/AlGaAs example, carrier leakage is not significant at room temperature. However, in material systems which do not have the luxury of large heterobarriers (for example, 630–680 nm emission AlInGaP), carrier leakage can be much more of a problem in view of the exponential dependence on barrier height through N_{p0} in Eq. (A2.14). Carrier leakage also increases rapidly with temperature, such that even lasers with relatively high barriers can be affected at high temperatures. Finally, if the cladding material has a high defect density, the reduced minority carrier lifetime can lead to high carrier leakage currents even for a small N_{p0}.

If the resistance in the p-cladding is substantial, an electric field will assist electron diffusion away from the active region, enhancing the leakage rate. If we include this as well as the existence of a contact (which is assumed to be a region of zero lifetime) a distance x_p away, it can be shown that a more general expression for the electron leakage current results [9]:

$$J_n = qD_n N_{p0}\left[\sqrt{\frac{1}{L_n^2} + \frac{1}{L_{nf}^2}}\, \coth\sqrt{\frac{1}{L_n^2} + \frac{1}{L_{nf}^2}}\, x_p + \frac{1}{L_{nf}}\right], \qquad (A2.17)$$

where

$$L_{nf} \equiv \frac{2kT}{q} \cdot \frac{\sigma_p}{J_{tot}}.$$

σ_p is the conductivity of the p-cladding region, and J_{tot} is the total diode current density. In analogy with the diffusion length, we can think of L_{nf} as the *drift* length, which decreases with increasing total current (as the electric field in the cladding increases). For low currents and high p-type conductivity, $L_{nf} \gg L_n$, and the carrier leakage current is dominated by the diffusion component considered above. However, for high currents and/or low p-type conductivity in the cladding, it is possible for $L_{nf} \ll L_n$, in which case the carrier leakage current becomes dominated by the drift component, which increases with the total current in addition to increasing with N_{p0}. More specifically, if $L_{nf} \ll L_n$, x_p, then $J_n \to q\mu_n N_{p0} J_{tot}/\sigma_p$. The double dependence on injection level leads to a higher sensitivity to temperature which can be severe in lasers with low heterobarriers [9]. If the contact is placed close to the active region such that $x_p \ll L_n$, L_{nf}, then the current reduces to $J_n \to qD_n N_{p0}/x_p$, becoming independent of both drift and diffusion lengths. Finally, it should be realized that for either $x_p \to 0$ or $L_{nf} \to 0$, the carrier leakage current does not really go to infinity but becomes limited by the rate at which carriers can be supplied to the cladding region (it becomes thermionic emission-limited rather than drift-diffusion-limited).

So far we have been considering a simple double heterostructure laser for which the active region fills the entire waveguide. In quantum-well lasers, the active region is restricted to the width of the well(s). Thus, carrier leakage has two meanings in this case. Carriers can leak into the separate confinement waveguiding layers as well as leaking out of the entire SCH waveguide region into the doped cladding layers. Carrier populations in the SCH regions lead to recombination that can be approximated using Eq. (A2.16) with the diffusion length replaced by the width of the SCH region, or

$$J_{SCH} = qL_{SCH} \frac{N_{SCH}}{\tau_n}. \tag{A2.18}$$

For $L_{SCH} = 1500$ Å, and $\tau_n = 5$ ns, we find $J_{SCH} \sim 50$ A/cm^2 per 10^{17} cm^{-3} of carrier density. This highlights the importance of maintaining low carrier populations in the waveguide regions of the laser. Of course, depending on the material quality of the SCH regions, the carrier lifetime may be longer or shorter affecting the carrier leakage current accordingly.

In Chapter 2, we introduce an effective recombination rate per unit active volume for carrier leakage, R_l. In terms of the total carrier leakage current, $J_l = J_n + J_p + J_{SCH}$, we can define

$$R_l = \frac{J_l}{qd}, \tag{A2.19}$$

where d is the active region thickness. If lateral carrier leakage is important then it shoud also be added to R_l with the lateral active width replacing d in Eq. (A2.19). For example, in ridge lasers carriers are free to diffuse laterally since no heterobarrier exists. This lateral diffusion component of R_l is discussed in Chapter 4.

A2.4   INTERNAL QUANTUM EFFICIENCY

We would finally like to consider how carrier leakage affects the internal quantum efficiency, η_i, introduced in Chapter 2. In practice, η_i is defined as the fraction of current above threshold which results in stimulated emission, or $\eta_i(I - I_{th}) = I_{st}$. The output power is then $P_0 = (h\nu/q) \cdot \eta_0 I_{st}$, where η_0 is the *optical* efficiency of the laser cavity introduced in Chapter 5. This gives $P_0 = (h\nu/q) \cdot \eta_i \eta_0 (I - I_{th})$. The slope of the PI curve is related to $\eta_i \eta_0$, from which η_i can be extracted experimentally. This *above-threshold* definition of η_i is slightly different than the one given in Chapter 2; however, the two definitions are the same if η_i is not a function of current beyond threshold.

If we specify the total current as the stimulated current plus the sum of n various other components (i.e. spontaneous, Auger, leakage, etc.), such that $I = I_{st} + \sum I_n$ and $I_{th} = \sum I_{n,th}$, then we can expand $1 - \eta_i$ and rearrange to obtain

$$\eta_i = \frac{I_{st}}{I - I_{th}} = 1 - \frac{\sum(I_n - I_{n,th})}{I - I_{th}}. \tag{A2.20}$$

In this form we clearly see that any currents which clamp at threshold along with the carrier density do not contribute to a reduction of η_i. This point is often misunderstood in the laser community. The common mistake made is to set $\eta_i = I_{rad}/I$, where I_{rad} is the radiative current in the active region. This definition originates from the concept of radiative efficiency in an LED and has nothing to do with the laser performance above threshold. To demonstrate this point, consider a laser in which 90% of the current at threshold is Auger current, while the rest is spontaneous emission current in the active region (there is no leakage current). Because both Auger and spontaneous currents clamp at threshold (i.e. $I_n = I_{n,th}$), Eq. (A2.20) reveals that we can still have $\eta_i = 100\%$ even though the radiative efficiency near threshold is only 10%. The true radiative efficiency in this case does not approach 100% until the stimulated emission current becomes much larger than threshold.

The main question is which currents *do* contribute to a reduction of η_i? The simple answer is those that continue to increase above threshold. Now because the carrier density, N, in the active region clamps at threshold, all currents which depend monotonically on N should have no effect on η_i, including R_{sp}, R_{nr}, as well as carrier leakage out of the active region, R_l (since N_{p0} defined in the last section should in principle clamp along with N). However, the clamping of the modal gain (which is what really clamps) does not always result in a complete clamping of the carrier density in all of the various regions of the laser.

For example, there is always some small compression of the gain with increased photon density (see Chapter 5) which requires additional carriers to restore the threshold modal gain. Also, spatial hole burning of the carrier density profile by the optical mode can result in even larger changes in the local carrier density. These second-order effects can allow R_{sp}, R_{nr}, and R_l to increase above their threshold values, reducing η_i according to Eq. (A2.20). However, these effects are typically not very significant.

The more important factor to consider is the incomplete clamping of carrier densities in the surrounding regions, N_{SCH} and N_{p0}, defined in Section A2.3. For example, one model of the SCH-quantum well pn-junction suggests that carriers in the well couple to a second carrier pool in the SCH region which is not clamped at threshold, but related to the well carriers through a set of capture lifetimes (see Chapter 5). The carrier density in the SCH region and corresponding current defined in Eq. (2.18) can therefore increase with injected current above threshold, while the gain remains fixed in the quantum well [10].

From another point of view, there is no guarantee that the quasi-Fermi levels in the SCH and cladding regions clamp along with the quasi-Fermi levels in the active region at threshold. As a result, recombination rates away from the active region can potentially reduce η_i, if significant carrier populations exist there and if the quasi-Fermi levels in those regions remain unclamped or only partially clamped above threshold. Unfortunately, realistic modeling of the extent of clamping throughout the lasing junction is a very complex undertaking. Thus, the extent to which such effects contribute to a reduction of η_i is hard to predict. Experimental values for η_i are commonly in the range of 70% to 80% but can often be as low as 40% or as high as 95%.

In Chapter 2, we specifically assume that the leakage rate of carriers out of the active region, R_l, depends monotonically on the carrier density in the active region. This is assumed mainly to highlight the fact that *in principle*, carrier leakage does not affect the internal quantum efficiency. In view of the above discussion, this may not always be the case since some carrier leakage currents may clamp or partially clamp and others may not. In any case, only the amount of leakage *beyond* threshold, $R_l - R_{l,th}$, reduces η_i. For leakage currents which partially clamp at threshold, this value will typically be much smaller than that suggested by the magnitude of R_l itself.

We can modify the derivation in Chapter 2 to include the above effects by first expanding $\eta_i(I - I_{th})$ in Eq. (2.31) into $(I - I_L) - (I_{th} - I_{L,th}) = (I - I_{th}) - (I_L - I_{L,th})$, where I_L is the current leakage which does not generate carriers in the active region (see Fig. 2.2). Then by adding recombination terms $R_l - R_{l,th}$ for carrier leakage and similarly for all other recombination rates, we can use Eq. (A2.20) to reduce the carrier density rate equation back to Eq. (2.31), with the new η_i encompassing all currents which do not clamp at threshold.

Finally, it is interesting to note that even if the carrier density clamps everywhere, the *drift* carrier leakage defined in Eq. (A2.17) will continue to increase above threshold due to its dependence on the total current density. Therefore, if drift carrier leakage is significant, it can lead to a noticeable reduction of η_i, even for a fixed N_{p0} above threshold.

REFERENCES

[1]   J.S. Blakemore, *Solid State Electron.*, **25**, 1067 (1982).

[2]   K. Unger, *Z. Physik*, **207**, 322 (1967).

[3]   S.W. Corzine, unpublished.

[4]   W.B. Joyce and R.W. Dixon, *Appl. Phys. Lett.*, **31**, 354 (1977).

[5]   K. Unger, *Phys. Stat. Solidi (b)*, **149**, K141 (1988).

[6]   N.G. Nilsson, *Phys. Stat. Solidi (a)*, **19**, K75 (1973).

[7]   V.C. Aguilera-Navarro, G.A. Estevez, and A. Kostecki, *J. Appl. Phys.*, **63**, 2848 (1988).

[8]   K.J. Ebeling, *Integrated Opto-electronics*, Chapter 9, Springer-Verlag, Berlin (1993).

[9]   D.P. Bour, D.W. Treat, R.L. Thornton, R.S. Geels, and D.F. Welsh, *IEEE J. Quantum Electron.*, **QE-29**, 1337 (1993).

[10]   H. Hirayama, J. Yoshida, Y. Miyake, and M. Asada, *IEEE J. Quantum Electron.*, **30**, 54 (1994).

READING LIST

G.P. Agrawal and N.K. Dutta, *Semiconductor Lasers*, 2d ed., Ch. 3, Van Nostrand Reinhold, New York (1993).

Introduction to Optical Waveguiding in Simple Double-Heterostructures

A3.1 INTRODUCTION

Starting from Maxwell's equations a wave equation (sometimes referred to as the Helmholtz equation), which is very analogous to Schrödinger's equation, can be derived:

$$\nabla^2 \mathscr{E} = \mu\varepsilon \frac{\partial^2 \mathscr{E}}{\partial t^2}, \tag{A3.1}$$

where ε is the dielectric constant and μ is the magnetic permeability. In this derivation it is assumed that ε is uniform in space, and both μ and ε can be complex, although for the semiconductor materials of interest, $\mu \approx \mu_0$. The imaginary part of ε, includes the gain or loss that can occur in these materials. We are searching for time-harmonic fields propagating in the z-direction, so we try

$$\mathscr{E}(x, y, z, t) = \hat{\mathbf{e}}_i E_0 U(x, y) e^{j(\omega t - \tilde{\beta} z)}, \tag{A3.2}$$

as a solution to Eq. (A3.1). The unit vector $\hat{\mathbf{e}}_i$ gives the polarization. E_0 has units of volts, $U(x, y)$ has units of per unit length, and it is usually assumed to be normalized such that $\int |U(x, y)|^2 \, dA = 1$. Plugging Eq. (A3.2) into (A3.1) and factoring out common terms, we find that the transverse amplitude function, $U(x, y)$, must satisfy

$$\nabla^2 U(x, y) + [\tilde{n}^2 k_0^2 - \tilde{\beta}^2] U(x, y) = 0, \tag{A3.3}$$

where the square of the index of refraction $\tilde{n}^2 = \varepsilon/\varepsilon_0$, the free-space propagation constant, $k_0^2 = \omega^2 \mu_0 \varepsilon_0 = (2\pi/\lambda)^2$, and both $\tilde{n}$ and the propagation constant in the z-direction, $\tilde{\beta}$, are complex. Expanding $\tilde{\beta} = \beta + j\beta_i$, we see from the form of Eq. (A3.2) that $\beta = \omega/v_p$, where v_p is the velocity of a phase front, or the phase velocity. We can also introduce an effective index of refraction, $\bar{n}$, so that $\beta = k_0 \bar{n} = \omega \bar{n}/c = 2\pi \bar{n}/\lambda$. In Chapter 2 we model how β_i contains the gain and internal loss terms in a laser.

A3.2 THREE-LAYER SLAB DIELECTRIC WAVEGUIDE

As shown in Fig. A3.1, the double heterostructure used in diode lasers provides a three-layer slab configuration in which each layer has a different index. More generally, in lasers we may be interested in regions where the index varies both transversely and laterally. Thus, we might modify Eq. (A3.3) to give

$$\nabla^2 U(x, y) + [\tilde{n}^2(x, y)k_0^2 - \tilde{\beta}^2]U(x, y) \approx 0. \qquad (A3.4)$$

However, as indicated by the $\approx$ symbol, this now only approximately satisfies Maxwell's equations, since our derivation assumed that ε (and therefore n) was uniform in space.

In the current one-dimensional slab case, we can obtain an exact solution by solving Eq. (A3.3) for uniform n in each of the three regions and then matching the boundary conditions at the interfaces. By noting that the ratio of the real to the imaginary part of the index is very large in all practical cases of interest, we can replace $\tilde{\beta}$ by β, with some assurance that the mode shape, $U(x, y)$, will not be significantly in error. We shall still use Eq. (A3.2) to include the gain or loss in the propagating mode.

The solution procedure is very similar to how we solved for the confined

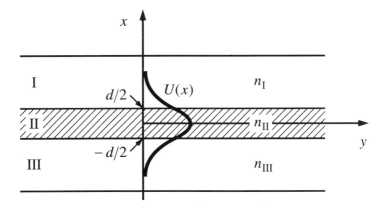

FIGURE A3.1 Schematic of a three-layer slab waveguide. Indices are assumed to be uniform in the z-direction.

states of the electron in a one-dimensional rectangular well in Appendix 1. Indeed, there is no difference in the form of the wave equation for the transverse electric field, Eq. (A3.3), and the time-independent Schrödinger's equation, Eq. (A1.5). Thus, the solutions will have the same form if the boundary conditions are analogous. As we shall see, they are analogous for the TE modes which are polarized in the y-direction, but a little different for the TM modes which are polarized in the x-direction.

A3.2.1   Symmetric Slab Case

For a symmetric three-layer slab guide ($n_I = n_{III}$), we follow the solution given in Appendix 1 for a one-dimensional potential well very closely. In the central region, we assume solutions of the form,

$$U_{II}(x) = \begin{cases} A \cos k_x x \text{ (symmetric solutions)} \\ A \sin k_x x \text{ (antisymmetric solutions)} \end{cases},$$ (A3.5)

In region I,

$$U_I(x) = Be^{-\gamma x}.$$ (A3.6)

After substituting Eqs. (A3.5) and (A3.6) into (A3.3) with $n = n_{II}$ and n_I, respectively, we find that

$$k_x^2 = k_0^2 n_{II}^2 - \beta^2,$$

and   (A3.7)

$$\gamma^2 = \beta^2 - k_0^2 n_I^2.$$

In region III, $U_{III} = Be^{\gamma x}$, but by symmetry in this case ($n_I = n_{III}$), we only need to use the single boundary condition at $x = d/2$ between regions II and I.

For the TE modes at $x = d/2$, we have that $U_{II} = U_I$ and $U'_{II} = U'_I$. These conditions derive from the requirements that both the tangential electric and magnetic fields, respectively, are equal at the boundary. For the symmetric solutions, this gives

$$A \cos \frac{k_x d}{2} = Be^{-\gamma d/2},$$ (A3.8a)

and

$$Ak_x \sin \frac{k_x d}{2} = B\gamma e^{-\gamma d/2}.$$ (A3.8b)

Dividing Eq. (A3.8b) by (A3.8a), we obtain the characteristic equation,

$$k_x \tan \frac{k_x d}{2} = \gamma.$$ (A3.9)

Similarly, for the antisymmetric solutions, we obtain $k_x \cot k_x d/2 = -\gamma$. Both can be included in a characteristic equation by recognizing the $\pi/2$ shift between the tan and cot functions. That is, after substituting for k_x and γ from Eqs.

(A3.7) and using $\beta = k_0 \bar{n}$,

$$\frac{k_0 d}{2} [n_{\text{II}}^2 - \bar{n}^2]^{1/2} = \tan^{-1} \left(\frac{\bar{n}^2 - n_{\text{I}}^2}{n_{\text{II}}^2 - \bar{n}^2} \right)^{1/2} + (m - 1) \frac{\pi}{2}, \qquad \text{(A3.10)}$$

where $m = 1, 2, 3, \ldots$ for the fundamental and higher-order modes. For the TM modes, the continuity of the tangential electric and magntic fields at the boundaries results in an additional factor of $(n_{\text{II}}^2/n_{\text{I}}^2)$ on the right side of Eq. (A3.9) and inside the brackets of the $\tan^{-1}$ function in Eq. (A3.10).

The solution to this transcendental equation is done graphically as was done for confined electrons in Appendix 1 and illustrated in Fig. A1.3. As might be expected, the results are analogous for this symmetrical slab case.

A3.2.2   General Asymmetric Slab Case

To add additional generality to the present case, we can repeat the above procedures for an antisymmetric three-layer slab waveguide. That is, $(n_{\text{I}} \neq n_{\text{III}})$. For the TE modes, we find that the characteristic equation, Eq. (A3.9), becomes

$$\tan k_x d = \frac{(\gamma_{\text{I}}/k_x) + (\gamma_{\text{III}}/k_x)}{1 - \gamma_{\text{I}} \gamma_{\text{III}}/k_x^2}, \qquad \text{(A3.11)}$$

where γ_{I} and γ_{III} are the decay constants in the upper and lower cladding regions, respectively, defined as in Eq. (A3.7). Equation (A3.11) also can be solved graphically, but it is convenient to define a normalized frequency, V, propagation parameter, b, and asymmetry parameter, a, in order to display the results. These normalized parameters are defined as follows:

$$V \equiv k_0 d (n_{\text{II}}^2 - n_{\text{III}}^2)^{1/2},$$

$$b \equiv \frac{\bar{n}^2 - n_{\text{III}}^2}{n_{\text{II}}^2 - n_{\text{III}}^2},$$

and $\qquad\qquad\qquad\qquad\qquad\qquad\qquad\qquad\qquad\qquad\qquad$ (A3.12)

$$a \equiv \frac{n_{\text{III}}^2 - n_{\text{I}}^2}{n_{\text{II}}^2 - n_{\text{III}}^2}.$$

Figure A3.2 gives plots of the normalized propagation parameter as a function of the normalized frequency for a range of normalized asymmetry parameters.

For TM modes we can use Figure A3.2 with some small error due to the neglected dielectric constant ratio that should multiply γ_{I} and γ_{III} in the dispersion relationship. The error becomes vanishingly small in weak dielectric guides.

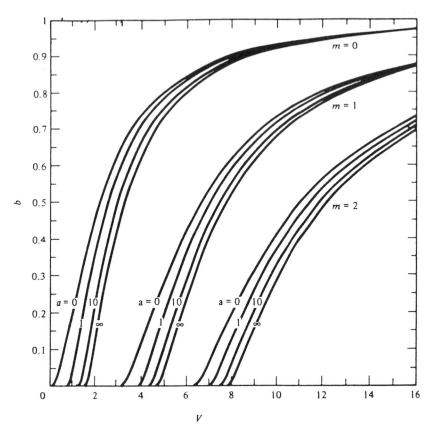

FIGURE A3.2   Normalized propagation parameter vs. normalized frequency for a range of asymmetries for the first three TE modes. After Kogelnik and Ramaswamy [1]. (Reprinted, by permission, from *Applied Optics*).

A3.2.3   Transverse Confinement Factor, Γ_x

The transverse confinement factor for the three-layer slab waveguide is defined as the fraction of the optical energy that is contained in the active slab region. As derived in Appendix 5, Eq. (A5.13), this fraction can be approximated as (neglecting $n/\bar{n}$)

$$\Gamma_x = \frac{\displaystyle\int_{-d/2}^{d/2} |U(x, y)|^2 \, dx}{\displaystyle\int_{-\infty}^{\infty} |U(x, y)|^2 \, dx}. \tag{A3.13}$$

For the symmetric case ($a = 0$), we can use Eqs. (A3.7) and (A3.5) in Eq. (A3.13)

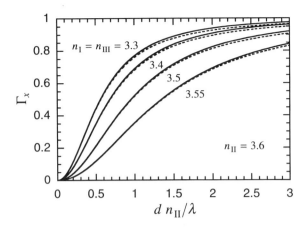

FIGURE A3.3   Comparison between the exact confinement factor (solid curve) and the approximate formula (A3.15) (dashed curve) for several values of the cladding index as a function of the guide thickness for the fundamental slab mode.

to get

$$\Gamma_x = \frac{1 + 2\gamma d/V^2}{1 + 2/\gamma d}. \tag{A3.14}$$

For the fundamental mode with relatively small index differences, Eq. (A3.14) can be approximated by

$$\Gamma_x \approx \frac{V^2}{2 + V^2}. \tag{A3.15}$$

Figure A3.3 compares the approximate formula to the exact confinement factor (neglecting $n/\bar{n}$) for several values of n_{III}.

A3.3   EFFECTIVE INDEX TECHNIQUE FOR TWO-DIMENSIONAL WAVEGUIDES

As indicated in Chapters 1 and 2, practical diode lasers usually employ waveguiding in the lateral y-direction as well as the transverse x-direction. These are referred to as either two-dimensional or channel waveguides. After Eq. (A3.4) it was suggested that exact analytic solutions to this two-dimensional problem depicted in Fig. A3.4 are impossible. The problems arise in matching the lateral boundary conditions for all values of x.

In the limiting case of very strong index discontinuities at the active–cladding interfaces, the field within the active region will fall to zero at the boundaries, and the boundary conditions can be met around the perimeter. That is, for

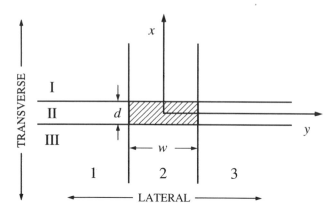

FIGURE A3.4   Waveguide cross section perpendicular to the z-direction. The transverse and lateral directions are divided into regions I, II, and III and 1, 2, and 3, respectively.

symmetric modes in region II-2 (the active region),

$$U(x, y) = U_0 \cos k_x x \cos k_y y, \qquad \text{(region II-2)} \qquad \text{(A3.16)}$$

and for antisymmetric modes, sine functions replace the cosines. Including both in sequence, the boundary conditions give $k_x d = \pi m_x$ and $k_y w = \pi m_y$, where the m_js are the respective mode numbers ($m_j = 1$ is the first symmetric mode in either direction, $m_j = 2$ is the first antisymmetric mode, etc.). Thus, the general solution for the symmetric modes is

$$U_{m_x m_y}(x, y) = U_0 \cos \frac{\pi m_x x}{d} \cos \frac{\pi m_y y}{w}. \qquad (\Delta n \to \infty) \qquad \text{(A3.17)}$$

Unfortunately, we cannot make the large index discontinuity assumption in most cases. (However, a semiconductor–air interface is sufficiently large to neglect fields outside the semiconductor provided it is thicker than a few wavelengths.) Thus, we must employ some other approximation. The most common of these is the *effective index technique*. This technique involves a sequential solution to the problem of Fig. A3.4. It is most accurate when the transverse slab mode solutions in regions 1, 2, and 3 are nearly the same, and when $w/d \gg 1$.

The effective index technique is the limiting form of a rigorous technique to match the fields in the lateral direction for all values of x along the interfaces between regions 1, 2, and 3 at $y = \pm w/2$ illustrated in Fig. A3.4. This rigorous technique uses a superposition of the simple (uniform-y) slab mode solutions along x derived in Section A3.2 to express the overall channel waveguide mode shape in each of the three lateral regions. To be completely rigorous, a complete set of slab modes along x must be used in each region (including the radiation modes) to accurately synthesize any arbitrary mode shape. Another consideration is that all components of the channel waveguide mode must propagate along z with the same phase velocity, or the same propagation constant, β.

To adjust the β's of the various slab modes in the expansion to the same value, we must add a y-component to the propagation direction. For example, if β_m is the propagation constant solution of the mth slab mode, and β is the propagation constant along z, then for a slab mode propagating along z, $\beta_m = \beta$ (as we assumed earlier). Now, if we tilt the propagation direction slightly toward y, then $\beta_m^2 = \beta^2 + k_{ym}^2$. Thus, k_{ym} can be used to make up the difference between the fixed β and each slab mode's β_m. For each slab mode in region 2, we therefore define $k_{ym} = \sqrt{\beta_{m2}^2 - \beta^2}$, which produces cosine and sine profiles along y (for $\beta_{m2} > \beta$). In regions 1 and 3, we anticipate that $\beta > \beta_{m1}$, β_{m3} so we define $\gamma_{ym1} = \sqrt{\beta^2 - \beta_{m1}^2}$ and $\gamma_{ym3} = \sqrt{\beta^2 - \beta_{m3}^2}$. This yields the familiar evanescent decaying solutions along y in these regions.

Adding the y profiles to each of the slab mode solutions along x, we can now express the overall channel waveguide mode as a weighted sum of the slab modes within each lateral region:

$$\mathscr{E}_1(x, y) = \sum_m B_{m1} U_{m1}(x) e^{\gamma_{ym1} y},$$

$$\mathscr{E}_2(x, y) = \sum_m A_m^e U_{m2}(x) \cos k_{ym} y + \sum_m A_m^o U_{m2}(x) \sin k_{ym} y,$$

$$\mathscr{E}_3(x, y) = \sum_m B_{m3} U_{m3}(x) e^{-\gamma_{ym3} y}.$$

B_{m1}, A_m^e, A_m^o, and B_{m3} are the weighting or *expansion* coefficients which are adjusted to accurately synthesize the overall channel waveguide mode in each of the lateral regions. $U_{m1}(x)$, $U_{m2}(x)$, and $U_{m3}(x)$ are the transverse mode shapes of the mth slab mode in regions 1, 2, and 3.

The final task is to match the boundary conditions at the planes between regions 1 and 2, and 2 and 3 using $\mathscr{E}_1$, $\mathscr{E}_2$, and $\mathscr{E}_3$. If the electric field is predominantly polarized along x (TM transverse mode), the matching of the tangential electric and magnetic fields is to a good approximation equivalent to matching $\mathscr{E}_x$ and $\partial \mathscr{E}_x / \partial y$ at each boundary (we neglect the small $\mathscr{E}_z$ and $\mathscr{E}_y$ components). If the lateral waveguide is symmetric, we can define laterally symmetric and antisymmetric solutions which require matching at only one boundary. For the symmetric solutions, we can set $B_{m1} = B_{m3} \equiv B_m$, $A_m^o = 0$, and $A_m^e \equiv A_m$. Matching these fields and their y-derivatives at the $y = w/2$ boundary yields

$$\sum_m A_m U_{m2}(x) \cos \frac{k_{ym} w}{2} = \sum_m B_m U_{m3}(x) e^{-\gamma_{ym} w/2}, \qquad \text{(A3.18a)}$$

and

$$\sum_m A_m U_{m2}(x) k_{ym} \sin \frac{k_{ym} w}{2} = \sum_m B_m U_{m3}(x) \gamma_{ym} e^{-\gamma_{ym} w/2}, \qquad \text{(A3.18b)}$$

where

$$k_{ym}^2 = \beta_{m2}^2 - \beta^2 \qquad \text{and} \qquad \gamma_{ym}^2 = \beta^2 - \beta_{m3}^2. \qquad \text{(A3.19)}$$

β_{m2} and β_{m3} are the mth slab mode propagation constant solutions in regions 2 and 3, respectively. The A_m and B_m expansion coefficients can only satisfy Eqs. (A3.18) at all values of x for discrete values of β, which yield the guided mode solutions of the overall channel waveguide.

If the electric field is instead predominantly polarized along y (TE transverse mode), as is more commonly the case, the matching of the tangential electric and magnetic fields across the $y = w/2$ boundary is approximately equivalent to matching $\varepsilon \mathscr{E}_y$ and $\partial \mathscr{E}_y / \partial y$. Thus for TE transverse modes, the dielectric constant ratio $\varepsilon_3(x)/\varepsilon_2(x)$ should appear on the right side of Eq. (A3.18a). This factor however can often be ignored without introducing much error.

To solve Eqs. (A3.18), we need to have as many equations as there are expansion coefficients. If M modes are included in the summation in region 2, and M modes in region 3, then there are $2M$ unknown coefficients overall, and two equations at each x location. To get enough equations along the lateral boundary then, we must evaluate the lateral boundary conditions at M values of x. The resulting system of equations can be solved uniquely only when the determinant of the matrix formed by the factors multiplying each expansion coefficient goes to zero. The zero crossing can be found numerically by scanning the value of β across some appropriate range. As already suggested, if the transverse slab mode shapes in the adjacent lateral regions are similar, only a few slab modes are required to give good results.

The effective index technique results when we use only one transverse slab mode of region 2 and region 3 to approximate the channel waveguide field. Because only one unknown coefficient exists in each region (two total), we only need to evaluate the boundary conditions at one point along x (usually at $x = 0$). In this limit, Eqs. (A3.18) reduce to Eqs. (A3.8) used earlier to solve for the one-dimensional slab modes. Thus, the solution to the lateral guide problem becomes identical to the transverse slab guide problem, except that Eqs. (A3.19) replace Eqs. (A3.7), or β_{m2} replaces k_{II} and β_{m3} replaces k_{III} ($= k_1$). More generally, the effective index of the slab mode in each lateral region is used in place of the index of the medium in the normalized parameters defined in Eqs. (A3.12).

In summary, the solution sequence for the effective index technique is as follows:

1. Three-layer slab problems are first solved across the smallest dimension (x-direction in Fig. A3.4) using n_I, n_{II}, n_{III}, as if the regions were uniform in the other longer (y) dimension. This is repeated in each of the three

lateral regions listed in Fig. A3.4 to produce $\bar{n}_1$, $\bar{n}_2$, and $\bar{n}_3$. The effective index in the central (#2) region is used for the transverse k_x and γ_x in the final solution.

2. The three effective indices thus obtained ($\bar{n}_1$, $\bar{n}_2$, and $\bar{n}_3$) are now used in a new three-layer slab problem for the other (y) direction. The result is a final effective index for the two-dimensional problem. This is used for the lateral k_y and γ_y. The net axial propagation constant, β, thus, as always satisfies

$$[k_0 n_{\mathrm{II}}]^2 = k_x^2 + k_y^2 + \beta^2. \tag{A3.20}$$

This effective index technique can be applied relatively easily by successive use of Fig. A3.2. Although this figure is strictly only for TE modes, it can be approximately used for TM modes if the index differences are small. The thinner transverse dimension determines the designation of TE or TM. The second half of the effective index procedure is really dealing with the opposite polarization from the first. However, if the indices are similar, the difference in the solution is negligible.

The overall solutions for the symmetric modes of $U(x, y)$ are identical to Eq. (A3.16) within the active region, region II-2. In region I-2 the y-dependent factor remains the same, but the x-dependent factor becomes a decaying exponential like Eq. (A3.6). That is,

$$U(x, y) = U_0 \cos \frac{k_x d}{2} e^{-\gamma_x(x - d/2)} \cos k_y y, \qquad \text{(region I-2)} \tag{A3.21}$$

where γ_x comes from the first step and k_y comes from the second step in the effective index technique. Note that the magnitude has been matched to the solution in region II-2. For region II-3, the solution in the x-direction remains from the first step and the lateral decay is found from the second step,

$$U(x, y) = U_0 \cos \frac{k_y w}{2} (\cos k_x x) e^{-\gamma_y(y - w/2)}. \qquad \text{(region II-3)} \tag{A3.22}$$

The functional forms of $U(x, y)$ in other regions should now be obvious.

The lateral confinement factor Γ_y can be approximated by the lateral equivalent to Eq. (A3.15). That is,

$$\Gamma_y \approx \frac{V_l^2}{2 + V_l^2}, \tag{A3.23}$$

where, $V_l = k_0 w(\bar{n}_2^2 - \bar{n}_3^2)^{1/2}$.

A3.4   FAR FIELDS

The waveguide field profiles given above are assumed to be uniform along the length of the laser as long as the cross section of the guide remains uniform. At the output facet this field emerges from the laser waveguide and diffracts freely into the surrounding dielectric (usually air). In most applications it is desirable to capture this output light into some other waveguide or detector. The amount of light actually captured depends on the size and shape of the beam at the cross section of the capturing object amongst other things. Thus, it is useful to predict the field profile as it extends beyond the output facet.

In diffraction theory we refer to the field emitted from the laser weaveguide as the *near field* and the diffracted field some distance away as the *far field*. The transition occurs at roughly w^2/λ, where w is some characteristic full width of the near-field pattern. In a real index-guided waveguide, the wavefronts are planar as they approach the emitting facet. They remain approximately planar in the near field, but begin to show noticeable curvature in the transition to the far field. After some distance into the far field, the wavefronts approach a spherical shape with a radius of curvature measured from the center of the emitted mode at the facet where the wavefronts are planar. The most narrow, planar wavefront location is termed the *beam waist* in Gaussian mode theory. Figure A3.5 describes the generation of the far field by the emitted near field.

To determine the far-field pattern, U_F, a distance R and angle θ_R from the

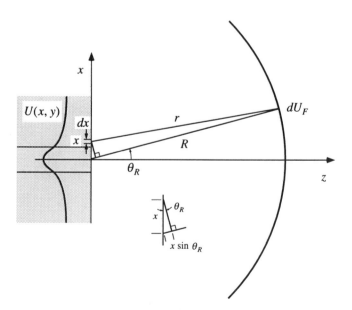

FIGURE A3.5   Illustration of near-field pattern, $U(x, y)$, and far-field element, dU_F, a distance R from the facet at the waveguide axis in the x-z plane.

origin as defined in Fig. A3.5, we consider each amplitude increment of the near-field pattern, $U(x, y)$, to be a radiating point source. Each point source or "spherical wavelet" then propagates a distance r to contribute a far-field element, dU_F. The total far-field amplitude at any given point is found by coherent addition of all wavelet contributions. The field of each spherical wavelet a distance r away from its source and weighted by the near-field amplitude within the increment $dx \, dy$ can be written as

$$dU_F = dx \, dy \, U(x, y) \frac{e^{-jkr}}{r} \left[\frac{j}{\lambda} \cos \theta_r \right], \tag{A3.24}$$

where $k = 2\pi/\lambda$ and θ_r is the angle between $\mathbf{r}$ and the z-axis (if the medium outside the laser is not air then $\lambda \to \lambda/n$). The added factors in square brackets are mathematical refinements to the intuitive wavelet concept originally conceived by Huygen. These factors introduced later by Fresnel (as well as Rayleigh and Sommerfeld) allow Huygen's principle to be applied to a wide range of scalar diffraction problems with excellent accuracy. The most significant of these terms is the *obliquity factor*, $\cos \theta_r$, which adds a directivity pattern to each spherical wavelet (it accounts for the reduced apparent area of each incremental emitter when viewed off-axis).

If $x, y \ll R$ for all x and y in which $U(x, y)$ has significant amplitude, we can approximate the distance from the increment to the measurement point as

$$r \approx R - x \sin \theta_x - y \sin \theta_y, \tag{A3.25}$$

where θ_x and θ_y are the angles measured from the z-axis toward the x- or y-axis, respectively. Since the phase factor is the most sensitive, we should use Eq. (A3.25) in the exponent of Eq. (A3.24). However in the denominator, we can set $r \approx R$ and also $\theta_r \approx \theta_R$. With these far-field approximations, we can sum over all wavelet contributions by integrating Eq. (A3.24) over the near-field emission plane to obtain

$$U_F(\theta_x, \theta_y) = \frac{j \cos \theta_R}{\lambda R} e^{-jkR} \iint U(x, y) e^{jk \sin \theta_x x} e^{jk \sin \theta_y y} \, dx \, dy, \tag{A3.26}$$

with $\cos \theta_R = \cos \theta_x \cos \theta_y / (1 - \sin^2 \theta_x \sin^2 \theta_y)^{1/2}$.

Concentrating on a single axis by setting $\theta_y = 0$, and taking the magnitude squared, we arrive at the angular power spectrum in the far field,

$$|U_F(\theta_x)|^2 = \frac{\cos^2 \theta_x}{\lambda^2 R^2} \left| \int U(x) e^{jk \sin \theta_x x} \, dx \right|^2, \tag{A3.27}$$

where $U(x) \equiv \int U(x, y) \, dy$. First of all we see that the far-field intensity drops as $1/R^2$, like a spherical wave. In addition, for small angles where $\sin \theta_x \approx \theta_x$

and $\cos \theta_x \approx 1$, we observe that $U_F(\theta_x)$ and $U(x)$ are Fourier transform pairs. In this approximation then, the far-field angular spectrum is just the Fourier transform of the near field, suggesting that a *narrow* emitting aperture (i.e. waveguide) leads to a *wide* angular distribution in the far field, and vice versa. For larger angles, the Fourier transform relationship breaks down, but the inverse relationship between the near-field and far-field beam width remains qualitatively valid.

With in-plane lasers, the transverse waveguide width is usually much smaller than the lateral width. This causes the transverse far-field pattern to have a much larger angular spread than the lateral far-field pattern. However, as long as $U(x, y)$ is real (i.e. has planar wave fronts), the far-field wave fronts will be spherical from the constant $\exp[-jkR]$ factor in Eq. (A3.26). In other words, even though the emitted power distribution is asymmetric, the output beam is *not* astigmatic. Thus, the emitted beam can be easily collimated into planar wave fronts with a simple spherical lens. However, the intensity pattern will be elliptical. To correct for this, the collimated beam can be refracted through a wedge with nonparallel planar surfaces to elongate one axis. The result is a circular collimated beam, but the procedure is rather inconvenient. With VCSELs, the transverse and lateral guide widths are usually the same, producing nice circular output beams in the far field.

REFERENCE

[1]   H. Kogelnik and V. Ramaswamy, *Appl. Opt.*, **8**, 1857 (1974).

READING LIST

H.A. Haus, *Waves and Fields in Optoelectronics*, Ch. 6, Prentice Hall, Englewood Cliffs, NJ (1984).

D.L. Lee, *Electromagnetic Principles of Integrated Optics*, Ch. 4, Wiley, New York (1986).

A. Yariv and P. Yeh, *Optical Waves in Crystals*, Ch. 11, Wiley, New York (1984).

J.W. Goodman, *Introduction to Fourier Optics*, Ch. 3, McGraw-Hill, New York (1968).

Density of Optical Modes, Blackbody Radiation, and Spontaneous Emission Factor

A4.1 OPTICAL CAVITY MODES

Figure A4.1 shows an optical cavity with dimensions d_x, d_y, and d_z. If we assume that the reflection coefficient at each boundary is real, then for each component of the propagation constant, k_j, the boundary condition for resonance is

$$2k_j d_j = 2m_j \pi, \qquad (A4.1)$$

where $j = x$, y, or z, and m_j is the respective mode number. This follows from the fact that the electric field of a resonant mode must replicate its phase after traversing a round-trip in the cavity. Thus, considering all three components, the magnitude of the propagation constant for some resonant mode of the cavity is given by

$$|\mathbf{k}|^2 = \pi^2 \left[\left(\frac{m_x}{d_x} \right)^2 + \left(\frac{m_y}{d_y} \right)^2 + \left(\frac{m_z}{d_z} \right)^2 \right], \qquad (A4.2)$$

where $|\mathbf{k}| = \omega n/c$.

Now just as in the case of electronic states, we wish to calculate the density of optical states or modes. The reasons are much the same. We would like to know the number of optical modes within some energy range, and rather than laboriously counting mode numbers, we would prefer to integrate a density of states. The process and the results are much the same, for there is little difference in concept between the electronic wavefunction and the normalized optical electric field. They both satisfy similar wave equations and boundary conditions as reviewed in Appendices 1 and 3.

However, there is one major difference between photons and electrons.

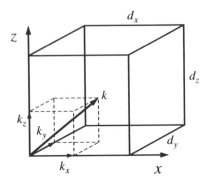

FIGURE A4.1 Optical cavity showing dimensions d_x, d_y, and d_z, and a superimposed k-vector decomposition for an optical mode.

Whereas only one electron can occupy an electronic state (after doubling the possible energy levels because of spin), an unlimited number of photons can occupy the same optical mode (again, after we have doubled the number of allowed k-values because of polarization). That is, there is no Pauli exclusion principle for photons. Thus, after we calculate the number of photon modes, we still have to determine how many photons occupy each mode to get the total number of photons.

Following the same procedure outlined after Eq. (A1.36), we now proceed to calculate the number of optical modes per unit volume per unit frequency. (We could do energy, $E = h\nu$, but let's have a little variety.) As in Appendix 1 we shall first consider *bulk* dimensions, a *spherical* coordinate system, but switch to *frequency* as the variable. The first step is to calculate the volume, $N_s(m)$, in mode number space as depicted in Fig. A1.10. Analogous to Eq. (A1.37), we determine the volume of the sphere, multiply by 2 for the two polarization states and divide by 8 to allow only positive mode numbers,

$$N_s(m) = \tfrac{4}{3}\pi m^3 \cdot 2 \cdot \tfrac{1}{8}. \tag{A4.3}$$

Now, in spherical coordinates, $k^2 = (2\pi\nu)^2/(c/n)^2 = \pi^2 m^2/d^2$, where n is the index of refraction, $m^2 = m_x^2 + m_y^2 + m_z^2$, and we have let $d^2 = d_x^2 = d_y^2 = d_z^2$. Solving for $m(\rightarrow 2\,dn\nu/c)$ and plugging into Eq. (A4.3),

$$N_s(\nu) = \frac{\pi}{3}\left(\frac{2\,dn}{c}\right)^3 \nu^3. \tag{A4.4}$$

Finally, we apply Eq. (A1.36) (set $\rho_0(\nu)\,d\nu = (1/V)(dN_s/d\nu)\,d\nu$), and use $V = d^3$ to obtain

$$\rho_0(\nu)\,d\nu = \frac{8\pi}{c^3} n^2 n_g \nu^2 \,d\nu, \tag{A4.5}$$

where we have defined a *group index*, $n_g = [n + v(\partial n/\partial v)]$, since the index of refraction can be frequency dependent.

If the dimensions of the optical cavity become small, say $d_j \leq 10\lambda$, the modes will not be so closely spaced, and it is advisable to use Eq. (A4.2) to count the modes within some range of wavelengths or frequencies. This situation is analogous to the electronic "quantum box." Because the states are widely spaced, the density of states just becomes a series of delta functions as each state is encountered. If only one dimension is small, we again have a similar situation to the quantum well for electrons as discussed in Appendix 1.

A4.2 BLACKBODY RADIATION

Equation (A4.5) gives us the desired density of optical states, but as mentioned above, any number of photons can occupy these states, so we need another piece of information to determine the photon density per unit frequency. One interesting boundary condition that can be applied to give a photon density is *thermal equilibrium*. In this case we can use the Maxwell–Boltzmann distribution to give the occupation probability of the optical states as a function of frequency. Maxwell–Boltzmann statistics dictate that a state with energy E will have a probability $\exp(-E/kT)$ of being occupied under thermal equilibrium at temperature T. Thus, we can calculate an average energy per state, $\langle E \rangle$, by taking a weighted average of the possible energies and their probability of occupation. Because light is quantized into photons of energy hv, the allowed energies are $E_j = jhv$, $j = 0, 1, 2, \ldots$. So we take the sum of allowed energies multiplied by the occupation probabilities and normalize this by dividing by the sum of the probabilities. That is,

$$\langle E \rangle = \frac{0e^{-0} + hve^{-hv/kT} + 2hve^{-2hv/kT} + \cdots}{1 + e^{-hv/kT} + e^{-2hv/kT} + \cdots}. \tag{A4.6}$$

Summing the infinite series and dividing, we get the average energy in each mode,

$$\langle E \rangle = \frac{hv}{e^{hv/kT} - 1}. \tag{A4.7}$$

The *blackbody radiation density*, $W(v)\, dv$, is now found by multiplying Eq. (A4.5) by Eq. (A4.7). That is, the equilibrium energy per unit volume per unit frequency is equal to the number of modes per unit volume per unit frequency times the average energy per mode under thermal equilibrium. Thus,

$$W(v)\, dv = \frac{\rho_0(v)hv}{e^{hv/kT} - 1}\, dv = \frac{8\pi n^2 n_g hv^3/c^3}{e^{hv/kT} - 1}\, dv. \tag{A4.8}$$

A4.3   SPONTANEOUS EMISSION FACTOR, β_{sp}

The spontaneous emission factor can be obtained from the density of optical modes per unit volume per unit frequency, Eq. (A4.5), by integrating over the cavity volume and the bandwidth of the spontaneous emission to find the number of optical modes that must contain the total spontaneous emission. Usually, it is assumed that the coupling to all modes is the same, so the reciprocal of this number of relevant modes is just the *fraction of energy going into each mode*, β_{sp}. Initially, making this assumption, the number of cavity modes in a bandwidth Δv_{sp} is found by integrating Eq. (A4.5) from v to $v + \Delta v_{sp}$ to be

$$N_{sp} = V_c \frac{8\pi}{c^3} n^2 n_g v^2 \Delta v_{sp}, \qquad (A4.9)$$

where V_c is the cavity volume. Taking the reciprocal and using $\Delta v_{sp}/v = \Delta \lambda_{sp}/\lambda$,

$$\beta_{sp} = \frac{\Gamma_c \lambda^4}{8\pi V n^2 n_g \Delta \lambda_{sp}}, \qquad (A4.10)$$

where $\Gamma_c = V/V_c$, and V is the active region volume. For typical values of parameters, $\beta_{sp} \sim 10^{-5}$ to 10^{-4}.

But the assumption that the spontaneous emission is uniformly distributed amongst the various cavity modes within $\Delta \lambda_{sp}$ is generally not true. In real devices, the emission spectrum follows a more bell-shaped curve as a function of both frequency and wavelength. If the mode of interest falls at the peak of the emission spectrum, then (A4.10) should be modified by setting $1/\Delta \lambda_{sp} \rightarrow$ (peak)/(area), where the area represents the total area under the emission spectrum, or the total emission rate. For example, if the emission spectrum follows a Lorentzian lineshape, we can set (peak)/(area) $= (2/\pi)/\Delta \lambda_{sp,FW}$, where the FW implies the full-width at half-maximum. Thus, an additional factor of $2/\pi$ should appear in Eq. (A4.10). For more complicated lineshapes, the ratio (peak)/(area) will be related to $1/\Delta \lambda_{sp}$, but the exact prefactor will depend on the specific lineshape.

Additional considerations involve the strength of the mode's wavefunction within the active region. The fraction of spontaneous emission represented by (A4.10) refers to the *modal* spontaneous emission rate which must account for the overlap of the mode of interest with the active region. In Appendix 5 we show how such modal averages are taken. We have intentionally set $1/V_c = \Gamma_c/V$ in Eq. (A4.10) to emphasize this modal averaging process. One interesting consequence of this substitution is that if the active region is very thin, and the mode for which we wish to calculate β_{sp} has a null at this point, then $\Gamma_c \approx 0$, from which it follows that $\beta_{sp} \approx 0$ as well. This would not be obvious if we had simply used $1/V_c$ in Eq. (A4.10).

Equation (A4.10) represents the semiclassical version of the spontaneous emission factor. In Chapter 4, a more useful version of the spontaneous emission factor derived from quantum mechanical considerations will be introduced. The quantum version is more general and simpler to evaluate.

READING LIST

J.T. Verdeyen, *Laser Electronics*, 2d ed., Ch. 7, Prentice Hall, Englewood Cliffs, NJ (1989).

Modal Gain, Modal Loss, and Confinement Factors

A5.1 INTRODUCTION

Within a laser cavity the gain and loss are not uniform throughout the volume occupied by the optical modes of interest. In fact, the gain region typically occupies only a few percent of the volume occupied by the optical modes, and the material absorption loss is typically very different in different regions of the cavity. Thus, we must develop some sort of overlap factor which gives the net gain or loss provided to an optical mode.

To be completely general, let's define the localized *material* gain as a function of all three dimensions of space: $g(x, y, z)$. To provide the net effect of $g(x, y, z)$ on the mode as a whole, a properly weighted average of the gain distribution must be taken throughout the entire cavity volume. Now from Chapter 4, we know that the gain varies according to the stimulated emission rate, which in turn is proportional to the square of the electric field. Thus, it seems reasonable to use the electric field pattern, $\mathscr{E}(x, y, z)$, of the appropriate resonant mode as our weighting function. The standard definition of a weighted average from classical or quantum mechanics leads then to the following definition of *modal gain*:

$$\langle g \rangle = \frac{\displaystyle\int \mathscr{E}^*(x, y, z) g(x, y, z) \mathscr{E}(x, y, z) \, dV}{\displaystyle\int |\mathscr{E}(x, y, z)|^2 \, dV}. \tag{A5.1}$$

Before applying this result to typical laser structures, it is instructive to examine a little more carefully the appropriateness of this somewhat ad hoc definition.

A5.2   CLASSICAL DEFINITION OF MODAL GAIN

To rigorously determine the appropriate weighting function to be used in defining the modal gain or loss experienced by a waveguide mode, we turn to a classical description of gain and loss in the cavity. If we define $w_E(x, y, z)$ as the energy density throughout the cavity, then gain and loss *per unit time* can be associated with the time rate of change of this local energy density, dw_E/dt. For example, if $dw_E/dt > 0$ at some point in the cavity, energy is being generated locally which indicates the presence of gain in the material. Integrating dw_E/dt over the entire cavity then allows us to determine the *total* energy generation rate. Defining the modal gain per unit time as the *fractional* generation rate of energy, we have

$$\langle G \rangle \equiv \frac{1}{W_E} \frac{dW_E}{dt} = \frac{\int dw_E/dt \, dV}{\int w_E \, dV}, \tag{A5.2}$$

where W_E is the total energy in the cavity. To check that this classical definition is in agreement with what we know from Chapter 2, we can express the total energy in terms of the photon density using $W_E = h\nu N_p V_p$. Equation (A5.2) then reduces to the familiar rate equation, $dN_p/dt = \langle G \rangle N_p$ where $\langle G \rangle = v_g \langle g \rangle$ and $\langle g \rangle$ is the modal gain (written explicitly as Γg in Chapter 2).

Equation (A5.2) is a rigorous classical definition of the modal gain per unit time. We now wish to express this definition in terms of the electric fields of a given waveguide mode. It can be shown using a variant of Poynting's theorem (which makes use of slowly time-varying phasors) that dw_E/dt is related to the local field strength through the imaginary part of the dielectric constant, ε_i. The exact relation (assuming $\mu = \mu_0$) is given by

$$\frac{dw_E}{dt} = \tfrac{1}{2}\omega \varepsilon_0 \varepsilon_i \mathscr{E} \cdot \mathscr{E}^*, \tag{A5.3}$$

where $\mathscr{E}$ is the electric field vector. Equating the imaginary parts of the dispersion relation, $\tilde{k}^2 = \omega^2 \mu_0 \varepsilon_0 \tilde{\varepsilon}$, we find that $\varepsilon_0 \varepsilon_i = k(x, y, z)g(x, y, z)/\omega^2 \mu_0$ where $g \equiv 2 \, \text{Im}\{\tilde{k}\}$. Decomposing the total field into the two counterpropagating fields within the cavity, $\mathscr{E}^+(x, y, z)$ and $\mathscr{E}^-(x, y, z)$, the above relation then becomes

$$\frac{dw_E}{dt} = \tfrac{1}{2}(kg/\omega\mu_0)|\mathscr{E}^+ + \mathscr{E}^-|^2. \tag{A5.4}$$

For notational convenience, it is to be understood that if the electric field vector has more than one component, then $|\mathscr{E}|^2$ implies $\mathscr{E} \cdot \mathscr{E}^*$. This expression can be used in the numerator of Eq. (A5.2). Now we need to replace the denominator.

To relate the total energy in the cavity to the electric fields, we start by writing down an integral relationship between the energy density and the Poynting vector taken over any cross-sectional area of the cavity:

$$v_g \int w_E \, dA = \tfrac{1}{2}(\beta/\omega\mu_0) \int [|\mathscr{E}_T^+|^2 + |\mathscr{E}_T^-|^2] \, dA. \tag{A5.5}$$

On the left-hand side, the group velocity of the mode, v_g, converts the integrated energy density to total power flowing through the cross-section. On the right-hand side, the sum of forward and backward time-averaged Poynting vectors defines the local power density which when integrated also yields the total power flowing through the cross-section. We have written the Poynting vectors explicitly in terms of the *transverse* electric field components and the effective propagation constant of the mode, β.

To determine the total energy, we need to integrate Eq. (A5.5) over the length of the cavity which we take as the z-direction. To include possible axial variations in the waveguiding structure, we let $v_g \to v_g(z)$ and $\beta \to \beta(z)$. Moving $v_g(z)$ to the right-hand side, integrating over z, and setting $dA \, dz \to dV$, we obtain

$$\int w_E \, dV = \int \frac{\tfrac{1}{2}\beta(z)/\omega\mu_0}{v_g(z)} [|\mathscr{E}_T^+|^2 + |\mathscr{E}_T^-|^2] \, dV. \tag{A5.6}$$

Using Eqs. (A5.4) and (A5.6) we can write $\langle G \rangle$ in terms of the electric fields. We can then use the group velocity in the active waveguide to convert the modal gain per unit time to modal gain per unit *active* length: $\langle g \rangle = \langle G \rangle / v_{ga}$. Setting $k = nk_0$ and $\beta(z) = \bar{n}(z)k_0$, we finally obtain

$$\langle g \rangle = \frac{\displaystyle\int gn|\mathscr{E}^+ + \mathscr{E}^-|^2 \, dV}{\displaystyle\int \frac{v_{ga}}{v_g(z)} \bar{n}(z)[|\mathscr{E}_T^+|^2 + |\mathscr{E}_T^-|^2] \, dV}. \tag{A5.7}$$

If the material gain, g, is dependent on the polarization of the electric field (which is true in quantum-well and quantum-wire active materials), then the numerator should be written more precisely as $g_x|\mathscr{E}_x|^2 + g_y|\mathscr{E}_y|^2 + g_z|\mathscr{E}_z|^2$ (where each field component represents a sum, $\mathscr{E}_i^+ + \mathscr{E}_i^-$).

It is interesting to note that the energy generation rate in the numerator of Eq. (A5.7) includes coherent effects such as standing waves, while the denominator does not. This is because gain in a medium is obtained specifically through interactions with the electric field. Hence, the energy generation rate in the cavity is concentrated at the peaks of the electric field standing wave pattern. The stored energy, on the other hand, exists in both the electric and magnetic fields continually shifting back and forth between them. As a result, the time-averaged energy distribution is independent of the standing wave pattern.

When there are no axial variations in group velocity or propagation constant, we can simplify Eq. (A5.7) by setting $v_{ga}/v_g(z) = 1$ and pulling $\bar{n}$ out of the integral:

$$\langle g \rangle = \frac{\int \bar{g} |\mathscr{E}^+ + \mathscr{E}^-|^2 \, dV}{\int [|\mathscr{E}_T^+|^2 + |\mathscr{E}_T^-|^2] \, dV}, \qquad \text{where } \bar{g} = g\frac{n}{\bar{n}}. \qquad (A5.8)$$

By defining an *effective* material gain, $\bar{g}$, in our weighted average (which is found by replacing the index with the effective index in the expression for gain given in Chapter 4), the rigorous definition of modal gain can be made to resemble our initial guess Eq. (A5.1). In fact, for TE modes which have no longitudinal electric field components, the correspondence is exact (aside from the incoherent sum in the denominator). For TM modes which do have longitudinal electric fields, there is still a slight difference between (A5.8) and (A5.1). For laser applications, the mode of interest is typically a TE mode which is very well confined to the active region, implying that $\bar{n} \approx n$, and hence $\bar{g} \approx g$. As a result, it is quite common to simply use (A5.1) with the real material gain, g, in defining the modal gain. This procedure typically introduces little error. To be completely rigorous, however, Eq. (A5.8) (or Eq. (A5.7) if there are axial variations in the waveguiding structure) should be used.

A5.3 MODAL GAIN AND CONFINEMENT FACTORS

Figure 2.6 illustrates a laser cavity in which the gain region intersects the optical mode over some portion of its width and length. Figure A5.1 gives a slightly more detailed version of a typical laser cavity. The ith component of the electric field in such a cavity can be written as

$$\mathscr{E}_i(x, y, z) = U_i(x, y)\sqrt{2} \cos \beta z, \qquad (A5.9)$$

where $U_i(x, y)$ is the normalized transverse electric field profile, derived in Appendix 3 for a three-layer slab, and $\beta \, (= 2\pi/\lambda_z \equiv 2\pi\bar{n}/\lambda)$ is the z-component of the propagation constant k. In writing (A5.9), we have neglected the imaginary part of β, and hence are ignoring the growth of the fields in the cavity (see Section A5.5). The simple cosine function also indicates that an infinite standing wave ratio has been assumed in the z-direction. Such a situation will actually exist only over ranges in the cavity where the two counterpropagating waves are equal in magnitude (see Fig. A5.3). For a cavity with large mirror reflectivities ($r > 0.9$), such as in a vertical cavity laser, this range extends over the entire cavity. For a laser with low mirror reflectivities ($r < 0.2$), an infinite standing wave ratio exists only over a small portion of the cavity. In either case, if the axial (z-direction) integration is over many wavelengths, the contribution from this factor averages out.

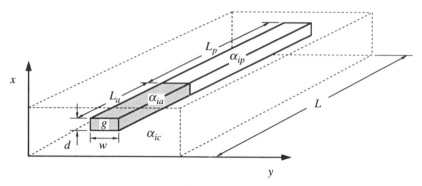

FIGURE A5.1   General laser cavity with active and passive axial sections.

If we assume that the group velocity and propagation constant are the same (or very similar) in both active and passive sections then we can use Eq. (A5.8) to define the modal gain. Plugging Eq. (A5.9) into Eq. (A5.8) and considering the $e^{+j\beta z}$ and $e^{-j\beta z}$ components of the $\cos \beta z$ separately in the denominator, we obtain

$$\langle g \rangle = \frac{\displaystyle\iiint_{xyz} \bar{g}(x, y, z) |U(x, y)|^2 \, 2 \cos^2 \beta z \, dx \, dy \, dz}{L \displaystyle\iint_{xy} |U(x, y)|^2 \, dx \, dy}. \tag{A5.10}$$

We have dropped the i subscript on the transverse field pattern, assuming that only one electric field component exists. If $U(x, y)$ is normalized, the denominator integral will be unity. However, we shall carry it along to include cases where it is not normalized.

In applying Eq. (A5.10) to the laser cavity in Fig. A5.1, a few further assumptions are usually made. For in-plane lasers, the gain can be assumed to be constant within the active region, which has transverse thickness d, width w, and length L_A. Therefore, the gain can be removed from the integral, replacing the limits of integration by these dimensions. Assuming the origin to be in the center of the active region, and also resetting $\bar{g} = gn/\bar{n}$ (where n is the refractive index of the active material), we have

$$\langle g \rangle = g \frac{n}{\bar{n}} \frac{\displaystyle\int_{-L_a/2}^{L_a/2} \int_{-w/2}^{w/2} \int_{-d/2}^{d/2} |U(x, y)|^2 \, 2 \cos^2 \beta z \, dx \, dy \, dz}{L \displaystyle\iint_{xy} |U(x, y)|^2 \, dx \, dy}. \tag{A5.11}$$

The integration over x and y yields a transverse confinement factor Γ_{xy} and a

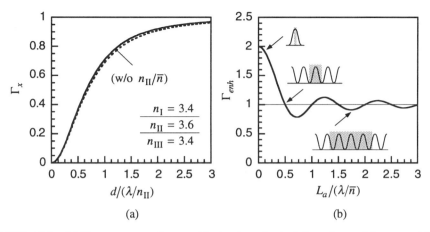

FIGURE A5.2 (a) Transverse confinement factor for a typical three-layer slab waveguide vs. normalized waveguide thickness. (b) Enhancement factor vs. normalized active length (where $\bar{n}$ is the effective index of the guide). The insets display the standing wave pattern in the cavity and its overlap with the active material.

transverse modal gain, $\langle g \rangle_{xy} = \Gamma_{xy} g$, which defines the incremental rate of growth of the fields in the active section (aside from any losses present). The integration over z gives an axial confinement factor Γ_z and the overall cavity modal gain:

$$\langle g \rangle = \Gamma_z \langle g \rangle_{xy} = \Gamma_{xy} \Gamma_z g, \tag{A5.12}$$

where

$$\Gamma_{xy} = \frac{n}{\bar{n}} \frac{\displaystyle\int_{-w/2}^{w/2} \int_{-d/2}^{d/2} |U(x, y)|^2 \, dx \, dy}{\displaystyle\iint_{xy} |U(x, y)|^2 \, dx \, dy}, \tag{A5.13}$$

$$\Gamma_z = \frac{1}{L} \int_{-L_a/2}^{L_a/2} 2 \cos^2 \beta z \, dz = \frac{L_a}{L} \left[1 + \frac{\sin \beta L_a}{\beta L_a} \right]. \tag{A5.14}$$

The ratio $n/\bar{n}$ in Eq. (A5.13) is usually close to unity and is typically neglected. The axial confinement factor can be further separated into a fill factor, $\Gamma_{fill} = L_a/L$, and an enhancement factor, $\Gamma_{enh} = 1 + \sin(\beta L_a)/\beta L_a$, such that $\Gamma_z = \Gamma_{fill} \Gamma_{enh}$. Figure A5.2 plots the one-dimensional transverse confinement factor, Γ_x, for a symmetric three-layer slab waveguide geometry (with and without $n/\bar{n}$), and Γ_{enh} in general.

A5.4 INTERNAL MODAL LOSS

Equation (A5.7) with g replaced by α_i gives the correct expression for the internal modal absorption loss, $\langle \alpha_i \rangle$. Unfortunately, it does not simplify quite as much as the modal gain, even if it is taken as constant within certain regions. This is because the passive loss has a value everywhere, unlike the gain. However, because we have already defined the transverse and axial confinement factors above, we can still construct an abbreviated form for the modal loss for this particular case, if we assume that the loss is constant within the active and passive channel waveguide regions as well as in all of the surrounding cladding material. We take the respective loss values as α_{ia}, α_{ip}, α_{ic}, as labeled in Fig. A5.1. Then using Eq. (A5.13) we can define *transverse* modal averages within both active and passive sections as

$$
\begin{aligned}
\langle \alpha_i \rangle_{xy}^a &= \Gamma_{xy} \alpha_{ia} + (1 - \Gamma_{xy}) \alpha_{ic}, \\
\langle \alpha_i \rangle_{xy}^p &= \Gamma_{xy} \alpha_{ip} + (1 - \Gamma_{xy}) \alpha_{ic}.
\end{aligned}
\tag{A5.15}
$$

The complete modal average over the entire cavity (assuming the group velocity is similar in both sections) is then found using Eq. (A5.14):

$$
\langle \alpha_i \rangle = \Gamma_z \langle \alpha_i \rangle_{xy}^a + (1 - \Gamma_z) \langle \alpha_i \rangle_{xy}^p.
\tag{A5.16}
$$

Combining these equations, we obtain:

$$
\langle \alpha_i \rangle = \Gamma \alpha_{ia} + \Gamma_{xy}(1 - \Gamma_z) \alpha_{ip} + (1 - \Gamma_{xy}) \alpha_{ic},
\tag{A5.17}
$$

where the cavity confinement factor, $\Gamma = \Gamma_{xy} \Gamma_z$. In some cases, the modal loss expression can become even more complicated than listed here. For example, the P and N cladding regions on either side of the active region may have different losses. Furthermore, in separate confinement lasers (including quantum-well lasers), there are additional waveguide (barrier) layers which have different losses than the active and cladding materials. In these cases the transverse confinement factor must be calculated for every layer that has a unique loss value. In VCSELs, the loss is often a function of the axial direction. In such cases, the axial confinement factor for the internal loss must be calculated by integrating Eq. (A5.14), including $\alpha_i(z)$ as well as additional axial variations of the standing wave pattern such as an exponential decay into a distributed Bragg reflector, as considered in Chapter 3. Because the exact form for modal loss is very cavity-specific, we will continue to refer to cavity modal loss as $\langle \alpha_i \rangle$ in most expressions.

A5.5   MORE EXACT ANALYSIS OF THE ACTIVE/PASSIVE SECTION CAVITY

Sections A5.3 and A5.4 make two simplifying assumptions in defining the various axial averages: (1) the propagation constant β in Eq. (A5.9) is purely real, and (2) the group velocities in the active and passive sections are the same. In this section we will remove these two assumptions. However, to retain relatively simple expressions we must assume that no reflections exist at the active-passive interface. For this case, $\bar{n}|\mathscr{E}_T|^2$ is preserved across the interface and multiple bounces in the cavity are eliminated. Standing wave effects are also ignored since the following analysis applies primarily to in-plane lasers. However, footnotes are provided to indicate where standing wave effects would alter the result. We will first examine the axial confinement factor and then describe the threshold condition and differential efficiency more carefully than considered in Chapter 2.

A5.5.1 Axial Confinement Factor

If we include the growth of the fields within the active section (i.e. we allow β to be complex in Eq. (A5.9)), the axial confinement factor (neglecting coherent effects) is no longer equal to the simple geometric fill factor, L_a/L. This can understood by examining Fig. A5.3 which illustrates the power flowing back and forth in a laser cavity assuming the active section provides net gain and the passive section provides loss. The area under the power flow curves divided by the group velocity in each section is proportional to the total energy in the cavity. From the figure it is evident that the fractional area contained in the active section is nontrivial when the fields display exponential growth and decay characteristics.

To determine the fractional energy contained in the active section (i.e. the axial confinement factor), we need to integrate the power curves along z. The functionality of each curve is provided in Fig. A5.3. Performing the integrations (neglecting coherent effects) and weighting the area in each section by the group velocity in accordance with Eq. (A5.7), the axial confinement factor becomes[1]

$$\Gamma_z = \frac{\dfrac{P_1 + P_2}{\ln\sqrt{1/R_1 R_2}} L_a/v_{ga}}{\dfrac{P_1 + P_2}{\ln\sqrt{1/R_1 R_2}} L_a/v_{ga} + \dfrac{P_2 - P_3}{\ln\sqrt{R_3/R_2}} L_p/v_{gp}}, \qquad (A5.18)$$

where

$$P_i \equiv \frac{1}{\sqrt{R_i}} - \sqrt{R_i} = \text{relative power out mirror } i. \qquad (A5.19)$$

Equation (A5.18) is equivalent to $\Gamma_z = W_E^a/W_E$, where W_E^a is the total energy within the active section. If the passive section has no loss, then the second term in the denominator (i.e. W_E^p normalized to the power at the crossing point) reduces to $(1/\sqrt{R_3} + \sqrt{R_3}) L_p/v_{gp}$.

Typically the net transverse modal gain in the active section (i.e. $\langle g \rangle_{xy} - \langle \alpha_i \rangle_{xy}^a$)

---

[1] If the active section is very short, Eq. (A5.18) can be multiplied by Γ_{enh} as defined in Eq. (A5.14) to approximately account for standing wave effects.

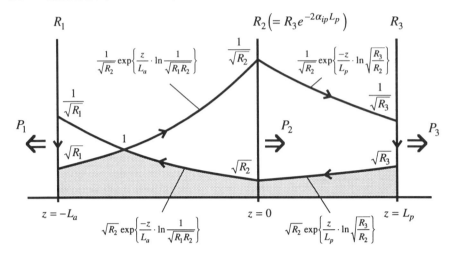

FIGURE A5.3 Illustration of power flow in a laser cavity with active and passive sections, assuming no reflections at the active-passive interface. R_2 defines the *effective* power reflectivity of the passive mirror (R_3) as seen by the active section at $z = 0$. The magnitudes of each power curve at the interfaces are indicated relative to the power level at the crossing point inside the active section. The functionality of each curve is also provided. In the active section, the threshold condition is used to express the *net* growth rate as: $\langle g \rangle_{xy} - \langle \alpha_i \rangle_{xy}^a \rightarrow \ln(1/\sqrt{R_1 R_2})/L_a$. In the passive section, the definition of R_2 is used to express the decay rate as $\alpha_{ip} \equiv \langle \alpha_i \rangle_{xy}^p \rightarrow \ln(\sqrt{R_3/R_2})/L_p$.

is larger than the loss in the passive section $\langle \alpha_i \rangle_{xy}^p$, implying a larger bowing of the power flow curves and hence a larger reduction of the area under the curves in the active section. As a result, the fill factor tends to be smaller than the geometric fill factor. However, the more linear the curves are, the more Eq. (A5.18) resembles the geometric fill factor. For example, linear curves imply low gain or high mirror reflectivities. If we set $R_i = 1 - T_i$ and examine the limits as $R_1, R_2, R_3 \rightarrow 1$, we find that $P_i \rightarrow T_i$, $\ln[1/R_1 R_2] \rightarrow T_1 + T_2$, and $\ln[R_3/R_2] \rightarrow T_2 - T_3$. The axial confinement factor in this limit reduces to:

$$\Gamma_z \rightarrow \frac{\bar{n}_{ga} L_a}{\bar{n}_{ga} L_a + \bar{n}_{gp} L_p}. \qquad \text{(roughly linear growth/decay)} \qquad \text{(A5.20)}$$

This limiting form is convenient to use even though it usually overestimates the axial confinement factor slightly. Note that the lengths are weighted by the group index. For monolithic active and passive sections, the group index does not vary much and the group indices can be cancelled out of the expression. However, if an external cavity is coupled to a laser, then $\bar{n}_{gp} \rightarrow 1$ and the inclusion of the group indices becomes important.

To include reflections at the active-passive interface in the derivation of the axial confinement factor, one must determine the power distribution throughout the entire cavity, and then calculate the fraction of energy contained in the active section.

A5.5.2 Threshold Condition and Differential Efficiency

With the axial confinement factor defined, we can now determine the threshold condition and differential efficiency of the active/passive section cavity shown in Fig. A5.3. The threshold condition can be found by setting the energy generation rate equal to the loss rates created by absorption and power coupled out of the cavity (i.e. $P_1 + P_3$). From Eqs. (A5.2) and (A5.12) we can write the total energy generation rate per unit time as $\Gamma_z v_{ga} \langle g \rangle_{xy} W_E$. Defining a similar term for the absorption losses per unit time and dividing by the total energy W_E, we obtain

$$\Gamma_z v_{ga} \langle g \rangle_{xy} = \Gamma_z v_{ga} \langle \alpha_i \rangle_{xy}^a + (1 - \Gamma_z) v_{gp} \langle \alpha_i \rangle_{xy}^p + \frac{P_1 + P_3}{W_E}. \qquad (A5.21)$$

The last term corresponds to the fractional mirror loss per unit time. Alternatively, by following the power curves through one round-trip of the cavity similar to the procedure outlined in Chapter 2 (see Section 2.5 and Eq. (2.22)), we obtain the more standard version of the threshold condition:

$$\langle g \rangle_{xy} L_a = \langle \alpha_i \rangle_{xy}^a L_a + \langle \alpha_i \rangle_{xy}^p L_p + \ln \frac{1}{\sqrt{R_1 R_3}}. \qquad (A5.22)$$

With this version, if we divide by L we obtain $\Gamma_z = L_a/L$ regardless of any exponential field growth in the cavity or differences in group velocity between the active and passive sections.[2] In contrast, Γ_z in Eq. (A5.21) is the actual axial energy confinement factor (A5.18). The two versions of the threshold condition are equivalent but appear different because the former equates the total generation and loss rates, while the latter simply equates the accumulated exponential growth and decay factors. So while the latter version (A5.22) is more convenient to use, it lacks a one-to-one correspondence with the generation and loss rates of the active/passive section laser.

For example, the $(1/L) \ln \sqrt{1/R_1 R_3}$ term in Eq. (A5.22) is usually associated with the mirror loss. However, with Eq. (A5.21) we can identify the true mirror loss rate per unit *active* length as: $\alpha_m = (P_1 + P_3)/(v_{ga} W_E)$. Using Eq. (A5.18) to set $W_E = W_E^a/\Gamma_z$, the true mirror loss becomes[3]

$$\alpha_m = \frac{P_1 + P_3}{P_1 + P_2} \frac{\Gamma_z}{L_a} \ln \frac{1}{\sqrt{R_1 R_2}}. \qquad (A5.23)$$

---

[2] In very short active sections, a more careful round-trip analysis including reflections at the active-passive interface(s) created by the gain discontinuity would yield $\Gamma_{enh} L_a/L$ instead of L_a/L.

[3] For very short active sections where standing wave effects are important, $\Gamma_z = \Gamma_{enh} W_E^a/W_E$ so that $\Gamma_z \to \Gamma_z/\Gamma_{enh}$ in the expression for the true mirror loss. In other words, standing wave enhancements to the gain do not affect the mirror loss.

With no passive section, $\alpha_m \to (1/L_a) \ln \sqrt{1/R_1 R_2}$. Thus, the mirror loss deviates from the standard definition only with the addition of a passive section. With no loss in the passive section, $P_2 = P_3$, $R_2 = R_3$, and $\alpha_m \to (\Gamma_z/L_a) \ln \sqrt{1/R_1 R_3}$. With high mirror reflectivities and similar group velocities, $\Gamma_z \to L_a/L$. Combining these restrictions we recover the standard definition: $\alpha_m \to (1/L) \ln \sqrt{1/R_1 R_3}$. In a VCSEL which typically has small losses and high mirror reflectivities, $\alpha_m \to \frac{1}{2}(T_1 + T_3)/(L_a + L_p v_{ga}/v_{gp})$.

Because $\Gamma_z \neq L_a/L$, we also know that the total generation rate is not exactly $\langle g \rangle_{xy} L_a/L$ and the total absorption loss rate is not exactly $\langle \alpha_i \rangle_{xy}^a L_a/L + \langle \alpha_i \rangle_{xy}^p L_p/L$ as one is tempted to conclude from Eq. (A5.22). This observation has no effect on the threshold value of $\langle g \rangle_{xy}$ since both Eqs. (A5.21) and (A5.22) predict the same value. However, it does become important when defining the differential efficiency which involves the ratio of these terms. Using Eqs. (A5.12) and (A5.23), the differential efficiency of the active/passive section cavity becomes

$$\eta_d = \eta_i \frac{\alpha_m}{\langle g \rangle} = \eta_i \frac{\ln \sqrt{1/R_1 R_2}}{\langle g \rangle_{xy} L_a} \frac{P_1 + P_3}{P_1 + P_2} = \eta_i \eta_{da} \eta_{dp}, \tag{A5.24}$$

where

$$\eta_{da} = \frac{\ln \sqrt{1/R_1 R_2}}{\langle g \rangle_{xy} L_a} = \frac{\ln \sqrt{1/R_1 R_2}}{\langle \alpha_i \rangle_{xy}^a L_a + \ln \sqrt{1/R_1 R_2}}, \tag{A5.25}$$

$$\eta_{dp} = \frac{P_1 + P_3}{P_1 + P_2} = \frac{(1 - R_1)/\sqrt{R_1} + (1 - R_3)/\sqrt{R_3}}{(1 - R_1)/\sqrt{R_1} + (1 - R_3 e^{-2\alpha_{ip} L_p})/\sqrt{R_3} e^{-\alpha_{ip} L_p}}. \tag{A5.26}$$

The active and passive section efficiencies, η_{da} and η_{dp}, were expanded using Eqs. (A5.22) and (A5.19) with $R_2 = R_3 e^{-2\alpha_{ip} L_p}$.

To more clearly see the high mirror reflectivity limit, we can rewrite Eq. (A5.26) using hyperbolic definitions and algebraic manipulations to obtain

$$\eta_{dp} = \frac{\sinh \frac{1}{2}[\ln \sqrt{1/R_1 R_3}]}{\sinh \frac{1}{2}[\langle \alpha_i \rangle_{xy}^p L_p + \ln \sqrt{1/R_1 R_3}]} \frac{\cosh \frac{1}{2}[\ln \sqrt{R_1/R_3}]}{\cosh \frac{1}{2}[\langle \alpha_i \rangle_{xy}^p L_p + \ln \sqrt{R_1/R_3}]}. \tag{A5.27}$$

For $R_1, R_3 \to 1$, $R_1 \sim R_3$, and $\alpha_{ip} L_p \ll 1$, we can replace the sinh functions by their arguments and set the cosh functions to one, which gives

$$\eta_{dp} \approx \frac{\ln \sqrt{1/R_1 R_3}}{\langle \alpha_i \rangle_{xy}^p L_p + \ln \sqrt{1/R_1 R_3}}, \tag{A5.28}$$

and

$$\eta_{da} \eta_{dp} \approx \frac{\ln \sqrt{1/R_1 R_3}}{\langle \alpha_i \rangle_{xy}^a L_a + \langle \alpha_i \rangle_{xy}^p L_p + \ln \sqrt{1/R_1 R_3}}. \tag{A5.29}$$

In this limit, the differential efficiency reduces to the ratio one would assume using the standard threshold condition (A5.22).

Another consideration is the fraction of power out of each facet. With $r_i = \sqrt{R_i}$ and $P_i = (1 - r_i^2)/r_i$ from Eq. (A5.19) we quickly find that

$$F_1 = \frac{P_1}{P_1 + P_3} = \frac{(1 - r_1^2)/r_1}{(1 - r_1^2)/r_1 + (1 - r_3^2)/r_3}, \tag{A5.30}$$

$$F_3 = \frac{P_3}{P_1 + P_3} = \frac{(1 - r_3^2)/r_3}{(1 - r_1^2)/r_1 + (1 - r_3^2)/r_3}. \tag{A5.31}$$

When F_1 is multiplied by $t_1^2/(1 - r_1^2)$ and F_3 is multiplied by $t_3^2/(1 - r_3^2)$ to account for possibly lossy mirrors, we obtain the same expression (3.30) as derived in Chapter 3. Thus, the fraction of power out is not affected by the lossy passive section.

In practice, Γ_z does not usually deviate substantially from the geometric fill factor. As a result, the terms comprising the standard version of the threshold condition (A5.22) usually represent reasonable approximations to the generation and loss rates, implying that the differential efficiency is usually well approximated by Eq. (A5.29). It is for this reason that we prefer to use the simpler more intuitive Eqs. (A5.22) and (A5.29) throughout this book. However, there are certain situations where Eq. (A5.29) fails to predict the differential efficiency accurately.

For example, with $R_1 = 0.1$ and $R_3 = 1$, the approximate differential efficiency (A5.29) is within 5% of the exact Eq. (A5.26) for $\alpha_{ip} L_p \leq 1$. For uncoated facets ($R_1 = R_3 = 0.3$), it overestimates the exact value by $\sim 10\%$ for $\alpha_{ip} L_p = 0.5$, and $\sim 30\%$ for a larger loss of $\alpha_{ip} L_p = 1$. For $R_1 = 1$ and $R_3 = 0.1$, the overestimate is close to 25% for $\alpha_{ip} L_p = 0.5$, and is almost 60% for $\alpha_{ip} L_p = 1$. In general, the approximation gets worse for smaller values of R_3 and larger values of $\alpha_{ip} L_p$, but is not significantly affected by the value of R_1. Hence, for $\alpha_{ip} L_p \leq 0.5$ and $R_3 \geq 0.3$, the approximate differential efficiency (A5.29) is generally fairly accurate.

It should be noted in closing this section that if there are reflections at the active-passive interface, then Γ_z and all dependent expressions will be different than given here.

A5.6 EFFECTS OF DISPERSION ON MODAL GAIN

Finally we would like to consider parenthetically the effects of material and waveguide dispersion on the modal gain. Equation (A5.8) reveals that $\langle g \rangle \propto 1/\bar{n}$. Physically, this dependence reflects the fact that the effective index controls the angle at which plane waves bounce down the guide (see Chapter 7). The smaller the effective index, the more bounces the plane waves make per unit length. As a result, the mode effectively "sees" more active material per unit length, which shows up as an enhancement in the modal gain per unit length. In other words, waveguide dispersion slows down the mode, allowing more stimulated transitions to be acquired into the mode per unit length. Thus, the modal gain per unit length is *enhanced by waveguide dispersion*. As for material dispersion, we find that (A5.8) contains no dependence on the overall group velocity, implying that $\langle g \rangle$ is *independent of material dispersion*.

The gain per unit time defined by $\langle G \rangle = v_g \langle g \rangle$, is inversely related to the group effective index since $\langle g \rangle$ is independent of $\bar{n}_g$. Now because $\bar{n}_g > n$ in most cases, we conclude that *waveguide and material dispersion reduce* the modal gain per unit time. The waveguide dispersion component of $\bar{n}_g$ is roughly compensated by the $1/\bar{n}$ implicit in $\langle g \rangle$. Neglecting waveguide dispersion, we still find that $\langle G \rangle$ is compromised by material dispersion. Physically this occurs because the local energy density in a dispersive medium is $\propto n_g n |E|^2$ (where n_g is the group index associated with material dispersion exclusively). Thus, an increase in the group index for a given energy density compromises the electric field strength, which reduces the local stimulated emission rate. The result is a reduced modal gain per unit time in the cavity. The modal gain per unit length does not suffer this consequence, because as n_g increases, the wave moves slower, allowing more stimulated transitions to be acquired per unit length. The reduction in field strength and the slowing down of the mode offset each other, leaving the modal gain per unit length independent of material dispersion, as concluded above.

Einstein's Approach to Gain and Spontaneous Emission

A6.1 INTRODUCTION

Equation (2.14) gives us the relationship between gain and the stimulated recombination rate,

$$R_{st} = v_g g N_p. \tag{A6.1}$$

As shown in Fig. 1.3, the net stimulated rate $R_{st} = R_{21} - R_{12}$, is the stimulated emission less the stimulated absorption of photons. Thus, we wish to calculate R_{st}, from which the gain, g, can be obtained. As also suggested in Chapters 1 and 2, the stimulated emission and absorption rates depend on the number of available electronic states and their probability of occupation in both the conduction and valence bands for the transitions to occur. The unknowns are the multiplicative rate constants. Once these are determined, we can calculate R_{st} and the gain, g.

Einstein gave us a technique to calculate these rates without delving too deeply into the details of the stimulated emission physics. His technique is to determine the desired rate constants under a particular set of boundary conditions. Once obtained, however, these constants are generally applicable to other situations. As shown in Fig. A6.1, the medium of interest is placed in a closed cavity, which has neither inputs nor outputs, and held under thermal equilibrium. Then, a dynamic balance equation can be set up which expresses the desired rates in terms of the equilibrium optical energy density, $W(v)$. Since this is known (from Appendix 4), the rate constants can be determined.

In this situation we must include all carrier recombination mechanisms in writing a dynamic balance equation. The nonradiative rates generate heat, which is naturally taken into account in this closed system. By one means or another, they must saturate in equilibrium. Thus, in equilibrium, for a pair of energies, E_2 and E_1, in the conduction and valence bands, respectively,

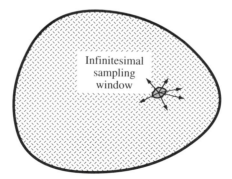

FIGURE A6.1 Schematic of an arbitrary closed cavity with negligible energy in or out. Contents must be in thermal equilibrium.

we have

$$\frac{dN_p}{dt} = 0 = R_{21} + R_{sp,21} - R_{12}, \tag{A6.2}$$

where the first two terms represent electrons recombining by either stimulated or spontaneous processes, respectively, and the last term represents electrons being generated by stimulated absorption. The 21 subscript on R_{sp} distinguishes this two-level spontaneous rate from the net recombination between two bands in a semiconductor as we have considered elsewhere. We could summarize Eq. (A6.2) by saying that the downward transition rate (i.e., conduction to valence band) must equal the upward transition rate.

The equilibrium occupation probability at some temperature is given by the Fermi function introduced in Appendix 2

$$f_i = \frac{1}{e^{(E_i - E_F)/kT} + 1}, \tag{A6.3}$$

where $i = 1$ or 2 for the involved transition energies in the valence or conduction bands, respectively. That is, $f_2 \equiv$ fraction of states filled at E_2, and $f_1 \equiv$ fraction of states filled at E_1. Figure A6.2 illustrates the various energy levels for reference.

In Appendix 1, we also defined a density of states, $\rho(E)$, to describe the distribution of states in a band. The number of states are equally distributed in k-space, but by integrating $\rho(E)$ over some energy range, the number of states in that range is obtained. As shown in Figure A6.2, the radiative recombination of an electron and hole involves states in the conduction and valence bands with the same k-vector. That is, both energy ($E_{21} = h\nu_{21}$) and momentum ($\hbar\mathbf{k}$-electron $\approx \hbar\mathbf{k}$-hole) conservation must be satisfied. (As discussed in Chapter 4, the photon momentum is negligible.) Thus, we can consider the *density of*

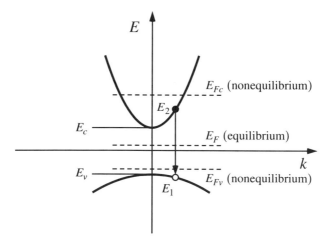

FIGURE A6.2   Energy vs. momentum schematic illustrating a transition between two energy states in the conduction and valence bands, respectively. Quasi-Fermi levels for both equilibrium and nonequilibrium carrier densities are also illustrated.

state pairs, or a reduced density of states, $\rho_r(E_{21})$, in calculating emission at E_{21} from electron–hole recombination. We shall explicitly derive $\rho_r(E_{21})$ for parabolic bands to obtain an analytic expression, however, the concept is entirely general.

With reference to Fig. A6.2 and using the results of Appendix 1, we can express the transition energy difference, $E_2 - E_1 = E_{21}$ as

$$E_{21} = E_g + \frac{\hbar^2 k^2}{2m_c^*} + \frac{\hbar^2 k^2}{2m_v^*} = E_g + E' \frac{m_v^* + m_c^*}{m_v^*}, \qquad (A6.4)$$

where $E' = \hbar^2 k^2 / 2m_c^*$ for parabolic bands. For relatively thick active regions (bulk), we found from Eq. (A1.39) in Appendix 1 that

$$\rho(E')\,dE' = \frac{1}{2\pi^2}\left[\frac{2m_c^*}{\hbar^2}\right]^{3/2}(E')^{1/2}\,dE'. \qquad (A6.5)$$

Solving for E' from Eq. (A6.4), and forming dE', we have

$$E' = \frac{m_v^*}{m_c^* + m_v^*}(E_{21} - E_g),$$

$$dE' = \frac{m_v^*}{m_c^* + m_v^*}\,dE_{21}. \qquad (A6.6)$$

Finally, plugging Eq. (A6.6) into Eq. (A6.5), we obtain the desired reduced density of states,

$$\rho_r(E_{21}) = \frac{1}{2\pi^2}\left[\frac{2m_r^*}{h^2}\right]^{3/2}(E_{21} - E_g)^{1/2}, \tag{A6.7}$$

where $m_r^* = m_c^* m_v^*/(m_c^* + m_v^*)$.

A6.2   EINSTEIN A AND B COEFFICIENTS

The general approach of Einstein was to assign rate constants to the three radiative processes appearing in Eq. (A6.2), with the assumption that these rates must be proportional to the carrier density. These were written empirically as $R_{sp,21} = A_{21}N_2$, $R_{21} = B_{21}W(v)N_2$, and $R_{12} = B_{12}W(v)N_1$. Generally speaking, the A rate constant is associated with spontaneous processes, while the B rate constants are associated with stimulated processes, and hence are weighted by the radiation *spectral* density, $W(v)$ (introduced in Appendix 4). In Einstein's day, most radiative transitions of interest took place between atoms with very isolated, sharp energy levels. The carrier densities therefore referred to the density of *atoms* with electrons in either energy level 1 or 2. In the current context, we must interpret these definitions somewhat differently because in semiconductors, the energy levels are neither isolated nor sharp.

To deal with the continuous nature of energy states in semiconductors, we restrict our attention to a differential population of state pairs existing between E_{21} and $E_{21} + dE_{21}$. Using the reduced density of states function derived above, the differential population available for producing downward transitions becomes

$$dN_2 = f_2(1 - f_1) \cdot \rho_r(E_{21})\,dE_{21}, \tag{A6.8}$$

where $dN_2 \equiv$ number of state pairs per unit volume between E_{21} and $E_{21} + dE_{21}$ available to interact with photons near E_{21}, in which the upper state is full and the lower state is empty. Similarly, the differential population available for producing upward transitions becomes

$$dN_1 = f_1(1 - f_2) \cdot \rho_r(E_{21})\,dE_{21}, \tag{A6.9}$$

where $dN_1 \equiv$ number of state pairs per unit volume between E_{21} and $E_{21} + dE_{21}$ available to interact with photons near E_{21}, in which the lower state is full and the upper state is empty.

Another factor we must consider in semiconductors is that the states in $\rho_r(E_{21})$ have a finite lifetime due to collisions with other electrons and phonons. As a result, a given differential population can actually appear over a small range of energies and thus interact with photons spanning some narrow energy range, ΔE_{21}. The probability of finding this population at energies away from E_{21} is characterized by some lineshape function, $\mathscr{L}(E - E_{21})$ which has a full-width half-maximum (FWHM) ΔE_{21} and is centered at E_{21} as shown in Fig. A6.3. The longer the lifetime of a given state, the narrower the spread in energy ΔE_{21}, and hence, the more chance there is of finding the state pair at E_{21}.

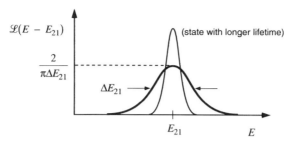

FIGURE A6.3 Plot of lineshape function vs. transition energy. We require that it be normalized so that its area equals one (the state must exist somewhere!). (The peak probability shown assumes a Lorentzian distribution.)

A direct consequence of the finite state lifetime is that when we consider interactions between our differential population and photons of a given energy, hv, we must somehow account for the fact that the population only spends a fraction of its time at that photon energy. In other words, we must weight our differential population by the probability of finding that population between photon energies hv and $hv + h\,dv$, which is given by $\mathscr{L}(hv - E_{21})h\,dv$. The differential population "seen" by photons at energy, hv, therefore becomes

$$dN_2 \rightarrow dN_2 \cdot \mathscr{L}(hv - E_{21})h\,dv,$$
$$dN_1 \rightarrow dN_1 \cdot \mathscr{L}(hv - E_{21})h\,dv. \tag{A6.10}$$

These are the forms for N_1 and N_2 that we must use when analyzing semiconductors.

With our differential populations defined, we can now make use of Einstein's *A* and *B* coefficients. The *differential* downward transition rates created by our differential dN_2 population can be written as

$$dR_{21} + dR_{sp,21} = \int \left[dN_2\,W(v)B_{21} + dN_2\,A_{21} \right]\mathscr{L}(hv - E_{21})h\,dv. \tag{A6.11}$$

The integral over photon frequency is necessary to include contributions that $W(v)$ makes over the full range of energies that dN_2 is spread over.

The differential transition rates on the left-hand side of (A6.11) have units of [volume^{-1} time^{-1}]. The population density, dN_2, has units of [volume^{-1}]. Therefore, the spontaneous recombination rate constant, A_{21}, has units of [time^{-1}] and is often expressed as an inverse spontaneous lifetime, $1/\tau_{sp}$. In atomic systems, A_{21} does represent the inverse of the spontaneous lifetime of energy level 2. However, in the present context, A_{21} is associated with the spontaneous lifetime (which we will denote as τ_{sp}^{21}) of only the differential population, dN_2, and is *not* equal to the entire band-to-band spontaneous lifetime, τ_{sp}. Later we will determine the relationship between the two-level lifetime τ_{sp}^{21} and τ_{sp}. For the stimulated term, we note that $W(v)$ represents the optical energy per unit volume per unit frequency. Therefore, the units of the stimulated recombination rate constant, B_{21}, are [(volume × frequency)/(energy × time)].

The differential upward transition rate created by our differential dN_1 population can be written as

$$dR_{12} = \int [dN_1 W(v)B_{12}]\mathcal{L}(hv - E_{21})h\,dv, \qquad (A6.12)$$

where B_{12} is the stimulated generation rate constant and has the same units as B_{21}.

A6.3 THERMAL EQUILIBRIUM

Einstein's approach allows us to relate the three rate constants appearing in Eqs. (A6.11) and (A6.12) in a straightforward manner. Under thermal equilibrium, we can set (A6.11) equal to (A6.12) according to (A6.2) which we assume holds for differential rates as well as the integrated rates. To remove the integrals over photon frequency, we note that equilibrium blackbody radiation is broadband and varies little over typical linewidths, ΔE_{21}, associated with the lineshape function. This allows us to treat the lineshape as a delta function, or $h\mathcal{L}(hv - E_{21}) \rightarrow \delta(v - v_{21})$. In other words, we simply evaluate all terms under the integral at v_{21}. The balance equation then reduces to

$$dN_2 W(v_{21})B_{21} + dN_2 A_{21} = dN_1 W(v_{21})B_{12}. \qquad (A6.13)$$

Rearranging, we obtain

$$\frac{dN_2}{dN_1} = \frac{W(v_{21})B_{12}}{W(v_{21})B_{21} + A_{21}}. \qquad (A6.14)$$

Alternatively, from Eqs. (A6.8) and (A6.9), using (A6.3) and $E_{21} = hv_{21}$, we find the simple result

$$\frac{dN_2}{dN_1} = \frac{f_2(1 - f_1)}{f_1(1 - f_2)} = e^{-hv_{21}/kT}. \qquad (A6.15)$$

Setting Eq. (A6.15) equal to (A6.14) and solving for $W(v_{21})$, we obtain

$$W(v_{21}) = \frac{A_{21}/B_{21}}{(B_{12}/B_{21})e^{hv_{21}/kT} - 1}. \qquad (A6.16)$$

From Appendix 4, the blackbody radiation formula which defines the spectral density of photons under thermal equilibrium is given by

$$W(v_{21}) = \frac{\rho_0(v_{21})hv_{21}}{e^{hv_{21}/kT} - 1}.$$ (A6.17)

Comparing Eq. (A6.16) with Eq. (A6.17), we see that both can be true for all temperatures only if the following two equalities hold:

$$B_{12} = B_{21},$$ (A6.18)

$$A_{21} = \rho_0(v_{21})hv_{21} \cdot B_{21}.$$ (A6.19)

Equation (A6.18) reveals that stimulated emission and stimulated absorption are truly complementary processes associated with the same rate constant. Perhaps more significantly, Einstein's approach establishes a fundamental link between stimulated and spontaneous emission processes through Eq. (A6.19). Thus, by analyzing the system under thermal equilibrium, Einstein's approach allows us to reduce the three differential rate constants to one independent constant, B_{21}. We could have designated A_{21} as the independent constant, however, too often this leads to the *incorrect* conclusion that B_{21} is inversely dependent on the density of optical modes, $\rho_0(v_{21})$. More correctly, we should view B_{21} as the rate constant of a single optical mode, and A_{21} as this rate constant multiplied by the equivalent spectral density that induces spontaneous emission into the full density of optical modes near v_{21}. It is interesting to note that the equivalent spectral density, $\rho_0(v_{21})hv_{21}$, implies one photon per optical mode. In Chapter 4, more insight into this observation is gained through a quantum mechanical perspective.

A6.4 CALCULATION OF GAIN

Now that we have established the connection between the three rate constants, we can leave the closed-system, thermal-equilibrium environment, and proceed to calculate the gain for a *monochromatic* radiation field under *nonequilibrium* conditions. Under forward bias, the Fermi level in the active region splits into two quasi-Fermi levels to reflect the nonequilibrium electron and hole densities. The splitting corresponds roughly to the applied voltage. The nonequilibrium carrier densities are then calculated from

$$N = \int \rho_c(E)f_c(E)\,dE \approx P = \int \rho_v(E)[1 - f_v(E)]\,dE,$$ (A6.20)

where we have assumed $N \approx P$ because of negligible doping in the active region. The factors, $\rho_c(E)$ and $\rho_v(E)$, refer to the densities of states in the conduction and valence bands, respectively. Also, $f_c(E)$ and $f_v(E)$ are the Fermi functions, in which E_F is replaced by the quasi-Fermi levels, E_{Fc} and E_{Fv}, for the conduction and valence bands, respectively.

To calculate the gain we turn to Eq. (A6.1) which allows us to relate the gain to the net stimulated emission rate. However, we must keep in mind that contributions to the gain at a particular frequency of radiation, v_0, will come from many differential populations distributed roughly over energies comparable to the lineshape width, ΔE_{21}. The gain contributed by each population can be written as

$$dg(hv_0) = \frac{dR_{st}}{v_g N_p}$$

$$= \frac{1}{v_g N_p} \int W(v) B_{21} [dN_2 - dN_1] \mathscr{L}(hv - E_{21}) h \, dv. \quad (A6.21)$$

The second equality is obtained by setting $dR_{st} = dR_{21} - dR_{12}$ and using Eqs. (A6.11) and (A6.12) with $B_{12} = B_{21}$. For a monochromatic field, $W(v) \rightarrow hv_0 N_p \delta(v - v_0)$, where the strength of the delta function is equal to the energy density of the field and v_0 is the frequency of the wave. With this substitution, the integral reduces to evaluating all photon frequency-dependent factors at v_0, and we are left with

$$dg(hv_0) = \frac{hv_0}{v_g} hB_{21}[dN_2 - dN_1] \mathscr{L}(hv_0 - E_{21}). \quad (A6.22)$$

The appearance of the lineshape function reminds us that the further away the differential population is in transition energy from the photon energy, the less contribution it makes to the gain at that frequency.

The total gain at hv_0 is found by integrating dg over all existing populations of state pairs which might possibly interact with the field. Expanding the differential populations in (A6.22) using Eqs. (A6.8) and (A6.9), simplifying the Fermi factors, and integrating over all possible transition energies, we obtain

$$g(hv_0) = \frac{hv_0}{v_g} h \int B_{21} \rho_r(E_{21})(f_2 - f_1) \mathscr{L}(hv_0 - E_{21}) \, dE_{21}. \quad (A6.23)$$

If the energy-dependent factors, $B_{21} \rho_r(E_{21})(f_2 - f_1)$, are slowly varying compared to the lineshape function, then we can set $\mathscr{L}(hv_0 - E_{21}) \rightarrow \delta(hv_0 - E_{21})$ and the gain expression reduces to

$$g_{21} \approx \frac{hv_{21}}{v_g} hB_{21} \rho_r(E_{21})(f_2 - f_1), \quad (\Delta E_{21} \rightarrow 0) \quad (A6.24)$$

where E_{21} is evaluated at the photon energy of interest (i.e., $g_{21} \equiv g(hv_0 = E_{21})$).

Equation (A6.23) is the central result of this section. It reveals that the gain is directly proportional to the rate constant B_{21}, the reduced density of states, and the Fermi probability factors. It is immediately apparent from (A6.24) that to achieve positive gain, we must create enough electrons and holes to allow $f_2 > f_1$. This places a condition on the quasi-Fermi levels which reduces to the requirement that the quasi-Fermi level separation be larger than the incident photon energy, or $\Delta E_F > h v_0$. Chapter 4 considers these issues in more detail.

To fully evaluate the gain, we still need to determine the rate constant, B_{21}. In lasers which use atomic transitions, a measurement of the spontaneous emission linewidth of a given transition can allow us to estimate A_{21} if the broadening of the line is dominated by the spontaneous emission lifetime, $\tau_{sp} = 1/A_{21}$. With this information, B_{21} can readily be determined using (A6.19). Thus, with Einstein's approach and this one simple measurement, the description of gain and spontaneous emission in atomic systems is completely self-contained.

Unfortunately the situation is not so simple in semiconductors since the spontaneous emission spectrum represents a superposition of transitions from all of our differential populations. The resulting broad emission spectrum prevents us from isolating the linewidth of just one differential population, and therefore prevents us from evaluating A_{21} (and hence, B_{21}) via direct experiment. The approach that must be followed in semiconductors is to estimate the transition rates using other more in-depth theories, and then relate the resulting expressions back to B_{21} and A_{21}. Chapter 4 details the theory required to evaluate the transition rates explicitly from a more fundamental quantum mechanical analysis. An explicit expression for B_{21} will be presented there.

We can alternatively express the gain in terms of the spontaneous rate constant, or the two-level lifetime, $\tau_{sp}^{21} = 1/A_{21}$. Using (A6.19) in (A6.24), the gain takes the form

$$g_{21} = \frac{A_{21}}{\rho_0(v_{21})} \frac{h}{v_g} \rho_r(E_{21})(f_2 - f_1)$$

$$= \frac{\lambda_0^2}{8\pi n^2 \tau_{sp}^{21}} h\rho_r(E_{21})(f_2 - f_1), \tag{A6.25}$$

where we have set $\rho_0(v_{21}) = 8\pi n^2/\lambda_0^2 v_g$ according to Eq. (A4.5), with $\lambda_0 \equiv c/v_{21}$. While this expression is equivalent to Eq. (A6.24), it is very deceptive for two reasons.

First of all, a common mistake in the literature is to equate τ_{sp}^{21} with the band-to-band spontaneous lifetime, τ_{sp}, incorrectly linking the gain with the carrier lifetime. In fact, some go even further by linking A_{21} with the overall spontaneous emission bandwidth (analogous to atomic transitions)—a procedure which is completely misguided in semiconductors, but nevertheless encouraged by writing the gain in terms of A_{21}.

The second problem is that when written in this way, one might conclude that the gain varies inversely with the optical mode density. Only upon more careful inspection does one realize that implicit in the two-level lifetime, τ_{sp}^{21}, is an inverse dependence on the mode density (i.e., the higher the mode density, the shorter the lifetime). As a result, the product $\rho_0(v_{21})\tau_{sp}^{21}$ (and hence the gain) becomes independent of the optical mode density—a conclusion which is obvious from Eq. (A6.24). Reduced optical mode densities possible in very small VCSEL structures (or "microcavities") have lead some researchers to conclude using (A6.25) that the gain is increased as a result—again, a conclusion which is misguided. For these reasons, Eq. (A6.24) which more appropriately defines the gain in terms of the single-mode stimulated rate constant, B_{21}, is preferable.

Another issue we need to resolve is whether we should use (A6.24) or the more complex (A6.23) to evaluate the gain. The time between collisions for electrons in typical semiconductors is on the order of 0.1 ps which leads to a FWHM of $\Delta E_{21} \approx 14$ meV (assuming a Lorentzian lineshape). At room temperature, this bandwidth is small enough that we can assume $f_2 - f_1$ to be roughly constant. Furthermore, B_{21} does not have a strong energy dependence. Therefore, our main concern lies with $\rho_r(E_{21})$. In "bulk" active regions, $\rho_r(E_{21})$ varies as $E^{1/2}$ and the rate of change can be neglected in comparison to the bandwidth of the lineshape function. In other words, at room temperature, (A6.24) can generally be used with bulk active regions.

For quantum wells, the reduced density of states can be found by using the reduced mass in Eq. (A1.45). It is zero up to the first allowed energy states in the conduction and valence bands where $n_z = 1$, i.e., $(E_{c1} - E_{h1})$. There it abruptly increases to

$$\rho_r(h\nu_{21}) = \frac{m_r^*}{\pi \hbar^2 d}, \qquad E_{21} > (E_{c1} - E_{h1}) \qquad \text{(A6.26)}$$

where it remains constant up to the point where $n_z = 2$. There it again increases by the same amount. Thus, it violates the assumptions about being slowly varying over the bandwidth ΔE_{21} made between Eq. (A6.23) and Eq. (A6.24). Therefore, we must use Eq. (A6.23). Unfortunately, the actual lineshape function is not well established. A Lorentzian is often used, but the results are somewhat nonphysical; so other, more complex functions have been developed to better fit the experimental data. The simple Lorentzian with an FWHM of ΔE_{21} takes the form

$$\mathcal{L}(E - E_{21}) = \frac{2/\pi \Delta E_{21}}{1 + 4(E - E_{21})^2/\Delta E_{21}^2}. \qquad \text{(A6.27)}$$

The integration of Eq. (A6.23) with Eq. (A6.27) will smooth the discontinuities in $\rho_r(E_{21})$ that exist in quantum-well (as well as quantum-wire and box) lasers. However, the plateau gain levels obtained by inserting Eq. (A6.26) in Eq. (A6.24) will be correct, as long as we are $> \Delta E_{21}$ away from a step edge. The numerical gain calculations in Chapter 4 illustrate this behavior more quantitatively.

A common feature of all active materials (regardless of the lineshape broadening or reduced dimensionality) is that the gain increases from an initial unpumped absorption level given by $g_{21}(f_2 = 0, f_1 = 1)$, to a transparency gain value given by $g_{21}(f_2 = f_1)$, finally approaching a saturation level of $g_{21}(f_2 = 1, f_1 = 0)$ (equal in magnitude to the unpumped absorption level) as more and

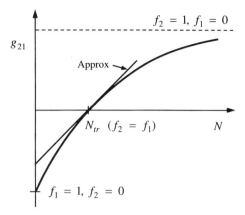

FIGURE A6.4 Illustration of gain vs. carrier density. Straight-line approximation valid over limited ranges.

more carriers are injected into the active region. In Fig. A6.4 we schematically illustrate this characteristic.

In Chapter 4 we shall find that this characteristic can be well approximated by a logarithmic function; however, for many situations only a small portion of the curve near and somewhat above the transparency point is of interest. In these cases, a straight-line approximation is often very useful. That is,

$$g_{21} = a(N - N_{tr}), \tag{A6.28}$$

where a is the differential gain, $\partial g/\partial N$, and N_{tr} is the transparency carrier density.

A6.5   CALCULATION OF SPONTANEOUS EMISSION RATE

In creating a given amount of gain in the semiconductor by increasing dN_2 relative to dN_1, we unfortunately end up creating a large amount of spontaneous emission over a relatively broad range of frequencies. This section deals with developing the relation for the spectrum of spontaneously emitted photons, which when integrated allows us to determine the total number of spontaneous photons being generated per second.

We start by defining the spontaneous emission within a small energy interval to be $R_{sp}^{h\nu}(h\nu)h\,d\nu$, where $R_{sp}^{h\nu}$ is the emission rate per unit energy per unit volume occurring at $h\nu$. As with the gain, we need to sum over all differential populations to determine the emission rate at a single frequency. The probability of dN_2 appearing at $h\nu$ is given by $\mathscr{L}(h\nu - E_{21})h\,d\nu$. Weighting dN_2 by this factor, multiplying by the spontaneous rate constant, A_{21}, and integrating over all state pairs, we obtain

$$R_{sp}^{hv}(hv)h\,dv = \int A_{21}[\mathscr{L}(hv - E_{21})h\,dv]\,dN_2. \tag{A6.29}$$

Canceling out $h\,dv$ on both sides, and expanding dN_2, we arrive at the desired result:

$$R_{sp}^{hv}(hv) = \int A_{21}\rho_r(E_{21})f_2(1 - f_1)\mathscr{L}(hv - E_{21})\,dE_{21}. \tag{A6.30}$$

Equation (A6.30) reveals that the spontaneous emission spectrum is smoothed in the same manner as the gain spectrum discussed above. It is interesting to note that while we must have $E_{21} > E_g$ to have a nonzero reduced density of states, it *is* possible for $hv < E_g$ since $\mathscr{L}(hv - E_{21})$ can be nonzero for $hv - E_{21} < 0$. Thus, spontaneous emission can actually be observed at energies $\approx \Delta E_{21}/2$ below the bandgap. This reflects the uncertainty in the energy of states at the band edge, which results from the finite lifetimes of electrons in those states.

Again if $A_{21}\rho_r(E_{21})f_2(1 - f_1)$ is slowly varying compared to the line-shape function, then we can set $\mathscr{L}(hv - E_{21}) \rightarrow \delta(hv - E_{21})$ and Eq. (A6.30) simplifies to

$$R_{sp}^{21} \approx A_{21}\rho_r(E_{21})f_2(1 - f_1), \qquad (\Delta E_{21} \rightarrow 0) \tag{A6.31}$$

where we have defined $R_{sp}^{21} \equiv R_{sp}^{hv}(hv = E_{21})$ in analogy with g_{21}. Comparing R_{sp}^{21} to g_{21} given in Eq. (A6.24), it is interesting to note that the two are quite similar. In fact, we can express the spontaneous emission at E_{21} in terms of the gain at E_{21} as follows:

$$R_{sp}^{21} = \frac{v_g}{h^2 v_{21}}\frac{A_{21}}{B_{21}}\frac{f_2(1 - f_1)}{(f_2 - f_1)}g_{21} = \frac{1}{h}\rho_0(v_{21})\cdot v_g n_{sp}g_{21}, \tag{A6.32}$$

where we have made use of the relation between the rate constants expressed in Eq. (A6.19). We have also introduced the *population inversion factor* which is defined as

$$n_{sp} = \frac{f_2(1 - f_1)}{(f_2 - f_1)} = \frac{1}{1 - e^{(hv_{21} - \Delta E_F)/kT}}. \tag{A6.33}$$

The popular usage of the sp subscript originates from the fact that n_{sp} was initially referred to as the spontaneous emission factor, but was later changed to eliminate conflict with β_{sp} which is also defined as the spontaneous emission factor. We now refer to n_{sp} as the population inversion factor because it is the semiconductor laser equivalent to the ratio $N_2/(N_2 - N_1)$ encountered in atomic laser physics. In atomic systems, when $N_2 - N_1 > 0$, the population is said to be *inverted*, the ratio $N_2/(N_2 - N_1)$ is positive, and optical gain is achieved. Similarly, when $n_{sp} > 0$, a population inversion is established in the semiconductor, indicating a net optical gain.

If the quasi-Fermi level separation in the semiconductor is known, then using Eq. (A6.32), we can determine the spontaneous emission rate if we know the gain, or we can determine the gain if the spontaneous emission rate is known. In Chapter 4 a more thorough investigation of this fundamental relationship between spontaneous emission and gain is provided.

Now we wish to determine the total spontaneous emission rate, R_{sp}, by integrating (A6.30) over all photon energies:

$$
\begin{aligned}
R_{sp} &= \int R_{sp}^{hv}(hv) h\, dv \\
&= \iint A_{21}\rho_r(E_{21}) f_2(1 - f_1)\mathscr{L}(hv - E_{21})\, dE_{21} h\, dv \\
&\approx \int A_{21}\rho_r(E_{21}) f_2(1 - f_1) \left[\int \mathscr{L}(hv - E_{21}) h\, dv \right] dE_{21} \\
&= \int A_{21}\rho_r(E_{21}) f_2(1 - f_1)\, dE_{21} = \int R_{sp}^{21}\, dE_{21}.
\end{aligned}
\tag{A6.34}
$$

The third equality is found by inverting the order of integration and pulling out all terms not dependent on the photon frequency. The approximate sign is used here because A_{21} is not completely independent of photon frequency ($A_{21} \propto v$, from Eq. (A6.19) with $\rho_0(v) \propto v^2$ and $B_{21} \propto v^{-2}$ (see Chapter 4)). However, in comparison to $\mathscr{L}(hv - E_{21})$, this dependence can be neglected. The integral in brackets then reduces to unity which leads to the fourth equality. In other words, the lineshape broadening has no effect on the total spontaneous emission rate, and we can simply integrate over the simplified R_{sp}^{21} defined in (A6.31).

Setting $A_{21} = 1/\tau_{sp}^{21}$ and $R_{sp} = N/\tau_{sp}$ in (A6.34), we can define the total spontaneous lifetime in terms of the local spontaneous lifetime through the following nontrivial relation:

$$
\tau_{sp} \equiv N \left[\int \frac{1}{\tau_{sp}^{21}} \rho_r(E_{21}) f_2(1 - f_1)\, dE_{21} \right]^{-1}
\tag{A6.35}
$$

Generally speaking, the term in brackets will tend to go as N^2 due to the double dependence on the quasi-Fermi levels (f_2 related to N and $1 - f_1$ related to P). As a result, the total spontaneous lifetime typically follows a $1/N$ dependence as assumed in Chapter 2.

In performing actual calculations of the total spontaneous emission rate, it is useful to replace A_{21} with the single-mode rate constant, B_{21}. Doing this, Eq. (A6.34) becomes

$$
R_{sp} = \int B_{21} hv_{21}\rho_0(v_{21})\rho_r(E_{21}) f_2(1 - f_1)\, dE_{21}.
\tag{A6.36}
$$

Thus, we see that the spontaneous emission rate includes both optical and electronic density of states functions.

By reducing the cavity size to dimensions on the order of the emission wavelength in the material, it is in principle possible to significantly alter the optical mode density which allows us to actually alter the spontaneous emission rate. An active field of research which studies these microcavity effects is attempting to reduce the spontaneous emission rate substantially. The motivation for this lies in the following relation:

$$R_{sp} = \eta_i \eta_r \frac{I}{qV}. \tag{A6.37}$$

That is, the total spontaneous emission rate represents the number of carriers lost to spontaneous recombination per second, and can therefore be equated with the radiative portion of the injected current. By minimizing R_{sp}, researchers hope to minimize the threshold current of certain types of lasers. In particular, VCSELs represent excellent candidates for such experiments due to their scalable geometry.

In concluding this appendix, it is useful to appreciate that the carrier density, the optical gain, and the radiative current in the active region can all be calculated from the quasi-Fermi levels using Eqs. (A6.20), (A6.23), (A6.36), and (A6.37). Thus, E_{Fc} and E_{Fv} *completely determine* all relevant parameters under nonequilibrium conditions. Furthermore, by invoking charge neutrality, we can find E_{Fc} for a given E_{Fv} using (A6.20), reducing the entire problem to one independent parameter. In other words, we can obtain gain vs. current, gain vs. carrier density, or current vs. carrier density, by scanning E_{Fv} over the appropriate ranges. The linear relationship of gain to carrier density, and the quadratic relationship of current to carrier density discussed in Chapter 2 represent *approximations* to the more rigorous nontrivial relationships derived in this appendix.

READING LIST

J.T. Verdeyen, *Laser Electronics*, 2d ed., Chs. 7 and 11, Prentice Hall, Englewood Cliffs, NJ (1989).

Periodic Structures and the Transmission Matrix

A7.1 INTRODUCTION

Distributed Bragg reflectors (DBRs) are important in many laser applications because (1) they can provide extremely high reflectivities, and (2) they can be used as frequency-selective filters. If the dielectric stack is completely periodic, the entire reflection spectrum can be determined exactly. If the dielectric stack is not perfectly periodic, then for practical purposes, only the peak reflectivity can be determined analytically. The main portion of this appendix deals with periodic stacks, first at the Bragg frequency to obtain the peak reflectivity, and then away from the Bragg frequency to determine the entire reflection spectrum. Correspondence is then made between the exact analysis and approximate Fourier and coupled-mode analyses. Finally, nonperiodic dielectric stacks are considered at the Bragg frequency.

A7.2 EIGENVALUES AND EIGENVECTORS

We begin with a homogeneous dielectric slab as shown in Fig. A7.1. Dividing up the slab into half-wave segments, we can assign normalized magnitudes to the forward- and backward-propagating fields at the ith plane:

$$\mathbf{e}_i \equiv \begin{bmatrix} A_i \\ B_i \end{bmatrix}. \tag{A7.1}$$

Here $\mathbf{e}_i$ is defined as a two-component field vector, comprised of a forward wave of normalized amplitude, A_i, and a backward wave of normalized amplitude, B_i.

473

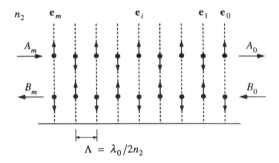

FIGURE A7.1 Propagation of fields through a uniform dielectric slab of index n_2, with reference planes placed at half-wavelength intervals. Arrows indicate the phase of the forward and backward waves.

Now because the planes are spaced by a half-wavelength, the phases of the waves must change by π as we move from one plane to the next, and therefore $\mathbf{e}_{i+1} = -\mathbf{e}_i$. We can also express this relation in terms of the half-wave phase delay transmission matrix introduced in Chapter 3:

$$\mathbf{e}_{i+1} = [\mathbf{T}_{\lambda/2}]\mathbf{e}_i = e^{j\pi}\mathbf{e}_i. \tag{A7.2}$$

Thus, the transmission matrix of a half-wave segment operating on any field vector is reduced to a multiplicative constant. The field at the input can be related to the field at the output as follows:

$$\mathbf{e}_m = [\mathbf{T}_{\lambda/2}]^m\mathbf{e}_0 = e^{jm\pi}\mathbf{e}_0. \tag{A7.3}$$

Thus, the input fields are simply equal to the output fields times the multiplicative constant taken to the mth power, from which the reflection and transmission through the structure can be immediately obtained.

For more complex dielectric structures, the **T**-matrix which relates the fields at one plane to the fields at the next is reduced to a multiplicative constant for only two specific field vectors. However, it can be shown that these two field vectors can be weighted and added to construct *any* desired field vector. In mathematical terms, the two fields are referred to as eigenvectors, and the corresponding multiplicative constants are the eigenvalues of the transmission matrix. For a reciprocal matrix with $T_{11}T_{22} - T_{12}T_{21} = 1$, the product of the eigenvalues of the two eigenvectors equals unity, allowing us to express them in the following manner:

$$[\mathbf{T}]\mathbf{e}_+ = e^{+\xi}\mathbf{e}_+ \quad \text{and} \quad [\mathbf{T}]\mathbf{e}_- = e^{-\xi}\mathbf{e}_-. \tag{A7.4}$$

Here $\mathbf{e}_+$ and $\mathbf{e}_-$ represent the two distinct eigenvectors of the **T**-matrix, and $e^{\pm\xi}$ are the corresponding eigenvalues. Written in exponential form, ξ plays the role of a propagation constant. However, the "propagation" we are referring to is from plane to plane, indicating that ξ is more appropriately defined as the *discrete* propagation constant relating the field vector at one plane to the field vector at the next. For a uniform slab with planes spaced by half-wavelengths, Eq. (A7.2) reveals that $\xi = j\pi$. In complex dielectric structures, $\mathrm{Re}\{\xi\} \neq 0$, and $\mathbf{e}_+$ and $\mathbf{e}_-$ grow and decay as we propagate through the stack.

By writing out the equations implicit in (A7.4), we can solve for the eigenvalues and eigenvectors, or the *eigensystem* in terms of the **T**-matrix components:

$$\begin{bmatrix} T_{11} & T_{12} \\ T_{21} & T_{22} \end{bmatrix} \begin{bmatrix} A_\pm \\ B_\pm \end{bmatrix} = e^{\pm\xi} \begin{bmatrix} A_\pm \\ B_\pm \end{bmatrix}, \tag{A7.5a}$$

$$e^{\pm\xi} = \tfrac{1}{2}(T_{11} + T_{22}) \pm \sqrt{\tfrac{1}{4}(T_{11} + T_{22})^2 - 1}, \tag{A7.5b}$$

$$\frac{A_\pm}{B_\pm} = \frac{T_{12}}{e^{\pm\xi} - T_{11}} = \frac{e^{\pm\xi} - T_{22}}{T_{21}}. \tag{A7.5c}$$

The eigenvalues are found by moving all terms in (A7.5a) to the left, setting the resulting determinant to zero, and applying $T_{11}T_{22} - T_{12}T_{21} = 1$ to simplify the square root term. For the eigenvectors, only the ratio of the two components is relevant (the absolute magnitude can be chosen arbitrarily). The two versions of A/B in (A7.5c) are found using the upper and lower equations in (A7.5a), respectively.

A7.3   APPLICATION TO DIELECTRIC STACKS AT THE BRAGG CONDITION

By filling roughly half of each segment defined in Fig. A7.1 with a different dielectric or index, n_1, we arrive at the quarter-wave distributed Bragg reflector introduced in Chapter 3, as indicated in Fig. A7.2. From Eqs. (3.43), the **T**-matrix coefficients of one period of this structure at the Bragg condition are symmetrical such that $T_{11} = T_{22}$, $T_{21} = T_{12}$, and $T_{11}^2 - T_{21}^2 = 1$. These relations simplify the eigensystem considerably:

$$e^{\pm\xi} = T_{11} \pm T_{21} \qquad \text{and} \qquad \frac{A_\pm}{B_\pm} = \pm 1. \tag{A7.6}$$

A convenient choice for the two eigenvectors is

$$\mathbf{e}_+ = \begin{bmatrix} 1 \\ 1 \end{bmatrix} \qquad \text{and} \qquad \mathbf{e}_- = \begin{bmatrix} 1 \\ -1 \end{bmatrix}. \tag{A7.7}$$

Now getting back to our objective, we want to know the input fields in terms of the output fields of the stack. Assuming no incoming wave from the right $(B_0 = 0)$, we can write the field vector at plane 0 as

$$\mathbf{e}_0 = \begin{bmatrix} 1 \\ 0 \end{bmatrix} = \tfrac{1}{2}(\mathbf{e}_+ + \mathbf{e}_-). \tag{A7.8}$$

To determine the field vector at plane 1, we apply the **T**-matrix to $\mathbf{e}_0$, which upon encountering the two eigenvectors is reduced to the two eigenvalues. Repeated application of the **T**-matrix allows us to determine the field vector at the ith plane:

$$\mathbf{e}_i = [\mathbf{T}]^i \mathbf{e}_0 = \tfrac{1}{2}(e^{+i\xi}\mathbf{e}_+ + e^{-i\xi}\mathbf{e}_-) = \begin{bmatrix} \cosh i\xi \\ \sinh i\xi \end{bmatrix}. \tag{A7.9}$$

The forward and backward wave components of the field vector are drawn according to Eq. (A7.9) at each plane in Fig. A7.2. Physically, the index discontinuities are feeding power from the forward to the backward wave, causing the forward wave to decay into the stack and the backward wave to grow from zero at the output.

It is apparent in Fig. A7.2 that the more periods there are, the closer the backward and forward wave are in magnitude at the input, and the higher the reflectivity of the stack. Evaluating Eq. (A7.9) at the mth plane and taking the reflected to incident amplitude ratio, we obtain

$$r_g = \frac{B_m}{A_m} = \tanh m\xi, \tag{A7.10}$$

or with $r_1 = \tanh \xi = T_{21}/T_{11}$ ($=$ the reflectivity of one period), we can write

$$r_g = r_1 m_{eff} \qquad \text{where} \qquad m_{eff} \equiv \frac{\tanh m\xi}{\tanh \xi}. \tag{A7.11}$$

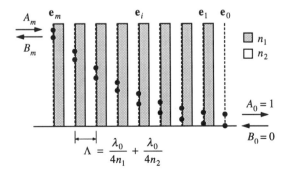

FIGURE A7.2   Forward and backward waves in a grating reflector.

Thus, the total reflectivity is just the reflectivity of one period times some *effective* number of mirror periods seen by the incident field, m_{eff}. As $m \to \infty$, m_{eff} saturates to a value of $m_{eff} \to 1/\tanh \xi \approx 1/\xi$ (for small ξ). From a physical viewpoint, the forward wave decays approximately exponentially into the mirror, and hence effectively "sees" only $1/\xi$ periods.

To further quantify the reflectivity, we must evaluate the discrete propagation constant. This is accomplished using Eqs. (3.43) to set $T_{11} = -(1 + r^2)/t^2$ and $T_{21} = -2r/t^2$, where r is the reflectivity of the 2–1 interface, and $t^2 = 1 - r^2$. With these substitutions, we have

$$e^\xi = T_{11} + T_{21} = -[(1 + r)/t]^2 = n_2/n_1. \tag{A7.12}$$

The first relation reduces to $-(1 + 2r)$ for small interface reflectivities. Solving for the discrete propagation constant, we find $\xi \approx j\pi + 2r$. In addition to the phase shift seen with the homogeneous slab, there is a real part to the discrete propagation constant which is responsible for the attenuation of the incident wave. The attenuation *per unit length* in this small r limit is just $2r/\Lambda$, or two reflections per half-wavelength—an intuitive result, indeed. Plugging the latter equality of (A7.12) into (A7.11), we find that the effective number of mirror periods at the Bragg frequency becomes

$$m_{eff} = \frac{\tanh\left[m\ln(n_2/n_1)\right]}{\tanh\left[\ln(n_2/n_1)\right]}. \tag{A7.13}$$

To determine the reflectivity, we expand (A7.10) into exponential form, and use (A7.12) for e^ξ:

$$r_g = \frac{(n_2/n_1)^m - (n_2/n_1)^{-m}}{(n_2/n_1)^m + (n_2/n_1)^{-m}} = \frac{1 - (n_1/n_2)^{2m}}{1 + (n_1/n_2)^{2m}}. \tag{A7.14}$$

As m approaches infinity, this ratio approaches one, with larger differences in the index accelerating the convergence (see Problem A7.1). In the last section of this appendix, we will develop a more generalized version of (A7.14), applicable to more general quarter-wave stacks at the Bragg condition.

A7.4 APPLICATION TO DIELECTRIC STACKS AWAY FROM THE BRAGG CONDITION

The above analysis corresponded to the quarter-wave stack at its peak reflectivity, where all interface reflections add in phase. With the thickness of each layer being exactly one quarter-wavelength, the T-matrix reduced considerably allowing us to simplify the description of the eigensystem. In the more general case away from the Bragg condition, $T_{11} \neq T_{22}$ and $T_{21} \neq T_{12}$, and the eigensystem becomes more complex. Rather than repeat the above procedure for this more general case (see Problem A7.2), we will explore an alternative approach which involves determining the overall grating matrix in terms of the single-period T-matrix.

We can relate the grating matrix to the eigenvectors and eigenvalues of the single-period T-matrix as follows:

$$[\mathbf{T}_g]\mathbf{e}_\pm = [\mathbf{T}]^m\mathbf{e}_\pm = e^{\pm m\xi}\mathbf{e}_\pm. \tag{A7.15}$$

This equation reveals that the eigenvectors of the grating matrix, $\mathbf{T}_g$, are in fact the same as the eigenvectors of the single period T-matrix, with the corresponding eigenvalues taken to the mth power. As a result, from the first equality in Eq. (A7.5c), we can set

$$\frac{T_{g12}}{e^{\pm m\xi} - T_{g11}} = \frac{T_{12}}{e^{\pm\xi} - T_{11}}. \tag{A7.16}$$

Subtracting the minus version of this equation from the plus version, we immediately obtain: $T_{g12} \sinh \xi = T_{12} \sinh m\xi$. If we now add the minus version to the plus version, we find: $(\sinh m\xi)(\cosh \xi - T_{11}) = (\sinh \xi)(\cosh m\xi - T_{g11})$. Solving for the grating matrix coefficients, we obtain

$$T_{g11} = \frac{\sinh m\xi}{\sinh \xi} T_{11} - \frac{\sinh m\xi \cosh \xi - \cosh m\xi \sinh \xi}{\sinh \xi},$$

$$T_{g12} = \frac{\sinh m\xi}{\sinh \xi} T_{12}. \tag{A7.17}$$

Identical relations exist for T_{g22} and T_{g21} using the second equality in Eq. (A7.5c). The overall reflectivity is then simply given by T_{g21}/T_{g11}, which becomes

$$r_g = \frac{T_{g21}}{T_{g11}} = \frac{T_{21}/T_{11}}{1 - \dfrac{\sinh(m-1)\xi}{T_{11}\sinh m\xi}}, \tag{A7.18}$$

where we have used the identity,

$$\sinh m\xi \cosh \xi - \cosh m\xi \sinh \xi = \sinh(m-1)\xi.$$

While Eq. (A7.17) provides us with the reflectivity of the grating, it is lacking in the sense that it does not provide us with any feel for how the reflectivity changes with frequency. Approximate expressions for r_g characterize the frequency dependence using the detuning parameter, $\delta = \beta - \beta_0$, introduced in

Chapter 3. Toward this end, we define a *generalized* detuning parameter, which is characterized by the asymmetry in the diagonal matrix coefficients:

$$\Delta \equiv j \frac{T_{22} - T_{11}}{T_{22} + T_{11}}. \tag{A7.19}$$

In the low interface reflectivity limit near the Bragg frequency, $\Delta \to \delta\Lambda = \pi(v - v_0)/v_0$ (see Eqs. (3.43) with no loss).

In order to use Δ in Eq. (A7.17), we first replace every sinh x function with (cosh x)(tanh x). The tanh x functions can then be replaced with the effective number of mirror periods, $m_{eff} = \tanh m\xi/\tanh \xi$, and we are left with only cosh x functions. By adding the plus and minus versions of Eq. (A7.5b), we can set

$$\cosh \xi = \tfrac{1}{2}(T_{11} + T_{22}) = \frac{T_{11}}{1 + j\Delta} = \frac{T_{22}}{1 - j\Delta}. \tag{A7.20}$$

The latter equalities make use of Eq. (A7.19). Using cosh $\xi = T_{11}/(1 + j\Delta)$ for T_{g11} and T_{g21}, and cosh $\xi = T_{22}/(1 - j\Delta)$ for T_{g22} and T_{g12}, Eqs. (A7.17) transform into

$$T_{g11} = (1 + jm_{eff}\Delta) \cosh m\xi,$$

$$T_{g21} = \frac{T_{21}}{T_{11}} m_{eff}(1 + j\Delta) \cosh m\xi, \tag{A7.21}$$

and

$$T_{g12} = \frac{T_{12}}{T_{22}} m_{eff}(1 - j\Delta) \cosh m\xi,$$

$$T_{g22} = (1 - jm_{eff}\Delta) \cosh m\xi. \tag{A7.22}$$

Using Eq. (A7.21) to define the grating reflectivity, $r_g = T_{g21}/T_{g11}$, we obtain

$$r_g = r_1 m_{eff} \frac{1 + j\Delta}{1 + jm_{eff}\Delta}, \tag{A7.23}$$

where $r_1 = T_{21}/T_{11}$, the reflectivity of one period.

In this version of r_g, the frequency dependence is more clearly visible. For example, at the Bragg frequency, $\Delta \to 0$ and Eq. (A7.23) reduces immediately to Eq. (A7.11). For small deviations from the Bragg condition, we can expand the denominator to first order in Δ to obtain

$$r_g \approx r_1 m_{eff}[1 - j(m_{eff} - 1)\Delta] \equiv r_1 m_{eff}[1 - j2\delta L_{eff}]. \tag{A7.24}$$

The latter equality, being the first-order expansion of $e^{-j2\delta L_{eff}}$, suggests that we can approximate the reflection phase deviation by a simple propagation delay associated with some *effective* length, or penetration depth. For small interface reflectivities near the Bragg frequency, $\Delta \approx \delta\Lambda$, $m_{eff} \gg 1$, and $L_{eff} \approx \frac{1}{2}m_{eff}\Lambda$. A more rigorous analysis which evaluates the exact phase slope at the Bragg frequency reveals that

$$L_{eff} \equiv -\frac{1}{2}\frac{d\phi_m}{d\delta}\bigg|_{\delta=0} = \frac{1}{2}m_{eff}\Lambda\left[\frac{1}{1+r^2} - \frac{1}{2m_{eff}}\right], \qquad (A7.25)$$

where m_{eff} is evaluated at the Bragg condition, and r is the interface reflectivity. For small r, m_{eff} becomes large and the term in brackets reduces to unity. Problem A7.3 examines the penetration depth concept applied to the energy distribution and absorption loss of a grating at the Bragg frequency. It is shown there that in the small r, large m limit, we can again set $L_{eff} = \frac{1}{2}m_{eff}\Lambda$. L_{eff} is equal to *half* the effective number of mirror periods seen by the field because the energy distribution and power go as the field squared, decaying twice as fast as the field.

To completely specify the grating reflectivity, we need to define explicit relations for r_1, Δ, and m_{eff} in terms of the matrix coefficients developed in Chapter 3. In Eqs. (3.43), two phase factors are used: $\phi_{\pm} = \tilde{\beta}_1 L_1 \pm \tilde{\beta}_2 L_2$. With no loss, we can use Eq. (3.46) to set $\phi_+ = \pi + \delta L$ and $\phi_- = 0$, where $\delta \equiv \beta - \beta_0$, and β is the average propagation constant of the grating defined in Eq. (3.44). For a lossless DBR, Eqs. (3.43) become

$$T_{11} = -\frac{1}{t^2}(e^{j\delta\Lambda} + r^2), \qquad T_{22} = -\frac{1}{t^2}(e^{-j\delta\Lambda} + r^2),$$

$$\qquad (A7.26)$$

$$T_{21} = -\frac{r}{t^2}(e^{j\delta\Lambda} + 1), \qquad T_{12} = -\frac{r}{t^2}(e^{-j\delta\Lambda} + 1).$$

From these relations we can immediately determine r_1 and Δ:

$$r_1 = \frac{T_{21}}{T_{11}} = 2r\frac{\cos(\delta\Lambda/2)e^{-j\delta\Lambda/2}}{1 + r^2 e^{-j\delta\Lambda}}, \qquad (A7.27)$$

$$\Delta = j\frac{T_{22} - T_{11}}{T_{22} + T_{11}} = \frac{\sin \delta\Lambda}{\cos \delta\Lambda + r^2}. \qquad (A7.28)$$

To help determine the eigenvalues and m_{eff}, we have

$$\tfrac{1}{2}(T_{11} + T_{22}) = -\frac{1}{t^2}(\cos \delta\Lambda + r^2), \qquad (A7.29)$$

$$\sqrt{\tfrac{1}{4}(T_{11} + T_{22})^2 - 1} = \frac{1}{t^2}[2r^2(1 + \cos \delta\Lambda) - \sin^2 \delta\Lambda]^{1/2}. \qquad (A7.30)$$

Using these relations in Eq. (A7.5b), we can calculate both ξ and m_{eff}. The reflection coefficient is now completely defined in terms of fundamental grating parameters: m, $r = (n_2 - n_1)/(n_2 + n_1)$, $t = \sqrt{1 - r^2}$, and $\delta\Lambda = (\beta - \beta_0)\Lambda$, where $1/\beta = \frac{1}{2}[1/\beta_1 + 1/\beta_2]$.

A7.5   CORRESPONDENCE WITH APPROXIMATE TECHNIQUES

In this section, we examine how the exact reflectivity reduces to common approximate expressions. We will consider two limiting cases. The first approximation involves neglecting r^2 and higher terms, while retaining all terms related to $\delta\Lambda$. In this case, the reflectivity reduces to the Fourier limit. The second approximation involves retaining terms up to r^2 and limiting the frequency deviation to $(\delta\Lambda)^2$ terms. In this case, the reflectivity reduces to the coupled-mode limit considered in Chapter 6.

A7.5.1   Fourier Limit

To analyze the grating reflector using Fourier analysis, we send an impulse function into the stack and track the resulting distribution of reflected impulse functions in time. Such an analysis gives us the impulse response of the grating. The Fourier transform of the impulse response should then correspond to the reflection spectrum. While this approach in principle is exact, in practice, the infinite number of impulse functions which appear back at the input after repeated bounces within a multilayer stack makes this approach mathematically intractable in the general case. However, if the interface reflections are small, any impulse functions which make double bounces or more before returning to the input will have negligible magnitude and can be ignored. In this limit, only a small uniform burst of impulse functions will return (assuming the transmission through each interface does not reduce the magnitude of the impulse). This square-shaped envelope of the impulse response when Fourier transformed leads to a $\sin x/x$ type reflection spectrum. It is this functionality we would like to reproduce using the exact expression (A7.23).

To approximate the simplified Fourier analysis, we neglect all but linear terms in r, simulating the single-bounce approximation. However, the frequency variation we retain completely, since this is accurately determined using the Fourier approach. With these approximations, (A7.27) and (A7.28) simplify to

$$r_1 \approx 2r \cos(\delta\Lambda/2)e^{-\delta\Lambda/2},$$
$$\Delta \approx \tan \delta\Lambda. \tag{A7.31}$$

The discrete propagation constant and the effective number of mirror periods found using (A7.29) and (A7.30) in (A7.5b) and (A7.11) simplify to

$$\frac{1}{2}(T_{11} + T_{22}) \approx -\cos \delta\Lambda,$$

$$\sqrt{\frac{1}{4}(T_{11} + T_{22})^2 - 1} \approx j \sin \delta\Lambda,$$

$$e^{\pm \xi} \approx -\cos \delta\Lambda \pm j \sin \delta\Lambda = e^{j\pi} e^{\mp j\delta\Lambda}, \tag{A7.32}$$

$$m_{eff} = \frac{\tanh m\xi}{\tanh \xi} \approx \frac{\tan m\delta\Lambda}{\tan \delta\Lambda}.$$

In the Fourier limit, the discrete propagation constant becomes simply the phase delay between mirror periods including detuning effects, $j\pi \mp j\delta\Lambda$.

Using these approximations in the expression for reflectivity (A7.23), we obtain

$$r_g \approx 2r \cos(\delta\Lambda/2)e^{-\delta\Lambda/2} \frac{\tan m\delta\Lambda}{\tan \delta\Lambda} \cdot \frac{1 + j \tan \delta\Lambda}{1 + j \tan m\delta\Lambda}. \tag{A7.33}$$

To reduce this expression to the Fourier limit, we move the cosines in the first fraction to the second fraction and convert the entire second fraction to exponential form. We then assume that the phase deviation *across one period* can be expanded to second order, such that $\cos(\delta\Lambda/2) \approx 1$ and $\sin \delta\Lambda \approx \delta\Lambda$. The accumulated phase deviation, $m\delta\Lambda$, across the entire grating may be large, so we leave this in general form. With $m\Lambda = L_g$ (the length of the grating), Eq. (A7.33) reduces to

$$r_g|_{Fourier} = 2mr \frac{\sin \delta L_g}{\delta L_g} e^{-j\delta L_g[(m - 1/2)/m]}. \tag{A7.34}$$

This result is exactly the result we would obtain from a simplified single-bounce Fourier analysis, revealing the characteristic $\sin x/x$ spectral dependence of weakly reflecting gratings. The $m - 1/2$ term in the phase factor results from our definition of L_g which places the output reference plane one half-period beyond the last reflection of the grating.

A7.5.2   Coupled-Mode Limit

The coupled-mode approach to be discussed in Chapter 6 analyzes the grating by solving the wave equation assuming the waves are sufficiently slowly varying that second-order derivatives of their magnitude can be neglected. This restriction implies that the interface reflectivities must not be too strong, and amounts to neglecting terms higher than r^2. Because the coupled-mode approach includes r^2 terms, it is much more accurate than the Fourier approach for strongly reflecting gratings near the Bragg condition. Another fundamental approximation of coupled-mode theory involves retaining only those Fourier components of the index variation which effectively couple forward- and backward-going waves at the Bragg condition. Away from the Bragg condition, other Fourier components of the index variation begin to contribute to the coupling between forward and backward waves. With these contributions neglected, we would expect that the coupled-mode approach is only good over a limited frequency range, in contrast to the Fourier approach. Therefore, to reproduce the coupled-mode approximation, we will limit the frequency deviation to $(\delta\Lambda)^2$.

Proceeding as before, we retain terms up to r^2 and $(\delta\Lambda)^2$, neglecting all others. We also neglect all products between r and $\delta\Lambda$. Equations (A7.27) and (A7.28) in this case simplify to

$$r_1 \approx 2r \equiv \kappa\Lambda,$$
$$\Delta \approx \delta\Lambda. \tag{A7.35}$$

Here we have introduced a new variable, $\kappa \equiv 2r/\Lambda$, which in coupled-mode theory represents the coupling per unit length between the forward and backward waves. For the square wave profile analyzed here, the *coupling constant*, κ, is just equal to two reflections per grating period. We will use κ in place of r for the rest of this section.

The discrete propagation constant and the effective number of mirror periods found using (A7.29) and (A7.30) in (A7.5b) and (A7.11) simplify to

$$\tfrac{1}{2}(T_{11} + T_{22}) \approx -(1 + (\kappa\Lambda)^2/2 - (\delta\Lambda)^2/2),$$
$$\sqrt{\tfrac{1}{4}(T_{11} + T_{22})^2 - 1} \approx [(\kappa\Lambda)^2 - (\delta\Lambda)^2]^{1/2} \equiv \sigma\Lambda,$$
$$e^{\pm\xi} \approx -(1 \mp \sigma\Lambda + (\sigma\Lambda)^2/2) \approx e^{j\pi}e^{\mp\sigma\Lambda}, \tag{A7.36}$$
$$m_{eff} = \frac{\tanh m\xi}{\tanh \xi} \approx \frac{\tanh m\sigma\Lambda}{\sigma\Lambda}.$$

For consistency with coupled-mode notation used in Chapter 6, we have introduced another new variable, $\sigma \equiv \sqrt{\kappa^2 - \delta^2}$. The discrete propagation constant is simply $j\pi \mp \sigma\Lambda$, to second order in $\sigma\Lambda$. Within the stopband ($|\delta| < \kappa$), $\sigma = \mathrm{Re}\{\xi/\Lambda\}$ and we can interpret it as the *decay* constant per unit length.

Using the approximations contained in (A7.35) and (A7.36), the general expression for reflectivity (A7.23) reduces to

$$r_g|_{coupled-mode} = \frac{\kappa}{\sigma} \tanh \sigma L_g \frac{1}{1 + j\dfrac{\delta}{\sigma} \tanh \sigma L_g}. \tag{A7.37}$$

In deriving this result, we have neglected $j\Delta$ in the numerator in comparison to one, and set $m\Lambda = L_g$. Equation (A7.37) is equivalent to the coupled-mode result, Eq. (6.37), aside from the $-j$ phase factor listed there, which arises from a difference in the choice of input reference plane (see Chapter 6 for details).

A7.6 GENERALIZED REFLECTIVITY AT THE BRAGG CONDITION

The analysis described above assumes that the overall grating matrix can be represented by the single-period T-matrix taken to the mth power. The disadvantage of this approach is that it assumes the grating is completely periodic with only two alternating index layers. In practice, the input and output layers usually have different refractive indices from those comprising the alternating layers of the grating. More generally, there might be a situation where many of the layers within the grating itself are different. In this section we would like to generalize the results of Section A7.3 to include such nonperiodic gratings. Unfortunately, this type of analysis is not possible away from the Bragg condition, so we focus our attention on determining the reflectivity at the Bragg frequency.

Breaking the grating down into its fundamental components, we find that there are three types of matrices we need to consider: (1) high-to-low interfaces (referenced from left to right), (2) low-to-high interfaces, and (3) quarter-wave phase delays. Denoting these by $\mathbf{T}_{HL}$, $\mathbf{T}_{LH}$, and $\mathbf{T}_{\lambda/4}$, we would like to consider their effect on the following vectors:

$$\mathbf{e}_+ = \begin{bmatrix} 1 \\ 1 \end{bmatrix} \quad \text{and} \quad \mathbf{e}_- = \begin{bmatrix} 1 \\ -1 \end{bmatrix}. \tag{A7.38}$$

Using Table 3.3, it can be shown that

$$\begin{aligned}
[\mathbf{T}_{HL}]\mathbf{e}_+ &= e^{+s}\mathbf{e}_+, & [\mathbf{T}_{HL}]\mathbf{e}_- &= e^{-s}\mathbf{e}_- \\
[\mathbf{T}_{LH}]\mathbf{e}_+ &= e^{-s}\mathbf{e}_+, & [\mathbf{T}_{LH}]\mathbf{e}_- &= e^{+s}\mathbf{e}_- \\
[\mathbf{T}_{\lambda/4}]\mathbf{e}_+ &= j\mathbf{e}_-, & [\mathbf{T}_{\lambda/4}]\mathbf{e}_- &= j\mathbf{e}_+
\end{aligned} \tag{A7.39}$$

where for the ith interface

$$s_i = \tfrac{1}{2}\ln\left[\frac{n_{Hi}}{n_{Li}}\right]. \tag{A7.40}$$

n_{Hi} is the high index of the ith interface, and n_{Li} is the low index. From Eqs. (A7.39), we see that $\mathbf{e}_+$ and $\mathbf{e}_-$ are actually the eigenvectors of both $\mathbf{T}_{HL}$ and $\mathbf{T}_{LH}$, suggesting that s_i is just the discrete *attenuation* constant of the ith interface. For a quarter-wave delay, $\mathbf{e}_+$ and $\mathbf{e}_-$ are not the eigenvectors, however, $\mathbf{T}_{\lambda/4}$ does transform one into the other. In Problem A7.4 the reader is asked to verify these relations.

Equations (A7.39) are all we need to propagate through any structure comprised of index discontinuities and multiples of quarter-wave segments. We are specifically interested in a grating whose index layers follow an HLHL sequence, and are each a quarter-wavelength thick. We begin by expanding the output field vector using Eq. (A7.8). Propagating through the first half-wavelength of the structure, assuming interface 0 is high–low, and interface 1 is low–high, we obtain

$$
\begin{aligned}
[\mathbf{T}_{LH}][\mathbf{T}_{\lambda/4}][\mathbf{T}_{HL}][\mathbf{T}_{\lambda/4}]\mathbf{e}_0 &= [\mathbf{T}_{LH}][\mathbf{T}_{\lambda/4}][\mathbf{T}_{HL}][\mathbf{T}_{\lambda/4}]\tfrac{1}{2}[\mathbf{e}_+ + \mathbf{e}_-] \\
&= [\mathbf{T}_{LH}][\mathbf{T}_{\lambda/4}]j\tfrac{1}{2}[e^{-s_0}\mathbf{e}_- + e^{+s_0}\mathbf{e}_+] \\
&= -\tfrac{1}{2}[e^{-(s_1+s_0)}\mathbf{e}_+ + e^{+(s_1+s_0)}\mathbf{e}_-]. \quad (A7.41)
\end{aligned}
$$

Thus, the s parameters of the two interfaces simply add together. Continuing the above procedure through N interfaces to the input, and taking the ratio of the reflected to incident wave amplitudes, we find

$$
|r_g| = \tanh \sum_0^N s_i, \quad (A7.42)
$$

where N is the number of quarter-wave layers in the grating. Using Eq. (A7.40) for s_i, we can alternatively express the reflectivity as

$$
|r_g| = \frac{1-b}{1+b} \quad \text{where} \quad b = \prod_0^N \left[\frac{n_{Li}}{n_{Hi}} \right]. \quad (A7.43)
$$

If n_L and n_H do not change throughout the grating, and there are $N = 2m$ quarter-wave layers with $n_{L0}/n_{H0} = 1$ (see Fig. A7.2), then $b = (n_L/n_H)^{2m}$, and (A7.43) reduces to the more limited (A7.14) derived earlier.

The tools developed here can also be applied to dielectric structures which may have half-wavelength spacings, as well as to structures which do not follow the HLHL alternating sequence (antireflection coatings are an example of this). The interested reader is invited to compare this method to the method discussed by Corzine, Yan, and Coldren (see reading list) which analyzes a wider range of dielectric structures using essentially the same approach.

READING LIST

S.W. Corzine, R.-H. Yan, and L.A. Coldren, *IEEE J. Quantum. Electron. Lett.*, **QE-27**, 2086 (1991).

D.I. Babic and S. W. Corzine *IEEE J. Quantum Electron.*, **QE-28**, 514 (1992).

P. Yeh, *Optical Waves in Layered Media*, Wiley, New York (1988).

PROBLEMS

A7.1 Design the grating mirrors of a VCSEL which meet the following requirements:

(i) The output coupler mirror must be as close to 99% peak power reflectivity as possible.

(ii) The back reflector must be greater than 99.9% peak power reflectivity.

Specify the mirror designs including number of periods and layer thicknesses for two material systems:

(a) $Al_{0.2}Ga_{0.8}As/AlAs$ alternating layers which provide a peak reflectivity at 0.87 μm

(b) $InP/InGaAsP(1.3$ μm$)$ alternating layers which provide a peak reflectivity at 1.55 μm

Table 1.1 may be helpful in establishing the proper design. For the back reflector, assume the grating terminates in the higher index material on both sides. For the output coupler, assume that one side terminates in the higher-index material, and that the other side terminates in air (in your output coupler design, take care to insure that the index follows the HLHL ordering sequence required for proper phasing of the reflected waves). Include a schematic of the index profile for all finished mirror designs.

If the region between the two mirrors of the VCSEL is one optical wavelength thick, how long would it take to grow both the short-wavelength and long-wavelength VCSELs assuming a typical growth rate of 1 μm/hour? Can you comment on the benefits of either the short- or long-wavelength material systems from this perspective?

A7.2 Derive the general expression for the grating reflectivity, Eq. (A7.23), using the techniques developed in Section A7.3. Specifically:

(i) Determine the eigenvectors using the second version of Eq. (A7.5c).

(ii) Expand the output vector in terms of the two eigenvectors.

(iii) Propagate the fields to the input plane.

(iv) Evaluate the reflected to incident amplitude ratio.

In defining the eigenvectors you may find it useful to set $e^{\pm\xi} = T_{11}(1 \pm \tanh\xi)/(1+j\Delta)$ and $T_{22} = T_{11}(1-j\Delta)/(1+j\Delta)$ (however, if you do use these relations, verify that they are indeed correct).

A7.3 The concept of a penetration depth not only applies to the linear phase deviation of the grating near the Bragg condition, but also to the energy distribution and absorption loss of the grating at the Bragg frequency.

(a) The energy penetration depth can be defined as the depth at which the incident energy taken as constant into the mirror is equal to the total energy integrated over the grating length. Setting the incident energy times the energy penetration depth equal to the summation of energy throughout all periods, we have

$$(A_m^* A_m + B_m^* B_m)L_{eff}^e = \sum_{i=1}^{m} (A_i^* A_i + B_i^* B_i)\Lambda. \qquad (A7.44)$$

Here the total energy in each period is approximated by the energy density at the left edge times the thickness of one period. This will be a good approximation if Λ is small compared to the length of the stack (m is large).

Using Eq. (A7.9) for A_i and B_i at the Bragg frequency, show that the energy penetration depth can be approximated by

$$L_{eff}^e = \frac{\tanh 2m\xi}{2\xi} \Lambda. \qquad (A7.45)$$

Hint: The summation can be approximated with an integral by setting $\Lambda \to dz$ and $i\Lambda \to z$.

Show that this expression is equivalent to $\frac{1}{2}m_{eff}\Lambda$ in the small r, large m limit.

(b) The absorption loss penetration depth can be defined as follows. If some small absorption loss, α, is distributed uniformly throughout the grating, the reflectivity of the grating will be reduced somewhat. If we model the reflector as a hard mirror recessed by some penetration depth into the lossy material, then the resulting round-trip attenuation is equal to $e^{-2\alpha L_{eff}^a} \approx 1 - 2\alpha L_{eff}^a$ (for small losses). The definition of L_{eff}^a is found by equating this effective reduction in *power* reflectivity with the true reduction.

By including a uniform loss in the phase terms, $\phi_\pm$, of Eqs. (3.43), show that at the Bragg frequency, $\Delta \approx -j\alpha\Lambda/2$ instead of zero. With this result and Eq. (A7.23), derive an approximate expression for L_{eff}^a, assuming the loss terms can be expanded to first order. Show that L_{eff}^a is equivalent to $\frac{1}{2}m_{eff}\Lambda$ in the small r limit.

(c) Derive the exact phase penetration depth, L_{eff}^p, given in Eq. (A7.25).

A7.4 Verify all relations in Eq. (A7.39). What are the corresponding relations for a half-wave phase delay? Answer this question by:

(a) directly using the half-wave matrix

(b) connecting two quarter-wave phase delays in series

Using the half-wave phase delay properties just derived, and other tools developed in Section A7.6, derive the reflectivity of a structure which is composed of:

  (i) a low-high interface
  (ii) a half-wavelength phase delay
  (iii) a high-low interface (where the high layer of both interfaces is the same)

Express the reflectivity both in terms of the three refractive indices and in terms of the individual interface reflectivities. Repeat the derivation assuming a quarter-wave phase delay instead. How do these expressions compare to the Fabry–Perot expressions derived in Chapter 3?

Electronic States in Semiconductors

A8.1 INTRODUCTION

In Appendix 1, a basic description of electronic states in periodic potentials was given. Important concepts such as energy bands, the electron effective mass, and the density of states were introduced. All of these concepts are essential to understanding the manner in which light interacts with semiconductor crystals. However, in order to provide a quantitative description of optical gain in these materials, a more complete description of the electronic states is required. In this appendix, the wavefunctions of the electrons required to evaluate the transition matrix element (see Appendix 10) will be considered in some detail. In addition, valence band-mixing effects related to the coupling of the heavy- and light-hole bands will be discussed. Finally the effects of strain on the subband structure of strained quantum wells will be treated. Some of the discussions will draw upon material introduced in Appendix 1.

A8.2 GENERAL DESCRIPTION OF ELECTRONIC STATES

The electron wavefunctions in the conduction and valence bands of the semiconductor are found by solving the Schrödinger equation which relates the system Hamiltonian, H_0, of the crystal lattice to the energy, E, of the electron. It can be written as

$$H_0 \psi = \left[\frac{\mathbf{p}^2}{2m_0} + V(\mathbf{r}) \right] \psi = E\psi, \tag{A8.1}$$

where $\mathbf{p}$ is the momentum operator, $\mathbf{r}$ is the position vector, m_0 is the free electron mass, ψ is the wavefunction of the electron, and $V(\mathbf{r})$ is the potential created by the crystal lattice. Due to the periodicity of $V(\mathbf{r})$, the solutions of the above equation are given by Bloch waves of the form

$$\psi = e^{i\mathbf{k}\cdot\mathbf{r}}u(\mathbf{k}, \mathbf{r}), \qquad (A8.2)$$

where $\mathbf{k}$ is the wavevector of the electron, and u is a Bloch function with the special property that it is periodic with the crystal lattice, and hence repeats itself in each unit cell of the crystal. In most practical cases, we never actually need to know the exact description of the Bloch function. What is important however is the *symmetry* properties of the Bloch functions in the conduction and valence bands, which will be considered in Section A8.3.

For analyzing localized electronic states such as those encountered in quantum wells and quantum wires, it is useful to consider linear combinations of the Bloch wave solutions in Eq. (A8.2). Using an arbitrary set of expansion coefficients, $A(k)$, we can express a spatially localized wavefunction as

$$\psi = \int A(k)e^{i\mathbf{k}\cdot\mathbf{r}}u(\mathbf{k}, \mathbf{r})\, d^3\mathbf{k} \approx u(0, \mathbf{r})\int A(k)e^{i\mathbf{k}\cdot\mathbf{r}}\, d^3\mathbf{k} \equiv F(\mathbf{r})u(\mathbf{r}). \quad (A8.3)$$

The description of localized states contained in (A8.3) is known as the envelope function approximation. The key assumption here is that within a given energy band, the Bloch function is not a strong function of $\mathbf{k}$ (at least in the proximity of the band edge) and can thus be approximately represented by the band edge ($\mathbf{k} = 0$) Bloch function, $u(0, \mathbf{r}) \equiv u(\mathbf{r})$. This allows us to pull it out of the expansion and define an envelope function, $F(\mathbf{r})$, whose Fourier spectrum is made up of the plane wave components of the solutions in (A8.2). Thus, our generalized *approximate* solutions in a given energy band consist of the band edge Bloch function multiplied by a slowly varying envelope function. We choose the two components to be normalized such that in Dirac notation, we have

$$\langle F|F\rangle \equiv \int_V F^*F\, d^3\mathbf{r} = 1, \qquad \langle u|u\rangle \equiv \frac{1}{V_{uc}}\int_{\text{unit cell}} u^*u\, d^3\mathbf{r} = 1. \quad (A8.4)$$

For the envelope functions, V is the volume of the crystal. For the Bloch functions, we only need to consider the volume of a single unit cell of the crystal, V_{uc}. Note that with our chosen definitions, the envelope functions have dimensions of $V^{-1/2}$, whereas the Bloch functions are dimensionless.

From symmetry considerations alone, one can deduce that the conduction band and valence band Bloch functions are orthogonal to each other. In Dirac notation, this orthogonality can be compactly written as

$$\langle u_c|u_v\rangle = \langle u_v|u_c\rangle = 0. \qquad (A8.5)$$

In Section A8.3, a more detailed discussion of the Bloch functions will reveal why this orthogonality exists. The envelope functions on the other hand can be determined explicitly. For example, in bulk material, the normalized envelope functions are simple plane waves:

$$F = \frac{1}{\sqrt{V}} e^{i\mathbf{k}\cdot\mathbf{r}}. \tag{A8.6}$$

The envelope functions in quantum-well material are considered in Appendix 1 and also later in this appendix.

We are particularly interested in the relationship between the energy, E, of the electron or hole given in (A8.1) and the electron's wavevector, $\mathbf{k}$, given in (A8.6). It is quite common to assume the bands to be parabolic in both the conduction and valence bands, allowing us to invoke the effective mass concept:

$$E_2(\mathbf{k}) = E_c + \frac{\hbar^2 k_2^2}{2m_c}, \qquad E_1(\mathbf{k}) = E_v - \frac{\hbar^2 k_1^2}{2m_v}, \tag{A8.7}$$

where $E_{c,v}$ are the band edge energies, $m_{c,v}$ are the effective masses in the two bands, and $k_{2,1}$ are the magnitudes of the wavevectors of a given electron or hole. However, these expressions are oversimplifications to reality. First of all, $E(\mathbf{k})$ is not the same along all directions of the crystal, and we cannot simply use k in place of $\mathbf{k}$. For example, the heavy-hole band in GaAs is known to be highly anisotropic as a function of the k-vector direction (in comparison, the conduction and light-hole bands are much more isotropic at energies near the band edge). Secondly, at energies away from the band edge, the band curvature does not remain perfectly parabolic, especially in the light-hole band. In quantum-well and quantum-wire material, the valence band becomes extremely nonparabolic. Thus, in defining the relations for gain, we must keep in mind that (A8.7) is not always a good approximation to reality.

A8.3   BLOCH FUNCTIONS AND THE MOMENTUM MATRIX ELEMENT

To define the electronic states more explicitly, we will first consider the Bloch functions of the various energy bands. The conduction and valence bands are illustrated in Fig. A8.1. The three valence bands are commonly known as the heavy-hole (HH), light-hole (LH), and split-off hole (SO) bands. We can view each energy band as originating from the discrete atomic energy levels of the isolated atoms that compose the crystal as introduced in Appendix 1. In this sense, the conduction band can be thought of as a remnant of an s atomic orbital, while the three valence bands are remnants of the three p atomic orbitals: p_x, p_y, and p_z. The corresponding Bloch functions for these orbitals are denoted here as u_s and u_x, u_y, and u_z. It is very useful to make this correspondence because the Bloch functions retain many of the symmetries that the atomic orbitals possess.

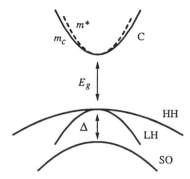

FIGURE A8.1 Typical III–V semiconductor band structure schematic illustrating the conduction (C), heavy-hole (HH), light-hole (LH), and split-off (SO) bands and their relative energy separations. The true (solid) and approximate (dashed) C bands are shown with their corresponding effective masses. (The relative scale of the four bands corresponds roughly to GaAs.)

For instance, the conduction band Bloch function, u_s, has even symmetry in all three directions within each unit cell, similar to the spherically symmetric s atomic orbital. In a similar manner, u_z has odd symmetry along z, but even symmetry in the other two directions, within each unit cell, similar to the p_z atomic orbital. From these two facts alone, we can state that the net odd symmetry along z must give $\langle u_s | u_z \rangle = 0$ (where the brackets indicate integration over a unit cell). However, operating on u_z with the momentum operator, p_z, inverts the symmetry along z and the momentum matrix element $\langle u_s | p_z | u_z \rangle$ is therefore, in general nonzero. From these and similar arguments we can immediately write down some useful symmetry relations and definitions [1]

$$\langle u_s | p_i | u_j \rangle = 0, \qquad \text{for } i \neq j \tag{A8.8}$$

$$\langle u_s | \mathbf{p} | u_i \rangle = \langle u_s | p_i | u_i \rangle \equiv M, \tag{A8.9}$$

$$\langle u_s | \mathbf{p} | \bar{u}_i \rangle = 0, \tag{A8.10}$$

where $i = x, y, z$, and u_i, $\bar{u}_i$ indicate spin-up and spin-down functions. The third relation comes from the fact that Bloch functions of opposite spin do not interact. The constant M is defined here as the basis function *momentum matrix element*. Thus, through simple symmetry arguments, the various momentum matrix elements between u_s and the three u_i can now all be related simply to one constant, M.

The valence band Bloch functions u_{hh}, u_{lh}, and u_{so} corresponding to the bands of Fig. A8.1 are usually written as linear combinations of the "basis" functions u_x, u_y, and u_z [2–4] in a manner analogous to the construction of hybrid orbitals in the study of molecular bonds. Spin degeneracy exists in all three bands, so we actually need to define six Bloch functions. Defining the electron's k-vector to be directed along z, the valence band Bloch functions can be written as

$$u_{hh} = -\frac{1}{\sqrt{2}}(u_x + iu_y), \qquad \bar{u}_{hh} = \frac{1}{\sqrt{2}}(\bar{u}_x - i\bar{u}_y),$$

$$u_{lh} = -\frac{1}{\sqrt{6}}(\bar{u}_x + i\bar{u}_y - 2u_z), \qquad \bar{u}_{lh} = \frac{1}{\sqrt{6}}(u_x - iu_y + 2\bar{u}_z), \quad \text{(A8.11)}$$

$$u_{so} = -\frac{1}{\sqrt{3}}(\bar{u}_x + i\bar{u}_y + u_z), \qquad \bar{u}_{so} = \frac{1}{\sqrt{3}}(u_x - iu_y - \bar{u}_z).$$

The prefactors are normalization constants which can have arbitrary phase (the phases chosen here are those of Broido and Sham [5]). For $\mathbf{k}$ directed along another direction we would have to redefine the above relations (however, cyclic permutation of x, y, and z do yield equivalent relations if we also include the direction of $\mathbf{k}$ in the permutation).

The above linear combinations of basis functions are known as the angular momentum representation. They are useful when we consider the spin–orbit interaction between the angular momentum of the p orbitals and the spin angular momentum of the electron. The spin–orbit interaction term is "diagonalized" in the above representation [2]. The SO band would be degenerate with the HH and LH bands if the spin–orbit interaction did not exist. As it is, the spin–orbit interaction partially removes the degeneracy, suppressing the SO band from the other two by a spin–orbit splitting energy, Δ, which in GaAs is equal to ~ 0.34 eV.

With the above description of the valence bands, we can obtain a more complete description of the transition matrix element. However, the magnitude of the basis function momentum matrix element is still an unknown. Therefore, we will close this section with an example use of the above relations to obtain an approximate expression for the magnitude of $|M|^2$.

Evaluation of $|M|^2$ in bulk material was first obtained by theoretically relating it to the *curvature* of the conduction band [3, 6]. Using a second-order perturbation technique known as the $k \cdot p$ method [1–3], we can express the conduction band effective mass along the electron's k-vector direction (which for the definitions of Eq. (A8.11) is the z-direction) as

$$\frac{1}{m_{cz}} = \frac{1}{m_0}\left[1 + \sum_{n \neq c} \frac{2}{m_0} \frac{|\langle u_c|p_z|u_n\rangle|^2}{E_c - E_n}\right], \qquad \text{(A8.12)}$$

where the summation sums over all n energy bands (not to be mistaken with energy subbands in quantum wells) of the crystal, and the E_n and E_c are the band edge energies of each band. From (A8.12) we see that the deviation of the conduction band effective mass from the free electron mass arises from the interaction between the conduction band and all other energy bands in the crystal.

It is interesting to note that due to the sign of the denominator, contributions from higher energy bands make the effective mass *heavier* and tend to flatten out the conduction band, while contributions from lower-lying energy bands tend to decrease the effective mass, increasing the curvature of the conduction band. In either case we find that the effect of a given band is to repel the conduction band away from it.

Also because of the denominator, only energy bands close in energy to the conduction band will contribute significantly to the summation. By neglecting all but the three valence bands in the summation, we can obtain an approximate closed-form expression for the conduction band effective mass using the relations given in (A8.8) through (A8.10) and (A8.11). Note that the HH Bloch function does not contain u_z, and hence its contribution to the sum is zero. Thus, summing over the LH and SO bands and using the energy separations defined in Fig. A8.1, we obtain

$$\frac{1}{m^*} = \left[1 + \frac{2|M|^2}{m_0} \left(\frac{2}{3} \frac{1}{E_g} + \frac{1}{3} \frac{1}{(E_g + \Delta)} \right) \right]. \tag{A8.13}$$

The approximate conduction band effective mass, m^*, is expected to be lighter than the true effective mass since the effects of any higher energy bands have been neglected in our approximation. The approximate conduction band (with curvature related to m^*) and the true conduction band are both illustrated in Fig. A8.1.

The true conduction band effective mass can be measured experimentally to a good degree of precision using cyclotron resonance techniques [7]. Thus, by assuming that m^* is close to the true effective mass, m_c, we can rearrange (A8.13) to obtain

$$|M|^2 = \left(\frac{m_0}{m^*} - 1 \right) \frac{(E_g + \Delta)}{2\left(E_g + \frac{2}{3}\Delta \right)} m_0 E_g. \tag{A8.14}$$

So we see that the simple description of the valence bands given in (A8.8) through (A8.10) and (A8.11) has led directly to a formula which can yield a rough estimate of $|M|^2$ [6]. And while the above relation is not exact, it does reveal that $|M|^2$ is roughly proportional to the *ratio* of the energy gap to the conduction band effective mass of the semiconductor [8].

We have derived Eq. (A8.14) as an exercise in using the valence band Bloch functions. It is a useful formula for materials which have not been fully characterized. However, in more common materials, such as GaAs, much more accurate methods of determining $|M|^2$ exist [9–13]. The inaccuracy of (A8.14) stems from the fact that the contribution from higher-lying energy bands can be significant, and thus, m^* is not always a very good approximation to the true effective mass, m_c. For example, in GaAs, m^* is approximately equal to $0.0502m_0$ compared to the true effective mass, $0.067m_0$. Thus, using the true effective mass in (A8.14) leads to an *under*estimation of the matrix element by about 26%. Many previous calculations have failed to recognize this correction (see Yan et al. [14] for a discussion of this) implying that both calculated spontaneous emission (and hence, calculated radiative current density) and optical gain will be underestimated by 26%, a significant factor. The most accurate estimates of $|M|^2$ have actually been obtained using electron spin resonance techniques [9–13]. In Table 4.1, we have tabulated the most accurately reported values of $|M|^2$ in several material systems commonly used in semiconductor laser applications.

A8.4 BAND STRUCTURE IN QUANTUM WELLS

With the Bloch functions now defined, we need to concentrate on defining and solving for the envelope functions in the conduction and valence bands.

A8.4.1 Conduction Band

For a nondegenerate energy band (aside from spin), such as the conduction band, it has been shown most notably by Luttinger and Kohn [4] (using a $k \cdot p$ formalism) that an "effective mass equation" or Schrödinger-like equation for the envelope function, F_2, as defined in (A8.3) can be derived. It can be given as

$$(H_c + V)F_2 = E_2(\mathbf{k})F_2, \tag{A8.15}$$

where the Hamiltonian for the conduction band is simply

$$H_c = -\frac{\hbar^2}{2m_c}\nabla^2 \rightarrow \frac{\hbar^2}{2m_c}(k_x^2 + k_y^2 + k_z^2). \tag{A8.16}$$

The arrow indicates the form of the Hamiltonian for plane wave solutions (when V is constant). The potential, V, in this case corresponds to the variation in the material band edge, and the total energy of the electron, $E_2(\mathbf{k})$, is measured relative to the bottom of the conduction band. What is extremely appealing about the effective mass equation as compared to Eq. (A8.1) is that the Bloch functions have been removed from the equation, and the effect of the periodic potential arising from the crystal lattice (and hence, the coupling to other energy bands) is now replaced by a conduction band "effective" mass, m_c. In this approximation, the quantum well created by the interfacing of three materials of different bandgap truly becomes a textbook particle-in-a-box problem with F as the wavefunction, and the material band edges as the potential, V. The solutions to Eqs. (A8.15) and (A8.16) for a quantum well are considered in Appendix 1.

A8.4.2   Valance Band

A8.4.2.1   Degenerate Effective Mass Equation The simplicity of the band structure in the conduction band of a quantum well relies on the assumption that the interaction with other energy bands is weak enough that we can treat it perturbatively by replacing that interaction with a conduction band effective mass. However, for bands degenerate in energy, the assumption of weak interaction is a poor one and cannot be used. Therefore, the effective mass equation (A8.15) must be modified to include the strong degenerate band interaction explicitly.

A modified derivation for the degenerate band effective mass equation has also been treated by Luttinger and Kohn [4]. In this case, we still obtain an effective mass equation for each degenerate band similar to (A8.15), however, as a result of the degeneracy, a coupling term is introduced which couples the equations together. For the degenerate HH and LH bands near the band edge (see Fig. A8.1), this implies that we must work with *four* coupled effective mass equations (we must include the spin degeneracy)! We can actually include as many energy bands in the coupled set of equations as we desire (two equations for each spin-degenerate band). For example, in addition to the HH and LH bands, Eppenga et al. [15, 16] have included the SO band as well as the conduction band in their four-band model, leading to eight coupled equations! However, interaction between the HH and LH bands is by far the strongest, and we do not pay a large penalty by neglecting the other bands. Only when we consider energy levels deep into the valence band (energies comparable to the spin–orbit splitting energy, Δ), do we need to include the SO band explicitly. In GaAs and InGaAs, $\Delta > 300$ meV, implying that for most gain calculations with these materials, we can neglect the SO band entirely. In InP, $\Delta = 100$ meV, however, the well material typically used in this system is closer to lattice-matched InGaAs where again Δ is closer to 300 meV.

The four coupled effective mass equations can be greatly simplified using a method first suggested by Kane [2] and later by Broido and Sham [5]. They pointed out than an appropriate linear combination of the four Bloch functions $(u_{hh}, u_{lh}, \bar{u}_{hh}, \bar{u}_{lh})$, into four new Bloch functions (u_A, u_B, u_C, u_D), decouples the four equations into two identical sets of two coupled equations. Thus, we actually only need to consider two coupled equations in our analysis. However, a price must be paid for this luxury. We must now restrict our attention to analyzing $E_1(k_{xy})$ in a given *plane* in the crystal (the direction of k_{xy} must be specified). Furthermore, the equations remain completely general only for the $\{100\}$ and $\{110\}$ planes. However, for the present purposes, these represent only minor restrictions.

The two coupled effective mass equations for the degenerate bands can be expressed as

$$(H_{hh} + V)F_{hh} + WF_{lh} = E_1(\mathbf{k})F_{hh}, \qquad (A8.17)$$

$$(H_{lh} + V)F_{lh} + W^{\dagger}F_{hh} = E_1(\mathbf{k})F_{lh}, \qquad (A8.18)$$

where $F_{hh, lh}$ are the heavy- and light-hole envelope functions corresponding to two new Bloch functions $u_{A, B}$ to be defined in the next section. These replace the generic valence band envelope function F_1 in this coupled case. $E_1(\mathbf{k})$ is the total energy of the hole measured from the valence band edge. Also, $W^{\dagger}$ is the Hermitian conjugate of W. The main difference between the degenerate (A8.17) and (A8.18) and nondegenerate (A8.15) effective mass equations lies in the coupling term, W. The energy, $E_1(\mathbf{k})$, is what we would like to solve for, but because of W, we must now solve two equations simultaneously to find it.

The form of the Hamiltonians, $H_{hh, lh}$, are also slightly different from (A8.16). Let us define k_z and k_t to be two perpendicular k-vector components, with k_z directed along a $\langle 1\,0\,0 \rangle$ direction, and k_t directed either along a $\langle 1\,0\,0 \rangle$ direction or a $\langle 1\,1\,0 \rangle$ direction within the k_x–k_y plane. With these definitions, we can write (assuming plane wave solutions)

$$H_{hh} = (\gamma_1 - 2\gamma_2)k_z^2 + (\gamma_1 + \gamma_2)k_t^2, \qquad (A8.19)$$

$$H_{lh} = (\gamma_1 + 2\gamma_2)k_z^2 + (\gamma_1 - \gamma_2)k_t^2, \qquad (A8.20)$$

where

$$\gamma_1 - 2\gamma_2 \equiv \frac{\hbar^2}{2m_{hh}}, \qquad \gamma_1 + 2\gamma_2 \equiv \frac{\hbar^2}{2m_{lh}}. \qquad (A8.21)$$

In writing $H_{hh, lh}$, we are assuming for convenience that $E_1(\mathbf{k})$ measures positive into the valence band. Note that the form of (A8.19) and (A8.20) is very similar to (A8.16). The only surprising feature is the different effective masses used along k_z and k_t. As we shall see in the next section, inclusion of the coupling term compensates for this apparent asymmetry in bulk material. However, in quantum-well material, this asymmetry produces a much lighter HH mass in the plane of the well than along the confinement axis. The material constants $\gamma_{1, 2}$ are referred to as the Luttinger parameters [17], and are easily related to the HH and LH effective masses, m_{hh} and m_{lh}, through (A8.21). A third Luttinger parameter, γ_3, exists in the coupling term, W. The coupling term takes a slightly different form when k_t is directed along $\langle 1\,0\,0 \rangle$ and $\langle 1\,1\,0 \rangle$ directions. We can define the two forms as

$$W = \sqrt{3}\,k_t(\gamma_2 k_t - i2\gamma_3 k_z) \qquad \text{for } \{1\,0\,0\} \text{ planes}, \qquad (A8.22)$$

$$W = \sqrt{3}\,k_t(\gamma_3 k_t - i2\gamma_3 k_z) \qquad \text{for } \{1\,1\,0\} \text{ planes}. \qquad (A8.23)$$

The Hermitian conjugate is given by

$$W^{\dagger} = \sqrt{3}\,k_t(\gamma_{2(3)} k_t + i2\gamma_3 k_z), \qquad (A8.24)$$

independent of whether or not k_z is complex, (i.e., we do not take the complex conjugate of the hermitian operator k_z in defining $W^\dagger$). For either of the above forms, it is interesting to note that with $k_t = 0$, the coupling term disappears, and the effective mass equations decouple! Thus, in a quantum well, we can independently solve each of the effective mass equations (A8.17) and (A8.18) for the quantized $E_{1n}(0)$ of the HH and LH valence bands, just as was done for the conduction band in the previous section. However, for finite k_t (i.e., away from the band edge), the equations become coupled and the solutions for $E_{1n}(k_t)$ become more complicated.

A8.4.2.2 Bulk Solutions As in the conduction band, our first step in solving the quantum-well problem is to find the bulk solutions within each material. To find a general relation for $E_1(\mathbf{k})$ in bulk material, it is convenient to cast the effective mass equations (A8.17) and (A8.18), into matrix form:

$$\begin{bmatrix} H_{hh} + V & W \\ W^\dagger & H_{lh} + V \end{bmatrix} \begin{bmatrix} F_{hh} \\ F_{lh} \end{bmatrix} = E_1(\mathbf{k}) \begin{bmatrix} F_{hh} \\ F_{lh} \end{bmatrix}. \tag{A8.25}$$

For the bulk solutions we take V to be a constant, V_0. Then in looking at Eqs. (A8.19) and (A8.20), we find that for a given k-vector, the 2×2 Hamiltonian matrix consists of simple constants. The eigenvalue problem in this case is easily solved for the eigenenergies. In general form, the bulk $E_1(\mathbf{k})$ relations for the HH and LH bands are given by

$$E_1(\mathbf{k}) - V_0 = \tfrac{1}{2}(H_{hh} + H_{lh}) \pm \tfrac{1}{2}[(H_{hh} - H_{lh})^2 + 4W^\dagger W]^{1/2}. \tag{A8.26}$$

Using Eqs. (A8.19) through (A8.24), $E_1(\mathbf{k})$ can be given explicitly in terms of $\mathbf{k}$ within any $\{1\,0\,0\}$ plane, by

$$E_1(\mathbf{k}) = \gamma_1(k_z^2 + k_t^2) \pm [4\gamma_2^2(k_z^2 + k_t^2)^2 + 12(\gamma_3^2 - \gamma_2^2)k_z^2 k_t^2]^{1/2}, \tag{A8.27}$$

where we have set $V_0 = 0$. A similar equation can be found for $E_1(\mathbf{k})$ in any $\{1\,1\,0\}$ plane. Figure A8.2 illustrates the constant energy contour curves of (A8.27) for both the HH band solution (negative root) and the LH band solution (positive root). From these curves, we see that the third Luttinger parameter, γ_3, can be related to the effective mass anisotropy along the $\langle 1\,0\,0 \rangle$ and $\langle 1\,1\,0 \rangle$ directions (because with $\gamma_3 = \gamma_2$, the contour curves would become circles, from (A8.27)). Looking along either k_z or k_t, (along a $\langle 1\,0\,0 \rangle$ direction), the cross-term in Eq. (A8.27) disappears and we simply have

$$E_1(\mathbf{k}) = (\gamma_1 \pm 2\gamma_2)k^2, \qquad \mathbf{k} \text{ directed along any } \langle 1\,0\,0 \rangle \text{ direction.} \tag{A8.28}$$

From the definition of the Luttinger parameters in (A8.21), we see that the standard HH and LH $E_1(\mathbf{k})$ relations are obtained from (A8.28). Thus,

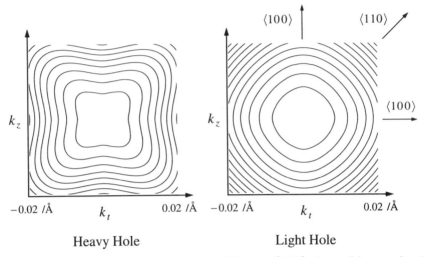

Heavy Hole   Light Hole

FIGURE A8.2   Contours of constant energy within any $\{100\}$ plane of k-space for the HH and LH bands in bulk GaAs. The energy spacing between each contour level is 0.5 meV for the HH band and 3 meV for the LH band.

contrary to what (A8.19) and (A8.20) may suggest, the effective masses along k_z and k_t are identical in bulk material (within $\{100\}$ planes).

To completely specify the bulk solutions, we also need to find the eigenvectors of (A8.25). With $E_{hh}(\mathbf{k})$ and $E_{lh}(\mathbf{k})$ given by the two roots of (A8.26), the eigenvectors, apart from a normalization constant, are found to be

$$\psi_1(\mathbf{k}, \mathbf{r}) = \begin{bmatrix} F_{hh} \\ F_{lh} \end{bmatrix} = e^{i\mathbf{k}\cdot\mathbf{r}} \begin{bmatrix} H_{lh} + V_0 - E_{hh} \\ -W^\dagger \end{bmatrix} \equiv e^{i\mathbf{k}\cdot\mathbf{r}} \begin{bmatrix} \Delta_{1h}(\mathbf{k}) \\ \Delta_{1l}(\mathbf{k}) \end{bmatrix}, \quad (A8.29)$$

$$\psi_2(\mathbf{k}, \mathbf{r}) = \begin{bmatrix} F_{hh} \\ F_{lh} \end{bmatrix} = e^{i\mathbf{k}\cdot\mathbf{r}} \begin{bmatrix} H_{lh} + V_0 - E_{lh} \\ -W^\dagger \end{bmatrix} \equiv e^{i\mathbf{k}\cdot\mathbf{r}} \begin{bmatrix} \Delta_{2h}(\mathbf{k}) \\ \Delta_{2l}(\mathbf{k}) \end{bmatrix}, \quad (A8.30)$$

where for either solution, the matrix notation implies

$$\psi = F_{hh} u_A + F_{lh} u_B. \quad (A8.31)$$

The validity of (A8.29) and (A8.30) can be checked by substituting them into (A8.18). In addition, the Hamiltonians in Eqs. (A8.19) through (A8.24) implicitly assumed plane wave solutions, so their inclusion in (A8.29) and (A8.30) is mandatory. Equation (A8.31) gives the wavefunctions in vector notation. The Bloch functions, $u_{A, B}$, are orthogonal to each other (analogous to two orthogonal unit position vectors in real space) and are given by linear combinations of the valence band Bloch functions defined in Section A8.3. We can write them as [18]

$$u_A = \frac{1}{\sqrt{2}} (\alpha u_{hh} - \alpha^* \bar{u}_{hh}), \tag{A8.32}$$

$$u_B = \frac{1}{\sqrt{2}} (\beta \bar{u}_{lh} - \beta^* u_{lh}). \tag{A8.33}$$

For $\{100\}$ planes, $\alpha = \beta = 1$. For $\{110\}$ planes, $\alpha = \exp[i3\pi/8]$, and $\beta = \exp[-i\pi/8]$.

A8.4.2.3 Quantum-well Solutions To solve the quantum-well problem, we again choose the quantum-well direction to be along z (this is not mandatory, but the effective mass equations would not decouple at the band edge otherwise (see (A8.22) through (A8.24)). In this case, k_z is directed along the confinement axis and k_t is the transverse k-vector component which lies in the plane of the well, as shown in Fig. A8.3(a). We now construct a general solution out of all bulk solutions that exist at a given energy within each material. From

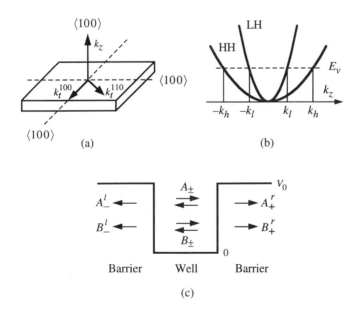

FIGURE A8.3 (a) Coordinate system to be used in the valence band model. The transverse or in-plane k-vector, k_t, can be directed along either a $\langle 100 \rangle$ or $\langle 110 \rangle$ direction, whereas the confinement axis must be along a $\langle 100 \rangle$ direction. (b) Illustration of the four plane wave states that exist at a given energy in the bulk valence band structure with $k_t = 0$. (c) Quantum-well potential and the eight coefficients which are to be used in solving the degenerate effective mass equation (see Eq. (A8.34)). The l and r superscripts indicate left and right coefficients.

Fig. A8.3(b), we see that in general, four plane wave solutions exist at a given energy [19]. The general solution in each region of Fig. A8.3(c) is then given by a linear combination of these four waves, or

$$\Psi = \sum A_{\pm} \psi_1(\pm k_{hh}, k_t, \mathbf{r}) + \sum B_{\pm} \psi_2(\pm k_{lh}, k_t, \mathbf{r}). \qquad (A8.34)$$

The sums are over the plus- and minus-going waves, and the $k_{hh, lh}$ are defined in Fig. A8.3(b). The four coefficients, $A_{\pm}$, $B_{\pm}$, are unknown constants. Equation (A8.34) is the valence band analog of (A1.6) in the conduction band. In this sense, Eq. (A8.27) is the more complicated valence band analog of Eq. (A1.7).] Both ψ_1 and ψ_2 are two-component vectors, from their definitions in (A8.29) and (A8.30). The general solution, Ψ, is also a two-component vector. Each component can be written as

$$F_{hh} = e^{ik_t r_t} [\sum A_{\pm} \Delta_{1h}(\pm k_{hh}, k_t) e^{\pm ik_{hh} z} + \sum B_{\pm} \Delta_{2h}(\pm k_{lh}, k_t) e^{\pm ik_{lh} z}], \quad (A8.35)$$

$$F_{lh} = e^{ik_t r_t} [\sum A_{\pm} \Delta_{1l}(\pm k_{hh}, k_t) e^{\pm ik_{hh} z} + \sum B_{\pm} \Delta_{2l}(\pm k_{lh}, k_t) e^{\pm ik_{lh} z}], \quad (A8.36)$$

where we have pulled out the common transverse plane wave component. Thus, in the case of the valence band, there are four unknown coefficients, $A_{\pm}$, $B_{\pm}$ in each region, as shown in Fig. A8.3(c), as opposed to just two, as was the case in the conduction band.

To find the coefficients, we match the general solutions within each region at both well–barrier interfaces. The boundary conditions for the degenerate effective mass equation involve matching the four following quantities [19] across the interface

$$F_{hh} \qquad \text{and} \qquad (\gamma_1 - 2\gamma_2) \frac{dF_{hh}}{dz} + \sqrt{3} \gamma_3 k_t F_{lh}, \qquad (A8.37)$$

$$F_{lh} \qquad \text{and} \qquad (\gamma_1 + 2\gamma_2) \frac{dF_{lh}}{dz} - \sqrt{3} \gamma_3 k_t F_{hh}. \qquad (A8.38)$$

With $k_t = 0$, the above boundary conditions are identical to those given in Appendix 1 for the nondegenerate case. In addition, for any k_t, if the Luttinger parameters are the same on both sides of the interface, the second boundary conditions reduce to the simple slope continuity conditions. The generalized slope continuity conditions (A8.37) and (A8.38) can be derived by integrating Eq. (A8.25) over an infinitesimal thickness which straddles the interface. However, we must first symmetrize the Hamiltonian which involves setting $(\gamma_1 \pm 2\gamma_2)k_z^2 \rightarrow k_z(\gamma_1 \pm 2\gamma_2)k_z$ in Eqs. (A8.19) and (A8.20), *and* setting $\gamma_3 k_z \rightarrow (\gamma_3 k_z + k_z \gamma_3)/2$ in Eqs. (A8.22) through (A8.24), before setting $k_z \rightarrow -i\partial/\partial z$, since γ_1, γ_2, γ_3 all depend on z. These symmetrizing substitutions guarantee that the Hamiltonian in Eq. (A8.25) remains hermitian for any arbitrary z-dependence of γ_1, γ_2, γ_3. It should be noted that the above boundary conditions hold for both $\{1\,0\,0\}$ and $\{1\,1\,0\}$ planes. Caution should be issued here that the above boundary conditions apply only when the Bloch functions of the two materials are similar (this similarity should be mirrored in the values of the Luttinger parameters).

Applying the four boundary conditions at each interface gives us a total of eight equations. There are four unknown coefficients in each of the three regions, or twelve in total. However, requiring the envelope functions to go to zero at infinity leaves us with a total of eight unknown coefficients as shown in Fig. A8.3(c). Thus, our problem is now completely specified and we can solve the eight homogeneous equations by numerically finding the roots of the 8×8 determinant.

The general procedure for obtaining $E_1(k_t)$ is then as follows: (1) find the $E_{1n}(0)$ of a particular HH or LH band edge energy level using the conventional method of Appendix 1; (2) increment k_t and guess at the new energy of the state; (3) find $k_{hh,\,lh}$ from the two $E_1(\mathbf{k})$ relations given in (A8.26) within each material (each material having its own V_0 and its own set of Luttinger parameters); (4) evaluate the Δ's within each material from their definitions in (A8.29) and (A8.30); (5) evaluate the 8×8 coefficient determinant; (6) if it is not equal to zero, use Newton's method to repeat the process until the energy root is found for that given k_t; and (7) increment k_t and repeat the entire process to find the new energy root, using an educated initial guess. The entire $E_{1n}(k_t)$ can then be traced in this way. The rate of convergence is very good for this type of problem. For example, for each k_t, the energy root can be found in typically 2–3 iterations (Fig. A8.4 was generated in less than 30 seconds on a Macintosh Quadra 650).

Our description of the valence subband structure is now complete and we can move on to an example. The one-dimensional quantum confinement in the conduction band gives rise to a set of parabolic subbands in the plane of the well. In the valence band, coupling between the HH and LH subbands changes the situation drastically, giving rise to a much more interesting band structure. Figure A8.4 shows the valence subband structure for an 80Å GaAs/Al$_x$Ga$_{1-x}$As quantum well with $x = 0.2$ in the barrier regions, calculated using the procedure outlined above. The subband structure is seen to be far from parabolic, and in some regions, the band curvature is even inverted, leading to a negative "local" hole mass.

With respect to predicting the optical gain achievable in a material with the subband structure shown in Fig. A8.4, we are particularly interested in the density of states of the subbands, also shown in Fig. A8.4. The density of states is important because it determines the relationship between carrier density and the quasi-Fermi level of the band as discussed in Appendix 2. From Fig. A8.4, we see that ρ_v is roughly $2.5\rho_c$ near the band edge, but rapidly becomes very large as mixing between the bands starts to become significant. The mismatch between ρ_c and ρ_v as well as the overall large ρ_v reduces the performance of the quantum well, increasing transparency levels and reducing the differential gain (ideally the DOS curve in Fig. A8.4 would be a straight line of magnitude one).

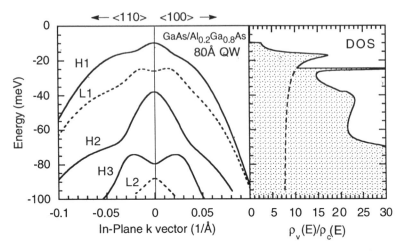

FIGURE A8.4 Plotted on the left is the valence subband structure of an 80Å GaAs/Al$_{0.2}$Ga$_{0.8}$As quantum well ($V_0 \approx 95$ meV). On the right is the total (solid curve) and H1 subband (dashed curve) density of states plotted relative to the density of states in the first conduction (C1) subband.

Because the energy bands are different along the $\langle 110 \rangle$ and $\langle 100 \rangle$ directions within the quantum well, the density of states at any given energy should in principle be calculated by averaging over all in-plane k-vector directions. Fortunately, this procedure can be approximated by calculating the density of states for just one set of energy bands which is some appropriate average between the $\langle 100 \rangle$ and $\langle 110 \rangle$ dispersion curves shown in Fig. A8.4. These average energy bands can be found by using a coupling term of the form $W_{ave} = (W_{100} + W_{110})/2$ in the band calculation. This approach is commonly known as the *axial* approximation [3]. We use it to calculate the density of states in Figs. A8.4 and A8.6 as well as for all quantum well gain calculations presented in Chapter 4. For the "bulk" barrier regions, the energy bands (the HH and LH effective masses) are found using a "spherical average" of the effective masses along all directions in the crystal. It basically involves replacing γ_2 with $(2\gamma_2 + 3\gamma_3)/5$ in Eq. (A8.21), and is known as the *spherical* approximation [19].

In the next section, we consider ways in which ρ_v can be modified, in beneficial ways, by the introduction of strain.

A8.4.3 Strained Quantum Wells

It was originally suggested by Yablonovich and Kane [20] and independently by Adams [21] that the introduction of compressive strain into the crystal lattice of a semiconductor could lead to enhanced performance in semiconductor lasers. To understand why, we need to examine the effects of

(a)

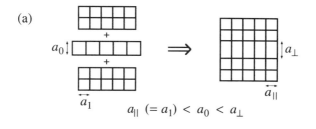

$a_{\parallel} \ (= a_1) < a_0 < a_{\perp}$

(b)

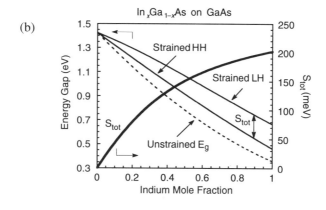

In$_x$Ga$_{1-x}$As on GaAs

(c)

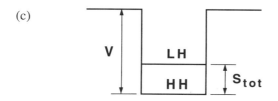

FIGURE A8.5 (a) Crystal lattice deformation under compressive strain. (b) Bulk band-gap of In$_x$Ga$_{1-x}$As, when the in-plane lattice constant is compressed to that of GaAs. (c) The modified potential profile in the valence band of a quantum well, when the well material is under compressive strain.

strain, particularly in a quantum well. Introducing compressive strain into a quantum well configuration is particularly simple; just grow the well layer out of a material with a *larger* native lattice constant than the barrier layers. Because the quantum-well layer is typically very thin, instead of forming misfit dislocations, the lattice actually compresses in the plane of the well to match that of the barrier layers. In addition, the lattice constant in the direction normal to the plane becomes elongated (in an effort to keep the volume of each unit cell the same), as shown in Fig. A8.5(a).

Because the energy gap of a semiconductor is related to its lattice spacing, we might expect that distortions in the crystal lattice should lead to alterations in the bandgap of the strained layer (putting aside for the moment the changes created simply by quantum confinement). In fact, there are two types of modifications which occur as discussed in more detail in Appendix 11. The first effect produces an upward shift in the conduction band as well as a downward shift in both valence bands, increasing the overall bandgap by an amount, H (which is positive for compressive strain and negative for tensile strain). The H indicates that this shift originates from the *hydrostatic* component of the strain. The second more important effect separates the HH and LH bands, each being pushed in opposite directions from the center by an amount, S. The S indicates that this shift originates from the *shear* component of the strain. Thus, the band edge degeneracy of the two valence bands is removed and two energy gaps must now be defined. The total strained bandgap can be written as $E_g + H \pm S$, where the upper sign refers to the C–LH bandgap, $E_g(LH)$, and the lower sign refers to the C–HH bandgap, $E_g(HH)$. In Fig. A8.5(b) we have plotted the unstrained bulk bandgap as well as the two compressively strained bulk bandgaps of InGaAs on a GaAs substrate (which has a smaller lattice constant than InGaAs). Note that S as defined above is positive for compressive strain, since $E_g(LH) > E_g(HH)$. For tensile strain, S would be negative and the bandgap ordering would be reversed.

The two energy shifts, H and S, increase linearly with the lattice constant mismatch, which in turn increases linearly with indium mole fraction. We should expect then that the bandgap difference, $E_g(LH) - E_g(HH)$, defined in the plot as S_{tot}, should be a linearly increasing function of indium mole fraction, since from the above discussion, $S_{tot} = 2S$. For small indium mole fractions, we see from the plot that this is true. However, as the indium mole fraction increases, S_{tot} begins to saturate. This is a result of the interaction between the LH band and the SO band. When taken into account, this interaction introduces a correction term into the expression for the strained LH bandgap such that to second order, $S_{tot} = 2S(1 - S/\Delta)$, where Δ is the spin–orbit splitting energy. A more detailed discussion of strained bandgaps is given in Appendix 11.

In a quantum well, the splitting of the HH and LH bands can have dramatic consequences, since the large nonparabolicity of the subband structure in Fig. A8.4 is a direct result of the HH and LH band mixing. If we place the strained bandgaps shown in Fig. A8.5(b) into a quantum well, the situation becomes as shown in Fig. A8.5(c), where the depth of the quantum well as seen by light holes is reduced by the splitting energy, S_{tot}. To predict the valence subband structure of the strained quantum well in Fig. A8.5(c), we simply need to add a potential offset to the effective mass equation describing the LH band in the well. Equation (A8.25) in the previous section now simply becomes

$$\begin{bmatrix} H_{hh} + V & W \\ W^\dagger & H_{lh} + V + S_{tot} \end{bmatrix} \begin{bmatrix} F_{hh} \\ F_{lh} \end{bmatrix} = E_1(\mathbf{k}) \begin{bmatrix} F_{hh} \\ F_{lh} \end{bmatrix}, \qquad (A8.39)$$

where V is zero inside the well and S_{tot} is zero outside the well. The procedure for solving (A8.39) in a strained quantum well is entirely analogous to the procedure presented in Section A8.4.2 for an unstrained quantum well. Thus, we can immediately turn to an example calculation using Eq. (A8.39).

For direct comparison, we take the GaAs/AlGaAs 80Å quantum well used in the example of Section A8.4.2 and simply add a bit of indium to the well layer. InGaAs has a larger native lattice constant than GaAs, thus sandwiched between two AlGaAs layers, it will be compressed in the plane of the well. For an indium mole fraction of 20%, the resulting HH–LH splitting energy, S_{tot}, is approximately 80 meV, or 1.4% compressive strain.

Figure A8.6 shows both the valence subband structure and the density of states of the 80 Å $In_{0.2}Ga_{0.8}As/Al_{0.2}Ga_{0.8}As$ strained quantum well calculated from Eq. (A8.39) [22]. Immediately we see that the LH bands have been pushed deep into the band (the full depth of the well is not shown). As a result, the band warping has been greatly reduced. Comparison with Fig. A8.4 reveals that the density of states in the strained quantum well is reduced significantly and matching between ρ_c and ρ_v is greatly improved. Both of these features translate into lower transparency levels and higher differential gain as calculations of gain presented in Chapter 4 reveal.

Equation (A8.39) not only applies to quantum wells, but can also be applied to bulk strained material. The $E(k)$ relations in bulk material are obtained by finding the eigenenergies of Eq. (A8.39) as we did for Eq. (A8.25) in the last section (see (A8.26) and (A8.27)) (the definitions for H and W given in Section

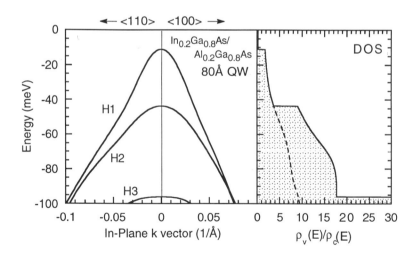

FIGURE A8.6 Plotted on the left is the valence subband structure of an 80 Å $In_{0.2}Ga_{0.8}As/Al_{0.2}Ga_{0.8}As$ strained quantum well ($V_0 \approx 175$ meV). The LH bands have been pushed further out of the well as a result of the strain and cannot be seen with the energy scale shown. On the right is the total (solid curve) and H1 subband (dashed curve) density of states plotted relative to the density of states in the first conduction (C1) subband.

A8.4.2 apply here as well). The general bulk solutions for the eigenenergies of (A8.39) can be obtained in closed form, but are somewhat messier than Eq. (A8.27). We leave it as an exercise for the reader to show that to first order in $\gamma_2 k^2/S$ (i.e., in the large-strain regime where $S \gg \gamma_2 k^2$), the eigenenergies of (A8.39) can be expressed as [23]

$$E_1(k_{\parallel}) = (\gamma_1 \mp \gamma_2)k_t^2 \pm S_{tot}/2, \qquad (k_z = 0) \qquad \text{(A8.40)}$$

$$E_1(k_{\perp}) = (\gamma_1 \pm 2\gamma_2)k_z^2 \pm S_{tot}/2. \qquad (k_t = 0) \qquad \text{(A8.41)}$$

The upper signs refer to the LH band solutions while the lower signs refer to the HH band solutions (we have thrown away any common shifts in the band edges to concentrate on the *difference* between the LH and HH bandgap energies). The parallel (in-plane) and perpendicular (normal to the plane) k-vectors refer to the notation used in Fig. A8.5(a). Note that the above relations are also obtained by setting $W = 0$ (as seen by combining (A8.19) and (A8.20) with (A8.39)). This makes sense because in the large-strain regime we are basically saying that $S \gg W$, allowing us to neglect the coupling between the bands altogether. The modified band structure given by (A8.40) and (A8.41) is shown to scale in Fig. A8.7. Thus, even in bulk material, strain serves to reduce the effective mass of the HH band dramatically within the "plane of compression." Perpendicular to the plane, it is interesting to note, however, that apart from the splitting of the HH and LH bands, the dispersion relation (A8.41) remains unchanged compared to (A8.28).

We have thus far not mentioned the effects of strain on the conduction band. The reason for this is that due to its relative isolation from other bands, the conduction band curvature remains relatively unaffected by the shifting energy gaps. Equation (A8.13) would suggest that the increase in the bandgap should increase the conduction band effective mass slightly. However, we must be careful here because the third parameter which ties into this equation, $|M|^2$,

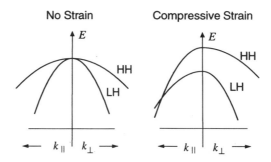

FIGURE A8.7   Effects of compressive strain on the bulk valence band structure. The relative band curvatures are drawn according to Eqs. (A8.40) and (A8.41), which are derived under the large-strain limit (the true dispersion curves would not cross each other as suggested in the figure).

does not necessarily remain constant (for example, it is conceivable that $|M|^2$ could increase proportionally with the bandgap, leaving m^* unaffected). In any case, the change in the conduction band curvature should be slight.

REFERENCES

[1]   D. Long, *Energy Bands in Semiconductors*, Interscience, New York (1968).

[2]   E.O. Kane, *J. Phys. Chem. Solids*, **1**, 82 (1956).

[3]   E.O. Kane, *J. Phys. Chem. Solids*, **1**, 249 (1957).

[4]   J.M. Luttinger and W. Kohn, *Phys. Rev.*, **97**, 869 (1955).

[5]   D.A. Broido and L.J. Sham, *Phys. Rev. B*, **31**, 888 (1985).

[6]   H.C. Casey, Jr. and M.B. Panish, *Heterostructure Lasers Part A: Fundamental principles*, Academic Press, Orlando, FL (1978).

[7]   J.S. Blakemore, *J. Appl. Phys.*, **53**, R123 (1982).

[8]   M. Asada, A. Kameyama, and Y. Suematsu, *IEEE J. Quantum Electron.*, **QE-20**, 745 (1984).

[9]   D.J. Chadi, A.H. Clark, and R.D. Burnham, *Phys. Rev. B*, **13**, 4466 (1976).

[10] C. Hermann and C. Weisbuch, *Phys. Rev. B*, **15**, 823 (1977).

[11] C. Hermann and C. Weisbuch, *Modern Problems in Condensed Matter Sciences, Volume 8; Optical Orientation*, ed. V.M. Agranovich and A.A. Maradudin, North-Holland, pp. 463–508 (1984).

[12] B. Jani, P. Gilbart, J.C. Portal, and R.L. Aulombard, *J. Appl. Phys.*, **58**, 3481 (1985).

[13] R.J. Nicholas, J.C. Portal, C. Houlbert, P. Perrier, and T.P. Pearsall, *Appl. Phys. Lett.*, **34**, 492 (1979).

[14] R.H. Yan, S.W. Corzine, and L.A. Coldren, *IEEE J. Quantum Electron.*, **QE-26**, 213 (1990).

[15] R. Eppenga, M.F.H. Schuurmans, and S. Colak, *Phys. Rev. B*, **36**, 1554 (1987).

[16] S. Colak, R. Eppenga, and M.F.H. Schuurmans, *IEEE J. Quantum Electron.*, **QE-23**, 960 (1987).

[17] J.M. Luttinger, *Phys. Rev.*, **102**, 1030 (1956).

[18] S.L. Chuang, *Phys. Rev. B*, **40**, 10379 (1989).

[19] L.C. Andreani, A. Pasquarello, and F. Bassani, *Phys. Rev. B*, **36**, 5887 (1987).

[20] E. Yablonovitch and E.O. Kane, *IEEE J. Lightwave Technol.*, **4**, 504 (1986).

[21] A.R. Adams, *Electron. Lett.*, **22**, 249 (1986).

[22] S.W. Corzine, R.H. Yan, and L.A. Coldren, *Appl. Phys. Lett.*, **57**, 2835 (1990).

[23] T.C. Chong and C.G. Fonstad, *IEEE J. Quantum Electron.*, **QE-25**, 171 (1989).

READING LIST

D. Ahn, S.L. Chuang, and Y.C. Chang, *J. Appl. Phys.*, **64**, 4056 (1988).

D. Ahn and S.L. Chuang, *IEEE J. Quantum Electron.*, **QE-26**, 13 (1990).

Fermi's Golden Rule

A9.1 INTRODUCTION

This appendix derives a general expression for the rate of decay from an initial quantum mechanical state into a continuum of final states in the presence of a harmonic perturbation. The resulting expression for the transition rate is referred to as *Fermi's Golden Rule*, since it is a general result applicable to many quantum mechanical systems. Here we concentrate on the interaction between light and matter.

In physical terms, radiative transitions occur because the oscillating field of the photon alters the oscillating phase of the electron wavefunction in such a way that it becomes similar to the oscillating phase of another electron wavefunction typically in a different energy band. This phase-matching in time results in a strong coupling between the two electron states, analogous to the coupling of waveguide modes considered in Chapter 6. In the latter case, the coupling causes the energy initially in one mode to be transferred back and forth between the two modes. If only two electron wavefunctions were involved in the coupling, the electron would also oscillate between the two states (known as *Rabi* oscillations in the quantum world). However in most cases, a *density* of states are coupled to the initial electron wavefunction, and as a result, the electron transforms with an exponential decay from its initial wavefunction to one of many resonant wavefunctions. This electromagnetically induced phase-matching and subsequent transformation of an electron from one state to another is the fundamental mechanism of radiative transitions. The following treatment quantifies these physical arguments.

A9.2 SEMICLASSICAL DERIVATION OF THE TRANSITION RATE

To characterize the electronic system and its interaction with light, we make use of two Hamiltonians: one that describes the electronic system in isolation,

H_0, and a second that describes a classical perturbation created by the electromagnetic field, $H'(t)$. The wavefunction in Dirac notation, $|\Psi(t)\rangle$, provides a description of the state of the electron. Its evolution in time is governed by Schrödinger's equation:

$$i\hbar \frac{d}{dt}|\Psi(t)\rangle = \{H_0 + H'(t)\}|\Psi(t)\rangle. \tag{A9.1}$$

The problem we wish to solve using (A9.1) is the process of absorption (stimulated emission is analogous to absorption and hence leads to the same result, while spontaneous emission is treated separately in Chapter 4). To model an absorption event, we assume the electron initially occupies some ground state of the system, $|\psi_0\rangle$. The presence of the time-varying field excites a change in the state of the electron. Under the appropriate conditions, the electron can be excited to any number of higher energy final states, $|\psi_s\rangle$, where the sth state has energy $\hbar\omega_s$ above the ground state. This situation is depicted in Fig. A9.1.

Which state the electron eventually occupies is unknown, so we hypothesize a time-dependent superposition of the initial and possible final states of the electron:

$$|\Psi(t)\rangle = c_0(t)|\psi_0\rangle + \sum_s c_s(t)e^{-i\omega_s t}|\psi_s\rangle, \tag{A9.2}$$

with the initial conditions:

$$c_0(0) = 1 \quad \text{and} \quad c_s(0) = 0. \tag{A9.3}$$

The time-dependence of the unperturbed wavefunctions, $e^{-i\omega_s t}$, is included explicitly to remove the rapidly oscillating phase from the time dependence of the expansion coefficients ($|\psi_0\rangle$ and $|\psi_s\rangle$ are assumed to be independent of time). Note that (A9.2) satisfies (A9.1) with no perturbation present since $H_0|\psi_s\rangle = \hbar\omega_s|\psi_s\rangle$ and $H_0|\psi_0\rangle = 0$.

To obtain a set of useful equations, we begin by substituting (A9.2) into (A9.1). To move forward we recognize that all states are orthogonal to each other such that $\langle\psi_0|\psi_s\rangle = 0$ and $\langle\psi_{s'\neq s}|\psi_s\rangle = 0$. Furthermore, we have

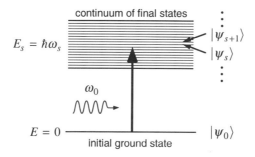

FIGURE A9.1 Model assumed for the process of absorption.

$\langle\psi_0|\psi_0\rangle = 1$ and $\langle\psi_s|\psi_s\rangle = 1$. If we multiply (A9.1) from the left with $\langle\psi_0|$, we obtain one equation for the coefficient c_0. Multiplying (A9.1) from the left with $\langle\psi_s|$, we obtain an additional set of equations, one for each c_s:

$$i\hbar\frac{dc_0}{dt} = H'_{00}c_0 + \sum_s H'_{0s}c_s e^{-i\omega_s t},$$

$$i\hbar\frac{dc_s}{dt} = H'_{s0}c_0 e^{i\omega_s t} + \sum_{s'} H'_{ss'}c_{s'} e^{-i(\omega_{s'} - \omega_s)t}, \tag{A9.4}$$

where the following shorthand notation is used:

$$H'_{s0} \equiv \langle\psi_s|H'|\psi_0\rangle. \tag{A9.5}$$

To simplify (A9.4) further, we need to express the time dependence of the perturbation explicitly. For a harmonic perturbation, we can set

$$H'(t) \rightarrow H'(e^{i\omega_0 t} + e^{-i\omega_0 t}), \tag{A9.6}$$

where ω_0 is the oscillation frequency of the incident electromagnetic wave. The amplitude of the wave is contained in H', however, we will not need to know the explicit details of H' for the purposes of this appendix.

The set of equations contained in (A9.4) is exact. The first approximation we make is known as the *rotating wave approximation* which involves ignoring all terms that oscillate at frequencies comparable to or greater than either ω_s or ω_0. The reasoning is that these terms will oscillate positive and negative much faster than the time scales involved with changes in $c(t)$ and hence, will average out to zero net contribution as time increases. Placing (A9.6) into (A9.4), the following combinations of angular frequencies appear: ω_0, $\omega_s \pm \omega_0$, $(\omega_{s'} - \omega_s) \pm \omega_0$. If we assume $\omega_s \sim \omega_0$ and $\omega_{s'} \sim \omega_s$, then all combinations except $\omega_s - \omega_0$ are comparable to or greater than either ω_s or ω_0.

Ignoring all terms in (A9.4) except those which contain the difference frequency between the electronic and electromagnetic frequencies, $\omega_s - \omega_0$, we obtain the central equations of motion:

$$i\hbar\frac{dc_0}{dt} = \sum_s H'_{0s}c_s e^{-i(\omega_s - \omega_0)t},$$

$$i\hbar\frac{dc_s}{dt} = H'_{s0}c_0 e^{i(\omega_s - \omega_0)t}. \tag{A9.7}$$

These equations are known as the Wigner–Weisskopf equations within the rotating wave approximation.

To solve (A9.7) we integrate the second equation from 0 to t, using the initial conditions expressed in (A9.3) to obtain

$$c_s(t) = -\frac{i}{\hbar} H'_{so} \int_0^t c_0(t') e^{i(\omega_s - \omega_0)t'} dt'.$$ (A9.8)

Inserting (A9.8) into the first equation of (A9.7), we obtain

$$\frac{dc_0(t)}{dt} = -\frac{1}{\hbar^2} \sum_s |H'_{so}|^2 \int_0^t c_0(t') e^{-i(\omega_s - \omega_0)(t-t')} dt',$$ (A9.9)

where we have made use of the fact that $H'_{0s} = (H'_{so})^*$. We can convert the sum over final states to an integral with the assumption that the states are so closely spaced that they form a continuous distribution characterized by a density of states function in energy:

$$\sum_s |H'_{so}|^2 \rightarrow \int |H'(E_s)|^2 \rho_f(E_s) \hbar \, d\omega_s.$$ (A9.10)

The matrix element $|H'(E_s)|^2$ represents an average of $|H'_{so}|^2$ over all final s states existing with energies close to E_s above the ground state. Also, the final density of states function, ρ_f, represents the total *number* of states per unit energy.

Substituting (A9.10) into (A9.9) and reversing the order of integration, we can identify one portion of the equation as the inverse Fourier transform of the matrix element–density of states product:

$$f(t - t') = \int_{-\infty}^{\infty} \frac{1}{\hbar} |H'(E_s)|^2 \rho_f(E_s) e^{-i(\omega_s - \omega_0)(t-t')} \, d\omega_s.$$ (A9.11)

With this definition, (A9.9) becomes simply

$$\frac{dc_0(t)}{dt} = -\int_0^t c_0(t') f(t - t') \, dt'.$$ (A9.12)

Under certain circumstances, this integro-differential equation can be solved exactly. Its solution depends critically on the Fourier transform of the product $|H'(E_s)|^2 \rho_f(E_s)$, represented by the time-domain response function, $f(t - t')$. We will examine three such solutions here, the first of these leading to the famed Fermi's Golden Rule.

A9.2.1   Case I: the Matrix Element-Density of Final States Product is a Constant

If the density of final states is distributed evenly over a large energy range, and if the matrix element does not vary over this range, we can set $|H'(E_s)|^2 \rho_f(E_s)$ to a constant. Taking the Fourier transform we find that $f(t - t')$ becomes a delta function in time, as is evident from the following relation:

$$\int_{-\infty}^{\infty} e^{-i(\omega_s - \omega_0)(t-t')} \, d\omega_s = 2\pi\delta(t-t').$$ (A9.13)

The integro-differential equation for this case reduces to a simple first-order differential equation

$$|H'(E_s)|^2 \rho_f(E_s) \rightarrow |H'|^2 \rho_f$$

$$f(t-t') = \frac{1}{\hbar}|H'|^2\rho_f \cdot 2\pi\delta(t-t')$$ (A9.14)

$$\frac{dc_0(t)}{dt} = -\frac{2\pi}{\hbar}|H'|^2\rho_f \cdot \int_0^t c_0(t')\delta(t-t') \, dt' = -\frac{\pi}{\hbar}|H'|^2\rho_f \cdot c_0(t).$$

The factor of 2π representing the total area under the delta function is reduced to π in the last equality because the time integration terminates in the center of the delta function such that only half of the area is included. The solution to (A9.14) is a decaying exponential:

$$c_0(t) = e^{-Wt/2} \rightarrow |c_0(t)|^2 = e^{-Wt},$$ (A9.15)

where the decay rate is given by

$$W = \frac{2\pi}{\hbar}|H'|^2\rho_f.$$ (A9.16)

We can also solve for the final state probability coefficients to determine where the electron ends up after making the transition. Using (A9.15) in (A9.8), we obtain

$$|c_s(t \rightarrow \infty)|^2 = \frac{|H'_{s0}|^2}{(\hbar W/2)^2 + (E_s - \hbar\omega_0)^2}.$$ (A9.17)

If we sum this probability over all final states (i.e., multiply (A9.17) by $\rho_f \, dE_s$ and integrate over all energies), the total probability will equal one. Thus, the electron will eventually appear somewhere in the continuum, we just don't know exactly where. The peak probability occurs at the state with energy $E_s = \hbar\omega_0$. The electron can appear at other states as well, however, the probability of this happening declines away from the peak with a distribution characterized by a Lorentzian with $\Delta E_{FWHM} = \hbar W$.

Equation (A9.17) also tells us that the electron only interacts with final states clustered around $E_s = \hbar\omega_0$. In other words, states that are not within $\hbar W/2$ of the resonant energy effectively do not participate in the transition. Therefore, our earlier assumption that $|H'|^2\rho_f$ must be constant over a large energy range can be refined to state that as long as $|H'(E_s)|^2\rho_f(E_s)$ is flat over energies for which (A9.17) has significant amplitude, the solutions of Case I can be applied.

This translates into evaluating $|H'(E_s)|^2 \rho_f(E_s)$ at $E_s = \hbar\omega_0$, and requiring that it be constant for energies $\gg \hbar W$. In semiconductor applications, $|H'|^2 \rho_f$ is usually a smooth enough function of E_s to easily meet this requirement. Therefore, interactions between light and semiconductor materials will typically induce exponentially decaying solutions for the initial state of the system.

Fermi's Golden Rule. Our ultimate objective is to obtain a transition rate that can be used in a rate equation. The above analysis allows us to determine how quickly photons are being removed from the electromagnetic field. For example, each transition event decays with the decay rate, W. If we resupply the ground state with a new electron every $1/W$ seconds, then the ground state will remain filled and every $1/W$ seconds on average, a photon will be absorbed. We can therefore view W as the rate at which photons are absorbed in the active region, or the transition rate. If N_p is the photon density and V_p is the mode volume, then $N_p V_p$ photons exist in a given mode. The rate at which these photons disappear can be expressed as

$$\frac{dN_p V_p}{dt} = -W. \tag{A9.18}$$

Dividing by the mode volume and introducing the confinement factor, the photon *density* rate equation becomes

$$\frac{dN_p}{dt} = -\frac{V}{V_p}\frac{W}{V} \equiv -\Gamma R_r, \tag{A9.19}$$

where $\Gamma(= V/V_p)$ is the optical confinement factor, and $R_r(= W/V)$ is defined as the radiative transition rate *per unit volume* of active material. If we absorb the $1/V$ into ρ_f, such that the final density of states is interpreted as a density per unit energy *and* volume (as is customary), then the transition rate per unit volume becomes

$$R_r = \frac{2\pi}{\hbar}|H'(E_s)|^2 \rho_f(E_s)|_{E_s - \hbar\omega_0}. \tag{A9.20}$$

This result and its interpretation as a transition rate is known as *Fermi's Golden Rule.* The evaluator bar reminds us that the product $|H'|^2 \rho_f$ must be evaluated at $E_s = \hbar\omega_0$ for reasons discussed in reference to Eq. (A9.17).

The derivation presented here assumes the electron is making an upward transition to model absorption. However, we could just as well assume the electron is making a downward transition by setting $\omega_s \to -\omega_s$, allowing us to model stimulated emission. This substitution requires that we use the opposite time harmonic in the perturbation Hamiltonian (A9.6), however, aside from this simple change the rest of the derivation is identical. Therefore, R_r has the same form for both absorption and stimulated emission. The difference between the upward and downward transition rates appears only when we include the probability of finding an electron in the initial state and no electron in the final state. The theory of gain given in Chapter 4 elaborates on this issue in some detail since the balance between absorption and stimulated emission provides the key to understanding optical gain in any material.

A9.2.2   Case II: the Matrix Element-Density of Final States Product is a Delta Function

Assume that only a single final state exists as opposed to a continuous density of final states, as might exist in an atomic transition from one energy level to another. We can model this situation by replacing the density of final states with a delta function in energy to represent the solitary energy level, E_f. To simplify matters, we will assume that $\omega_f = \omega_0$ such that perfect resonance between the electronic and electromagnetic systems exists (see Problem A9.2 for the more general solution). Taking the Fourier transform of the delta function, we find that $f(t - t')$ is a constant independent of time, which allows us to again solve (A9.12) exactly. Summarizing the results, we have

$$\rho_f(E_s) \rightarrow \frac{1}{\hbar} \delta(\omega_s - \omega_f)$$

$$f(t - t') = \frac{1}{\hbar^2} |H'|^2 \quad \text{(assuming } E_f = \hbar\omega_0) \tag{A9.21}$$

$$\frac{dc_0(t)}{dt} = -\frac{1}{\hbar^2} |H'|^2 \int_0^t c_0(t') \, dt' \rightarrow c_0(t) = \cos \Omega t, \quad \Omega^2 = \frac{1}{\hbar^2} |H'|^2.$$

Using (A9.8) with $\omega_f = \omega_0$, the final state probability amplitude is given by $c_f(t) = \sin \Omega t$. In words, when the electron interacts with only one possible final state, the probability amplitude oscillates back and forth sinusoidally between the initial and final states. The oscillation frequency depends on the magnitude of interaction, $|H'(E_f)|^2$. Physically, the energy is continually shifting back and forth between the electromagnetic field and the electron.

In a more general sense, we can view $f(t - t')$ as a *memory* function that represents how strongly previous events are coupled to current changes in $c_0(t)$. With only one final state to interact with, the system has infinite memory ($f(t - t')$ is independent of time), and the interaction continues indefinitely in the same periodic fashion. This periodic exchange of energy between the electromagnetic field and the electron is characteristic of *Rabi* oscillations, which occur whenever the coupled system exhibits a strong memory of interaction.

In contrast, the system considered in Case I (Fermi's Golden Rule) has no memory of previous events ($f(t - t')$ is a delta function in time). The existence of numerous *equally probable* transition pathways destroys the system memory, such that the interaction only depends on the present state of the electron. From another point of view, the probability amplitudes of the various pathways combine in such a way as to make the process irreversible causing the initial state to decay exponentially toward a final state (rather than oscillating back and forth indefinitely). The next case bridges the gap between Case I and Case II, producing a system with finite memory and oscillatory decaying exponential solutions.

A9.2.3   Case III: the Matrix Element-Density of Final States Product is a Lorentzian

This situation can occur when analyzing the spontaneous emission process of a two-level system placed in a highly resonant cavity. In general, the analysis of spontaneous emission leads to an equation identical to Eq. (A9.12), with the exception that the density of final electronic states is interpreted as the density of free-space optical modes, and the field strength within the matrix element is set equal to the vacuum-field strength of the free-space mode at E_s. When the two-level system is placed in a cavity, the vacuum-field strengths of the free-space optical modes become enhanced near the cavity resonances, which in turn enhances the matrix element. For a highly resonant cavity, this enhancement has a Lorentzian lineshape about each resonance.

With $|H'(E_s)|^2 \rho_f(E_s)$ set to a Lorentzian, its Fourier transform, $f(t - t')$, becomes a decaying exponential in time. In other words, the memory of the system decays exponentially as it recedes away from the present. One way of viewing this is that the spontaneously emitted photon only has a finite lifetime in the cavity before it escapes, limiting the memory of the system to a finite duration. In any case, (A9.12) can again be solved exactly, however, the solution is not as obvious as it was in Case I or II (see Problem A9.3). Summarizing, we have

$$|H'(E_s)|^2 \rho_f(E_s) \rightarrow |H'|^2 \rho_f K \cdot \frac{1}{1 + (2\Delta\omega\tau_p)^2}$$

$$f(t - t') = \frac{KW_0}{4\tau_p} e^{-|t-t'|/2\tau_p} \tag{A9.22}$$

$$\frac{dc_0(t)}{dt} = -\frac{KW_0}{4\tau_p} \int_0^t c_0(t') e^{-|t-t'|/2\tau_p} \, dt', \quad \text{where} \quad W_0 = \frac{2\pi}{\hbar} |H'|^2 \rho_f.$$

In the above equations, $\Delta\omega \equiv \omega_s - \omega_0$, τ_p is the photon lifetime of the cavity (note that $\Delta\omega_{FWHM}\tau_p = 1$), K is the enhancement of the vacuum-field strength within the cavity at the resonance peak ($K \propto \tau_p$), and W_0 is the decay rate found in Case I (with ρ_f in this case representing the density of free-space optical modes). Using the initial conditions expressed in Eq. (A9.3), the function satisfying (A9.22) can be written as

$$c_0(t) = \frac{p_+ e^{p_- t} - p_- e^{p_+ t}}{p_+ - p_-}, \quad \text{where} \quad p_\pm = -\frac{1}{4\tau_p} [1 \pm (1 - 4\tau_p KW_0)^{1/2}]. \tag{A9.23}$$

If the memory of the system characterized by the photon lifetime, τ_p, is very short such that $4\tau_p K W_0 \ll 1$, the solution reduces to

$$p_+ \approx -\frac{1}{2\tau_p}, \quad p_- \approx -\tfrac{1}{2}K W_0, \quad \text{and} \quad |c_0(t)|^2 \approx e^{-K W_0 t}. \quad \text{(A9.24)}$$

The last equality is valid for all but very short times, which can be verified from (A9.23) and the fact that $p_+ \gg p_-$. In words, when the photon escapes much quicker than the time it takes to make a transition (represented by the inequality $4\tau_p K W_0 \ll 1$), the initial state decays exponentially as was found in Case I for a system with no memory. From another point of view, the inequality $4\tau_p K W_0 \ll 1$ allows us to treat $|H'|^2 \rho_f$ as a constant for energy ranges $\gg \hbar K W_0$, making this case equivalent to Case I. However in the present case, the decay rate is enhanced by the vacuum-field strength enhancement inside the cavity, K. This enhancement of the decay or emission rate was first postulated by Purcell many years ago.

If we go to the other extreme and assume that the system has a strong memory (the photon lifetime in the cavity is very long) such that $4\tau_p K W_0 \gg 1$, the roots become

$$p_\pm \approx -\frac{1}{4\tau_p} \pm i \left(\frac{K W_0}{4\tau_p} \right)^{1/2}. \quad \text{(A9.25)}$$

Complex roots suggest that $|c_0(t)|^2$ exhibits damped oscillatory behavior. Rabi oscillations of this nature are again the result of the system having a strong memory. In other words, the lifetime of the fields in the cavity is long enough to induce additional absorption and re-emission of electromagnetic energy. However, contrary to Case II, $|c_0(t)|^2$ does eventually reduce to zero at a decay rate $\approx 1/2\tau_p$ (which corresponds to the decay rate of the field *amplitude* within the cavity).

Case III is perhaps the most interesting theoretically because it bridges the gap between the first two cases showing clearly how the oscillating solution transforms to the exponentially decaying solution as the photon lifetime of the cavity is adjusted from infinity to zero, respectively.

READING LIST

H. Kroemer, *Quantum Mechanics*, Ch. 19, Prentice Hall, Englewood Cliffs, NJ (1994).
H.M. Lai, P.T. Leung, and K. Young, *Phys. Rev. A*, **37**, 1597 (1988).
E.M. Purcell, *Phys. Rev.*, **69**, 681 (1946).

PROBLEMS

A9.1   Determine $f(t - t')$ in Case I assuming $|H'|^2 \rho_f$ is constant for energies within $E_s \pm \Delta E/2$, but *zero* everywhere else. Using this expression for

$f(t - t')$, describe qualitatively the range over which the solution for $c_0(t)$ is expected to behave like a decaying exponential. In your description, be sure to consider

(i) ΔE in relation to W, and

(ii) E_s in relation to $\hbar\omega_0$.

Qualitative plots of $f(t - t')$ superimposed on the expected dependence of $c_0(t')$ may prove useful in developing your answer.

A9.2 Solve for $c_0(t)$ and $c_s(t)$ in Case II without assuming that $E_f = \hbar\omega_0$. Plot both as a function of time, assuming the energy difference $(E_f - \hbar\omega_0)^2$ is equal to

(a) $|H'|^2$ and

(b) $12\,|H'|^2$.

Can you make a comment about the significance of tuning E_f to $\hbar\omega_0$?

A9.3 Using the initial conditions in Eq. (A9.3), derive Eq. (A9.23) in Case III assuming the solution is the sum of two exponentials with constant coefficients. Numerically plot $c_0(t)$ as a function of t/τ_p for $4\tau_p K W_0$ equal to

(a) 100,

(b) 1, and

(c) 0.01.

Plot any other cases that may seem interesting to you. Can you explain this behavior?

Transition Matrix Element

A10.1 GENERAL DERIVATION

As derived in Chapter 4 the matrix element $|H'_{21}|^2$ can be written in terms of a *transition matrix element* $|M_T|^2$. We begin by repeating this equation, Eq. (4.21),

$$|H'_{21}|^2 = \left(\frac{q \mathscr{A}_0}{2m_0}\right)^2 |M_T|^2, \qquad \text{where} \qquad |M_T|^2 \equiv |\langle u_c|\hat{\mathbf{e}} \cdot \mathbf{p}|u_v\rangle|^2 |\langle F_2|F_1\rangle|^2. \tag{A10.1}$$

As discussed in Appendix 8, the momentum matrix element, $|M|^2$ can be estimated from experiment. We now need to determine $|M_T|^2$ in terms of $|M|^2$. The difference between the two matrix elements is that $|M|^2$ determines the transition probability between u_s and the basis functions (u_x, u_y, u_z, or collectively u_i), whereas $|M_T|^2$ determines the transition probability between $u_c (= u_s)$ and the valence band Bloch functions (u_{hh}, u_{lh}, u_{so}, or collectively u_v). By expanding the u_v in terms of the u_i using Eqs. (A8.11), we can express $|M_T|^2$ in terms of $|M|^2$. Before we do this, however, we need to discuss spin degeneracy and how to include it here.

In Appendices 1 and 8, a simple factor of 2 for spin degeneracy was included in the definition of the density of states function. However, there are subtleties involved which are often overlooked when we simply include a factor of 2 in our equations. For example, to include the spin degeneracy in our evaluation of $|M_T|^2$, we must obviously sum over both $u_c \rightarrow u_v$ and $\bar{u}_c \rightarrow \bar{u}_v$ transitions. However, what is not so obvious is that in our sum we must also include $\bar{u}_c \rightarrow u_v$ and $u_c \rightarrow \bar{u}_v$ transitions! This is necessary because the LH and SO valence band Bloch functions are made up of both spin-up *and* spin-down basis functions, as seen from their definitions in Eqs. (A8.11). Therefore, a total of *four* transitions must be considered, as shown in Fig. A10.1. Because spin is already accounted for in the density of states function, we will sum over these transitions

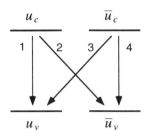

FIGURE A10.1   The four possible transitions between the spin-degenerate C and V bands, which must be considered when estimating the transition matrix element.

and then divide by 2 to remove the spin degeneracy. Thus, the reader should be aware that spin degeneracy has been removed from all expressions for the transition matrix element $|M_T|^2$ derived in this appendix.

Summing over the four transitions shown in Fig. A10.1, the transition matrix element defined in Eq. (A10.1) becomes

$$|M_T|_v^2 = \frac{1}{2} \sum_{u_c, \bar{u}_c} \sum_{u_v, \bar{u}_v} |\langle u_c | \hat{\mathbf{e}} \cdot \mathbf{p} | u_v \rangle|^2, \qquad \text{(A10.2)}$$

where the factor of 1/2 is to remove the spin degeneracy. We have set the envelope function overlap integral, $|\langle F_2 | F_1 \rangle|^2$, equal to unity because for the moment we will be interested in transitions between two bulk plane wave electron states. Later on, we will return to a more general form of Eq. (A10.2) which does include the envelope function overlap integrals.

To simplify Eq. (A10.2), we can first of all replace the dot product between the unit polarization vector and the electron momentum operator, $\hat{\mathbf{e}} \cdot \mathbf{p}$, with the expansion, $e_x p_x + e_y p_y + e_z p_z$. Then by using the selection rules given in Eqs. (A8.8) through (A8.10), in combination with the expansions of the valence band Bloch functions given in Eqs. (A8.11), we can reduce the expression for $|M_T|^2$ to a very simple form. To aid the reader in following the derivation, we give here the intermediate step in simplifying $|M_T|^2$ for the three valence band transitions:

$$|M_T|_{hh}^2 = \tfrac{1}{4}|M|^2 \{|-e_x - ie_y|^2 + 0 + 0 + |e_x - ie_y|^2\},$$

$$|M_T|_{lh}^2 = \tfrac{1}{12}|M|^2 \{|2e_z|^2 + |e_x - ie_y|^2 + |-e_x - ie_y|^2 + |2e_z|^2\}, \quad \text{(A10.3)}$$

$$|M_T|_{so}^2 = \tfrac{1}{6}|M|^2 \{|-e_z|^2 + |e_x - ie_y|^2 + |-e_x - ie_y|^2 + |-e_z|^2\}.$$

Each of the terms within brackets corresponds to one of the four spin-degenerate transitions (the ordering of terms from left to right corresponds to the numbering shown in Fig. A10.1). Note that in every case, the first term is equal to the fourth and the second term is equal to the third (leading to the standard factor of 2 for spin degeneracy).

To make the final expression as general as possible, we first replace every occurrence of $e_x^2 + e_y^2$ with the equivalent expression, $1 - e_z^2$ (since $\hat{\mathbf{e}}$ is a unit vector). This substitution places everything in terms of e_z. We can then interpret e_z as the component of $\hat{\mathbf{e}}$ which is parallel to the electron k-vector, since $\mathbf{k}$ is directed along z (an assumption made in defining (A8.11)). In other words, we can set $e_z = \hat{\mathbf{k}} \cdot \hat{\mathbf{e}}$, where $\hat{\mathbf{k}}$ is a unit vector directed along $\mathbf{k}$. Using these substitutions, we find that

$$|M_T|_v^2/|M|^2 = \begin{cases} \frac{1}{2}(e_x^2 + e_y^2) = \frac{1}{2}(1 - |\hat{\mathbf{k}} \cdot \hat{\mathbf{e}}|^2) & \text{for HH band,} & \text{(A10.4)} \\ \frac{1}{6}(e_x^2 + e_y^2 + 4e_z^2) = \frac{1}{2}(\frac{1}{3} + |\hat{\mathbf{k}} \cdot \hat{\mathbf{e}}|^2) & \text{for LH band,} & \text{(A10.5)} \\ \frac{1}{3}(e_x^2 + e_y^2 + e_z^2) = \frac{1}{3} & \text{for SO band.} & \text{(A10.6)} \end{cases}$$

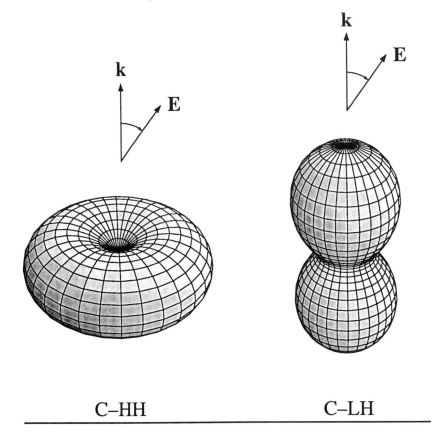

C–HH   C–LH

FIGURE A10.2 Dependence of the transition strength, $|M_T|^2$, on angle between the electron's k-vector and the incident electric field vector, $\mathbf{E}$, for C–HH and C–LH transitions (C–SO transitions are independent of angle). For C–HH transitions, $|M_T|^2$ is zero when $\mathbf{E} \| \mathbf{k}$ and becomes a maximum of $\frac{1}{2} \times |M|^2$ when $\mathbf{E} \perp \mathbf{k}$. For C–LH transitions, when $\mathbf{E} \| \mathbf{k}$, $|M_T|^2$ has a peak value of $\frac{2}{3} \times |M|^2$ and is reduced to $\frac{1}{6} \times |M|^2$ when $\mathbf{E} \perp \mathbf{k}$.

The *relative* transition strengths given in Eqs. (A10.4) through (A10.6) allow us to relate the transition matrix element, $|M_T|^2$, needed in our gain calculations to the experimentally measurable matrix element, $|M|^2$. Note that the use of a dot product has allowed us to drop any reference to a coordinate system, and hence, drop the constraint that the electron k-vector be directed along z (in other words, the physics does not lie in the coordinate system we choose, but in the relative orientation between the field polarization $\hat{e}$ and the electron k-vector).

To examine the dependence of Eqs. (A10.4) through (A10.6) on the field polarization in a more visual fashion, we have plotted the relative transition strengths for C–HH and C–LH transitions in Fig. A10.2 as a function of the angle between the electron k-vector and the electric field polarization, $\hat{e}$. These three-dimensional renderings reveal that the strength of interaction between each electron plane wave state and photon is highly polarization dependent. However, the striking features of Fig. A10.2 do not reveal themselves in bulk material because photons of a given polarization interact with a great number of electrons, all with k-vectors pointing in different directions. The average over all these interactions transforms the interesting shapes in Fig. A10.2 into uniform spheres. In fact, the average of $|\hat{k}\cdot\hat{e}|^2$ for $\hat{k}$ sweeping over all three dimensions is equal to $1/3$. Thus, for all three valence band transitions, the bulk material transition matrix element is just equal to $1/3 \times |M|^2$ (spin excluded) for any electric field polarization.

A10.2   POLARIZATION-DEPENDENT EFFECTS

The derivation presented above assumed plane wave states for the envelope functions, which then led to the polarization dependence illustrated in Fig. A10.2. In quantum-confined structures, the envelope functions are typically constructed from two (or more) plane wave states. The magnitude squared of the transition matrix element in (A10.2) will, in general, then contain cross-terms between the various plane waves that make up the confined state. In the following discussion we will ignore these cross-terms, making the analysis simpler. Later, we will see that in quantum wells, the conclusions derived here are consistent with the band-mixing model for transitions near the band edge.

Neglect of the cross-terms in (A10.2) implies that we can treat the plane waves that make up the confined states as independent from each other. In this approximation, each plane wave's k-vector direction will then have a corresponding polarization dependence similar to that derived in Section A10.1 and shown in Fig. A10.2. Near the band edge in quantum-confined structures, the k-vectors are quantized along certain directions, and the situation will be as shown in Fig. A10.3 for a typical quantum well and quantum wire. In the quantum well, all k-vectors point along the same axis and the polarization dependence is simply proportional to Fig. A10.2. However, in the quantum-wire

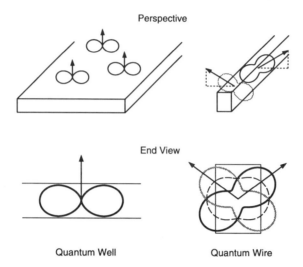

Perspective

End View

Quantum Well   Quantum Wire

FIGURE A10.3 Illustration of how quantum confinement in a quantum well and a quantum wire serves to "polarize" the momentum of band edge electrons along certain directions. The C–HH transition strength is superimposed on each electron's k-vector in the quantum-well case. The end view of the quantum well suggests that C–HH interactions with light are strongest when the light is polarized in the plane of the well. In the quantum-wire case, the C–LH transition strength is used. The end view shows two possible k-vector directions for electrons. The average C–LH transition strength is indicated by the dashed curve.

case, we must average over the polarization dependence of each plane wave (as indicated by the dashed curve in the lower right side of Fig. A10.3). To quantify the *average* polarization dependence, we choose our coordinate system along the confinement axis (axes) of the structure. It is then possible to evaluate the average transition strength along the three orthogonal field polarizations, $\hat{\mathbf{e}}_x$, $\hat{\mathbf{e}}_y$, and $\hat{\mathbf{e}}_z$, simply by replacing $\hat{\mathbf{k}}$ in Eqs. (A10.4) through (A10.6) with some appropriate *average* k-vector direction, $\hat{\mathbf{k}}_{ave}$.

We leave it to the reader to justify that $\hat{\mathbf{k}}_{ave}$ is obtained simply by finding the average direction of all allowed k-vectors within the *first octant* of our coordinate system (if we included all octants in our average, $\hat{\mathbf{k}}_{ave}$ would always be zero!). Below we list $\hat{\mathbf{k}}_{ave}$ for bulk and various quantum-confined structures for band edge states (where "band edge" implies that the total k-vectors are simply equal to the quantized k-vectors):

$$\hat{\mathbf{k}}_{ave} = (1/\sqrt{3})(\hat{\mathbf{k}}_x + \hat{\mathbf{k}}_y + \hat{\mathbf{k}}_z), \quad \text{bulk}$$
$$\hat{\mathbf{k}}_{ave} = \hat{\mathbf{k}}_z, \quad \text{quantum well } (L_z)$$
$$\hat{\mathbf{k}}_{ave} = (1/\sqrt{2})(\hat{\mathbf{k}}_x + \hat{\mathbf{k}}_z), \quad \text{quantum wire } (L_x = L_z) \quad \text{(A10.7)}$$
$$\hat{\mathbf{k}}_{ave} = (1/\sqrt{5})(\hat{\mathbf{k}}_x + 2\hat{\mathbf{k}}_z). \quad \text{quantum wire } (L_x = 2L_z)$$

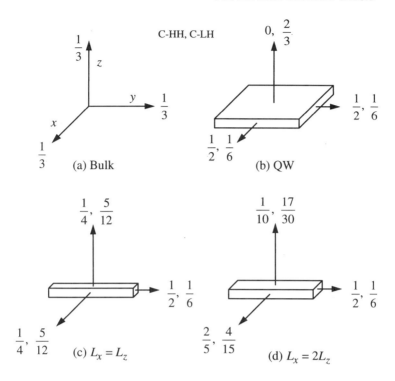

FIGURE A10.4 Relative band edge transition strengths for various quantum confinement structures. The coordinates referred to in the text are indicated in (a). The magnitude of the transition matrix element is found by multiplying the relative numbers by $|M|^2$. For example, with light polarized along the wire direction of (c), the band edge C–LH transition strength $|M_T|^2 = \frac{1}{6} \times |M|^2$.

The prefactors are normalization constants since $\hat{k}_{ave}$ is a unit vector. The last equation was obtained assuming $k \propto 1/L$ (which is exactly true only for an infinitely deep well).

Figure A10.4 illustrates the band edge transition strengths for the three orthogonal polarizations in the four structures listed above, obtained by substituting the $\hat{k}_{ave}$ defined for each structure into (A10.4) through (A10.6). When multiplied by $|M|^2$, the numbers in Fig. A10.4 give the magnitude of the transition matrix element, $|M_T|^2$, for each particular case (spin excluded). Note that the sum of the transition strengths over the three polarizations for each type of transition is always equal to $|M|^2$, as is the case in bulk material. Thus, the "total" transition strength for band edge transitions is always conserved. The difference is that in quantum-confined structures, a *redistribution* of the transition strength among the three polarizations occurs due to the nonuniform distribution of k-vector directions.

To treat arbitrary polarizations, we can always break the field up into the three orthogonal components shown in Fig. A10.4. The transition matrix element is then simply given by the trigonometric sum of the three components, or

$$|M_T|_v^2 = |M|^2 \sum_i e_i^2 S_i^v, \qquad i = x, y, z \tag{A10.8}$$

where the S_i^v are the transition strengths determined from Fig. A10.4. As an example use of (A10.8), we examine the polarization-dependent characteristics of a quantum-wire structure. From Fig. A10.4(c), the *ratio* of the transition matrix element between the C–LH and C–HH transitions is $\frac{1}{3}$ when $\hat{e}$ is parallel to the wire and $\frac{5}{3}$ when $\hat{e}$ is perpendicular to the wire. In general, from Fig. A10.4(c) and Eq. (A10.8), we have

$$\frac{|M_T|_{lh}^2}{|M_T|_{hh}^2} = \frac{\frac{1}{3}\cos^2\theta + \frac{5}{6}\sin^2\theta}{\cos^2\theta + \frac{1}{2}\sin^2\theta}, \tag{A10.9}$$

where θ is the angle between $\hat{e}$ and the axis of the wire.

A10.3 INCLUSION OF ENVELOPE FUNCTIONS IN QUANTUM WELLS

As shown in Appendix 8 quantum confinement changes the valence band structure through the interaction and mixing of the envelope functions. That is, the general valence band wavefunction in (A8.31) consists of both HH and LH envelope function components. Thus, the transition matrix element in Eq. (A10.1) must be modified and is now expressed as

$$|M_T|_v^2 = \sum_{u_c, \bar{u}_c} |\langle u_c|\hat{e}\cdot\mathbf{p}|u_A\rangle\langle F_2|F_{hh}\rangle + \langle u_c|\hat{e}\cdot\mathbf{p}|u_B\rangle\langle F_2|F_{lh}\rangle|^2. \tag{A10.10}$$

Equivalent terms summing over the spin degenerate counterparts of u_A and u_B (the u_C and u_D Bloch functions) increase the sum by a factor of two. This is accounted for by the removal of the factor of $\frac{1}{2}$ appearing in Eq. (A10.2) (throughout this appendix, the spin degeneracy is removed from the transition matrix element and included in the reduced density of states function in Chapter 4).

The transition matrix element in (A10.10) can be placed in a more elegant form by following a procedure similar to that outlined in Section A10.1. However, in the present case we must average the transition matrix element over all in-plane k-vector directions to remove the cross-term which results from squaring Eq. (A10.10). With the help of Eqs. (A8.32), (A8.33), and (A8.11), the transition matrix element reduces to

$$|M_T|_v^2 = |M|^2[\tfrac{2}{3}|\langle F_2|F_{lh}\rangle|^2], \qquad \text{TM } (\hat{e}\parallel\hat{z}) \tag{A10.11}$$

$$|M_T|_v^2 = \frac{|M|^2}{2}[|\langle F_2|F_{hh}\rangle|^2 + \tfrac{1}{3}|\langle F_2|F_{lh}\rangle|^2], \qquad \text{TE } (\hat{e}\perp\hat{z}) \tag{A10.12}$$

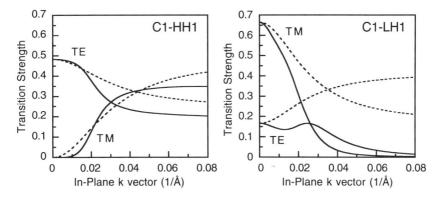

FIGURE A10.5 Relative transition strengths for both TE and TM light polarization for the two lowest subband transitions in an unstrained $GaAs/Al_{0.2}Ga_{0.8}As$ 80Å QW. The dashed curves represent what one would calculate assuming parabolic subbands. The transition strength as plotted here is defined as $|M_T|^2/|M|^2$ (bulk value is 1/3).

where z is assumed to be the quantization direction. At the band edge (where $F_{lh} = 0$ for HH states, and $F_{hh} = 0$ for LH states), the above expressions give the same transition strengths as those in Fig. A10.4 assuming the overlap integrals are close to unity, (which is true to within 5–10% typically, and would be true exactly if the effective mass in the C and V bands were identical). However, as we move away from the band edge, band-mixing occurs such that both $F_{hh, lh}$ are present in any one wavefunction, altering the transition strengths from those shown in Fig. A10.4. Sample calculations of the transition strength which illustrate this effect are plotted as a function of transverse k-vector, k_t in Fig. A10.5.

In finding the envelope functions and evaluating the overlap integrals numerically, we must make sure they are properly normalized. Normalization of the wavefunctions is obtained through the following relations

$$F_i(\text{norm}) = \frac{F_i}{\sqrt{N_i}}, \qquad \text{where} \qquad N_{hh, lh} = \langle F_{hh} | F_{hh} \rangle + \langle F_{lh} | F_{lh} \rangle,$$

$$N_2 = \langle F_2 | F_2 \rangle. \qquad (A10.13)$$

The above envelope functions refer to the functions along the confinement direction, and hence the brackets indicate integration along z. The in-plane envelope functions are simple plane waves. Thus, we require **k** conservation in the plane of the well, which then yields an in-plane overlap integral of unity, justifying our conversion of a volume integral into an integral along z only.

READING LIST

S.W. Corzine, R-H. Yan, and L.A. Coldren, Optical gain in III–V bulk and quantum well semiconductors, Ch. 1 in *Quantum Well Lasers*, ed. P.S. Zory, Jr., Academic Press, San Diego (1993).

Strained Bandgaps

In this appendix, the details of how strain affects the bandgap of III–V semiconductors is considered. To provide some background, we will begin by reviewing concepts of stress and strain in a crystal lattice [1].

A11.1 GENERAL DEFINITIONS OF STRESS AND STRAIN

A crystal lattice which feels an external force will react by distorting in some fashion. The force per unit area, or stress, is usually defined by a *stress* tensor, σ_{ij}, as depicted in Fig. A11.1. *Shear* components of stress ($i \neq j$) will cause the crystal to rotate unless equal and opposite components exist (for example, if $\sigma_{23} = \sigma_{32}$). If equal and opposite components do exist, the shear stress will deform a cubic lattice into a nonrectangular shape (i.e., the crystal axes will become nonorthogonal). In typical semiconductor applications, these types of deformations are rare, and for the present purposes, we will assume that $\sigma_{ij} = 0$ for $i \neq j$. *Normal* components of stress ($i = j$) will cause the crystal to expand or contract along the crystal axes, but in contrast to shear stress, the deformed lattice remains rectangular. The six faces of a cubic crystal can be acted upon by three normal forces: σ_{11}, σ_{22}, and σ_{33}. Because we are only considering normal components, the notation can be abbreviated to σ_1, σ_2, and σ_3. As suggested in Fig. A11.1, the σ_i are defined as positive for outward directed forces.

The next consideration involves the mathematical description of the lattice distortion. The strained state of the lattice is usually defined by a *strain* tensor, ε_{ij}. Each component of the strain tensor defines some aspect of the distortion of the lattice away from its unstrained shape. Restricting our attention to the effects of normal forces on the lattice, we only need to consider the three diagonal components of the strain tensor: ε_{11}, ε_{22}, and ε_{33} (since the crystal axes remain orthogonal to each other). Again using the abbreviated notation, these can be written as ε_1, ε_2, and ε_3. The ε_i measure the fractional increase ($\varepsilon_i > 0$) or decrease ($\varepsilon_i < 0$) of the crystal lattice along the ith axis as illustrated

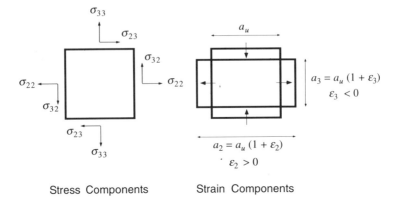

Stress Components   Strain Components

FIGURE A11.1 Components of stress tensor, σ_{ij}, and examples of the abbreviated strain tensor, ε_i, within the y–z plane.

in Fig. A11.1. For example, if the description of the lattice distortion is given as $\varepsilon_1 = 0$, $\varepsilon_2 = 0.01$, and $\varepsilon_3 = -0.01$, then we know that the lattice is undistorted along x, is larger by 1% along y, and is smaller by 1% along z, as qualitatively illustrated in the figure within the y–z plane.

With the stress and strain tensors defined, we now need to relate them to predict how a particular stress or set of external forces leads to a particular strain or lattice deformation. In a uniform material, the strain is proportional to the magnitude of the applied stress, as long as we remain within the elastic limits of the material. Hooke's law in an isotropic medium expresses this scalar relation as $\sigma = C\varepsilon$, where the constant, C, is Young's modulus. In a crystal, one type of stress may lead to more than one type of strain, implying that the different components of both stress and strain tensors are potentially related. The more general form of Hooke's law excluding shear components of stress can be written as

$$\begin{bmatrix} \sigma_1 \\ \sigma_2 \\ \sigma_3 \end{bmatrix} = \begin{bmatrix} C_{11} & C_{12} & C_{12} \\ C_{12} & C_{11} & C_{12} \\ C_{12} & C_{12} & C_{11} \end{bmatrix} \begin{bmatrix} \varepsilon_1 \\ \varepsilon_2 \\ \varepsilon_3 \end{bmatrix}. \tag{A11.1}$$

The C_{ij} are referred to as the *elastic stiffness coefficients* or the *elastic moduli*. The expression written here assumes the crystal has cubic symmetry such that all off-diagonal elements are equal, and all diagonal elements are equal, leaving us with only two elastic moduli, denoted C_{11} and C_{12}. For crystals with less symmetry, more C_{ij} components may need to be specified (if we include shear stress, an additional C_{44} is required to complete the description of stress and strain in cubic crystals). In common semiconductors, C_{11}, $C_{12} > 0$ with $C_{11} > C_{12}$, and both are usually described in units of 10^{11} dyn/cm^2.

Some relevant examples using Eq. (A11.1) are shown in Fig. A11.2. In the

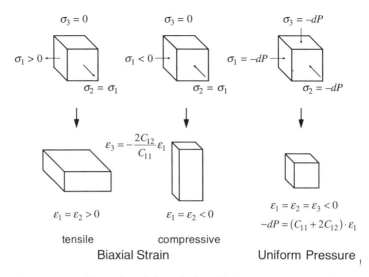

FIGURE A11.2   Examples of the relationship between stress and strain.

first two examples shown on the left, stress is applied to all four x and y faces of the cube such that $\sigma_1 = \sigma_2$, and no stress is applied to the z faces, such that $\sigma_3 = 0$. For both of these cases, the crystal is under *biaxial* strain. When the stress is directed outward ($\sigma_1 > 0$), the resulting strain is referred to as biaxial *tensile* strain, while inward stress ($\sigma_1 < 0$) gives rise to biaxial *compressive* strain. The resulting lattice deformation can be calculated using (A11.1). By symmetry, with $\sigma_1 = \sigma_2$, the strain in both x and y directions must be equal and we can set $\varepsilon_1 = \varepsilon_2$. With $\sigma_3 = 0$ and $\varepsilon_1 = \varepsilon_2$, the first and third equations in (A11.1) reduce to

$$\sigma_1 = C_{11}\varepsilon_1 + C_{12}\varepsilon_1 + C_{12}\varepsilon_3,$$
$$0 = C_{12}\varepsilon_1 + C_{12}\varepsilon_1 + C_{11}\varepsilon_3. \tag{A11.2}$$

These equations are valid for both tensile and compressive biaxial strain. The relationship between the strain components is immediately obtained from the lower equation:

$$\varepsilon_3 = -\frac{2C_{12}}{C_{11}}\varepsilon_1. \tag{A11.3}$$

Since $C_{11}, C_{12} > 0$, we conclude that under biaxial strain, the lattice deformation along z will be opposite to the deformation along either x or y, as depicted in Fig. A11.2. This effect is similar to squeezing a balloon—as the sides are compressed, the top and bottom expand as the balloon attempts to maintain the same volume of air. Using (A11.3) in the first equation of (A11.2), we then find

$$\sigma_1 = C_{11}\varepsilon_1[1 + C_{12}/C_{11} - 2(C_{12}/C_{11})^2]. \tag{A11.4}$$

This equation gives us the absolute measure of the stress required to achieve a given strain. Note that if $C_{12} = 0$, the strain perpendicular to the stress plane (i.e., along z) reduces to zero, and Hooke's law reduces to the scalar relation: $\sigma_1 = C_{11}\varepsilon_1$.

The example shown on the right side of Fig. A11.2 corresponds to a uniform stress applied equally to all sides of the crystal. If we define this inward-directed stress as a differential pressure change, dP, surrounding the crystal, then we can set $\sigma_1 = \sigma_2 = \sigma_3 = -dP$. Adding up all three equations in (A11.1), we immediately find

$$-3\,dP = (C_{11} + 2C_{12})(\varepsilon_1 + \varepsilon_2 + \varepsilon_3). \tag{A11.5}$$

Neglecting cross-terms between the ε_i, the fractional change in the *volume* of the crystal lattice is given by

$$dV/V \approx \varepsilon_1 + \varepsilon_2 + \varepsilon_3. \tag{A11.6}$$

Therefore, (A11.5) allows us to determine the change in the crystal volume in response to a change in the uniform pressure surrounding the crystal. This relationship will be used later to relate the experimentally measured pressure dependence of the bandgap to the strain of the lattice.

A11.2 RELATIONSHIP BETWEEN STRAIN AND BANDGAP

With an understanding of how a particular stress leads to a particular strain, we are left with the task of determining how a particular strain affects the bandgap of the semiconductor, the main topic of this appendix. Pikus and Bir [2] in 1959 provided the fundamental theory necessary to describe how lattice deformations affect the Hamiltonian which describes the interaction between the three valence bands: the heavy-hole (HH), light-hole (LH), and split-off (SO) bands. Their analysis involved transforming the coordinate system describing the crystal potential and the hole wavefunctions into a new deformed coordinate system, making use of the strain tensor. To first order in the strain, it was shown that a Hamiltonian identical in form to the Luttinger–Kohn (LK) valence band Hamiltonian described in Appendix 8, must be added to account for the strain. This new Pikus and Bir or *strain* Hamiltonian can be found directly from the LK Hamiltonian using the following substitutions:

$$k_i k_j \rightarrow \varepsilon_{ij},$$

$$\gamma_1 \rightarrow a,$$

$$-2\gamma_2 \rightarrow b, \tag{A11.7}$$

$$-2\sqrt{3}\gamma_3 \rightarrow d.$$

In these substitutions, the k_i are the various wavevectors appearing in the LK Hamiltonian and the γ_i are the Luttinger parameters, both of which are discussed in Appendix 8. The ε_{ij} are the components of the strain tensor describing the lattice deformation and a, b, and d are *lattice deformation potentials* which relate shifts in the valence bands to the strain tensor. An additional strain-dependent spin–orbit interaction Hamiltonian [3, 4] also exists, however, the deformation potentials associated with this interaction are generally more than an order of magnitude smaller [3] than those associated with the Pikus and Bir Hamiltonian, and will therefore be neglected here.

In this appendix, only normal components of the stress tensor are considered and we can set $\varepsilon_{ij} = 0$ for $i \neq j$. Furthermore, if we concentrate on biaxial strain then we can also set $\varepsilon_1 = \varepsilon_2$. Under these conditions, the 6×6 strain Hamiltonian [4, 5] relating the three twofold spin-degenerate valence bands is block-diagonalized into two identical 3×3 Hamiltonians (the interaction between states of opposite spin is removed). We can write the 3×3 strain Hamiltonian as

$$
\begin{array}{ccc}
HH & LH & SO
\end{array}
$$
$$
\begin{bmatrix}
H - S & 0 & 0 \\
0 & H + S & \sqrt{2}S \\
0 & \sqrt{2}S & H + \Delta
\end{bmatrix}
\begin{array}{l}
HH \\
LH \\
SO
\end{array}
\tag{A11.8}
$$

where

$$
H = a(\varepsilon_1 + \varepsilon_2 + \varepsilon_3), \tag{A11.9}
$$

$$
S = b(\tfrac{1}{2}(\varepsilon_1 + \varepsilon_2) - \varepsilon_3), \tag{A11.10}
$$

and Δ is the spin–orbit energy which separates the SO band from the HH and LH bands. The energies, H and S, are related to the strain tensor through the a and b deformation potentials (the deformation potential, d, appears only when shear components of stress are considered).

Equation (A11.9) reveals that H is proportional to the change in volume of the crystal lattice created by the strain (compare with (A11.6)). It is referred to as the *hydrostatic* component of the strain. Because H appears with the same sign in every diagonal term of the Hamiltonian (A11.8), we conclude that changes in the crystal volume result in a rigid shift of all three valence bands either up or down in energy, depending on whether the volume change is positive or negative.

Equation (A11.10) defines the *shear* component of the strain, S. The shear strain energy is proportional to the asymmetry in the strain parallel and perpendicular to the stress plane (the shear *strain* should not be confused with shear *stress* which is zero in this case). The dependence of the Hamiltonian on S is more complex than its dependence on H. The HH band is isolated from the other two bands, however, the LH and SO bands are coupled through the

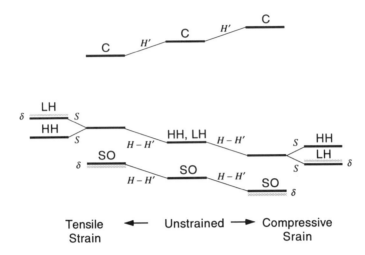

FIGURE A11.3 Qualitative band energy shifts of the conduction band and three valence bands for biaxial compressive and tensile strain. The magnitude of the energy shift is indicated next to each shift. The gray energies include LH–SO band coupling.

$\sqrt{2}S$ term. If $S \ll \Delta$, the LH–SO band coupling can be neglected, and the dominant contribution from S shows up in the HH and LH diagonal terms, where it is seen to split the HH and LH bands in opposite directions.

The strain Hamiltonian is written in terms of the hole energy, such that positive energy moves further into the valence band. For biaxial compressive strain, H and S are both positive, indicating an increase in the bandgap. For biaxial tensile strain, H and S are both negative. Figure A11.3 illustrates the band edge shifts of the three valence bands for both types of biaxial strain. The conduction band experiences a shift due to the hydrostatic component of strain only. In fact, we can define a separate H' with a corresponding deformation potential a' associated with the movement of the conduction band. However, it can be difficult to experimentally separate H' from the total bandgap shift [6]. A common approach is to interpret H in the strain Hamiltonian (and deformation potential a) as the total energy shift of the bandgap resulting from hydrostatic strain, and not worry about how the total shift is divided up between the conduction and valence bands. In fact, to estimate strained bandgaps, the division of H into H' and $H - H'$ is irrelevant. For heterointerface applications (including strained QWs), the division of H into H' and $H - H'$ can indicate how the conduction band offset, Q_c, is affected by strain. However, in practice Q_c is measured experimentally by other means, implying that again we do not need H'. As a result, the strained bandgap and its lineup with other material bandgaps is typically defined using H and measured values of Q_c.

To quantify the energy shifts depicted in Fig. A11.3, we need to evaluate H and S. The strain components within the stress plane, ε_1 and ε_2, are easily identified as the fractional change in the in-plane lattice constant of the strained material. The strain perpendicular to the stress plane, ε_3, was discussed earlier in reference to Fig. A11.2, and can be related to ε_1 and ε_2 using Eq. (A11.3). If we assume the in-plane lattice constant, $a_\parallel$, of the strained material is strained to match the substrate lattice constant such that $a_\parallel = a_{sub}$, we can define the strain as

$$\varepsilon \equiv \frac{a_{native} - a_{sub}}{a_{native}}, \tag{A11.11}$$

where a_{native} is the native unstrained lattice constant of the strained material. The various strain components are related to this definition as follows:

$$\varepsilon_1 = \varepsilon_2 = -\varepsilon \quad \text{and} \quad \varepsilon_3 = \frac{2C_{12}}{C_{11}}\varepsilon. \tag{A11.12}$$

The *lattice-mismatch parameter*, ε, is defined negative to the in-plane strain components by convention such that $\varepsilon > 0$ for compressive strain, and $\varepsilon < 0$ for tensile strain. Plugging these into the definitions of H and S, we find

$$H = (-a) \cdot 2 \frac{C_{11} - C_{12}}{C_{11}} \varepsilon, \tag{A11.13}$$

$$S = (-b) \cdot \frac{C_{11} + 2C_{12}}{C_{11}} \varepsilon. \tag{A11.14}$$

The parentheses around the deformation potentials are used because in common semiconductors, $a, b < 0$. The parentheses therefore enclose positive numbers, and the sign of ε determines the sign of both H and S.

The change in the HH, LH, and SO bandgaps for $S \ll \Delta$ can be approximated by the diagonal terms of Eq. (A11.8). However, for larger strains, the matrix must be diagonalized. Since the HH band is isolated, the problem is reduced to finding the eigenvalues of the 2×2 submatrix which couples the LH and SO bands. Performing this procedure, the bandgap shifts become

$$\Delta E_{HH} = H - S,$$
$$\Delta E_{LH} = H + S - \delta,$$
$$\Delta E_{SO} = H + \Delta + \delta, \tag{A11.15}$$
$$\delta = \tfrac{1}{2}\Delta\{[1 - 2(S/\Delta) + 9(S/\Delta)^2]^{1/2} - (1 - S/\Delta)\} \approx 2S^2/\Delta.$$

The energy, δ, represents an additional repulsion between the LH and SO bands which increases approximately quadratically with S and hence is positive for both tensile and compressive strain. This LH–SO band-coupling energy shift is indicated by the gray energy levels in Fig. A11.3. The band edge energy shifts in (A11.15) are all relative to the unstrained direct bandgap of the material, E_{g0}. Once E_{g0} is added to the energy shifts, the strained bandgaps are completely defined. The material parameters that we must know to predict the strained bandgaps (aside from the unstrained bandgap and lattice-mismatch parameter) are a, b, C_{11}, C_{12}, and Δ.

Values for all of the required strain parameters can be found in the literature. However, the hydrostatic deformation potential, a, is a bit more difficult to track down. The division of a between the valence and conduction band is not well standardized. As a result, a sometimes refers to the conduction band shift and sometimes the valence band shift, or sometimes the sum of these two (which is what we are interested in). Unfortunately, it is often difficult to know which definition is being used in any given citation. For this reason it is common to estimate a based on measurements of the pressure dependence of the bandgap. The strain example given earlier in reference to Fig. A11.2 showed that a uniform pressure is related to the volume change as follows:

$$-3 \, dP = (C_{11} + 2C_{12})(\varepsilon_1 + \varepsilon_2 + \varepsilon_3). \tag{A11.16}$$

Furthermore, for a uniform pressure we can set $S = 0$, implying that the shift in the HH and LH bandgaps is caused solely by H. Thus, the differential change in bandgap in response to a differential change in pressure is simply

$$dE = a(\varepsilon_1 + \varepsilon_2 + \varepsilon_3). \tag{A11.17}$$

Dividing (A11.17) by (A11.16), we find

$$a = -\tfrac{1}{3}(C_{11} + 2C_{12})\frac{dE}{dP}. \tag{A11.18}$$

Because $dE/dP > 0$, we conclude that $a < 0$, as stated earlier. This formula allows us to determine a for any bandgap once the pressure dependence of the gap is known. The pressure dependence of the energy gap is commonly quoted in units of 10^{-6} eV/bar or sometimes 10^{-6} eV cm^2/kg (1 bar $= 1.019$ kg/cm^2). In defining a this way, a useful conversion factor to keep handy is

$$10^{-6} \text{ eV/bar} \times 10^{11} \text{ dyn/cm}^2 = 0.1 \text{ eV}. \tag{A11.19}$$

Strain-related parameters of III–V semiconductor materials are summarized in Table A11.1. The strain parameters related to shear stress, d and C_{44}, are included for completeness but are not required in the calculation of biaxially strained bandgaps. The hydrostatic deformation potential, a, was calculated from other parameters in the table using (A11.18); the values of dE/dP in the table correspond to the direct bandgap change with pressure. The first five entries were taken primarily from Adachi [7, 8] while the rest were taken from

TABLE A11.1   Strain Parameters in III–V Semiconductors.

Material	Lattice Constant $a(Å)$	Deformation Potentials (eV)			Elastic Moduli (10^{11} dyn/cm^2)			(10^{-6} eV/bar)	
		a	b	d	C_{11}	C_{12}	C_{44}	dE/dP	$\Delta(eV)$
GaAs	5.6533	-8.68	-1.7	-4.55	11.88	5.38	5.94	11.5	0.34
InAs	6.0583	-5.79	-1.8	-3.6	8.329	4.526	3.959	10.0	0.371
AlAs*	5.6611	-7.96	-1.5	-3.4	12.02	5.70	5.89	10.2	0.30
GaP*	5.4512	-9.76	-1.5	-4.6	14.12	6.253	7.047	11.0	0.10
InP	5.8688	-6.16	-2.0	-5.0	10.22	5.76	4.60	8.5	0.10
AlP*	5.4635	-8.38	-1.75	-4.8	13.2	6.3	6.15	9.75	0.10
GaSb	6.0959	-8.28	-1.8	-4.6	8.842	4.026	4.322	14.7	0.8
InSb	6.4794	-7.57	-2.0	-4.8	6.47	3.65	3.02	16.5	0.98
AlSb*	6.1355	2.04	-1.35	-4.3	8.769	4.341	4.076	-3.5	0.75

* Indirect gap.

Landolt–Bornstein [9]. Complete data for AlP could not be found. Thus, blank entries in this row were filled with the average value of the other two phosphides.

If we wanted to know, for example, the strained bandgap of InGaAs grown on GaAs as a function of indium mole fraction, we could determine the strain parameters for the InGaAs ternary by linear interpolation between the GaAs and InAs values listed in Table A11.1. For most ternaries this procedure works well. For quaternaries, Vegard's law [7] can be applied to provide reasonable estimates.

A11.3   RELATIONSHIP BETWEEN STRAIN AND BAND STRUCTURE

The energy shifts discussed in this appendix refer to the band edge shifts exclusively. To predict the band structure away from the band edge, the strain Hamiltonian must be added to the LK Hamiltonian and the combination must be diagonalized. In Appendix 8, this combined Hamiltonian uses the band edge of the HH band as the zero-energy reference. Thus, $H - S$ should be subtracted from all diagonal terms of the strain Hamiltonian. The net result once the strain Hamiltonian is diagonalized is that the difference between the HH and LH band edges should be added to the LH diagonal term in the LK Hamiltonian. In Appendix 8, this total splitting energy between the HH and LH bands was defined as S_{tot}. Using Eq. (A11.15), we can write

$$S_{tot} \equiv E_{LH} - E_{HH} = 2S - \delta \approx 2S(1 - S/\Delta), \qquad \text{(A11.20)}$$

where the latter approximation expands δ to second order in S.

In Appendix 8, the LK Hamiltonian used for the subband structure calculations ignores any coupling to the SO band. Chao and Chuang [5] have analyzed the consequences of this approximation and found that significant differences do exist between the subband structure with and without SO band coupling. However, their calculations without SO band coupling assume $S_{tot} = 2S$ (i.e., they ignore the SO band coupling entirely). By assuming $S_{tot} = 2S - \delta$, the SO band coupling can at least be partially included since we obtain more accurate estimates of the LH subband energies at the zone center (i.e., at the subband edges). This procedure should improve the accuracy of strained subband structure calculations which ignore SO band coupling in the LK Hamiltonian.

REFERENCES

[1] D.R. Lovett, *Tensor Properties of Crystals*, Ch. 4, Adam Hilger, New York (1989).

[2] G.E. Pikus and G.L. Bir, *Sov. Phys.-Solid State*, **1**, 1502 (1960).

[3] L.D. Laude, F.H. Pollak, and M. Cardona, *Phys. Rev. B*, **3**, 2623 (1971).

[4] T.B. Bahder, *Phys. Rev. B*, **41**, 11992 (1990).

[5] C.Y.P. Chao and S.L. Chuang, *Phys. Rev. B*, **46**, 4110 (1992).

[6] D.D. Nolte, W. Walukiewicz, and E.E. Haller, *Phys. Rev. Lett.*, **59**, 501 (1987).

[7] S. Adachi, *J. Appl. Phys.*, **53**, 8775 (1982).

[8] S. Adachi, *J. Appl. Phys.*, **58**, R1 (1985).

[9] Landolt–Börnstein, in *Numerical Data and Functional Relationships in Science and Technology*, edited by O. Madelung, Vols. 17a and 22a, New Series, Group III. Springer-Verlag, Berlin (1982).

Threshold Energy for Auger Processes

A12.1 CCCH PROCESS

For the CCCH process in Fig. 4.15, we can write the momentum and energy conservation laws (i.e., initial = final) as

$$\mathbf{k}_1 + \mathbf{k}_2 = \mathbf{k}_3 + \mathbf{k}_4, \tag{A12.1a}$$

$$\Delta E_1 + \Delta E_2 = -(E_g + \Delta E_3) + \Delta E_4, \tag{A12.1b}$$

where the $\mathbf{k}$'s are vectors in k-space, and E_c has been used as the energy reference level in the latter equation. Rearranging the energy conservation law, we find

$$\Delta E_4 - E_g = \Delta E_1 + \Delta E_2 + \Delta E_3. \tag{A12.2}$$

From Eq. (4.76), it follows that the most probable transition corresponds to the minimum possible energy of state 4. This minimum value for ΔE_4 is referred to as the *threshold* energy, E_T, of the Auger process.

To determine the threshold energy, we need to minimize (A12.2). If we assume that all bands are parabolic, we can set $\Delta E_i \propto \mathbf{k}_i \cdot \mathbf{k}_i / m_C$ for the three conduction band states and $\Delta E_3 \propto \mathbf{k}_3 \cdot \mathbf{k}_3 / m_H = \mu \mathbf{k}_3 \cdot \mathbf{k}_3 / m_C$, for the valence band where $\mu = m_C/m_H$. At the minimum, ΔE_4 will be independent of variations in any of the k-vectors. The differential of any one term is $d(\mathbf{k}_i \cdot \mathbf{k}_i) = \mathbf{k}_i \cdot d\mathbf{k}_i + d\mathbf{k}_i \cdot \mathbf{k}_i$. Taking the total differential of Eq. (A12.2) and setting it to zero, we obtain

$$\mathbf{k}_4 \cdot d\mathbf{k}_4 = \mathbf{k}_1 \cdot d\mathbf{k}_1 + \mathbf{k}_2 \cdot d\mathbf{k}_2 + \mu \mathbf{k}_3 \cdot d\mathbf{k}_3 = 0, \tag{A12.3}$$

$$d\mathbf{k}_3 = d\mathbf{k}_1 + d\mathbf{k}_2 - d\mathbf{k}_4. \tag{A12.4}$$

The second equation follows from taking the differential of the constraint (A12.1a). Replacing $d\mathbf{k}_3$ in (A12.3) using (A12.4), we find

$$(\mathbf{k}_1 + \mu\mathbf{k}_3)\cdot d\mathbf{k}_1 + (\mathbf{k}_2 + \mu\mathbf{k}_3)\cdot d\mathbf{k}_2 - \mu\mathbf{k}_3\cdot d\mathbf{k}_4 = 0. \tag{A12.5}$$

The last term is zero when $\mathbf{k}_3\|\mathbf{k}_4$, since $\mathbf{k}_4\cdot d\mathbf{k}_4 = 0$ from (A12.3). The remaining two terms are zero for any $d\mathbf{k}_1$ and $d\mathbf{k}_2$ when $\mathbf{k}_1 = \mathbf{k}_2 = -\mu\mathbf{k}_3$. Substituting this into Eq. (A12.1a), we can determine $\mathbf{k}_4$. To summarize, we have

$$\mathbf{k}_1 = \mathbf{k}_2 = -\mu\mathbf{k}_3, \tag{A12.6}$$

$$\mathbf{k}_4 = -(1 + 2\mu)\mathbf{k}_3, \tag{A12.7}$$

where $\mu = m_C/m_H$. Thus, the most probable Auger transition occurs when all four k-vectors are colinear, with $\mathbf{k}_3$ pointing in the opposite direction to the rest. The Auger transitions in Fig. 4.14 reflect this conclusion. Using (A12.6) and (A12.7) in Eq. (A12.1b) and solving for k_3^2, we obtain

$$k_3^2 = \frac{1}{(1 + 2\mu)(1 + \mu)} k_g^2, \tag{A12.8}$$

where k_g is corresponds to $E_g = \hbar^2 k_g^2/2m_C$. The energies of the four states involved are then easily found using (A12.8):

$$\Delta E_1 = \Delta E_2 = \mu\Delta E_3 = \frac{\mu^2}{(1 + 2\mu)(1 + \mu)} E_g, \tag{A12.9}$$

$$\Delta E_4 = \frac{1 + 2\mu}{1 + \mu} E_g. \tag{A12.10}$$

The threshold energy for the CCCH process is therefore given by

$$E_T \equiv \Delta E_4 = \frac{1 + 2\mu}{1 + \mu} E_g = \frac{2m_C + m_H}{m_C + m_H} E_g. \tag{A12.11}$$

The most probable CCCH Auger transition is now completely defined.

A12.2 CHHS AND CHHL PROCESSES

For the CHHS process in Fig. 4.14, we again start by writing the momentum and energy conservation laws:

$$\mathbf{k}_1 + \mathbf{k}_2 = \mathbf{k}_3 + \mathbf{k}_4, \tag{A12.12a}$$

$$-(\Delta E_1 + \Delta E_2) = E_g + \Delta E_3 - (\Delta_{so} + \Delta E_4). \tag{A12.12b}$$

In this case, E_v has been used as the energy reference level in the latter equation. Rearranging the energy conservation law, we find

$$\Delta E_4 - (E_g - \Delta_{so}) = \Delta E_1 + \Delta E_2 + \Delta E_3. \qquad (A12.13)$$

Again, we wish to minimize ΔE_4 to maximize the transition probability. Using the same procedure as outlined above, the minimum k-vectors are found to be

$$\mathbf{k}_1 = \mathbf{k}_2 = -\mathbf{k}_3/\mu_H, \qquad (A12.14)$$

$$\mathbf{k}_4 = -(1 + 2/\mu_H)\mathbf{k}_3, \qquad (A12.15)$$

where $\mu_H = m_C/m_H$. Again, all four k-vectors are colinear, with $\mathbf{k}_3$ pointing in the opposite direction to the rest. The energies of the four states involved are found to be

$$\mu_S \Delta E_1 = \mu_S \Delta E_2 = \Delta E_3$$

$$= \frac{1/\mu_S}{(1 + 2/\mu_H)(1 + 2/\mu_H - 1/\mu_S)} (E_g - \Delta_{so}), \qquad (A12.16)$$

$$\Delta E_4 = \frac{(1 + 2/\mu_H)}{(1 + 2/\mu_H - 1/\mu_S)} (E_g - \Delta_{so}), \qquad (A12.17)$$

where $\mu_S = m_C/m_S$. The threshold energy for the CHHS process is therefore given by

$$E_T \equiv \Delta E_4 = \frac{2m_H + m_C}{2m_H + m_C - m_S} (E_g - \Delta_{so}). \qquad (A12.18)$$

The equations for the CHHL process are found by setting $\Delta_{so} \to 0$ and $m_S \to m_L$.

Langevin Noise

This appendix gives a more detailed account of the Langevin noise sources used in Chapter 5. It is divided into three main sections. The first section considers basic properties of Langevin noise sources and covers the general definition and evaluation of the correlation strength between two Langevin noise sources. The second section considers the specific Langevin noise correlations between: (1) the photon density and carrier density Langevin noise sources, (2) the photon density and output power Langevin noise sources, and (3) the photon density and phase Langevin noise sources. The final section makes use of the specific correlation strengths to evaluate the noise spectral densities of the photon density, output power, and carrier density, using the formulas developed in Chapter 5. Practical approximations to these lengthy expressions are also discussed.

A13.1 PROPERTIES OF LANGEVIN NOISE SOURCES

A13.1.1 Correlation Functions and Spectral Densities

One of the defining characteristics of a Langevin noise source, $F(t)$, is its completely random nature. In fact, the best analogy to $F(t)$ is a random number generator that generates a new number between $\pm \infty$ every Δt seconds in the limit of $\Delta t \to 0$. In mathematical terms, this characteristic is described as a *memoryless* process which means that the value of $F(t)$ at time t has absolutely no correlation with any previous value $F(t - \tau)$ including $\tau \to 0^+$. Now because $F(t)$ is just as often positive as it is negative, the average value over time is zero: $\langle F(t) \rangle = 0$.[1] Furthermore, since $F(t)$ and $F(t - \tau)$ fluctuate randomly

---

[1] The brackets $\langle \ \rangle$ actually refer to a statistical average over many similarly prepared systems at the same time t. However, we can obtain the same result by averaging a single system over extended time intervals if the statistical processes involved are both *stationary* and *ergodic*, which we assume to be the case here. Thus, whether we define $\langle \ \rangle$ as a statistical average or time average is a matter of conceptual convenience. In the time domain, it is often convenient to think in terms of a time average, whereas in the frequency domain, a statistical average is usually more appropriate.

relative to each other, they have the same sign just as often as they have opposite signs over time. As a result, the average of the *product* of the two over time t is also zero: $\langle F(t)F(t - \tau)\rangle = 0$. The only exception is when $\tau = 0$, in which case the product is always positive: $\langle F(t)F(t)\rangle = \langle F(t)^2\rangle = \infty$ (the magnitude is infinite because $F(t)^2$ can take on any value between 0 and ∞, which when averaged is infinite). As a function of τ then, the *correlation function* $\langle F(t)F(t - \tau)\rangle$ displays a delta function-like behavior. Generalizing this result, we can define the correlation function between any two (memoryless) Langevin noise sources as

$$\langle F_i(t)F_j(t - \tau)^*\rangle = S_{ij} \cdot \delta(\tau). \tag{A13.1}$$

The proportionality constant S_{ij} defines the *correlation strength* between the two noise sources (it has units of (seconds) × (fluctuating variable units)^2). When $i = j$, S_{ii} defines the *auto*correlation strength. When $i \neq j$, S_{ij} defines the *cross*-correlation strength which is nonzero only if the fluctuations of one noise source are in some way correlated with the fluctuations of the other noise source. The complex conjugate is included in the definition of the correlation function to account for possibly complex Langevin noise sources (such that at $\tau = 0$, the autocorrelation function reduces to $\langle |F_i(t)|^2\rangle$).

It is interesting to point out that according to Eq. (A13.1), a Langevin noise source has an infinite mean-square value: $\langle |F_i(t)|^2\rangle = \infty$. However, this would only be observable by a detection system which had infinite bandwidth. In practice, the fluctuations observed on an oscilloscope are never infinite because they are limited by the system rise time or measurement bandwidth. The actual *measured* mean-square noise is found from the overlap of the measurement bandwidth with the *spectral density* of the noise source as worked out in Eqs. (5.103)–(5.105) in Section 5.5.

We can relate the spectral density to the correlation function by examining the *frequency domain* correlation function as follows:

$$\langle F_i(\omega)F_j(\omega')^*\rangle = \left\langle \int F_i(\tau)e^{-j\omega\tau}\,d\tau \cdot \int F_j(t)^*e^{j\omega't}\,dt \right\rangle$$

$$= \iint \langle F_i(t + \tau)F_j(t)^*\rangle e^{-j\omega\tau}e^{-j(\omega - \omega')t}\,dt\,d\tau$$

$$= \int \langle F_i(t + \tau)F_j(t)^*\rangle e^{-j\omega\tau}\,d\tau \cdot \int e^{-j(\omega - \omega')t}\,dt$$

$$= \int \langle F_i(t)F_j(t - \tau)^*\rangle e^{-j\omega\tau}\,d\tau \cdot 2\pi\delta(\omega - \omega')$$

$$\equiv S_{ij}(\omega) \cdot 2\pi\delta(\omega - \omega'). \tag{A13.2}$$

The first step in this derivation uses Eq. (5.102) to transform to the time domain. The second step sets $\tau \to t + \tau$ (with $d\tau = d\tau$) and regroups, confining the statistical average (see footnote 1) to the statistically varying processes. The third and fourth steps assume that $\langle F(t + \tau)F(t)^* \rangle$ only depends on the *relative* time delay τ between the two functions and not on the absolute time t. This characteristic is described as a *stationary* process and is not limited to Langevin noise sources but applies to all statistical processes considered in this book. For stationary processes then, we can separate out the integration over t in the third step, and shift the time origin to $t \to t - \tau$ in the fourth step. The fourth step also recognizes the integration over t to be a delta function of strength 2π.

The final step in Eq. (A13.2) defines the correlation strength of the frequency domain delta function as the spectral density $S_{ij}(\omega)$ (compare with Eq. (5.104)), where

$$S_{ij}(\omega) = \int \langle F_i(t)F_j(t - \tau)^* \rangle e^{-j\omega\tau} \, d\tau. \tag{A13.3}$$

In words, the spectral density is just the Fourier transform of the correlation function. This fundamental relation is known as the *Wiener–Khinchin* theorem and it applies to all types of stationary processes (not just Langevin noise). For example, when discussing linewidth in Section 5.5.6, we stated that the power spectrum was equal to the Fourier transform of the electric field autocorrelation function. Equation (A13.2) is the proof of that statement. However, here we are interested in applying this result to Langevin noise.

Plugging the Langevin noise correlation Eq. (A13.1) into Eq. (A13.3), we find

$$S_{ij}(\omega) = S_{ij} \int \delta(\tau)e^{-j\omega\tau} \, d\tau = S_{ij}. \tag{A13.4}$$

Thus, the Langevin noise spectral density is simply equal to the correlation strength and is independent of frequency—it is a "white" noise source (this is in fact true for any memoryless, stationary process). Since the Langevin noise spectral density and correlation strength can be used interchangeably, we adopt a common notation for both:

$$\langle F_i F_j \rangle \equiv S_{ij}(\omega) = S_{ij}. \tag{A13.5}$$

In some cases it is more meaningful to interpret $\langle F_i F_j \rangle$ as the noise spectral density while in other cases the correlation aspect is more relevant. In this appendix, we will generally refer to $\langle F_i F_j \rangle$ as the correlation strength. Note that both the spectral density and the correlation strength have units of (seconds) × (fluctuating *time domain* variable units)^2.

A13.1.2   Evaluation of Langevin Noise Correlation Strengths

The Langevin noise sources discussed here and in Chapter 5 are based on a shot noise model advanced by McCumber [1] and others such as Lax [2] as a method of simplifying the rigorous quantum description of noise in lasers. Within this model, the laser noise is assumed to originate from shot noise associated with the discrete random flow of particles into and out of the carrier and photon reservoirs. It can be shown that the spectral density of shot noise is constant and *proportional to the average rate of particle flow*. With this in mind, consider a reservoir where particles are flowing into and out of the reservoir by a host of discrete random processes. In the Langevin formalism, each of these discrete processes contributes shot noise to the overall noise in the reservoir. Therefore, to determine the total Langevin noise spectral density or correlation strength $\langle F_i F_i \rangle$, we simply sum over all shot noise contributions—or over all rates of particle flow into and out of reservoir i. To determine the cross-correlation strength $\langle F_i F_j \rangle$ between two reservoirs i and j, we sum only over particle flows which affect *both* reservoirs simultaneously. However in this case, when one reservoir gains a particle $(F_i > 0)$, the other loses a particle $(F_j < 0)$. Hence the product $\langle F_i F_j \rangle$ is always negative, and the noise between the two reservoirs is said to be *negatively correlated*.

Within the shot noise Langevin model then, the correlation strengths between the various Langevin noise sources are found by simple inspection of the rates into and out of the various reservoirs:

$$\langle F_i F_i \rangle = \sum R_i^+ + \sum R_i^-, \tag{A13.6}$$

$$\langle F_i F_j \rangle = -\left[\sum R_{ij} + \sum R_{ji} \right]. \tag{A13.7}$$

The reservoir Langevin noise sources, F_i and F_j, as well as the rates of particle flow into, R_i^+, out of, R_i^-, and between the two reservoirs, R_{ij} and R_{ji} are all in units of numbers per unit time. For single-sided spectral densities, an additional factor of 2 would be required on the RHS of these definitions.

As an example use of Eq. (A13.6), let's consider a simple and perhaps familiar example: shot noise in a detector. The shot noise associated with the current generated by discrete random absorption events[2] is given by: $\langle i_N(t)^2 \rangle = 2qI\Delta f$, where I is the average current, and Δf is the bandwidth of the detection system. We can arrive at this same result by defining the current noise in terms of a Langevin noise source: $i(t) = I + F_I(t)$. After converting to pure numbers per unit time (i.e., I/q and F_I/q), we can use Eq. (A13.6) to determine the spectral density of the noise or the correlation strength $\langle F_I F_I \rangle$:

$$\frac{1}{q^2} \langle F_I F_I \rangle = \frac{I}{q} \rightarrow \langle F_I F_I \rangle = qI. \tag{A13.8}$$

---

[2] If the absorbed photons do not arrive in an entirely random fashion, it is possible to generate a noise current smaller than the standard shot noise level. Such a nonrandom, or sub-shot-noise-limited photon stream can be generated by semiconductor lasers (see Section 5.5.4 for details).

Multiplying this double-sided spectral density by $2\Delta f$ (see Eq. (5.106)) gives the shot noise expression for $\langle i_N(t)^2 \rangle$. The main point of this exercise is to show the importance of first converting to numbers per unit time in order to get the correct proportionality constant (in this case, q).

A13.2   SPECIFIC LANGEVIN NOISE CORRELATIONS

A13.2.1   Photon Density and Carrier Density Langevin Noise Correlations

Figure 5.1 displays all rates into and out of both carrier and photon reservoirs. The associated density Langevin noise sources for each reservoir, F_N and F_P, when converted to units of numbers per unit time are VF_N and $V_p F_P$. Using Fig. 5.1 in combination with Eqs. (A13.6) and (A13.7), we immediately obtain

$$V_p^2 \langle F_P F_P \rangle = N_p V_p / \tau_p + (R_{21} + R_{12} + R'_{sp})V, \tag{A13.9}$$

$$V^2 \langle F_N F_N \rangle = \eta_i I/q + (R_{sp} + R_{nr} + R_{21} + R_{12})V, \tag{A13.10}$$

$$V_p V \langle F_P F_N \rangle = -(R_{21} + R_{12} + R'_{sp})V. \tag{A13.11}$$

The current term $\eta_i I/q$ in (A13.10) may be smaller or larger depending on the noise characteristics of the current pumping source [3].

We can simplify the correlation strengths using the following substitutions. From the steady-state relations, we can set $N_p V_p / \tau_p = (R_{21} - R_{12} + R'_{sp})V$ in Eq. (A13.9). From the fundamental relations (4.32) and (4.48), we can also set $R_{12} = R_{21} - v_g g N_p$ and $R_{21} = R'_{sp} N_p V_p$. Finally with $R_{sp} - R'_{sp} + R_{nr} = \eta_i I_{th}/qV$ from Eq. (5.18′), we obtain

$$\langle F_P F_P \rangle = 2\Gamma R'_{sp} N_p \left[1 + \frac{1}{N_p V_p} \right], \tag{A13.12}$$

$$\langle F_N F_N \rangle = \frac{2R'_{sp} N_p}{\Gamma} \left[1 + \frac{1}{2N_p V_p} \right] - \frac{v_g g N_p}{V} + \frac{\eta_i (I + I_{th})}{qV^2}, \tag{A13.13}$$

$$\langle F_P F_N \rangle = -2R'_{sp} N_p \left[1 + \frac{1}{2N_p V_p} \right] + \frac{v_g g N_p}{V_p}. \tag{A13.14}$$

For lasers putting out milliwatt power levels above threshold, the photon number in the cavity is typically $\sim 10^5$ or more. Thus it is generally safe to assume $N_p V_p \gg 1$. For this reason, Eqs. (5.127) through (5.129) are written without the terms in square brackets.

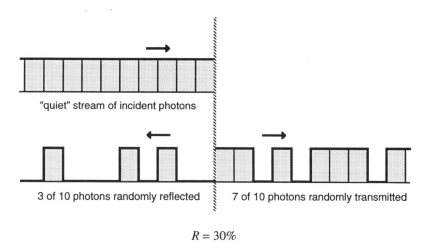

$$R = 30\%$$

FIGURE A13.1   Illustration of partition noise created by a partially reflecting mirror. The key point is that a negative correlation exists between the partition noise reflected and transmitted at the mirror facet.

A13.2.2   Photon Density and Output Power Langevin Noise Correlations

In Chapters 2 and 5, we simply used $P_0 = (h\nu V_p v_g \alpha_m F)N_p$ or equivalently $P_0 = (\eta_0 h\nu V_p/\tau_p)N_p$ to convert from photon density to output power. However, when considering the *noise* inside and outside of the cavity we must be careful. To understand why, consider a perfectly uniform stream of photons incident on a partially reflecting mirror as sketched in Fig. A13.1. While on average 30% of the photons are reflected, each individual photon must either be completely transmitted or completely reflected. This random division into reflected and transmitted photons leads to *partition* noise in the stream of both reflected and transmitted photons.

The partition noise reflected back into the cavity is accounted for in $\langle F_P F_P \rangle$ by the term $N_p V_p/\tau_p$ in Eq. (A13.9), which is equivalently the shot noise created by photons escaping the cavity. The partition noise transmitted outside the cavity however is different from that reflected back in. In fact, it is the exact inverse of the reflected noise, as Fig. A13.1 reveals. When added to the other noise contributions, the overall noise associated with the power outside the cavity is not the same as the overall noise associated with the photon density inside the cavity. This perhaps subtle point was first brought out by Yamamoto [4] using more fundamental quantum mechanical arguments.[3]

---

[3] Yamamoto describes the partition noise as vacuum-field fluctuations incident on the mirror facet from outside the cavity. In this case, the vacuum fields transmitted into the cavity have a negative phase relationship to those reflected off of the mirror facet and this is what provides the negative correlation between the facet noise inside and outside of the cavity. Despite this difference in interpretation, Yamamoto's derivation produces the same result for output power noise considered in this appendix (however, Yamamoto's derivation requires a high mirror reflectivity resonator).

We can again use the Langevin method to determine the effect of this *negativity correlated* partition noise by treating the stream of output photons as another reservoir with its own associated Langevin noise source (similar to the detector current considered initially). Adapting Eq. (5.118), we have for the output power fluctuations:

$$\delta P(t) = (\eta_0 h v V_p/\tau_p) N_{p1}(t) + F_0(t). \qquad (A13.15)$$

As we did in deriving Eqs. (5.124) and (5.125), we convert to the frequency domain, multiply both sides by $\delta P(\omega')^*$, take the time average, and integrate over ω' to obtain

$$S_{\delta P}(\omega) = (\eta_0 h v V_p/\tau_p)^2 S_{N_p}(\omega) + 2 \operatorname{Re}\{(\eta_0 h v V_p/\tau_p)\langle N_{p1} F_0 \rangle\} + \langle F_0 F_0 \rangle. \qquad (A13.16)$$

The first term in this equation is what we would naively expect the relationship to be. However, the partition noise at the mirror facet creates two additional noise contributions which are important to consider, particularly at high output powers.

Using Eq. (5.121) for $N_{p1}(\omega)$, we can set

$$\langle N_{p1} F_0 \rangle = \frac{H(\omega)}{\omega_R^2} [(\gamma_{NN} + j\omega)\langle F_P F_0 \rangle + \gamma_{PN}\langle F_N F_0 \rangle]. \qquad (A13.17)$$

Thus, to evaluate (A13.16) we need to know the various correlations between the three noise sources, F_0, F_P, and F_N. First of all, there is no correlation between the carrier noise and the phenomenon depicted in Fig. A13.1. Therefore, we can immediately set $\langle F_N F_0 \rangle = 0$. For the other two correlation strengths we can apply Eqs. (A13.6) and (A13.7) using F_0/hv and $V_p F_P$ to obtain

$$\langle F_0 F_0 \rangle = \eta_0 N_p V_p/\tau_p \cdot (hv)^2 = h v P_0, \qquad (A13.18)$$

$$\langle F_P F_0 \rangle = -\eta_0 N_p V_p/\tau_p \cdot (hv/V_p) = -P_0/V_p, \qquad (A13.19)$$

$$\langle F_N F_0 \rangle = 0. \qquad (A13.20)$$

Using these in combination with Eqs. (A13.12) through (A13.14), we can evaluate $S_{N_p}(\omega)$ (5.125) and $\langle N_{p1} F_0 \rangle$ (A13.17) from which we can ultimately obtain the output power spectral density function $S_{\delta P}(\omega)$ (A13.16). The complete expression is deferred to the next section where we evaluate all three photon density, output power, and carrier density noise spectral density functions.

A13.2.3 Photon Density and Phase Langevin Noise Correlations

The phase Langevin noise source $F_\phi(t)$ and its related correlation strengths are found by studying the electric field in the laser cavity. To begin, we assume the noise on the electric field contains fluctuations in-phase and out-of-phase with the average field as shown in Fig. A13.2. Defining the in-phase and out-of-phase random fields as $\Delta\mathscr{E}_r(t)$ and $\Delta\mathscr{E}_i(t)$, the total instantaneous field and associated *complex* Langevin noise source become

$$\mathscr{E}(t) = \mathscr{E}_0 + \Delta\mathscr{E}_r(t) + j\Delta\mathscr{E}_i(t),$$
$$\tilde{F}(t) = F_r(t) + jF_i(t). \tag{A13.21}$$

Here $F_r(t)$ models the in-phase field fluctuations and $F_i(t)$ models the out-of-phase field fluctuations.

The power and phase fluctuations and their associated Langevin noise sources can be related to the in-phase and out-of-phase noise components as follows:

$$\mathscr{E}^*\mathscr{E} \approx \mathscr{E}_0^2 + 2\mathscr{E}_0\Delta\mathscr{E}_r \quad \text{and} \quad \angle\mathscr{E} \approx \Delta\mathscr{E}_i/\mathscr{E}_0,$$
$$F_P(t) \approx 2\sqrt{N_p}F_r(t) \quad \text{and} \quad F_\phi(t) \approx F_i(t)/\sqrt{N_p}. \tag{A13.22}$$

The approximate equalities are valid for $\Delta\mathscr{E}_r, \Delta\mathscr{E}_i \ll \mathscr{E}_0$. For the latter relations, the fluctuating portion of $\mathscr{E}^*\mathscr{E}$ translates into photon density fluctuations, $F_P(t)$, while $\mathscr{E}_0^2$ translates into average photon density, N_p (and hence, $\mathscr{E}_0$ translates into $\sqrt{N_p}$). Now if we assume the spectral density of the in-phase and out-of-phase components of the field noise are equal in magnitude (i.e., $\langle F_i F_i\rangle = \langle F_r F_r\rangle$), we obtain

$$\langle F_r F_r\rangle = \frac{1}{4N_p}\langle F_P F_P\rangle = \frac{\Gamma R'_{sp}}{2}, \tag{A13.23}$$

$$\langle F_\phi F_\phi\rangle = \frac{1}{N_p}\langle F_i F_i\rangle = \frac{1}{N_p}\langle F_r F_r\rangle = \frac{\Gamma R'_{sp}}{2N_p}. \tag{A13.24}$$

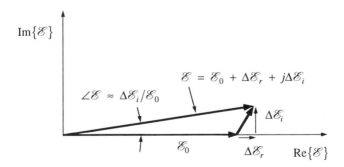

FIGURE A13.2 Vector illustration of the relationship between the instantaneous field magnitude and the quadrature (in-phase and out-of-phase) noise components.

The first equation uses Eq. (A13.12) to set $\langle F_P F_P \rangle = 2\Gamma R'_{sp} N_p$ assuming $N_p V_p \gg 1$.

To determine the cross-correlations, we note that $\langle F_\phi F_P \rangle = 2\langle F_i F_r \rangle$ from Eq. (A13.22). Assuming $\Delta \mathscr{E}_i(t)$ and $\Delta \mathscr{E}_r(t)$ are completely uncorrelated, we can set $\langle F_i F_r \rangle = 0$. As for $\langle F_\phi F_N \rangle$, we note that phase fluctuations due to changes in carrier density are accounted for separately with the linewidth enhancement factor, α (see Eq. (5.138)). Thus, the inherent phase fluctuations which exist even when $\alpha = 0$ are unrelated to carrier noise. From these arguments, we conclude that

$$\langle F_\phi F_P \rangle = \langle F_\phi F_N \rangle = 0. \tag{A13.25}$$

Finally it should be noted that a difference appears between the phase fluctuations inside the cavity and outside the cavity as discussed by Yamamoto [4]. However, the difference in this case is relatively minor and will not be considered here (it is only important to consider at frequencies comparable to or greater than $1/\tau_p$).

A13.3   EVALUATION OF NOISE SPECTRAL DENSITIES

A13.3.1   Photon Noise Spectral Density

For reference, the spectral density of the photon density noise derived in Chapter 5 is given by

$$S_{N_p}(\omega) = \frac{|H(\omega)|^2}{\omega_R^4} [(\gamma_{NN}^2 + \omega^2)\langle F_P F_P \rangle + 2\gamma_{NN}\gamma_{PN}\langle F_P F_N \rangle + \gamma_{PN}^2 \langle F_N F_N \rangle]. \tag{5.125}$$

Using the rate coefficients defined in Eq. (5.35) and the Langevin noise correlation strengths defined in Eqs. (A13.12) through (A13.14), the photon density noise becomes

$$S_{N_p}(\omega) = \frac{N_p}{V_p} \tau_p \cdot \frac{a'_1 + a'_2 \omega^2}{\omega_R^4} |H(\omega)|^2, \tag{A13.26}$$

where

$$a'_1 = \frac{a'_2}{\tau_{\Delta N}^2} + \omega_R^2 \frac{2}{\tau_{\Delta N}\tau_p} + \omega_R^4 \left[\frac{\eta_i(I + I_{th})}{I_{st}} + 1 \right],$$

$$a'_2 = \frac{8\pi(\Delta\nu)_{ST} N_p V_p}{\tau_p} \left[1 + \frac{1}{N_p V_p} \right],$$

and $(\Delta\nu)_{ST} = \Gamma R'_{sp}/4\pi N_p$, $I_{st} = qN_p V_p/\tau_p$. To simplify the expression, we have neglected the dependence of the single-mode spontaneous emission rate on N

by setting $1/\tau'_{\Delta N} \to 0$. Also, in defining a'_1 we have set $v_g a N_p/\tau_p = \omega_R^2$. Thus within a'_1, the resonance frequency is *by definition*: $\omega_R^2 \equiv v_g a N_p/\tau_p$.

The three terms comprising a'_1 form a power series in ω_R^2, or equivalently in N_p. At very low powers the first term dominates, while at moderate-to-high powers the last ω_R^4 term dominates. In fact at very high powers, the ω_R^4 term in a'_1 dominates the entire numerator of (A13.26) and its coefficient in square brackets reduces to 2 (for a shot noise-limited current source). At very high powers, we can also set $|H(\omega)|^2 \approx (1 + \omega^2\tau_p^2)^{-1}$ if we neglect gain compression (with gain compression, τ_p is replaced by γ/ω_R^2). In this limit then, $S_{N_p}(\omega) \approx (2N_p/V_p) \cdot \tau_p/(1 + \omega^2\tau_p^2)$. Using Eq. (5.105), we can estimate the mean-square photon number fluctuation. Integrating $S_{N_p}(\omega)V_p^2$ over all frequencies, we obtain $\langle|\delta N_p(t)V_p|^2\rangle = N_p V_p$. This result reveals that in the limit of high powers, the statistics of the photon number inside the cavity converges toward a Poisson distribution, indicative of a *coherent* state (in the language of quantum optics) [4].

A13.3.2   Output Power Noise Spectral Density

For reference, the spectral density of the output power noise derived earlier is given by

$$S_{\delta P}(\omega) = (\eta_0 h v V_p/\tau_p)^2 S_{N_p}(\omega) + 2 \, \text{Re}\{(\eta_0 h v V_p/\tau_p)\langle N_{p1}F_0\rangle\} + \langle F_0 F_0\rangle. \quad (A13.16)$$

Using the expression for $\langle N_{p1}F_0\rangle$ given in Eq. (A13.17) and the Langevin noise correlation strengths defined in Eqs. (A13.18) through (A13.20), the output power noise becomes

$$S_{\delta P}(\omega) = h v P_0 \cdot \left[\frac{a_1 + a_2\omega^2}{\omega_R^4} |H(\omega)|^2 + 1 \right], \quad (A13.27)$$

where

$$a_1 = \frac{8\pi(\Delta v)_{ST} P_0}{h v}\left[1 + \frac{1}{N_p V_p}\right]\frac{1}{\tau_{\Delta N}^2} + \eta_0\omega_R^4\left[\frac{\eta_i(I + I_{th})}{I_{st}} - 1\right]$$
$$- 2\eta_0\omega_\Delta^2\left[\omega_R^2 + \frac{1}{\tau_{\Delta N}\tau_p}\right],$$

$$a_2 = \frac{8\pi(\Delta v)_{ST} P_0}{h v}\left[1 + \frac{1}{N_p V_p}\right] - 2\eta_0\omega_R^2\left[\frac{\Gamma a_p}{a} + \frac{4\pi(\Delta v)_{ST}}{\omega_R^2\tau_p}\right],$$

and $(\Delta v)_{ST} = \Gamma R'_{sp}/4\pi N_p$, $I_{st} = q P_0/\eta_0 h v$. To simplify the expression, we have neglected the dependence of the single-mode spontaneous emission rate on N by setting $1/\tau'_{\Delta N} \to 0$. Also, in defining a_1 and a_2, we have set $v_g a N_p/\tau_p = \omega_R^2$. Thus within a_1 and a_2, the resonance frequency is *by definition*: $\omega_R^2 \equiv v_g a N_p/\tau_p$. The remainder of terms comprising the exact expression for the resonance frequency are sectioned off into the variable: $\omega_\Delta^2 \equiv \gamma_{NP}\gamma_{PN} + \gamma_{NN}\gamma_{PP} - v_g a N_p/\tau_p$.

Within a_1, the first term is independent of power and dominates at very low powers. The second term is $\propto P_0^2$ and dominates at moderate-to-high powers even though $\eta_i(I + I_{th})/I_{st}$ approaches one at high powers (for a shot noise-limited current source). The third ω_Δ^2 term can be neglected for all but very small power levels since the difference between $v_g a N_p/\tau_p$ and the actual ω_R^2 is usually negligible. Within a_2, the first term is independent of power, but nevertheless dominates for all but very high powers. The second term is $\propto P_0$, so it eventually becomes comparable to the first term. However, gain compression limits ω_R^2 to a maximum value, such that the second term never actually grows larger than the first term. Within the square brackets of the second term, $\Gamma a_p/a$ dominates for all but very small powers since $4\pi(\Delta v)_{ST}/\omega_R^2\tau_p$ is $\propto 1/P_0^2$.

In comparing $S_{\delta P}(\omega)$ to $S_{N_p}(\omega)$, note that in addition to the slightly more complex frequency coefficients, a new "$+1$" factor appears in $S_{\delta P}(\omega)$. This factor insures that the output power spectral density never drops below the shot noise limit of hvP_0 for all frequencies (unless the current source has sub-shot noise characteristics). Because the shot noise is included implicitly in $S_{\delta P}(\omega)$, it is in principle not necessary to explicitly add a shot noise term to the noise current in a photodetector [4]. However in practice, the partition noise created by the random loss of photons in getting from the laser to the detector contributes a shot noise-like term $\propto (1 - \eta_{det})$ that should be added to the detector current noise (see Eq. (5.137) for details).

The expression for $S_{\delta P}(\omega)$ given in Chapter 5 (5.130) neglects the ω_Δ^2 term of a_1, and the second term within square brackets of the ω_R^2 term in a_2. The first set of square brackets in both a_1 and a_2 are also set equal to one, assuming $N_p V_p \gg 1$ above threshold.

A13.3.3 Carrier Noise Spectral Density

For reference, the spectral density of the carrier density noise derived in Chapter 5 is given by

$$S_N(\omega) = \frac{|H(\omega)|^2}{\omega_R^4} [\gamma_{NP}^2 \langle F_P F_P \rangle - 2\gamma_{PP}\gamma_{NP} \langle F_P F_N \rangle + (\gamma_{PP}^2 + \omega^2) \langle F_N F_N \rangle]. \quad (5.124)$$

Using the rate coefficients defined in Eq. (5.35) and the Langevin noise correlation functions defined in Eqs. (A13.12) through (A13.14), the carrier density noise becomes

$$S_N(\omega) = \frac{8\pi N_p^2}{\Gamma^2 \tau_p^2 \omega_R^4} (\Delta v)_{ST} \cdot (1 + \delta)|H(\omega)|^2, \quad (A13.28)$$

where

$$\delta = -\frac{\varepsilon N_p/n_{sp}}{(1 + \varepsilon N_p)^2} \left[1 + \varepsilon N_p - \frac{\varepsilon N_p}{2} \left\{ \frac{\eta_i(I + I_{th})}{I_{st}} + 1 \right\} \right]$$

$$+ \omega^2 \tau_p^2 \left[1 + \frac{1}{2n_{sp}} \left\{ \frac{\eta_i(I + I_{th})}{I_{st}} - 1 \right\} \right],$$

and $(\Delta v)_{ST} = \Gamma R'_{sp}/4\pi N_p$, $I_{st} = qN_p V_p/\tau_p$. To simplify the expression for this case, we have set the gain equal to the loss ($\Gamma v_g g = 1/\tau_p$) and assumed $N_p V_p \gg 1$. We have also written out a_p explicitly using Eq. (5.32) and set $R'_{sp} V \tau_p \to n_{sp}$. For practical uses, it turns out that δ does not need to be considered. For example at low powers, the entire first term reduces to $-\varepsilon N_p/n_{sp}$ which is negligible in comparison to one ($n_{sp} \sim 1$ above threshold). At very high powers, it reaches a maximum of $-1/4n_{sp}$ when $\varepsilon N_p = 1$ ($\eta_i(I + I_{th})/I_{st} \to 1$ at high powers for a shot noise-limited current source). Thus, gain compression can reduce the low frequency carrier noise by as much as 25% at very high powers, however, we neglect this contribution in Chapter 5. The second term is small in comparison to one for $\omega \ll 1/\tau_p$, so unless we are interested in very high frequency carrier noise, we can neglect it as well.

The expression for $S_N(\omega)$ given in Chapter 5 (5.124) neglects δ in (A13.28) and sets $\omega_R^2 = v_g a N_p/\tau_p$.

REFERENCES

[1] D.E. McCumber, *Phys. Rev.*, **141**, 306 (1966). See also: C. Harder, J. Katz, S. Margalit, J. Shacham, and A. Yariv, *IEEE J. Quantum Electron.*, **QE-18**, 333 (1982).

[2] M. Lax, *Phys. Rev.*, **160**, 290 (1967).

[3] Y. Yamamoto and S. Machida, *Phys. Rev. A*, **35**, 5114 (1987).

[4] Y. Yamamoto and N. Imoto, *IEEE J. Quantum Electron.*, **QE-22**, 2032 (1986).

PROBLEMS

A13.1 Assume the output power of a laser with a given noise spectral density, S_i, is split by a beam splitter and fed into two photodetectors. Use the Langevin method to determine the detected noise spectral densities. Start by defining reservoirs for the incident and detected photon streams, with associated Langevin noise sources (F_i, F_{d1}, and F_{d2}). For the incident photon reservoir, assume an in-flow from the laser, I, and two out-flows, ηI and $(1 - \eta)I$. For the detector reservoirs assume only in-flows from the incident photon reservoir (a flow chart is useful here). Next, define all auto- and cross-correlation strengths between the three Langevin noise sources, considering the in-flow correlation term in the incident photon reservoir to be S_i instead of shot noise, I (assume photon number flow rates in all reservoirs for convenience). Finally, using a small-signal analysis, show that the spectral density of the first detected photon stream $\langle I_{d1} I_{d1} \rangle$ is equal to the product $\langle (\eta F_i + F_{d1})(\eta F_i + F_{d1}) \rangle$, and then evaluate the product for both detectors. Interpret how this result applies to Eq. (5.137). How and why does the beam splitter affect the detected noise?

A13.2 For the arrangement described in Problem A13.1, show that the spectral density of the *sum* of the detected photon streams reproduces the noise of the incident photon stream, independent of η. What spectral density does the *difference* between the two detected photon streams yield in general, and for $\eta = 0.5$? Can you use the results for the sum and difference spectral densities to suggest a measurement technique which calibrates the measured intensity noise of a laser to the shot noise floor? This type of configuration is in fact known as a *balanced detector pair* and is commonly employed to determine whether or not a laser is operating below the standard quantum limit for noise.

Derivation Details for Perturbation Formulas

In Chapter 6 several calculations involve deriving an approximate solution to a perturbation on a known waveguide problem. This typically involves inserting a trial field into the wave equation, dropping out second-order terms, multiplying through by a complex conjugate of a transverse-mode eigenfunction of the unperturbed problem, integrating over the cross section, and using modal orthogonality as well as other arguments to drop additional terms, so that a simple analytic formula can be derived.

During the course of these calculations we repeatedly come upon a collection of terms composed of the transverse-mode perturbation acted upon by the wave equation. The first such instance is given in Eq. (6.8), which is the result just after multiplying times U^* and integrating over the cross section. That is,

$$2\beta\Delta\beta \int |U|^2 \, dA = \int \Delta\varepsilon k_0^2 |U|^2 \, dA$$

$$+ \int [(\nabla_T^2 \Delta U)U^* + \varepsilon k_0^2 \Delta U U^* - \beta^2 \Delta U U^*] \, dA. \quad (6.8)$$

The terms in question are contained in the second integral on the right side of the equation (second line). We need to show that the integral is zero or negligible in the cases of interest.

First, we multiply the complex conjugate of the transverse wave equation, Eq. (6.5), by ΔU to get

$$\Delta U(\nabla_T^2 U^*) + \Delta U[\varepsilon^*(x, y, z)k_0^2 - \beta^{*2}]U^* = 0. \quad (A14.1)$$

Thus, the last two terms in the last integral in Eq. (6.8) can be replaced by $-\Delta U(\nabla_T^2 U^*) + (\varepsilon - \varepsilon^*)k_0^2 U^* - (\beta^2 - \beta^{*2})U^*$. If ε and β were real, then we

553

would only need $-\Delta U(\nabla_T^2 U^*)$. Making the replacement,

$$\int [(\nabla_T^2 \Delta U)U^* + \varepsilon k_0^2 \Delta U U^* - \beta^2 \Delta U U^*] \, dA$$

$$= \int [(\nabla_T^2 \Delta U)U^* - \Delta U(\nabla_T^2 U^*)] \, dA$$

$$+ \int [(\varepsilon - \varepsilon^*)k_0^2 \Delta U U^* - (\beta^2 - \beta^{*2})\Delta U U^*] \, dA. \quad \text{(A14.2)}$$

Now, the first integral on the right is identically zero because

$$\int [(\nabla_T^2 \Delta U)U^* - \Delta U(\nabla_T^2 U^*)] \, dA = \int \nabla_T \cdot [(\nabla_T \Delta U)U^* - \Delta U(\nabla_T U^*)] \, dA$$

$$= \oint_\infty \hat{\mathbf{e}}_n \cdot [(\nabla_T \Delta U)U^* - \Delta U(\nabla_T U^*)] \, ds \equiv 0.$$

$$\text{(A14.3)}$$

The latter equality uses Green's Theorem to convert the integral over the cross-sectional area to a line integral around the perimeter of the cross section (in this case, at infinity). The vector $\hat{\mathbf{e}}_n$ is the unit vector normal to the contour of integration. The contour integral at infinity is zero because both U and ΔU must vanish at infinity for any guided mode. The second integral on the right side of Eq. (A14.2) also is identically zero for ε and β real. In fact, even for complex ε and β, it tends to be negligible in comparison to the first integral on the right side of Eq. (6.8) in most cases, since $\Delta U \ll U$, and because the loss or gain of the unperturbed problem can usually be chosen to be sufficiently small. Only in the extreme case where the gain provides most of the waveguiding effect (as in gain-guided lasers) will this term be nonnegligible. In this rare case, the use of the $\Delta\beta$ formula is questionable.

READING LIST

H.A. Haus, *Waves and Fields in Optoelectronics*, Ch. 6, Prentice Hall, Englewood Cliffs, NJ (1984).

The Electro-Optic Effect

For many tunable lasers and photonic integrated circuits it is desirable to change the index of refraction by the application of a dc (or rf) electric field. In certain crystals that do not possess inversion symmetry this is possible through what is known as the electro-optic effect. Because the index changes virtually instantaneously with the field, this effect can be used in very high modulation bandwidth modulators and tunable filters.

The electro-optic effect is somewhat more complex than one might at first expect, because the change in index is generally different for different polarizations of the optical field for a given applied dc field orientation, and moreover, it is usually associated with anisotropic crystals that have different indexes for different optical polarizations initially [1]. Thus, it is necessary to use the full dielectric tensor in relating the displacement, $\mathbf{D}$, to the electric field, $\mathscr{E}$, of the lightwave. That is, the displacement might not be parallel to the electric field for all orientations. Also, the applied dc or rf field generally has still another orientation, so we have three orientations to keep track of in this process.

The displacement field is related to the optical electric field via the dielectric tensor, $\mathbf{D} = \varepsilon \mathscr{E}$, or

$$
\begin{bmatrix} D_x \\ D_y \\ D_z \end{bmatrix} = \begin{bmatrix} \varepsilon_{xx} & \varepsilon_{xy} & \varepsilon_{xz} \\ \varepsilon_{yx} & \varepsilon_{yy} & \varepsilon_{yz} \\ \varepsilon_{zx} & \varepsilon_{zy} & \varepsilon_{zz} \end{bmatrix} \begin{bmatrix} \mathscr{E}_x \\ \mathscr{E}_y \\ \mathscr{E}_z \end{bmatrix}. \tag{A15.1}
$$

If we choose axes to diagonalize the matrix, then we have found the principle dielectric axes of the medium. Switching to the optical index, we then have

$$
\begin{bmatrix} D_x \\ D_y \\ D_z \end{bmatrix} = \varepsilon_0 \begin{bmatrix} n_{xx}^2 & 0 & 0 \\ 0 & n_{yy}^2 & 0 \\ 0 & 0 & n_{zz}^2 \end{bmatrix} \begin{bmatrix} \mathscr{E}_x \\ \mathscr{E}_y \\ \mathscr{E}_z \end{bmatrix}. \tag{A15.2}
$$

555

We continue to allow for anisotropy of the index, but we have defined an index ellipsoid with axes along the material's principle axes. When the field is polarized along one of these directions, $\mathbf{D}$ will be parallel to $\mathscr{E}$. In most materials at least two of these indexes are equal. These are called the *ordinary* index, while the third is called the *extraordinary* index. Rays with their electric field aligned along these directions are also referred to as the ordinary or extraordinary rays. By convention we label n_{zz} as the extraordinary index, and the z-axis is called the *optic* axis. Materials with two such indexes are called *birefringent*. In the III–V semiconductor materials considered in this book, the index is isotropic when no additional dc or rf field is applied. Thus, all of the indexes are the same with no applied field. However, with the application of a field, these materials become birefringent also.

The general index ellipsoid from Eq. (A15.2) is

$$\frac{x^2}{n_{xx}^2} + \frac{y^2}{n_{yy}^2} + \frac{z^2}{n_{zz}^2} = 1. \tag{A15.3}$$

The index of any propagating wave can be found by constructing a plane perpendicular to its k-vector that passes through the origin of the ellipsoid. The wave's electric field direction is then constructed as a line from the origin on this plane. The index is given by the distance to the surface of the ellipsoid. Figure A15.1 illustrates this construction. Note that for GaAs or InP this ellipsoid is just a sphere in the absence of any applied dc or rf field.

With the application of a dc or rf field to an electro-optic material this ellipsoid is distorted, so that generally off-diagonal terms appear in (A15.2). The electro-optic tensors are defined from the perturbation to each of the terms in the index ellipsoid. That is,

$$\Delta\left(\frac{1}{n^2}\right)_{ij} = \sum_k r_{ijk}\mathscr{E}_k, \tag{A15.4}$$

where $\mathbf{r}$ is the linear electro-optic tensor and $\mathscr{E}_k$ is the applied dc electric field. Three subscripts are necessary to account for the relative orientation of the crystal axis, the optical electric field, and the dc electric field. It is also possible that quadratic effects are important, so we can also define a similar equation where $\mathscr{E}$ is replaced by $\mathscr{E}^2$ and $\mathbf{r}$ is replaced by $\mathbf{s}$, the quadratic electro-optic tensor. Quadratic effects are important for wavelengths near the absorption edge in bulk and MQW waveguides, where an electric field can move the effective absorption edge [2, 3]. Also, in depletion regions of pn-junctions quadratic effects become dominant if a significant carrier density is being depleted at the same time [4]. We will come back to this topic after completing the linear effect.

In order to simplify the notation somewhat, an abbreviated subscript notation is usually introduced, so that the three-dimensional tensor can be

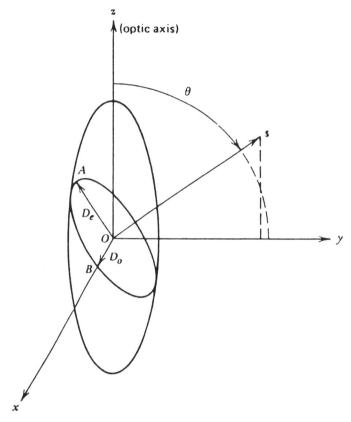

FIGURE A15.1   Plot of index ellipsoid [1] (From *Optical Waves in Crystals*, A. Yariv and P. Yeh, Copyright © 1984 John Wiley & Sons, Inc. Reprinted by permission of John Wiley and Sons, Inc.)

expressed in two dimensions. The notation involves using numbers from 1 to 6 for the first two subscripts, where $1 \equiv xx$, $2 \equiv yy$, $3 \equiv zz$, $4 \equiv yz$, $5 \equiv xz$, and $6 \equiv xy$. The final subscript, which refers to the applied dc or rf electric field orientation also is given a numerical subscript, where $1 \equiv x$, $2 \equiv y$, and $3 \equiv z$. Thus, Eq. (A13.4) becomes

$$\Delta\left(\frac{1}{n^2}\right)_i = \sum_{j=1}^{3} r_{ij}\mathscr{E}_j, \qquad i = 1, 2, \ldots, 6 \qquad (A15.5)$$

With the application of a dc or rf field, the distorted index ellipsoid takes on the form,

$$\left(\frac{1}{n^2}\right)_1 x^2 + \left(\frac{1}{n^2}\right)_2 y^2 + \left(\frac{1}{n^2}\right)_3 z^2 + 2\left(\frac{1}{n^2}\right)_4 yz + 2\left(\frac{1}{n^2}\right)_5 xz + 2\left(\frac{1}{n^2}\right)_6 xy = 1,$$

$$(A15.6)$$

where the additional terms are solely due to (A15.5) and the original terms may also be changed if nonzero perturbation terms are generated by (A15.5). To determine the change in index for a propagating optical wave, we again must diagonalize the index matrix, or equivalently, find the new principle axes for the distorted index ellipsoid. This generally involves a rotation in space of the original axes. Once this new principle axis set (x', y', z') is found, the index ellipsoid will again have the form of (A15.3). The coefficient of x'^2 will give the reciprocal of the square of the perturbed index, $n_{x'}$, for an optical field polarized along the x'-axis, and similarly for the other components.

In the case of GaAs, InP, or their alloys, the process of finding the perturbed principle axes and the related indexes is somewhat simplified, since for crystals of this zinc blende class (*cubic-$\bar{4}3m$*) only r_{41}, r_{52}, and r_{63} are nonzero for x, y, and z aligned with the crystal axes. Also, these all have the same value, so we set them all equal to r_{41}. Thus, for a dc field in the z-direction, or $[0\,0\,1]$, Eq. (A15.6) becomes

$$\frac{x^2 + y^2 + z^2}{n_0^2} + 2r_{41}\mathscr{E}_z xy = 1, \qquad (A15.7)$$

where n_0 is the unperturbed index for the isotropic semiconductor in all directions. That is, in the x–y plane, the index ellipsoid is somewhat squashed in the first and third quadrants and somewhat stretched in the second and fourth quadrants. To put Eq. (A15.7) in the form of (A15.3), it can be shown that a rotation of the coordinate system about the z-axis by $45°$ is required. That is, $z = z'$, and

$$x = x' \cos 45° + y' \sin 45°,$$
$$y = -x' \sin 45° + y' \cos 45°, \qquad (A15.8)$$

so that the new (primed) axes become $x' = [1\,\bar{1}\,0]$ and $y' = [1\,1\,0]$. Plugging this into (A15.7), we obtain

$$\left(\frac{1}{n_0^2} - r_{41}\mathscr{E}_z\right)x'^2 + \left(\frac{1}{n_0^2} + r_{41}\mathscr{E}_z\right)y'^2 + \left(\frac{1}{n_0^2}\right)z^2 = 1. \qquad (A15.9)$$

Thus,

$$\frac{1}{n_{x'}^2} = \frac{1}{n_0^2} - r_{41}\mathscr{E}_z \quad \text{and} \quad \frac{1}{n_{y'}^2} = \frac{1}{n_0^2} + r_{41}\mathscr{E}_z. \qquad (A15.10)$$

For $r_{41}\mathscr{E}_z \ll 1/n_0^2$, we can approximately solve for the desired perturbed indices,

$$n_{x'} = n_0 + \frac{n_0^3}{2} r_{41}\mathscr{E}_z,$$
$$n_{y'} = n_0 - \frac{n_0^3}{2} r_{41}\mathscr{E}_z. \qquad (A15.11)$$

The result of the above calculation is that the index for an optical field polarized along the $[1\bar{1}0]$ direction (i.e., x'-direction) will be increased by $(n_0^3/2)r_{41}\mathscr{E}_z$, while the index for an optical field polarized along the $[110]$ direction (i.e., y'-direction) will be decreased by $(n_0^3/2)r_{41}\mathscr{E}_z$. There will be no change in index for a component of optical field along the direction of the applied field $[001]$. Thus, for a waveguide on a (001) wafer aligned perpendicular to the natural (110) cleavage planes in these III–V materials, the application of a surface-normal field will cause the index of a TE mode to change, but it will not change for a TM mode.

In GaAs, r_{41} varies from 1.1 to 1.5×10^{-12} m/V in the 0.9 to 1.3 μm wavelength range. In other materials the electro-optic coefficient can be much larger than this. Of course, it is really the product $n^3 r_{ij}$ which gives the best measure of the index change for a given applied field. In GaAs, $n^3 r_{41} \sim 55 \times 10^{-12}$ m/V. Thus, for fields ~ 400 kV/cm, obtainable in *pin* undoped regions (as well as in the depletion regions of *pn*-junctions), $\Delta n/n \sim 0.001$. Outside of the semiconductor regime, the most popular material for photonic integrated circuits is lithium niobate, $LiNbO_3$. In this *trigonal-3m* material several terms in the electro-optic tensor are nonzero. At 1.3 μm, $r_{13} = -r_{23} \sim 7$; $r_{22} = -r_{12} = -r_{61} \sim 3.3$; $r_{33} \sim 30$; and $r_{51} = r_{42} \sim 26 \times 10^{-12}$ m/V. The ordinary and extraordinary indices of refraction are 2.22 and 2.14, respectively. Also, $n_0^3 r_{33} \sim 328 \times 10^{-12}$ m/V.

In certain III–V device configurations the quadratic electro-optic effect is larger than the linear effect. This can happen if the wavelength of the lightwave is close to the absorption edge in bulk or quantum-well material. When a field is applied the absorption edge moves to longer wavelengths via either the Franz–Keldysh (FK) in bulk or quantum-confined Stark effect (QCSE) in MQWs. Since the index of refraction is decreasing roughly as $1/\lambda$ above the absorption edge, the application of a field also increases the index at some wavelength in this region. The combined effect is nearly quadratic with electric field [4]. Also, if free carriers are depleted when a field is applied, e.g., in the depletion region of a *pn*-junction, the index will increase due to the removal of the index reduction associated with the existence of free carriers as well as the shift in absorption edge due to a removal of band filling. This effect again is approximately quadratic with the applied field.

Analogous to the linear effect, the quadratic index change in all cases can be written as

$$\Delta n_q = \frac{n_0^3}{2} s \mathscr{E}_z^2. \tag{A15.12}$$

The difficulty in these cases is that the quadratic coefficient, s, also decreases with increasing distance away from the absorption edge, and therefore, it is difficult to effectively parameterize the problem as with the linear electro-optic effect. With the QCSE, associated with absorption due to excitons in MQWs, the index shift per unit field change can be larger than in bulk GaAs, but the effect also decays more rapidly with increasing wavelength. For example, the effect becomes less than the linear effect ~ 40 nm from the absorption edge, whereas in bulk materials the quadratic effect due to the absorption edge alone is larger than the linear effect 100 nm away. The depletion of charge also only works effectively in bulk materials, since the charge will tend to screen the excitons in quantum wells. Thus, it is unclear whether or not quantum wells can offer a significant practical advantage in electro-optic modulators.

Figure A15.2 plots the phase modulation available in a doped GaAs waveguide under reverse bias. This device has a *pn*-junction in the center of the doped waveguide, and the waveguide is oriented along the [0 1 1] direction so that the quadratic effects add to the linear effect. A separate curve is plotted for each physical effect. The net phase shift $\Delta\phi$, in a device of length L, is obtained from the plotted phase-shift efficiency, η_{ps}, from

$$\Delta\phi = \eta_{ps} L(\Delta V), \tag{A15.13}$$

where ΔV is the applied voltage shift.

In all cases the proximity of the operating wavelength to the absorption edge is limited by loss. That is, although the amount of index change available increases as the absorption edge is approached, so does the loss and the loss change. As a consequence, it is important to limit the residual loss and loss change to some values when comparing different kinds of index modulation. As defined in Chapter 2, the ratio of the changes in the real to the imaginary

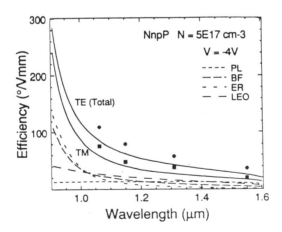

FIGURE A15.2 Calculated and measured (data) phase-shift efficiency vs. wavelength with $L_m = 740$ μm, and equal *n*- and *p*-type doping to the center of the 0.25 μm thick GaAs waveguide [5]. The AlGaAs cladding material contains 40% Al. PL, BF, ER and LEO refer to the plasma, band-filling, electro-refractive, and linear electro-optic effect respectively. All except LEO are approximately quadratic with applied reverse bias in this configuration.

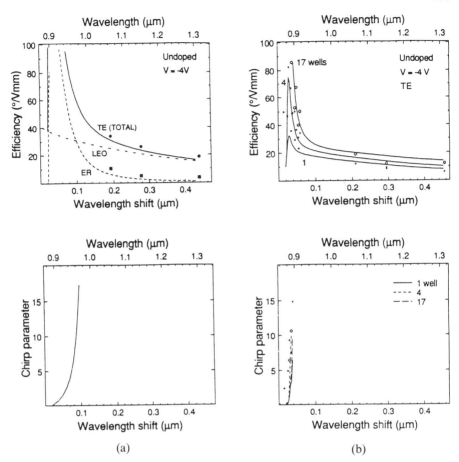

FIGURE A15.3   Calculated and measured phase-shift efficiency (top) and chirp (bottom) for undoped material vs. wavelength deviation from the zero-bias absorption edges at 0.87 μm for bulk and 0.855 μm for the MQW waveguides [5, 6]. The waveguide regions (MQW separate-confinement region) are 0.25 μm wide clad by $Al_{0.4}Ga_{0.6}As$. The bulk guide is undoped GaAs; the MQW region has either 1, 4, or 17 GaAs wells, 10-nm-thick, separated by $Al_{0.2}Ga_{0.8}As$ barriers. In all cases the phase-shift efficiency and chirp are calculated between 0 and −4 V. For the bulk case (a) the TE mode measurements are given by circles; the TM mode by squares. The calculated net quadratic electro-refractive (ER) effect is shown with the short dashed line, the net linear electro-optic (LEO) effect is shown with the long dashed line, and the total with a solid line. For the MQW case (b) only the TE mode is measured.

↳ more accurate to look at detuning in energy rather than wavelength.

parts of the refractive index is called the chirp parameter, α. Thus for good phase modulation with low loss this parameter should be large. As a rule of thumb, we prefer α > 10 for reasonably low-loss phase modulation. Figure A15.3 shows the calculated and measured phase modulation efficiency and the chirp parameter for undoped bulk and MQW waveguides for comparison.

REFERENCES

[1] A. Yariv and P. Yeh, *Optical Waves in Crystals*, Chs. 4 and 7, Wiley, New York (1984).

[2] A. Alping and L.A. Coldren, *J. Appl. Phys.*, **61**, 2430 (1987).

[3] D.A.B. Miller, D.S. Chemla, and S. Schmitt-Rink, *Phys. Rev. B*, **33**, 6976 (1985).

[4] J.G. Mendoza-Alvarez, L.A. Coldren, A. Alping, R.H. Yan, T. Hausken, K. Lee, and K. Pedrotti, *IEEE J. Lightwave Technol.*, **6**, 793 (1988).

[5] T. Hausken, Ph.D. thesis, ECE Technical Report #90-18, Ch. 2, UCSB (1990).

[6] T. Hausken, Ph.D. thesis, ECE Technical Report #90-18, Ch. 4, UCSB (1990).

READING LIST

A. Yariv, *Optical Electronics*, 4th ed. Ch. 9, Saunders College Publishing, Philadelphia (1991).

A. Yariv and P. Yeh, *Optical Waves in Crystals*, Ch. 7, Wiley, New York (1984).

Solution of Finite Difference Problems

A16.1 MATRIX FORMALISM

In Section 7.3.8, the finite-difference technique was introduced to solve for the effective index and field profiles of an arbitrary channel waveguide structure. The scalar wave equation was discretized to provide the following linear matrix equation:

$$\frac{U_j^{i-1}}{\Delta X^2} + \frac{U_{j-1}^i}{\Delta Y^2} - \left(\frac{2}{\Delta X^2} + \frac{2}{\Delta Y^2} - (n_j^i)^2\right)U_j^i + \frac{U_{j+1}^i}{\Delta Y^2} + \frac{U_j^{i+1}}{\Delta X^2} = \bar{n}^2 U_j^i. \quad \text{(A16.1)}$$

where U_j^i and n_j^i are the normalized electric field and refractive index at the grid point (i, j) for $i = 0, 1, 2, \ldots, I + 1$ and $j = 0, 1, 2, \ldots, J + 1$. Also, $\Delta X = k_0 \Delta x$ and $\Delta Y = k_0 \Delta y$ are the normalized coordinate steps between grid points, and $\bar{n}$ is the effective index of the waveguide mode.

To complete the problem we need to specify boundary conditions. While different choices exist, the simplest choice is simply to set the fields to zero around the border of the computational window. This approximation is valid as long as we are far enough away from the guiding layers that the true field solutions are essentially zero. Experimentation with the size of the computational window can determine the validity of this approximation. So our boundary conditions are $U_j^0 = U_j^{I+1} = U_0^i = U_{J+1}^i = 0$, and we are left solving for the fields within the window. In other words, we need to solve all equations for $i = 1$ to I and $j = 1$ to J, or a total of $I \times J$ coupled equations.

Defining coefficients for terms with common j indices, Eq. (A16.1) becomes

$$\frac{U_j^{i-1}}{\Delta X^2} + [bU_{j-1}^i + a_j^i U_j^i + bU_{j+1}^i] + \frac{U_j^{i+1}}{\Delta X^2} = \bar{n}^2 U_j^i, \quad \text{(A16.2)}$$

where

$$a_j^i = (n_j^i)^2 - \frac{2}{\Delta X^2} - \frac{2}{\Delta Y^2}.$$

$$b = \frac{1}{\Delta Y^2}.$$

We can compact the y-direction into matrix notation by defining a vector which encompasses y for each x position, i:

$$\mathbf{U}^i = \begin{bmatrix} U_1^i \\ \vdots \\ U_J^i \end{bmatrix}. \tag{A16.3}$$

Then, by vertically listing all J equations (A16.2) for a given i, we can group common i indices into a matrix-difference equation along x:

$$\mathbf{BU}^{i-1} + \mathbf{A}^i\mathbf{U}^i + \mathbf{BU}^{i+1} = \bar{n}^2\mathbf{U}^i, \tag{A16.4}$$

where

$$\mathbf{A}^i = \begin{bmatrix} a_1^i & b & 0 & \cdots & 0 & 0 \\ b & a_2^i & b & 0 & & 0 \\ 0 & b & a_3^i & b & 0 & \vdots \\ \vdots & 0 & & \ddots & & 0 \\ 0 & & 0 & b & a_{J-1}^i & b \\ 0 & 0 & \cdots & 0 & b & a_J^i \end{bmatrix},$$

$$\mathbf{B} = \frac{1}{\Delta X^2}\mathbf{I},$$

and $\mathbf{I}$ is the $J \times J$ identity matrix. The second b does not appear in the top and bottom rows of the $\mathbf{A}^i$ matrix because $U_0^i = 0$ and $U_{J+1}^i = 0$ (this is in fact the reason we need to set the fields to zero at the boundaries).

Now writing all I equations (A16.4) along x in matrix form, we obtain

$$\begin{bmatrix} \mathbf{A}^1 & \mathbf{B} & 0 & \cdots & 0 & 0 \\ \mathbf{B} & \mathbf{A}^2 & \mathbf{B} & 0 & & 0 \\ 0 & \mathbf{B} & \mathbf{A}^3 & \mathbf{B} & 0 & \vdots \\ \vdots & 0 & & \ddots & & 0 \\ 0 & & 0 & \mathbf{B} & \mathbf{A}^{I-1} & \mathbf{B} \\ 0 & 0 & \cdots & 0 & \mathbf{B} & \mathbf{A}^I \end{bmatrix} \begin{bmatrix} \mathbf{U}^1 \\ \mathbf{U}^2 \\ \mathbf{U}^3 \\ \vdots \\ \mathbf{U}^{I-1} \\ \mathbf{U}^I \end{bmatrix} = \bar{n}^2 \begin{bmatrix} \mathbf{U}^1 \\ \mathbf{U}^2 \\ \mathbf{U}^3 \\ \vdots \\ \mathbf{U}^{I-1} \\ \mathbf{U}^I \end{bmatrix}, \tag{A16.5}$$

where again the second **B** does not appear in the top and bottom rows because $\mathbf{U}^0 = 0$ and $\mathbf{U}^{I+1} = 0$. We can write this equation symbolically as

$$\mathbf{AU} = \bar{n}^2 \mathbf{U} \qquad \text{with} \qquad \mathbf{U} = \begin{bmatrix} \mathbf{U}^1 \\ \vdots \\ \mathbf{U}^I \end{bmatrix}. \tag{A16.6}$$

This is the matrix equation to be solved for the eigenvalue, $\bar{n}^2$, and eigenvector, **U**. In this compact form, **A** represents an $I \times I$ matrix with elements that are themselves $J \times J$ matrices, and **U** is an I length vector with components that are themselves J length vectors. To solve the equation, we expand each element of **A** into a $J \times J$ block, so that **A** becomes an $(IJ) \times (IJ)$ matrix with scalar elements. Likewise, we expand the component vectors of **U**, so that **U** becomes an (IJ) length vector with scalar components. Various matrix methods are then available for determining the eigenvalues and eigenvectors.

A16.2 ONE-DIMENSIONAL DIELECTRIC SLAB EXAMPLE

To illustrate the basic numerical technique, we consider the most simple slab waveguide problem. In this case the scalar wave equation correctly describes the TE mode with its electric field polarized along the y-direction as illustrated in Fig. A16.1.

In this case the eigenvalue equation (A16.5) reduces to

$$\begin{bmatrix} \left(n_1^2 - \dfrac{2}{\Delta X^2} \right) & \dfrac{1}{\Delta X^2} & 0 & 0 \\ \dfrac{1}{\Delta X^2} & \left(n_2^2 - \dfrac{2}{\Delta X^2} \right) & \dfrac{1}{\Delta X^2} & 0 \\ 0 & & \ddots & \\ 0 & 0 & \dfrac{1}{\Delta X^2} & \left(n_I^2 - \dfrac{2}{\Delta X^2} \right) \end{bmatrix} \begin{bmatrix} U_1 \\ U_2 \\ \vdots \\ U_I \end{bmatrix} = \bar{n}^2 \begin{bmatrix} U_1 \\ U_2 \\ \vdots \\ U_I \end{bmatrix},$$

$$\tag{A16.7}$$

where we have simplified the notation in this one-dimensional case, in which there is no variation of the index in the y-direction, by letting, $n_i = n_j^i$, and $U_i = U_j^i$, since j only takes on a single value.

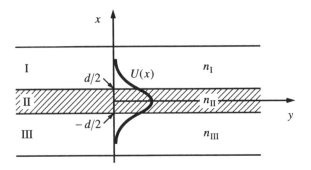

FIGURE A16.1 Schematic of dielectric slab waveguide assuming propagation in the z-direction.

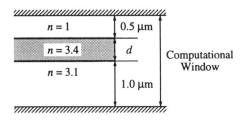

FIGURE A16.2 One-dimensional slab waveguide example.

Figure A16.2 shows an illustrative one-dimensional problem. To demonstrate the importance of the grid size, we solve the problem for several different finite-difference step sizes and two different slab thicknesses. The results are given in Tables A16.1 and A16.2.

TABLE A16.1 Results of Finite Difference Calculation of Slab Waveguide in Fig. A16.2.*

Δx (μm)	Air Grid Pts	Guide Grid Pts	Sub. Grid Pts	Matrix Size	Effective Index	Error (%)
0.25	2	2	4	8 × 8	3.315 23	1.081
0.125	4	4	8	16 × 16	3.292 13	0.376
0.0625	8	8	16	32 × 32	3.283 29	0.107
0.0312 5	16	16	32	64 × 64	3.280 70	0.028

* Slab waveguide thickness $d = 0.5$ μm (single transverse mode). Exact effective index is $\bar{n}_0 = 3.279\,790$ (for $\lambda_0 = 1.3$ μm).

TABLE A16.2   Results of Finite Difference Calculation of Slab Waveguide in Fig. A16.2.†

Δx (μm)	Air Grid Pts	Guide Grid Pts	Sub. Grid Pts	Matrix Size	Effective Indices	Error (%)
0.25	2	4	4	10×10	3.36583	0.242
					3.27223	1.234
0.125	4	8	8	20×20	3.36046	0.082
					3.24507	0.394
0.0625	8	16	16	40×40	3.35849	0.021
					3.23582	0.108
0.03125	16	32	32	80×80	3.35792	0.006
					3.23322	0.028

† Slab waveguide thickness $d = 1.0$ μm (two modes allowed). Exact effective indices are $\bar{n}_0 = 3.357\,718$ and $\bar{n}_1 = 3.232\,331$ (for $\lambda_0 = 1.3$ μm).

READING LIST

B.P. Flannery, S.A. Teukolsky, and W.T. Vetterling, *Numerical Recipes: Art of Scientific Computing*, Cambridge University Press, Cambridge, UK (1986).

Optimizing Laser Cavity Designs

A17.1 GENERAL APPROACH

The focus of this appendix will be on how to minimize the laser current given by either Eq. (2.36) or Eq. (5.19):

$$I = I_{th} + \frac{q}{hv} \frac{P_0}{\eta_d}. \tag{A17.1}$$

The cavity parameters we can use to minimize this equation include the active, passive, and cavity lengths (L_a, L_p, and L), the mean mirror reflectivity $R = \sqrt{r_1 r_2}$, the cross-sectional area of the active region, A, and the optical confinement factor, $\Gamma = \Gamma_{xy} \Gamma_z$. Other more "independent" cavity parameters include the active and passive section internal losses α_{ia} and α_{ip}, the injection efficiency of the active region η_i, and the fraction of light coupled out of the desired mirror facet, F. These latter parameters are referred to as independent since they can usually be optimized regardless of other design choices. Figure A5.1 illustrates the relevant cavity geometry for a typical in-plane laser.

In order to write Eq. (A17.1) more explicitly in terms of the cavity parameters, we first define a threshold current density per *unit volume*, J_v, such that $I_{th} = J_v A L_a$. From Eq. (3.31), we can also set $1/\eta_d = [1 + \alpha_i L/\ln(1/R)]/(\eta_i F)$. With these substitutions, the laser current becomes

$$I = J_v A L_a + I_P \left(1 + \frac{\alpha_i L}{\ln(1/R)} \right), \tag{A17.2}$$

where

$$I_P = \frac{q}{hv} \frac{P_0}{\eta_i F}. \tag{A17.3}$$

Equation (A17.3) translates the output power into injected current, including the penalty in current when η_i and F are less than unity. In designing the laser, we should always attempt to maximize η_i by clever design of the carrier confinement region, and maximize F by setting all mirrors aside from the output coupler as close to unity as possible within practical limits.

If we had the flexibility to vary every cavity parameter, our optimization procedure would inevitably produce a design with *unity* mirror reflectivity and *zero* cavity length, since this design eliminates internal losses and creates the smallest possible mode volume. However, in practice one or more constraints are placed on the cavity design. For example, we might have a phase-shifter passive region with a fixed length, or we might want to maintain a desired mode spacing implying a fixed overall cavity length. Or else we may only want to optimize one cavity parameter for any given combination of other parameters, such as optimizing the number of quantum wells to use in the active region.

In the following, we will summarize the relevant equations for many (but

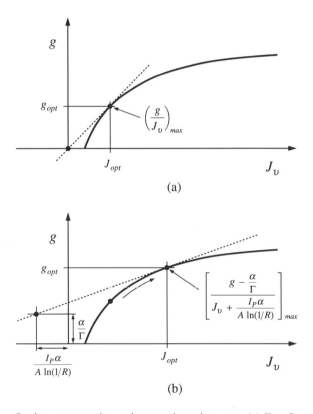

(a)

(b)

FIGURE A17.1 Optimum operating points on the gain curve. (a) For Case A1 and Case E the knee of the gain curve represents the optimum design. (b) For all other cases, the optimum design slides away from the knee depending on the internal loss and the desired output power. The internal loss α in (b) takes on different forms for each particular design constraint. Also note that the plot is *material* gain vs. *volume* current density.

not all) of the constraints typically encountered in designing lasers. The first three cases optimize the *active* length of an in-plane laser for (A) a fixed total cavity length, (B) a fixed passive length, and (C) a fixed passive-to-active length ratio. The final two cases optimize the number of quantum wells to use in (D) an in-plane laser, and (E) a VCSEL. For each case, three equations are provided. The first equation expresses the laser current using some form of the *current-to-gain conversion factor*. The second equation expresses the parameter to be optimized in terms of the optimum operating point on the gain curve g_{opt} as illustrated in Fig. A17.1. The third equation expresses the optimum reflectivity for any given choice of fixed parameters.

A17.2   SPECIFIC CASES

A17.2.1   Case A1: Optimize L_a for a Fixed L ($\alpha_{ia} = \alpha_{ip}$)

Using the threshold gain condition for an in-plane laser (with $\Gamma_{enh} = 1$), we can replace L_a in Eq. (A17.2) with $L_a = [\alpha_i L + \ln(1/R)]/(\Gamma_{xy}g)$. Rearranging, we obtain

$$I = \left\{ \left(\frac{J_v}{g}\right) \frac{A}{\Gamma_{xy}} \ln(1/R) + I_P \right\} \left(1 + \frac{\alpha_i L}{\ln(1/R)} \right), \tag{A17.4}$$

$$L_a|_{opt} = \frac{\alpha_i L + \ln(1/R)}{\Gamma_{xy}g_{opt}}, \tag{A17.5}$$

$$R|_{opt} = \exp\left\{ -\sqrt{I_P \alpha_i L \frac{\Gamma_{xy}}{A} \Big/ \left(\frac{J_v}{g}\right)_{min}} \right\}. \tag{A17.6}$$

The second and third equations will be explained in a moment. For now let's concentrate on the first equation. First of all, note that the laser current does not explicitly depend on L_a. The only effect L_a has on the laser current lies implicitly in the first term, J_v/g. Thus, to minimize the laser current we should adjust L_a so as to minimize J_v/g. This is accomplished by using L_a to adjust the threshold gain to the optimum operating point on the gain curve as illustrated in Fig. A17.1a. This optimum operating point, g_{opt}, maximizes the slope, g/J_v, and hence minimizes J_v/g. Once we know g_{opt} from the gain curve of the active material, we can use Eq. (A17.5) to determine the optimum L_a.

For a fixed R, that's all there is to our optimization procedure. However, if we have the freedom to adjust R, then Eq. (A17.4) reveals a tradeoff in the dependence on R. The first (threshold) term suggests that we should set $R \rightarrow 1$, while the last (power out) term suggests that we should set $R \rightarrow 0$. The minimum for the sum of these terms occurs when the two terms are equal (e.g., $(ax + b/x)_{min} \rightarrow x_{min} = \sqrt{b/a}$). Solving for R we obtain Eq. (A17.6). Thus, for a fixed cavity length, L, and a given power out, I_P, we can determine the optimum R to use from Eq. (A17.6). This value of R can then be used in Eq. (A17.5) to determine the optimum L_a.

A17.2.2   Case A2: Optimize L_a for a Fixed L ($\alpha_{ia} \neq \alpha_{ip}$)

For this more general version of Case A1, we must consider the dependence of the internal losses on the active length. With $\alpha_i L = \alpha_{ia} L_a + \alpha_{ip} L_p$ and L fixed, we must first set $L_p = L - L_a$. Defining $\Delta\alpha = \alpha_{ia} - \alpha_{ip}$, we can set $\alpha_i L = \Delta\alpha L_a + \alpha_{ip} L$ in Eq. (A17.2). Collecting terms involving L_a and using the threshold gain condition (again with $\Gamma_{enh} = 1$) to set $L_a = [\alpha_{ip} L + \ln(1/R)]/(\Gamma_{xy} g - \Delta\alpha)$, we obtain

$$I = \left\{ \left[\frac{J_v + \dfrac{I_P \Delta\alpha}{A \ln(1/R)}}{g - \dfrac{\Delta\alpha}{\Gamma_{xy}}} \right] \frac{A}{\Gamma_{xy}} \ln(1/R) + I_P \right\} \left(1 + \frac{\alpha_{ip} L}{\ln(1/R)} \right), \qquad (A17.7)$$

$$L_a \big|_{opt} = \frac{\alpha_{ip} L + \ln(1/R)}{\Gamma_{xy} g_{opt} - \Delta\alpha}, \qquad (A17.8)$$

$$R \big|_{opt} = \exp\left\{ - \sqrt{ I_P \alpha_{ip} L \frac{\Gamma_{xy}}{A} \bigg/ \left[\frac{J_v + \dfrac{I_P \Delta\alpha}{A \ln(1/R)}}{g - \dfrac{\Delta\alpha}{\Gamma_{xy}}} \right]_{min} } \right\}. \qquad (A17.9)$$

When the active and passive losses are not equal, the minimum ratio in Eq. (A17.7) no longer corresponds to the *knee* in the gain curve suggested in Fig. A17.1(a). Instead, the optimum operating point is shifted along the gain curve with the *origin* of the line tangent to the gain curve being shifted up by $\Delta\alpha/\Gamma_{xy}$ and to the left by $I_P \Delta\alpha/A \ln(1/R)$, as illustrated in Fig. A17.1(b). When $\alpha_{ia} > \alpha_{ip}$ ($\Delta\alpha$ positive), the optimum operating point slides *up* the gain curve favoring a shorter active section than Case A1. When $\alpha_{ia} < \alpha_{ip}$ ($\Delta\alpha$ negative), the optimum operating point slides *down* the gain curve favoring a longer active section. Both of these operating point shifts reflect the tendency to reduce the total internal losses. As we increase output power, the shift in the operating point away from the knee in the gain curve becomes more extreme, reflecting the increasing importance of reducing $\alpha_i L$ as opposed to maintaining a low threshold current.

Once we have located the optimum operating point on the gain curve, g_{opt}, which now depends on the desired output power *and* mirror reflectivity, we can determine the optimum L_a using Eq. (A17.8). If we wish to optimize R in addition to L_a, we again need to minimize Eq. (A17.7) with respect to R. However the dependence of g_{opt} on R complicates the matter. If we use an initial guess for R to determine g_{opt} and assume that the term in square brackets in Eq. (A17.7) is relatively insensitive to changes in R then we can concentrate on the two $\ln(1/R)$ factors outside the square brackets. Minimizing the current with respect to these factors in the same manner as Case A1, we obtain Eq. (A17.9). The optimum R obtained in this way is not exact but should be reasonably close for small output powers. At high output powers it may be necessary to iterate through the procedure once or twice to obtain more accurate estimates of both R and g_{opt}.

A17.2.3   Case B1: Optimize L_a for a Fixed L_p ($L_p = 0$)

For this case, we have no passive section and we are essentially asking what the optimum cavity length is. We can first of all set $L_a = L$ and $\alpha_{ia} = \alpha_i$. Collecting terms in Eq. (A17.2) involving L and using the threshold gain condition (with $\Gamma_{enh} = 1$) to set $L = \ln(1/R)/(\Gamma_{xy}g - \alpha_i)$, we obtain

$$I = \left[\frac{J_v + \dfrac{I_p \alpha_i}{A \ln(1/R)}}{g - \dfrac{\alpha_i}{\Gamma_{xy}}} \right] \frac{A}{\Gamma_{xy}} \ln(1/R) + I_P, \qquad (A17.10)$$

$$L|_{opt} = \frac{\ln(1/R)}{\Gamma_{xy} g_{opt} - \alpha_i}, \qquad (A17.11)$$

$$R|_{opt} = 1. \qquad (A17.12)$$

With no internal losses, the optimum operating point is again at the knee of the gain curve as we found in Case A1. However with finite internal losses, the operating point slides *up* the gain curve in favor of shorter cavity lengths which reduce $\alpha_i L$ as illustrated in Fig. A17.1(b)—the shift in operating point increasing with increasing output power. This situation is similar to Case A2.

Once g_{opt} is determined, the optimum cavity length can be found using Eq. (A17.11). If we also wish to optimize R, we should recognize that the second $1/\ln(1/R)$ term present in both Case A1 and Case A2 does not appear in Eq. (A17.10). As a result, the optimum R is found by setting $R \to 1$. Using this in Eq. (A17.11), we find that the optimum cavity length is found by setting $L \to 0$. This optimum ($L = 0$, $R = 1$) cavity design defines the lowest possible required current, however, it does not represent a practical design (although VCSELs venture toward this limit). Thus, it does not make sense to *simultaneously* optimize the cavity length *and* the mirror reflectivity when $L_p = 0$, since the best design will always approach $L \to 0$ and $R \to 1$. At some point, practical limits on either one of the parameters will define a design constraint which can be used to optimize the other parameter.

A17.2.4   Case B2: Optimize L_a for a Fixed L_p ($L_p \neq 0$)

For this more general version of Case B1, we must separate out the fixed portion of the internal losses. We begin by setting $\alpha_i L = \alpha_{ia} L_a + \alpha_{ip} L_p$. Collecting terms in Eq. (A17.2) involving L_a and using the threshold gain condition (with $\Gamma_{enh} = 1$) to set $L_a = [\alpha_{ip}L_p + \ln(1/R)]/(\Gamma_{xy}g - \alpha_{ia})$, we obtain

$$I = \left\{ \left[\frac{J_v + \dfrac{I_P \alpha_{ia}}{A \ln(1/R)}}{g - \dfrac{\alpha_{ia}}{\Gamma_{xy}}} \right] \frac{A}{\Gamma_{xy}} \ln(1/R) + I_P \right\} \left(1 + \frac{\alpha_{ip}L_p}{\ln(1/R)} \right), \quad \text{(A17.13)}$$

$$L_a|_{opt} = \frac{\alpha_{ip}L_p + \ln(1/R)}{\Gamma_{xy}g_{opt} - \alpha_{ia}}, \quad \text{(A17.14)}$$

$$R|_{opt} = \exp\left\{ -\sqrt{I_P \alpha_{ip} L_p \frac{\Gamma_{xy}}{A} \bigg/ \left[\frac{J_v + \dfrac{I_P \alpha_{ia}}{A \ln(1/R)}}{g - \dfrac{\alpha_{ia}}{\Gamma_{xy}}} \right]_{min}} \right\}. \quad \text{(A17.15)}$$

The operating point on the gain curve is the same as Case B1 with α_{ia} used to define the shift away from the knee in the gain curve. The optimum active length is found using the determined g_{opt} and Eq. (A17.14). For this more general case, we can define an optimum R which does not reduce to unity as it did in Case B1. The residual loss in the passive section $\alpha_{ip}L_p$ prevents us from removing the internal loss completely, implying that the optimum R must always be less than unity for finite output powers. For zero output power corresponding to threshold, $I_P = 0$ and the optimum $R \rightarrow 1$ as it does in every other case. As with Case A2, Eq. (A17.15) must be iterated a couple of times (updating g_{opt} on each iteration) to determine an accurate estimate of the optimum R at high output powers.

A17.2.5   Case C: Optimize L_a for a Fixed L_p/L_a

For this case we set $\alpha_i L = (\alpha_{ia} + \alpha_{ip}L_p/L_a)L_a$. Collecting terms in Eq. (A17.2) involving L_a and using the threshold gain condition (with $\Gamma_{enh} = 1$) to set $L_a = \ln(1/R)/(\Gamma_{xy}g - \alpha_{ia} - \alpha_{ip}L_p/L_a)$, we obtain

$$I = \left[\frac{J_v + \dfrac{I_P(\alpha_{ia} + \alpha_{ip}L_p/L_a)}{A \ln(1/R)}}{g - \dfrac{\alpha_{ia} + \alpha_{ip}L_p/L_a}{\Gamma_{xy}}} \right] \frac{A}{\Gamma_{xy}} \ln(1/R) + I_P, \quad \text{(A17.16)}$$

$$L_a|_{opt} = \frac{\ln(1/R)}{\Gamma_{xy}g_{opt} - \alpha_{ia} - \alpha_{ip}L_p/L_a}, \quad \text{(A17.17)}$$

$$R|_{opt} = 1. \quad \text{(A17.18)}$$

In this case, the shift away from the knee of the gain curve is governed by the internal loss term $\alpha_{ia} + \alpha_{ip}L_p/L_a$. The optimum active length is found using

Eq. (A17.17) once g_{opt} is determined. The optimum R in this case again reduces to unity due to the absence of the second $1/\ln(1/R)$ term, as we found in Case B1. Thus, here again it does not make sense to consider optimizing *both* the active length and the mirror reflectivity. We should simply be aware that the best design is found by approaching $L_a \to 0$ and $R \to 1$.

A17.2.6   Case D: Optimize N_w for Fixed L_a and L_p (In-Plane Laser)

For in-plane quantum-well lasers we can set $A = w N_w d_1$, where w is the lateral active width, N_w is the number of quantum wells, and d_1 is the well width. We can also set $\Gamma_{xy} = \Gamma_y \Gamma_1 N_w$, where Γ_1 is the transverse optical confinement per well. For the internal loss, we can define a material free-carrier loss per well α_{ia1} and write the total active loss as $\alpha_{ia} = \alpha_{ia0} + \Gamma_y \Gamma_1 \alpha_{ia1} N_w$ with $\alpha_{i0} L = \alpha_{ia0} L_a + \alpha_{ip} L_p$. Collecting terms in Eq. (A17.2) involving N_w and using the threshold gain condition to set $N_w = [\alpha_{i0} L + \ln(1/R)]/[\Gamma_y \Gamma_1 L_a (g - \alpha_{ia1})]$, we obtain

$$I = \left\{ \left[\frac{J_v + \dfrac{I_P \Gamma_y \Gamma_1 \alpha_{ia1}}{w d_1 \ln(1/R)}}{g - \alpha_{ia1}} \right] \frac{w}{\Gamma_y} \frac{d_1}{\Gamma_1} \ln(1/R) + I_P \left(1 + \frac{\alpha_{i0} L}{\ln(1/R)} \right) \right\}, \quad \text{(A17.19)}$$

$$N_w|_{opt} = \frac{\alpha_{i0} L + \ln(1/R)}{\Gamma_y \Gamma_1 L_a (g_{opt} - \alpha_{ia1})}, \quad \text{(A17.20)}$$

$$R|_{opt} = \exp\left\{ - \sqrt{I_P \alpha_{i0} L \frac{\Gamma_y \Gamma_1}{w d_1} \Bigg/ \left[\frac{J_v + \dfrac{I_P \Gamma_y \Gamma_1 \alpha_{ia1}}{w d_1 \ln(1/R)}}{g - \alpha_{ia1}} \right]_{min}} \right\}. \quad \text{(A17.21)}$$

If α_{ia1} is small, we can simply choose the number of quantum wells to align the threshold gain to the knee of the gain curve using Eq. (A17.20). The optimum R for a given cavity length can also be readily determined using Eq. (A17.21). If the free-carrier loss in each quantum well is not negligible, then the optimum operating point slides up the gain curve away from the knee, favoring fewer wells. This shift is illustrated in Fig. A17.1(b). The optimum R and g_{opt} must also be iterated once or twice to obtain accurate estimates of both.

Another issue we can resolve using Eq. (A17.19) is how to optimize the transverse and lateral optical confinement. Clearly it is beneficial to maximize the transverse optical confinement per well per unit well width, Γ_1/d_1, so we should work hard at providing a tightly confining transverse optical waveguide—a factor of 2 improvement in confinement translates into a factor of 2 reduction in the threshold current. For the lateral confinement, we can typically assume a simple three-layer waveguiding structure of width w. To minimize w/Γ_y, we use Eq. (A3.23) in Appendix 3 to obtain an approximate expression for Γ_y. Forming the ratio we obtain

$$\frac{w}{\Gamma_y} \approx w + \frac{2}{k_0^2 \Delta \bar{n} w} \rightarrow w_{opt} \approx \frac{1}{k_0}\sqrt{\frac{2}{\Delta \bar{n}}}, \tag{A17.22}$$

where $\Delta \bar{n}$ is the difference between the transverse effective index of the central guiding region and the transverse effective index of the lateral cladding region. For large w, we have good optical confinement and $w/\Gamma_y \approx w$, implying that we should attempt to minimize w. However as w becomes very small, we lose optical confinement and w/Γ_y actually begins to increase with further decreases in width. The minimum in w/Γ_y occurs at w_{opt} when the two terms comprising w/Γ_y are equal. Hence we should keep $w \geq w_{opt}$. For a small index difference of 0.01 and a wavelength of 1 μm, $w_{opt} = 2.25$ μm. For a larger index difference of 0.1, $w_{opt} = 0.7$ μm. Thus, in practice w_{opt} is quite small and for most cases we should simply attempt to make w as small as possible.

From a more general perspective, we can picture the ratios d_1/Γ_1 and w/Γ_y as the *mode widths* in the transverse and lateral dimensions just as V/Γ defines the mode volume. In fact we can define the cross-sectional area of the mode as $A_p = A/\Gamma_{xy}$. In this context, we see from Eq. (A17.19), or any one of Eqs. (A17.4)–(A17.16), that it is not the cross-sectional area of the active region itself but more fundamentally the cross-sectional area of the optical mode that is important in minimizing the threshold current of the laser.

A17.2.7 Case E: Optimize N_w for Fixed L (VCSEL)

For a quantum-well VCSEL, the active length becomes $L_a = N_w L_1$, where L_1 is the well width. We must also include standing wave enhancements such that $\Gamma = \Gamma_{enh}\Gamma_{xy}$. For this case we will also make the simplifying assumption that $\Gamma_{enh}\Gamma_{xy}\alpha_{ia1}N_w L_1$ is much smaller than other internal losses in the cavity, which is generally a very good assumption. Within this approximation $\alpha_i L$ can be assumed independent of N_w. Replacing L_a in Eq. (A17.2) with $L_a = [\alpha_i L + \ln(1/R)]/(\Gamma_{enh}\Gamma_{xy}g)$ and rearranging, we obtain

$$I = \left\{\left(\frac{J_v}{g}\right)\frac{A}{\Gamma_{enh}\Gamma_{xy}}\ln(1/R) + I_P\right\}\left(1 + \frac{\alpha_i L}{\ln(1/R)}\right), \tag{A17.23}$$

$$N_w|_{opt} = \frac{\alpha_i L + \ln(1/R)}{\Gamma_{enh}\Gamma_{xy}g_{opt}L_1}, \tag{A17.24}$$

$$R|_{opt} = \exp\left\{-\sqrt{I_P\alpha_i L \frac{\Gamma_{enh}\Gamma_{xy}}{A}\bigg/\left(\frac{J_v}{g}\right)_{min}}\right\}. \tag{A17.25}$$

As with Case A1, we find that to minimize the laser current we simply want to minimize J_v/g. This is accomplished by choosing the number of quantum wells which align the threshold gain with the knee of the gain curve as illustrated in Fig. A17.1(a). In practice, the design of the VCSEL starts with the evaluation of the internal losses of the cavity which we should attempt to minimize since the total laser current, I, scales directly with $\alpha_i L$. Once this is known we can use Eq. (A17.25) to determine the optimum R for a desired output power. Finally we can use Eq. (A17.24) to find the number of wells to use.

The only difficulty in evaluating Eq. (A17.25) lies in the implicit dependence of Γ_{enh} on N_w which is in principle an unknown quantity before we know R. To get around this snag, we can use an educated guess for N_w to estimate a value to use for Γ_{enh}. Evaluating Eqs. (A17.25) and (A17.24) to determine R and N_w, we can determine whether our initial guess for N_w was accurate. If our first guess was not accurate, we can update Γ_{enh} and reevaluate Eqs. (A17.25) and (A17.24) to obtain more accurate estimates of R and N_w.

A17.2.8 Summary of Cases A through E

The general form of the laser current, optimum active length (Cases A through C only), and optimum reflectivity for all cases considered above can be

TABLE A17.1 Summary of Optimum Design Parameters.

	Constraint		α	$\alpha_0 L$
	Vary L_a			
Fixed L	$(\alpha_{ia} = \alpha_{ip})$		0	$\alpha_i L$
	$(\alpha_{ia} = \alpha_{ip} + \Delta\alpha)$		$\Delta\alpha$	$\alpha_{ip} L$
Fixed L_p	$(L_p = 0)$		α_i	0
	$(L_p \neq 0)$		α_{ia}	$\alpha_{ip} L_p$
Fixed L_p/L_a			$\alpha_{ia} + \alpha_{ip}\dfrac{L_p}{L_a}$	0
	Vary N_w			
In-plane laser			$\Gamma_y \Gamma_1 \alpha_{ia1} N_w$	$\alpha_{i0} L$
VCSEL	$(\alpha_{ia1} \approx 0)$		0	$\alpha_i L$

written as

$$I = \left\{ \left[\frac{J_v + \dfrac{I_p \alpha}{A \ln(1/R)}}{g - \dfrac{\alpha}{\Gamma_{xy}}} \right] \frac{A}{\Gamma_{enh} \Gamma_{xy}} \ln(1/R) + I_p \right\} \left(1 + \frac{\alpha_0 L}{\ln(1/R)} \right), \qquad \text{(A17.26)}$$

$$L_a|_{opt} = \frac{\alpha_0 L + \ln(1/R)}{\Gamma_{xy} g_{opt} - \alpha}, \qquad \text{(A17.27)}$$

$$R|_{opt} = \exp\left\{ - \sqrt{I_p \alpha_0 L \frac{\Gamma_{enh} \Gamma_{xy}}{A} \left[\frac{J_v + \dfrac{I_p \alpha}{A \ln(1/R)}}{g - \dfrac{\alpha}{\Gamma_{xy}}} \right]_{min}} \right\}. \qquad \text{(A17.28)}$$

The only difference between the various cases lies in the terms α and $\alpha_0 L$. α is the loss whose loss-length product scales with the design variable, while $\alpha_0 L$ is the fixed background loss of the cavity. The specific forms for these two terms are summarized in Table A17.1 for all Cases A through E. Also, $\Gamma_{enh} = 1$ for all cases other than the VCSEL. Note that when $\alpha = 0$, the optimum design occurs at the knee in the gain curve (see Fig. A17.1a). When $\alpha_0 L = 0$, the optimum reflectivity is one and the optimum active length is zero.

A17.3   OPTIMUM OPERATING POINT ON THE GAIN CURVE

The knee of the gain curve (where $g/J = dg/dJ$) which defines the optimum operating point for many design constraints can be determined for the gain curve fits given in Chapter 4: $g = g_0 \ln[(J + J_s)/(J_{tr} + J_s)]$. If the linearity parameter J_s is zero, the solution to $g/J = dg/dJ$ is given by

$$J_{opt} = e J_{tr}, \qquad g_{opt} = g_0,$$

and $\qquad\qquad\qquad\qquad\qquad\qquad\qquad\qquad\qquad\qquad$ $(J_s = 0)$ $\qquad$ (A17.29)

$$\left(\frac{J_v}{g} \right)_{min} = \frac{e J_{tr}}{g_0 L_z}.$$

The well width L_z appears in the denominator since J_v is a current per unit volume. Thus, for many designs, we simply want to make sure that the threshold gain is near g_0.

When the linearity parameter is not zero, we cannot solve for the knee in the gain curve explicitly, however we can still write

$$J_{opt} = (J_{tr} + J_s) e^{1/(1 + J_s/J_{opt})} - J_s,$$

$\qquad\qquad\qquad\qquad\qquad\qquad\qquad\qquad\qquad\qquad\qquad$ $(J_s \neq 0)$ $\qquad$ (A17.30)

$$g_{opt} = \frac{g_0}{1 + J_s/J_{opt}}.$$

In the first equation, J_{opt} appears on both sides. To solve this, we can use $J_{opt} = eJ_{tr}$ as a first guess on the right-hand side and iterate a couple of times. Once we know J_{opt}, we can readily find g_{opt}.

The knee in the gain curve can alternatively be found graphically. One method is demonstrated in Fig. A17.1(a). Simply plot the gain curve and draw a tangent line from the origin. A much more insightful method however is to plot J_v/g directly as a function of material gain. This is shown in Fig. A17.2 for the three GaAs based active materials considered in Chapter 4.

The plot in Fig. A17.2 reveals a number of insights about the three active materials. For example, the penalty for operating too close to transparency or too far into gain saturation is immediately clear. Also in comparing the minimum J_v/g for the three active materials we immediately see that the InGaAs QW provides about a factor of 2 improvement over the GaAs QW, while bulk GaAs is only 25% more than the GaAs QW. In addition, we can immediately determine where the minimum J_v/g occurs and how broad the minimum is. The plot therefore allows us to quickly determine the optimum threshold material gain as well as the tolerance we have in choosing other operating points. The relevant features of the three active materials can be summarized as follows:

$$\left(\frac{J_v}{g}\right)_{min} = \begin{array}{l} 1.4 \text{ mA/}\mu\text{m}^2 \text{ at } 1250 \text{ cm}^{-1} \ (650\text{--}2100 \text{ cm}^{-1}) \quad \text{(InGaAs QW)} \\ 2.9 \text{ mA/}\mu\text{m}^2 \text{ at } 1450 \text{ cm}^{-1} \ (750\text{--}2500 \text{ cm}^{-1}) \quad \text{(GaAs QW)} \quad \text{(A17.31)} \\ 3.8 \text{ mA/}\mu\text{m}^2 \text{ at } 1000 \text{ cm}^{-1} \ (450\text{--}1900 \text{ cm}^{-1}) \quad \text{(bulk GaAs)} \end{array}$$

The range over which J_v/g remains within 20% of the minimum value is

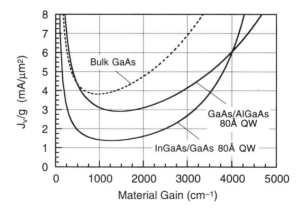

FIGURE A17.2 Plot of the calculated current-to-gain conversion factor as a function of material gain for three GaAs based active materials. For the bulk GaAs, an 80Å thickness is assumed (for details on these materials see Section 4.6). For $\eta_i < 1$, the J_v/g axis should be divided by η_i.

indicated in parentheses for each case. So, for example, while designing an InGaAs QW laser to have 1250 cm^{-1} threshold material gain provides the lowest J_v/g, we could alternatively pick the threshold gain anywhere in the range $650\text{--}2100 \text{ cm}^{-1}$ and not pay a high price. Thus in general, the design of the laser threshold gain is extremely flexible as long as we stay within certain boundaries.

The reason J_v/g is relevant is that it is the only active material-dependent term in the expression for threshold current. For example in Cases A1 and E, the threshold component of the current is given by

$$I_{th} = \left(\frac{J_v}{g}\right) \frac{A}{\Gamma_{enh}\Gamma_{xy}} [\alpha_i L + \ln(1/R)]. \tag{A17.32}$$

For a $10 \times 10 \text{ μm}^2$ VCSEL with $\Gamma_{enh} = 1.8$, $\Gamma_{xy} = 1$, $\alpha_i L = 0.3\%$, $\ln(1/R) = 0.6\%$, the coefficient relating I_{th} to J_v/g is 0.5 μm^2. Thus, aside from surface recombination and the internal quantum efficiency, threshold currents of $\sim 0.75 \text{ mA}$, $\sim 1.5 \text{ mA}$, or $\sim 2 \text{ mA}$ can be expected if we use InGaAs QW, GaAs QW, or bulk GaAs active regions. This is true as long as we choose the number of QWs such that the threshold material gain falls somewhere within the J_v/g minimum plotted in Fig. A17.2. When $\eta_i < 1$, the actual J_v/g is related to the theoretical estimate via: $J_v/g = (J_v/g)_{theory}/\eta_i$. Thus, threshold currents estimated using $(J_v/g)_{theory}$ in Eq. (A17.32) should be divided by η_i to account for a less-than-unity internal quantum efficiency.

A17.4   SHIFTED OPTIMUM OPERATING POINTS ON THE GAIN CURVE

For cases A1 and E, the minimum J_v/g determines the optimum laser design. However, for all other cases we wish to minimize $[J_v + I_P\alpha/A \ln(1/R)]/(g - \alpha/\Gamma_{xy})$, where α takes on different forms depending on the particular case, as summarized in Table A17.1. For such cases, the best approach is to plot the ratio as a function of material gain in order to determine the optimum gain operating point. Figure A17.3 illustrates this procedure for a single quantum-well in-plane laser at different output powers with an assumed loss of 10 cm^{-1}. At threshold (0 mW), the internal loss shifts the minimum to the right by $\sim \alpha/\Gamma_{xy}$. For finite output powers the minimum is pushed to even higher gains and higher values. This shift toward higher gains was qualitatively shown in Fig. A17.1. However, with Fig. A17.3 we can quantitatively estimate the best operating gain to use.

For example, if we are trying to pick a cavity length (Case B) which minimizes the laser current at 20 mW of output power, Fig. A17.3 tells us that we should adjust the cavity length to provide a $\sim 2800 \text{ cm}^{-1}$ threshold material gain. Using parameters in the caption and Eq. (A17.11), we find $L = 110 \text{ μm}$. The value of the minimum can then be used to evaluate the laser current via Eq. (A17.26) with $\alpha = 10 \text{ cm}^{-1}$ and $\alpha_0 L = 0$ (assuming there is no passive section). Using the parameters given in the caption, we find $I_P = 15.81 \text{ mA}$. Taking

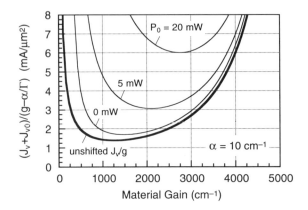

FIGURE A17.3   Plots of the shifted current-to-gain conversion factor for different output powers as a function of material gain for the InGaAs/GaAs 80Å QW. In the vertical axis caption, $J_{v0} = I_p \alpha / A \ln(1/R)$. The assumed parameters are: $\alpha = 10 \text{ cm}^{-1}$, $\Gamma_{xy} = 4\%$, $R = 0.32$, $A = 2 \, \mu\text{m} \times 80\text{Å}$, and $h\nu/q = 1.265 \text{ mW/mA}$. For $\eta_i < 1$ and/or $F < 1$, the power levels should be interpreted as $P_0/\eta_i F$ and the vertical axis should be divided by η_i.

6 mA/μm^2 as the minimum value of the curve, we obtain $I = 18.54$ mA from Eq. (A17.26).

We can summarize the general laser optimization procedure as follows. First we must determine the gain vs. current curve (either theoretically or experimentally) and apply a curve fit such as the two or three parameter fits discussed in Chapter 4. To optimize the laser for a given set of constraints, we use Table A17.1 to define both α and $\alpha_0 L$. With the output power chosen, we then plot $[J_v + I_p \alpha / A \ln(1/R)]/(g - \alpha/\Gamma_{xy})$ as a function of material gain, and choose an operating gain for our laser which minimizes this curve. We can then adjust either the active length or the quantum well number to achieve this material threshold gain. The laser current is then found from Eq. (A17.26). If we want to simultaneously optimize the mirror reflectivity, then we should evaluate Eq. (A17.28) and compare it to the value of R assumed initially. If different, then we should plot $[J_v + I_p \alpha / A \ln(1/R)]/(g - \alpha/\Gamma_{xy})$ again using the optimized value of R. The new minimum in the curve defines a new design with a new optimum R. We should iterate until the new value of R is close to the old value. When the value of R is self-consistent, we have found the design which requires the minimum laser current.

A17.5   OTHER DESIGN CONSIDERATIONS

A17.5.1   High-Speed Designs

For high-speed lasers, in addition to minimizing the laser current, we also want to maximize the differential gain in order to enhance the relaxation resonance

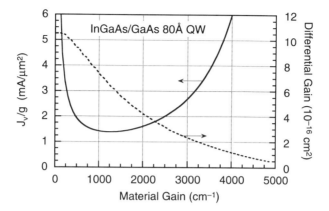

FIGURE A17.4 Plot of the current-to-gain conversion factor and differential gain as a function of material gain for the InGaAs/GaAs 80Å QW.

frequency as discussed in Chapters 2 and 5 (see Eq. (2.56) for example). However, it turns out that the maximum differential gain occurs at transparency which from Fig. A17.2 is far from ideal for minimizing the threshold current. Thus a tradeoff exists between obtaining low threshold current and high differential gain.

Figure A17.4 plots both J_v/g and dg/dN as a function of material gain for the InGaAs QW active material. This plot immediately quantifies the tradeoff between the two. For example, we can see that reducing the threshold material gain from 2000 cm^{-1} to 1000 cm^{-1} has little effect on the threshold current, while it increases the differential gain by close to a factor of 2. Reducing the threshold material gain much below 500 cm^{-1} leads to large increases in the threshold current without much increase in differential gain. Thus, we conclude that a reasonable compromise operating point exists somewhere in the range 500–1000 cm^{-1}.

A17.5.2 Heating Effects

In general, the internal losses are an independent parameter of the cavity which we should attempt to minimize. However in VCSELs there is a tradeoff involved with the amount of internal loss associated with doping. For low doping the internal losses are low and therefore a lower laser current is required to achieve a given output power. However, lower doping can also mean higher resistance in getting current to the active region, which results in heating of the device. For cw operation, heating effects limit the maximum output power possible in a VCSEL, and therefore we should increase the doping to minimize heating and increase the maximum output power. This tradeoff in design of the doping profile depends on the particular geometry of the VCSEL and its associated thermal impedance. However, generally speaking the doping should be sufficient to limit the series voltage drop to less than 1–2 V, regardless of how much internal losses are generated by the doping profile.

Heating effects can also be a problem for in-plane lasers, causing them to die prematurely or degrade rapidly. The maximum output power is also dependent on how hot the laser gets (in addition to other factors such as catastrophic optical damage at the mirror facets). The heat generation is related to the current *density* injected into the laser. So in addition to minimizing the total current, keeping the total current density below a certain level can also be important. While the design procedures discussed in this appendix do minimize the total current, such designs might not actually be the best, particularly if tens of kA/cm^2 are required for a laser which does not have adequate heat sinking. For these cases, it is best to use the maximum current density as a cap for the design and to modify the operating gain accordingly if the optimum design calls for too much current density. Chapter 8 considers these issues further.

READING LIST

P.W.A. McIlroy, A. Kurobe, and Y. Uematsu, *IEEE J. Quantum Electron.*, **QE-21**, 1958 (1985).

Index